Drug Discovery and Development

Ramarao Poduri
Editor

Drug Discovery and Development

From Targets and Molecules to Medicines

Springer

Editor
Ramarao Poduri
Pharmaceutical Sciences and Natural Products
Central University of Punjab
Bathinda, Punjab, India

ISBN 978-981-15-5533-6 ISBN 978-981-15-5534-3 (eBook)
https://doi.org/10.1007/978-981-15-5534-3

This Springer imprint is published by the registered company Springer Nature Singapore Pte Ltd.
The registered company address is: 152 Beach Road, #21-01/04 Gateway East, Singapore 189721, Singapore

Preface

Drug discovery is a complex process. The reason of which are attributed to unmet medical needs, different lifestyles, development of multidrug resistance, lack of clear cut understanding of the life processes, etc. Communities are increasingly aware of risk benefits and adverse effects of drugs and health cost.

Drug discovery was in place ever since the humans started its existence on the planet with records dating back to *Charka Samhita* and *Ebers Papyrus*. Initially, it was by trial and error method of screening vast array of natural products and discovered the useful and harmful effects of plant and animal materials. With the advancement of knowledge, reductionist approach, and to find the cause and effect relationship began the era of testing natural products on whole animals (in vivo), and animal organs and tissues (in vitro), followed by cells and subcellular organelles. With a better insight of structure and life processes (anatomy, physiology, and pathology), the effect of natural products was evaluated on chemical neurotransmission, which led to the concept of a "receptive substance" and "magic bullets." The concurrent developments in the areas of physical, biophysical, and chemical sciences paved the way to the easy availability of pure chemicals (including the stereoisomers) from either natural or synthetic sources. This led to the concept of drug–receptor interaction, target identification, hit and lead development, and screening of small molecules in vitro and in vivo resulting in the discovery and development of specific, selective efficacious drugs. The realization of physicochemical and ADME properties of chemicals produced long-acting medicines and optimization of their dosage.

However, limited knowledge in biological processes, especially the differences in physiological processes in experimental animals and humans, resulted in lack of effect and/or increased adverse reactions/increased attrition rate of new chemical entities. The implementation of stringent regulatory procedures and the introduction of regulatory toxicity testing in different phases of clinical trials prolonged both the drug discovery and development phases. This also escalated the drug approvals worldwide by the regulatory agencies along with increasing the cost of producing drugs.

The past few decades saw developments in genetics, epigenetics, toxicogenomics, molecular biology, nanotechnology, understanding the downstream signaling pathways, rDNA technology, monoclonal antibodies, stem cells, and biomarkers. These newer technologies transformed the face of drug discovery and shifted the emphasis from small molecules to biologics.

With the advent of computers, CADD (in silico) and IoT, a paradigm shift was observed in drug discovery and health care industry. Molecular simulations and modeling of various biological processes are being carried out. Adoption of digital technologies, artificial intelligence, and virtual, augmented, and extended reality and platform technologies are poised to change the entire gamut of drug discovery.

The introduction of potent medicines acting on off-targets and incomplete understanding of various physiological processes have increased the safety concerns of the approved products. Post-market survey and pharmacovigilance play a pivotal role in minimizing the adverse effects. Even after taking these precautions, the probability of serious adverse effects can occur. The risk benefit analysis needs to be assessed by sound statistical analyses of the preclinical and clinical data.

Historically, drug discovery has been led by chemistry and pharmacology. Now it is conducted through collaborative efforts of multidisciplinary experts who understand disease mechanisms, identify potential targets for therapeutic intervention, and evaluate potential drug candidates. The cordial interaction between academia (basic research), industry (clinical research and product development), regulatory authorities (review and approval), and last but not the least the end user is essential for successful drug discovery and development. Based on this concept, we have assembled experts from several disciplines who would highlight both the basic and applied aspects of the process for discovering, developing, testing, and regulation of new drugs. The proposed book will be useful for both entry-level research scholars and professionals who are engaged in the complex process of Drug Discovery and Development.

Bathinda, Punjab, India Ramarao Poduri

Contents

Chapter 1
Historical Perspective of Drug Discovery and Development

Ramarao Poduri

1.1 Introduction

Drug discovery and development is as old as of human civilization. They were curious from the beginning and observed keenly the nature for their survival and evolution. From their comprehension, identified the plants for their food and identified the ones which produce cure, toxic or poisonous effects to humans. Curare, a mixture of plant extracts, was applied on the arrows to immobilize the animals during the hunting in olden days in South America and later it was developed into muscle relaxant, d-tubocurarine [1]. The pharmacological basis of action of curare and the neurotransmitter acetylcholine role at neuromuscular junction was reported by Bernard and Dale, respectively [2, 3]. By trial and error method, humans tested various natural products on self and identified the useful and poisonous substances. The drug discovery evolved from natural products, synthetic, biotechnological to biopharmaceutical drugs. The concept of reductionism (drug receptor and theories [4–7]) to multiple paradigm concept transformed the drug discovery and development into a complex process [8]. This chapter will provide an overview and future scenario of history of drug discovery.

1.2 Early History and Natural Products: Drug Discovery

Ever since the existence of humans on this planet, they are exploring the nature for their daily requirements including medicines. The history of the use of medicinal herbs by Neanderthals, civilizations in Mesopotamia, Egypt and Rome is reviewed

R. Poduri (✉)
Department of Pharmaceutical Sciences and Natural Products, Central University of Punjab, Bathinda, Punjab, India

R. Poduri (ed.), *Drug Discovery and Development*,
https://doi.org/10.1007/978-981-15-5534-3_1

[9]. The drug discovery was influenced by the social and cultural traditions. Initially the knowledge regarding the medicinal uses of plants was transmitted through the word of the mouth from one generation to another or paintings on the caves. Majority of plant products which were depicted in the paintings are related the substances acting on brain. The identification of the medicinal use of plants was part of magic and religion [10]. It was believed that disease was due to possession demon and used to treat demon. Greeks thought that the drugs remove the imbalance in the humours which cause the disease. Greeks contributed to the rational development of the herbal drugs. Galen (130–200 AD) writings were well acclaimed (galenicals) [11]. Chinese and Arabs also contributed to the growth of the herbal drugs [12].

The connection between the west and India was mainly due to the silk road. *Atharvaveda* [13] contained the basic concepts of Indian system of medicine, *Ayurvedic* medicine. Charaka and Sushruta wrote *samhitas* dealing with drugs and surgical procedures, respectively. As per *Ayurveda*, the diet, lifestyle and environment cause the humoral imbalance, leading to the disease. The concept of personalized treatment is advocated in *Ayurveda. Ayurveda* and evidence-based medicine need to be amalgamated [14].

Several drugs have been discovered from the compounds obtained from natural products. In the early nineteenth century, several compounds were isolated from plants which were chemically modified into drugs. The first alkaline substance was isolated from the morphine in 1817 by Serturner. Acetylsalicylic acid was synthesized from salicylic acid (extracted from willow tree) to reduce the gastric irritation by Gerhardt [15]. Number of currently approved drugs are either plant products or derivatives of the natural products. A detailed account of the role played by natural products is discussed in review [12]. However, the interest in the plant products declined in the later part of the last century due to decrease in the activity of the active component on isolation or tedious extraction procedures. Recently, Harvey and co-workers reported that the use of genomic and metabolomic studies resulted in identification of antimicrobial agents from plants. The advances in the pharmacological screening methods renewed the interest in natural product [16]. The Nobel Prize in Physiology or Medicine 2015 was awarded to the discovery of two natural products: avermectin [17] and artemisinin [18] revived interest again in natural products [19]. The important contribution of natural products in drug discovery is reviewed [20].

1.3 Serendipity and Drug Discovery and Development

"The word serendipity was first coined by Horace Walpole in a letter written to his friend Sir Horace Mann in 1754. Walpole was impressed by a fairy tale he had read about the adventures of 'The Three Princes of Serendip'(an ancient name of Ceylon, now known as Sri Lanka) who were making discoveries by accidents and sagacity, of things which they were not in quest of…" [21]. Accidental discoveries played crucial role in science. Drug discovery and development had several examples of serendipity. Louis Pasteur, an excellent experimental researcher, made several discoveries by critical observation and said, "*in the field of observation chance only favours the prepared mind*" (see Ref. [22]). The classical example of serendipity is the discovery of penicillin. Alexander Fleming observed that the fungal contaminant inhibited the bacterial growth. The saga of antibiotics started by a chance [23].

> *Traditionally, serendipity discoveries are understood as accidental findings made when the discoverer is in quest for something else...... Serendipity is more than just the irrational part of certain scientific discoveries. It cannot be denied that there are "accidental" aspects in serendipity that may sometimes be crucial* [24].

Kubinyi [22] reported the list of serendipitous discoveries in drug research. The role of the fluoride ions in the stimulation of the adenylyl cyclase leading to the formation of cAMP was surprising and could not be explained for the next two decades [25]. This started new era of ligand-receptor mediated transmembrane signal mechanisms.

Serendipitous discoveries were made in preclinical as well as clinical phases of the drug discovery. More number of discoveries by chance were made in the clinical studies. Nearly a quarter of the medicines which are prescribed are the derivatives of serendipitously discovered chemicals [26]. Serendipity played a significant role in the discovery of numerous drugs currently being used for various lifestyle diseases. Several drugs which are used for the treatment of different types of cancers are discovered by chance observation, viz., artemisinin (antimalarial), acetylsalicylic acid (rheumatism), etoposide (cathartic), leucovorin (growth factor), metformin (antidiabetic), rapamycin (antimicrobial), streptozotocin (antibiotic), thalidomide (morning sickness) and vinblastine (antidiabetic) [27].

Some of the cardiovascular drugs are due to chance observations. Discovery of dicoumarol was an accidental observation that the cattle were dying due to internal haemorrhage after feeding on sweet clover [28]. The application of the venom from the Brazilian pit viper lowered blood pressure by Sergio Ferreira and Sir John Vane lead to understand the mechanism of renin angiotensin system [29]. "*The phenotype-based small-molecule discovery approaches are beginning to complement the more established target-based approaches to cardiovascular drug discovery*" [30].

The discovery of Viagra from NO was serendipity and it was well documented in the reviews [31, 32]. Initially, the research related to NO was with respect to its role in environmental chemistry and toxicology. Later, its role in the function of immune system and as a signalling molecule was reported [32]. Zopal and co-workers showed that NO can be used to treat acute respiratory distress syndrome [33]. The major therapeutic application of NO function is sildenafil. Pfizer originally thought that sildenafil can treat hypertension. However, it later turned out to be useful to treat erectile dysfunction. The discoveries associated with NO functions in human body have many unexpected and serendipitous outcomes.

Ban [34] reviewed the serendipitous discovery of drugs, chloral hydrate, lithium, meprobamate, chlordiazepoxide, chlorpromazine, imipramine and iproniazid which are acting on central nervous system. Serendipitous discoveries were the start of an era of psychopharmacological drug discovery and development. Later, the focus is to develop drugs having more selectivity for therapeutic targets. This resulted in less side effects and increased safety [35].

Serendipity also played role in the discovery of drugs for treating infectious diseases. The case study of levamisole indicates that choosing the right animal model can play role in successful development of drug. The metabolite of the test compound by the chickens modified to levamisole [36]. Acyclovir, used for the treatment of herpes simplex virus (HSV) infections, could be considered as serendipitous discovery [37].

Serendipity played a pivotal role in several discoveries of natural sciences. Lenox advocated for observations rather than seeing the things in laboratory training. *In every course, the student's modes of both observation and discovery should be examined, questioned, and shared with classmates* [38]. Serendipitous or chance discoveries along with the rational discoveries will contribute to provide solutions to the unmet needs of the humankind.

1.3.1 Chemistry-Stereochemistry and Allostery: Drug Discovery and Development

The discoveries in the nineteenth century in chemistry (organic chemistry) paved way for the rapid drug discovery. The concept of isomerism (Friedrich Wohler), synthesis of urea from inorganic material (Wohler) and synthesis of the mauveine, synthetic dye (William Perkin), started a new era of organic compounds as medicinal agents and in helping to understand how the drugs interact with biological systems. Louis Pasteur studied the impact of stereochemical aspects of the molecules action on the bacteria (see Ref. [39]).

The initiation of pharmacological response involves the formation of a complex between the drug/ligand and its site of action or receptor [7]. This interaction is dependent not only on the chemical structure of the drug/ligand which intern control the physicochemical properties but also spatial arrangement of the functional groups in the drug molecule. The drug-receptor interaction is highly dependent on the stereochemistry [40–43]. Further the stereochemistry of the drug will affect the pharmacokinetics and clinical pharmacology of the drug and thereby alter the concentration of the drug in the biosphere and the effect. The l-norepinephrine is 100 times more potent when compared to d-norepinephrine in producing effects at alpha-adrenoceptors due to its stereoisomerism. Easson and Stedman [44] proposed three-point interaction with the adrenoceptors. Based upon the interactions of the ligands and receptors the structure activity relationships for various classes of drugs were proposed.

Chirality is ubiquitous in the small molecules (ligands) and macromolecules of the biological systems. Majority of the drugs are racemic mixture and the enantiomers of the chiral drugs though have same physical and chemical properties differ in pharmacological and biological activities [45]. The other enantiomer (s) pharmacodynamically may have partial agonistic-, antagonistic- or no-activity. On the other hand, it may produce a different activity or toxicity [45, 46]. In addition, the stereoisomers may have different pharmacokinetic characteristics (absorption, distribution, metabolism, and excretion (ADME) [45]. The thalidomide incidence causing the teratogenicity is good example for interconversion of the isomers in the biological systems [47, 48].

Srinivas and co-workers reviewed the issues, considerations, and regulatory requirements related to enantiomer drug development [49]. The USDFA guidance document (https://www.fda.gov/regulatory-information/search-fda-guidance-documents/development-new-stereoisomeric-drugs) was issued in May 1992. As a result of the unexpected toxicity the regulatory approval process has become stringent and time consuming. Atropisomerism may give rise to geometrical isomers, diastereoisomers, or enantiomers and can interconvert may cause in design and development of new drugs and their regulations [50, 51].

The concept of allosterism was proposed as early as 1965 as the Monod–Wyman–Changeux model to explain the mechanisms involved in the action of ligands with bacterial enzymes [52]. The interest in allosterism was developed with the clinical success of the benzodiazepines [53]. Several classes of allosteric drugs affecting various biological targets were identified by both academia and pharmaceutical industry involved in allosteric drug discovery [54].

Majority of the drugs and endogenous substances act on the receptors present on the cell surface. Initial studies on characterization of drug receptor interactions are based on endogenous ligands with orthosteric site. Subsequently it is observed that the drugs can interact with other binding sites on the receptors which are known as allosteric sites. The binding to these sites can result in increased selectivity and decreased toxicity. Molecular mechanisms involved in allosteric modulation of receptor activity played a pivotal role in drug discovery and development due to innumerable number of possibilities to obtain selective drugs for various targets and to minimize the toxic effects [55, 56].

G protein coupled receptors (GPCRs) reported to interact allosterically with number of ligands. Kenakin and Laurence [56] reviewed allosteric interactions involving GPCR mechanisms and their effects. The authors discussed the functional selectivity (biased agonism and biased antagonism) and its potential therapeutic applications. The explosion of technology that has enabled observation of diverse 7TMR behaviour has also shown how drugs can have different efficacies depending upon how the ligand interacts and can elicit different effects [57].

The computational methods [58] played critical role in predicting the drug-receptor interactions [7] and effects and contributed lot to the allosterism-based drug discovery. The computational methods to identify the allosteric interactions in drug discovery and the role it can play in predicting the drug resistance and selectivity are reviewed [59–61].

Allostery in endogenous proteins produces disease and contributes to drug discovery and development. Better understanding of disease symptoms at the molecular level and specific altered allosteric interactions can leads to innovative and safe therapies [62].

1.4 Chemical Biology and Drug Discovery and Development

The identification of correct lead via assessing its biological/pharmacological properties, and understanding of structure activity relationships are essential for successful drug development. Initially the properties of the leads were determined mainly by pharmacological methods. Pharmacological receptor characterization was done on the end organ responses. In general, in vitro and in vivo pharmacological screening methods are used to test the ligand-mediated effects. Small molecules are screened for activity using the receptor protein by either conventional or high-throughput screen methods [63]. Receptor binding assays played prominent role in the identification of the target and ligand interactions which paved ways to drug discovery [64]. Target identification, identification of transmembrane signal mechanisms and downstream pathways that mediate the effects have important roles in drug discovery [65]. Target identification, deconvolution and validation are reviewed [66].

Genetics and genetic modifications [67] greatly influenced in understanding the post drug receptor interactions and downstream pathways. CRISPR-Cas gene editing is "ready to have immediate impact in real world drug discovery" [68]. CRISPR-Cas help in identifying the target which is critical for successful drug discovery. The technology will help in to switch off /knock out the specific genes and obtain the desired mutations. Variations in the genetic composition lead to alterations in the metabolic enzymes and thereby the sojourn of the drug in the body. Hence the pharmacokinetic parameters [69] vary accordingly [70]. Genome-Wide Association Studies (GWASs) identified the human genes which are specifically associated with diseases and the targets for drug discovery [71]. Amalgamation of the network biology/ pharmacology with genetics will play prominent role in successful drug discovery and drug repurposing [72]. Genetics played a critical role in the drug development (see Ref. [73]).

Advances in biotechnology and rDNA technology produced human insulin and erythropoietin in cell culture and the monoclonal antibodies started new era of biological drugs [74]. The bioavailability of the macromolecules is very poor and require the technologies of drug delivery. Nanotechnology provides solution to site-specific delivery and increases bioavailability. The biological/toxicological properties of these engineered nanomaterials are different from the bulk materials. The safety and toxicity properties need to be tested other than conventional methods [75, 76].

Once the successful preclinical evaluation of the investigational new completed, the clinical trials (Phase I–Phase IV) and pharmacovigilance studies are carried out.

The regulatory sciences and regulations evolved and analysis and interpretation of the data are carried out by statistical methods.

1.5 Future Drug Discovery and Development

The cost of marketing a new drug is billions of US$ and it is increasing year after year due to high attrition rate in Phase III clinical trials. To overcome the failures of the drugs in various stages of clinical studies, increasing the reproducibility of the results of biomedical experiments, a better understanding of the processes in the disease states and gaining experience to manage the translational research in academic institutions are prime requirements [77].

Disruptive technologies are helpful to understand the downstream pathways involved in normal and disease state. The advent of virtual reality for the drug discovery has significant advantages for screening of the novel drugs [78]. New virtual reality methodologies are being used in the drug development. Artificial intelligence provides critical support in data mining, curation and management of the drug discovery big data [83]. AI can provide solution to drug discovery intricacies. Judicious use of these methodologies will provide better results [79]. Human induced stem cell technology is very useful for cell-based drug discovery to screen the novel molecules for lifestyle diseases [80, 81]. Organoids which are derived from stem cells resemble the original organs, and are used in drug discovery and development [82]. Multiple assays are required to determine the efficacy, selectivity, and safety of compounds during early stages of drug development to reduce attrition rates and lead to successful drug discovery [80]. Digital technologies will enhance faster and successful drug discovery to meet the unmet needs of humans.

References

1. Gray TC. The use of D-Tubocurarine chloride in anaesthesia. Ann R Coll Surg Engl. 1947;1:191–203.
2. Bernard C. Lessons on the effects of toxic and medicinal substances (in French). Paris: J.B. Baillière; 1857. p. 369–80.
3. Dale HH. The effect of certain esters and ethers of choline and their relation to muscarine. J Pharmacol Expt Ther. 1914;6:147–90.
4. Langley JN. On the contraction of muscle, chiefly in relation to the presence of "receptive" substances: part I. J Physiol. 1907;31:347–84.
5. Langley JN. Preliminary notice of experiments on the physiological action of jaborandi. Br Med J. 1875;1:241–2.
6. Maehle A-H. "Receptive substances": John Newport Langley (1852-1925) and his path to a receptor theory of drug action. Med Hist. 2004;48:153–74.
7. Poduri R, Jagadeesh G. The concept of receptor and molecule interaction in drug discovery and development. Chapter 3 of present book; 2020

8. Medina-Franco JL, Giulianotti MA, Welmaker GS, Houghten RA. Shifting from single to the multiple paradigms in drug discovery. Drug Discov Today. 2013;18:495–501.
9. Sneader W. Drug discovery. A history. Sussex: Jhon Wiley & Sons; 2005.
10. Powell MA. Drugs and pharmaceuticals in ancient Mesopotamia. In: Jacob I, Jacob W, editors. The healing past. Pharmaceuticals in the biblical and rabbinic world. Leiden: E.J. Brill; 1994. p. 54–67.
11. Longrigg J. Greek rational medicine. Philosophy and medicine from Alcmaeon to the Alexandrians. London: Routledge; 1993.
12. Newman DJ, Cragga GM, Snaderb KM. The influence of natural products upon drug discovery. Nat Prod Rep. 2000;17:215–34.
13. Feuerstein G, Kak S, Frawley D. In search of the cradle of civilization: new light on ancient India. Wheaton, IL: Quest Books; 1995.
14. Patwardhan B. Bridging ayurveda with evidence-based scientific approaches in medicine. EPMA J. 2014;5:19. https://doi.org/10.1186/1878-5085-5-19.
15. Gerhardt CF. Untersuchungen über die wasserfrei organischen Saüren. Ann Chem Pharm. 1853;87:149–79.
16. Harvey AL, Edrada-Ebel R, Quinn RJ. The re-emergence of natural products for drug discovery in the genomics era. Nat Rev Drug Discov. 2015;14:111–29.
17. Burg RW, Miller BM, Baker EE, Birnbaum J, Currie SA, Hartman R, Kong Y-L, Monaghan RL, Olson G, Putter I, Tunac JB, Wallick H, Stapley EO, Oiwa R, Ōmura S. Avermectins, new family of potent anthelmintic agents: producing organism and fermentation. Antimicrob Agents Chemother. 1979;15:361–7.
18. Tu YY, Ni MY, Zhong YR, Li LN, Cui SL, Zhang MQ, Wang XZ, Liang XT. Studies on the constituents of Artemisia annua L. (author's transl). Yao Xue Xue Bao. 1981;16:366–70.
19. Efferth T, Zacchino S, Georgiev MI, Liu L, Wagner H, Panossian A. Nobel prize for artemisinin brings phytotherapy into the spotlight. Phytomedicine. 2015;22:A1–3.
20. Singh IP, Ahmad F, Chatterjee D, Bajpai R, Sengar N. Natural products: drug discovery and development. Chapter 2 of the present book; 2020.
21. Roberts RM. Serendipity: accidental discoveries in science. New York: John Wiley; 1989.
22. Kubinyi H. Chance favours the prepared mind-from serendipity to rational drug design. J Receptor Signal Trans Res. 1999;19:15–39.
23. Lobanovska M, Pilla G. Penicillin' discovery and antibiotic resistance: lessons for the future? Yale J Biol Med. 2017;90(1):35–145.
24. Garcia P. Discovery by serendipity: a new context for an old riddle. Found Chem. 2009;11:33–42.
25. Sternweis PC, Gillman AG. Aluminium: a requirement for the activation of the regulatory component of adenylate cyclase by fluoride. Proc Natl Acad Sci U S A. 1982;79:4888–91.
26. Hargrave-Thomas E, Yu B, Reynisson J. Serendipity in anticancer drug discovery. World J Clin Oncol. 2012;3:1–6.
27. Prasad S, Gupta SC, Aggarwal BB. Serendipity in cancer drug discovery: rational or coincidence? Trends Pharmacol Sci. 2016;37:435–50.
28. Mueller RL, Scheidt S. History of drugs for thrombotic disease: discovery, development, and directions for the future. Circulation. 1994;89:432–49.
29. Douglas SA, Ohlstein EH, Johns DG. Techniques: cardiovascular pharmacology and drug discovery in the 21st century. Trends Pharmacol Sci. 2004;25:225–33.
30. Schlueter PJ, Peterson RT. Systematizing serendipity for cardiovascular drug discovery. Circulation. 2009;120:255–63.
31. Li JJ. Laughing gas, Viagra, and Lipitor: the human stories behind the drugs we use. New York: Oxford University press; 2006.
32. Marletta MA. Serendipity in discovery: from nitric oxide to Viagra. Proc Am Philos Soc. 2017;162:189–201.

33. Rossaint R, Lewandowski K, Zapol WM. Our paper 20 years later: inhaled nitric oxide for the acute respiratory distress syndrome—discovery, current understanding, and focussed targets of future applications. Intensive Care Med. 2014;40:1649–58.
34. Ban TA. The role of serendipity in drug discovery. Dialogues Clin Neurosci. 2006;8:335–44.
35. Robinson E. Psychopharmacology: from serendipitous discoveries to rationale design, but what next? Brain Neurosci Adv. 2018;2:1–11.
36. Campbell WC. Serendipity and new drugs for infectious disease. ILAR J. 2005;46:352–6. https://doi.org/10.1093/ilar.46.4.352.
37. De Clercq E. Curious discoveries in antiviral drug development: the role of serendipity. Med Res Rev. 2015;35:698–719.
38. Lenox RS. Educating for the serendipitous discovery. J Chem Educ. 1985;62:282–5.
39. Chast F. A history of drug discovery. In: Wermuth CG, editor. The practice of medicinal chemistry. 3rd ed. Amsterdam: Elsevier; 2008.
40. Ariens EJ. Stereochemistry: a source of problems in medicinal chemistry. Med Chem Rev. 1986a;6:451–66.
41. Ariens EJ. Chirality in bioactive agents and its pitfalls. Trends Pharmacol Sci. 1986b;7:200–5.
42. Ariens EJ. Stereochemistry in the analysis of drug-action. Part II. Med Chem Rev. 1987;7:367–87.
43. Ariens EJ. Nonchiral, homochiral and composite chiral drugs. Trends Pharmacol Sci. 1993;14:68–75.
44. Easson LH, Stedman E. CLXX studies on the relationship between chemical constitution and physiological activity. Biochem J. 1933;27:1257–66.
45. Campo VL, Lílian SC, Bernardes LSC, Carvalho I. Stereoselectivity in drug metabolism: molecular mechanisms and analytical methods. Curr Drug Metab. 2009;10:188–205.
46. Nilos ME, Gan J, Schlenk D. Effects of chirality on toxicity. In: Ballantyne B, Marrs TC, Syversen T, Casciano DA, Sahu SC, editors. General, applied and systems toxicology: Wiley; 2011. https://doi.org/10.1002/9780470744307.gat031.
47. Brocks DR. Drug disposition in three dimensions: an update on stereoselectivity in pharmacokinetics. Biopharm Drug Dispos. 2006;27:387–406.
48. Waldeck B. Three-dimensional pharmacology, a subject ranging from ignorance to overstatements. Pharmacol Toxicol. 2003;93:203–10.
49. Srinivas NR, Barbhaiya RH, Midha KK. Enantiomeric drug development: issues, considerations, and regulatory requirements. J Pharm Sci. 2001;90:1205–15.
50. Clayden J, Moran WJ, Edwards PJ, LaPlante SR. The challenge of atropisomerism in drug discovery. Angew Chem Int Ed. 2009;48:6398–401.
51. Joshi G, Kaur M, Kumar R. Dynamic axial chirality in drug design and discovery: Introduction to atropisomerism, classification, significance, recent trends and Challenges. Chapter 4 of present book; 2020.
52. Monod J, Wyman J, Changeux J-P. On the nature of allosteric transitions: a plausible model. J Mol Biol. 1965;12:88–118.
53. Mohler HF, Fritschy JM, Rudolph U. A new benzodiazepine pharmacology. J Pharmacol Exp Ther. 2002;300:2–8.
54. Wenthur CJ, Gentry PR, Mathews TP, Lindsley CW. Drugs for allosteric sites on receptors. Annu Rev Pharmacol Toxicol. 2014;54:165–84.
55. Christopolous A. Allosteric binding sites on cell surface receptors novel therapeutic targets for drug discovery. Nat Drug Rev. 2002;1:198–210.
56. Kenakin T, Laurence JM. Seven transmembrane receptors as shapeshifting proteins: the impact of allosteric modulation and functional selectivity on new drug discovery. Pharmacol Rev. 2010;62:265–304.
57. Gaddam RR, Vikram A. Biased agonism: renewing GPCR's targetability for the drug discovery. Chapter 5 of present book; 2020.
58. Bharatam PV. Computer-aided drug design. Chapter 6 of present book; 2020.

59. Amamuddy OS, Veldman W, Manyumwa C, Khairallah A, Agajanian S, Oluyemi O, Verkhivker GM, TastanBisho O. Integrated computational approaches and tools for allosteric. Drug Discovery Int J Mol Sci. 2020;21:847. https://doi.org/10.3390/ijms21030847.
60. Andre AS, Ribeiro T, Ortiz V. A chemical perspective on allostery. Chem Rev. 2016;116:6488–502.
61. Wagner JR, Lee CT, Durrant JD, Malmstorm RD, Feher VA, Amaro RE. Emerging computational methods for the rational discovery of allosteric drugs. Chem Rev. 2016;116:6370–90.
62. Nussinov R, Tsai C-J. Allostery in disease and in drug discovery. Cell. 2013;153:293–306.
63. Kumar NVS. Pharmacological screening—drug discovery. Chapter 7 of present book; 2020.
64. Bylund DB, Enna SJ. Receptor binding assays and drug discovery. Adv Pharmacol. 2018;82:21–33.
65. Schenone M, Dančík V, Wagner BK, Clemons PA. Target identification and mechanism of action in chemical biology and drug discovery. Nat Chem Biol. 2013;9:232–40.
66. Gullapall S. Drug target identification and validation. Chapter 8 of present book; 2020.
67. Poduri A, Khanna A. Genetics and drug discovery. Chapter 9 of present book; 2020.
68. Fellmann C, Benjamin G, Gowen BG, Lin PC, Doudna JA, Corn JE. Cornerstones of CRISPR–Cas in drug discovery and therapy. Nat Rev Drug Discov. 2017;16:89–101.
69. Sharma P, Patel N, Prasad B, Varma M. Pharmacokinetics: theory and application in drug discovery and development. Chapter 11 of present book; 2020.
70. McCarthy AD, Kennedy JL, Middleton LT. Pharmacogenetics in drug development. Philos Trans R Soc B. 2005;360:1579–88.
71. Visscher PM, Wray NR, Zhang Q, Sklar P, McCarthy MI, Brown MA, Yang J. 10 years of GWAS discovery: biology, function, and translation. Am J Hum Genet. 2017;101:5–22.
72. Nabirotchkin S, Peluffo AE, Rinaudo P, Yu J, Hajj R, Cohen D. Next-generation drug repurposing using human genetics and network biology. Curr Opin Pharmacol. 2020;51:78–92. https://doi.org/10.1016/j.coph.2019.12.004.
73. Duggar SA, Platt A, Goldstein DB. Drug development in the era of precision medicine. Nat Rev Drug Dev. 2018;17:183–96.
74. Malaviiya AN, Mehra NK. A fascinating story of the discovery & development of biologicals for use in clinical medicine. Indian J Med Res. 2018;148:263–78.
75. Saifi, M.A., Poduri, R. and Godugu, C. (2020). Nanomedicine: implications of nanotoxicology. Chapter 13, the present book.
76. Udupa, V. and Rao, K.S. (2020). Regulatory toxicology testing of pharmaceuticals. Chapter 12, the present book.
77. Poduri R. (2020). Pharmaceutical industry, academia, regulatory authorities and end user collaboration in successful drug discovery and development. Chapter 17 of present book.
78. Liu XH, Wang T, Lin J-P, Wu M-B. Using virtual reality for drug discovery: a promising new outlet for novel leads. Expert Opinion Drug Discov. 2018;13:1103–14.
79. Álvarez-Machancoses Ó, Fernández-Martínez JL. Using artificial intelligence methods to speed up drug discovery. Expert Opin Drug Discovery. 2019;14:1–9.
80. Herholt A, Galinski S, Geyer PE, Rossner MJ, Wehr MC. Multiparametric assays for early drug discovery. Trends Pharmacol Sci. 2020;41:318–35. https://doi.org/10.1016/j.tips.2020.02.005.
81. Kumar AHS. Discovery and development of stem cells for therapeutic applications. Chapter 12 of present book; 2020.
82. Takahashi T. Organoids for drug discovery and personalized medicine. Annu Rev Pharmacol Toxicol. 2019;59:447–62.
83. Zhu H. Big data and artificial intelligence modeling for drug discovery. Annu Rev Pharmacol Toxicol. 2020;60(1):573–89.

Chapter 2
Natural Products: Drug Discovery and Development

Inder Pal Singh, Furkan Ahmad, Debanjan Chatterjee, Ruchi Bajpai, and Neha Sengar

2.1 Drug Discovery and Development

The classical drug discovery and development is one of the most challenging efforts that directly benefit the mankind by providing new medicines. The process is long time consuming and cost intensive despite the recent technological advances. It is estimated that 1 out of 10,000 molecules synthesized or isolated is finally approved for human use. The cost of developing a new drug is close to 1.5–2 billion US dollars and takes up to 15 years from discovery stage to drug approval. The major cost is for clinical trials as compared with the preclinical drug discovery process, making a careful strategization of initial drug discovery process all the more important. Therefore it is more important to identify the possible dropouts at an early stage. The three major reasons for compounds failing to be successful drugs are toxicity, lack of efficacy, and lack of bioavailability [1]. Currently, the medicinal chemists have the advantage of guiding principles that have developed over the decades of experiences in order to design and optimize lead compounds. The first guidelines introduced as Lipinski's rule of five states that orally active drug should obey these five criterion: molecular weight less than 500 Da, calculated partition coefficient clog*P* less than 5, hydrogen bond donors less than 5, hydrogen bond acceptors less than 10, and number of rotatable bonds more than 10. Several other parameters were later added to the five proposed by Lipinski. These drug design guidelines were later followed by several others such as quantitative estimate of drug-likeness (QED). These computation models provide target molecules based on desirable ADMET (absorption, distribution, metabolism, excretion, and toxicity) properties and these models are fairly able to predict the above properties. However, some of the most

I. P. Singh (✉) · F. Ahmad · D. Chatterjee · R. Bajpai · N. Sengar
Department of Natural Products, National Institute of Pharmaceutical Education and Research (NIPER), SAS Nagar, Punjab, India
e-mail: ipsingh@niper.ac.in

R. Poduri (ed.), *Drug Discovery and Development*,
https://doi.org/10.1007/978-981-15-5534-3_2

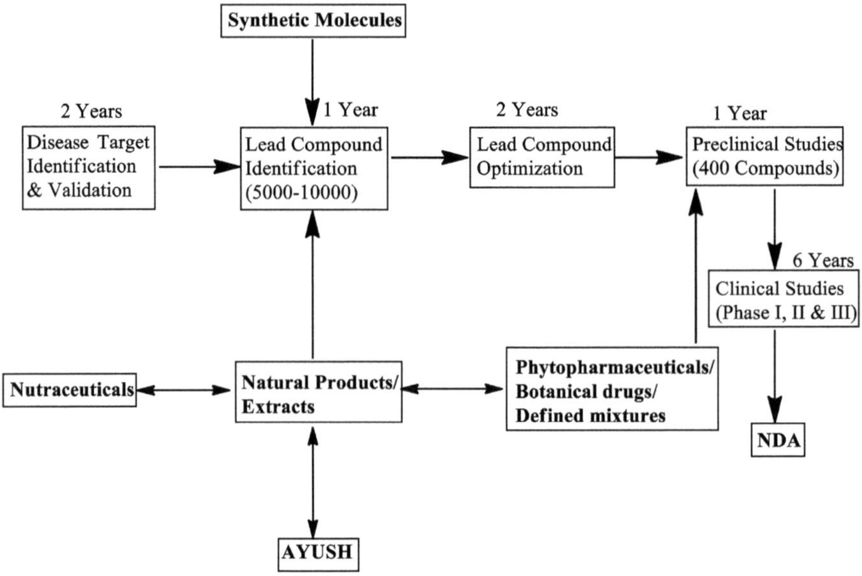

Fig. 2.1 Drug discovery and development process stages with approximate time lines [3–5]

important natural product drugs such as taxol, amphotericin B, vancomycin, and several others do not follow Lipinski's rule of five [2].

The process of drug discovery and development proceeds through several stages as shown in Fig. 2.1. Discovery begins with target identification, choosing a specific target involved in the biochemical pathway by understanding the disease condition at genetic, protein, and cellular level. This is followed by validating the target by confirming that it is actually involved in the disease and can interact with the drug. Several thousand molecules (synthetic, natural, semisynthetic, natural-mimics) are evaluated for interaction with the drug target and one or more lead molecules are selected. The promising compounds are evaluated for early safety studies. ADMET properties of each promising compound are studied. The potential drug molecules should be absorbed in the bloodstream, distributed to the site of action, metabolized, excreted from the body, and should be nontoxic. Lead optimization involves modifications in molecular structure to obtain more effective and safe molecules. A number of structural motifs can impart desirable properties to lead molecules in order to improve ADMET properties. The leads thus obtained are subjected to preclinical studies to determine if the drug is safe for human testing through a battery of in vitro and in vivo assays. Large quantities of drug are required for the development process. Therefore the scale-up of the drug is required at this stage to start with the clinical trials. Investigational New Drug (IND) application is filed before the trial can begin. The application includes the results of preclinical work, chemical structure, its manufacturing details, and any side effects. Phase I clinical trial is performed on small group of healthy volunteers in order to see the safety of drug in humans and study pharmacokinetics and pharmacodynamics. Phase II trials are conducted on a small group of patients to evaluate efficacy and to see possible short-term side effects. Phase III trials are conducted on larger number of patients (1000–5000) to generate statistically significant data on safety and efficacy of the drug. This is the most expensive part of drug discovery and development process. New Drug Application is submitted after the completion of three phases of clinical trials. Phase IV trials after the approval of drug as larger number of patients use the drug involve monitoring for any adverse reactions and long-term safety of the drug [3].

2.2 Natural Products for Human Health

Historically, plants have been used since millennia for the treatment of many diseases and illnesses. Natural products (NPs) have long been a traditional source of drug molecules and modern drug discovery programs have relied extensively on natural sources in the past. Natural product (NP) chemistry has produced a huge diversity of secondary metabolites from a vast array of natural resources including plants, microorganisms, and animals both from terrestrial and marine origin. Several of these NPs are successful drugs or are drug candidates. The use of NPs as traditional medicines across several ancient civilizations is well documented in history.

The proteins, fats, nucleic acids, and carbohydrates known as primary metabolites are essential for living organisms and are the products of primary metabolism. In contrast, secondary metabolites are not essential for growth and development of an organism and are biosynthesized through several fundamental processes such as photosynthesis, glycolysis, and Krebs cycle. The most important building blocks involved in biosynthesis of secondary metabolites are acetyl coenzyme A, shikimic acid, and mevalonic acid resulting in an infinite diversity through several biosynthetic pathways, and it is the secondary metabolites that have played a major role in drug discovery efforts in the past [3].

Despite the enormous contribution of NPs in drug discovery, the last century being the most productive source of leads for drugs, however, there was a diminished interest in NPs in the pharmaceutical industry in the late 1980s and 1990s. This was in spite of the fact that there were major technological advancements in chemistry and biology. The combinatorial chemistry and high-throughput screening failed to show the anticipated impact or success. The pharma industry still faces challenges for the discovery and development of new drugs and requires major changes and new paradigms. It requires concerted efforts from biologists, synthetic and medicinal chemists, structure biologists, biotechnologists, computational experts, and clinicians to meet the challenges for new drug discovery and development. Discovery and development of natural products as modern drugs is a hugely challenging task. Besides being successful drugs, natural products also form the basis of several traditional medicine systems such as Ayurveda, Sidha, Unani, Homeopathy, Traditional Chinese medicines (TCM), Kampo, Tibetan medicine, etc. [6]. A new drug class regulated in India is phytopharmaceuticals which contain partially purified extracts from traditionally used medicines with defined minimum four constituents. NPs also play a great role in beneficial effects of nutraceuticals. Aspects of natural products in modern drug discovery and the development of Ayurveda, Siddha, and Unani (ASU) drugs is covered in this chapter.

2.3 Natural Products in Modern Drug Discovery and Development

Drug discovery and development of natural products is a challenging task, which may be based on folklore, traditional or alternative medicine, and is entirely different from developing herbal remedies or dietary supplements or nutraceuticals. The process of finding and discovering a new and effective drug from Mother Nature involves isolating and purifying a single active molecule, characterizing the molecule followed by the entire process as shown in Fig. 2.1. Many drugs that are available in the market were discovered from natural sources; the approved drugs are either NPs or botanical drugs (defined mixtures) or are natural product derivatives. David Newman and Gordon Cragg, in a series of reviews over the last 20 years, have analyzed the role that NPs have played as sources of new drugs. In cancer alone, about 175 small molecules were approved during the time period from 1940s to 2014 and out of these 85 (49%) were either natural products or their derivatives. The other area where NPs have played a prominent role is the infectious diseases. During the period 1981–2014, a total of 174 approved drugs for cancer included 17 NPs, one botanical drug (defined mixture), and 38 NP derivatives. Besides these, 33 were biological macromolecules. NPs have also made enormous contribution in antibacterial area: 140 approved drugs included 11 NPs and 71 NP derivatives. NPs drugs were also approved as immunosuppressant (5), immunostimulant (3), hypocholesterolemic (4), antiparasitic (2), and one each for several other disease conditions. Four botanical drugs (defined mixtures) were also approved during this time period. This analysis by Newman and Cragg clearly provides an optimistic scenario for the important role of natural products in modern drug discovery [7].

Natural resources play a major role as starting materials for drug discovery. It has been estimated that nearly 75,000 species of higher plants exist and only 10% have been used in traditional medicine. Only 1–5% of these have been studied scientifically. Scaffold diversity is one of the most important features of a compound library that determines its success in screening hits when identifying bioactive molecules. Structural diversity in natural products is enormous underlining their importance in drug discovery and development. NPs often provide selective and specific biological activities based on mechanisms of action, examples are the HMG-CoA reductase inhibition by lovastatin and tubulin-assembly by paclitaxel. Pharmacologically active compounds from plants and microbes represent an important pipeline for a new investigational drug [8].

2.3.1 Historical Perspectives

Natural products have been used since ancient times for the treatment and prevention of many diseases. The earliest natural products were described on clay tablets in cuneiform from Mesopotamia era (2600 B.C.), oils from *Cupressus sempervirens* (Cypress) and Commiphora species (myrrh); these are still in use to treat coughs, colds, and inflammation. The Ebers Papyrus (2900 B.C.) is an Egyptian medical text compilation, which has 700 plant-based drugs ranging from gargles to ointments. The first and most famous well-known example of a natural product as a drug is morphine that was isolated from *Papaver somniferum* L. (opium poppy). Morphine was commercialized by Merck in 1826. Heroin (diacetylmorphine) was prepared by boiling crude morphine (isolated from the plant *P. somniferum*) with acetic anhydride. Codeine is also easily obtainable by simple methylation of morphine. The first NP-based semisynthetic pure drug was Aspirin (anti-inflammatory agent). It was derived from the natural product salicin isolated from the bark of the willow tree *Salix alba* L. and introduced by Bayer in 1899. Another example is penicillin from *Penicillium notatum* discovered by Fleming in 1929. Digitoxin is cardiotonic glycoside isolated from *Digitalis purpurea* L. (foxglove), which is frequently used in congestive heart failure. Paclitaxel (Taxol), the most widely used drug for breast cancer, was isolated from the bark of *Taxus brevifolia* (Pacific Yew). Pilocarpine, an alkaloid containing a histidine ring, was isolated from *Pilocarpus jaborandi* (Rutaceae). Pilocarpine has been shown to be effective in chronic glaucoma. Pilocarpine is also used to treat dry mouth (xerostomia) resulting from radiation therapy in cancer patients. Quinine and artemisinin are well-known anti-malarial drugs [9].

2.3.2 Sources of Natural Products

Natural products (NPs) with therapeutic potential can be obtained from various natural resources that include plants, animals, marine organisms, bacteria, yeasts, molds, fungi, etc. Therefore, it becomes very important to do a thorough literature search so that proper natural resource is selected. Most of the discoveries in the past from natural products have come out of serendipity. The selection can be based on phenotypic screening or target-based screening. There has been an explosion of data in biomedical sciences in the last 3–4 decades, and the main focus of drug discovery has been on drug targets using rational drug design strategies. Drug discovery in the past was mainly driven by serendipity with limited knowledge on molecular mechanisms of disease. The important factors to consider in choosing a starting point (plant/NP) are the history, folklore, medicinal use, availability, literature on pharmacology, toxicology, etc. Huge number of internet resources, retrieval databases, books, and primary literature are available to aid the process of selection of right resource for the intended purpose.

2.3.2.1 Plants

Plants produce a variety of different classes of compounds/secondary metabolites with diverse biological properties. In the last seven decades, several plant-derived compounds have been approved as antineoplastic agents; these include vinblastine, vincristine, vinorelbine, etoposide, teniposide, paclitaxel, docetaxel, topotecan, and irinotecan. Plants have provided useful drugs for many other disease conditions. NPs isolated and identified from both higher and lower plants will undoubtedly continue to provide useful drugs (Fig. 2.2).

2.3.2.2 Animals

The skin of poisonous Ecuadorian frog is a source of Epibatidine which is ten times more potent than morphine and has led to a new class of analgesics. Teprotide from the extract of Brazilian viper led to the development of enalopril and captopril, which are effective angiotensin-converting enzyme (ACE) inhibitors and have proved to be anti-hypertensive drugs (Fig. 2.3).

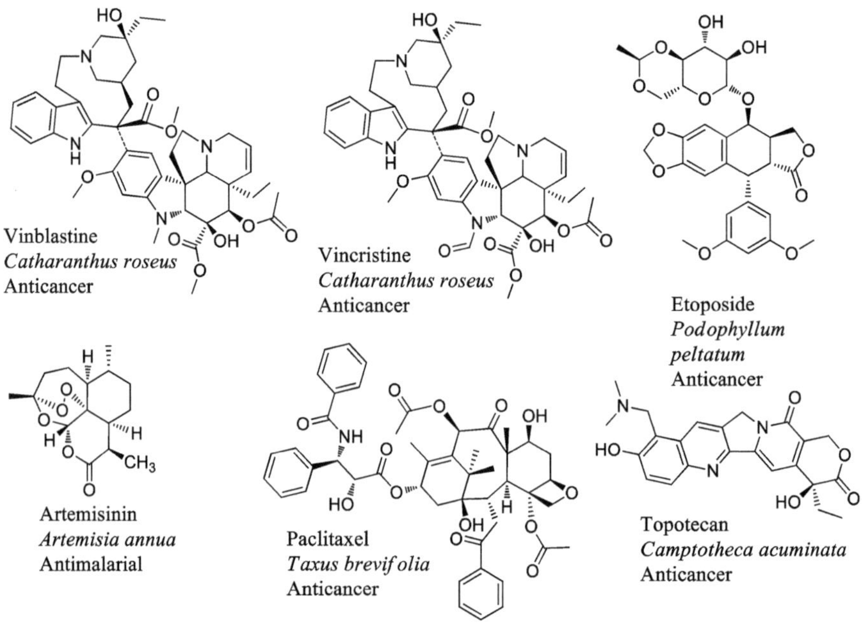

Fig. 2.2 Drugs from plants

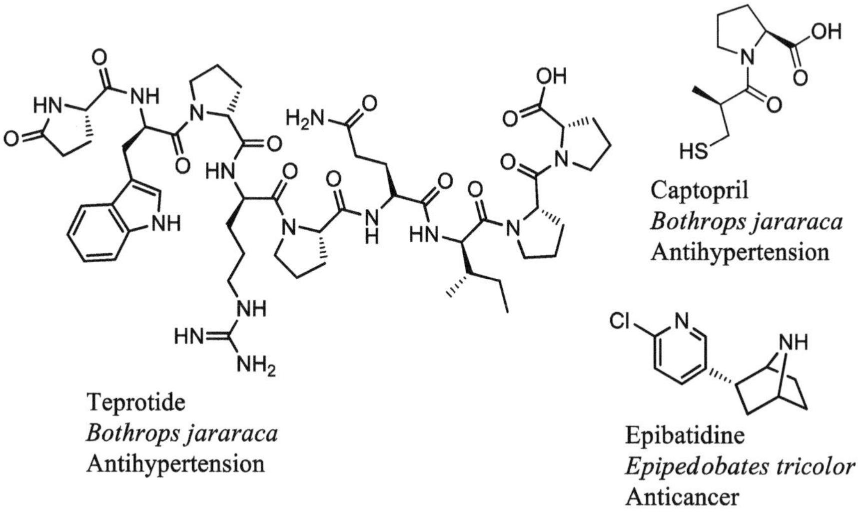

Fig. 2.3 Drugs from animals

2.3.2.3 Marine

Marine environment has proven to be a very rich source of bioactive secondary metabolites with unusual skeletons not encountered in terrestrial environment. More than 70% of earth surface is covered by oceans which have biodiversity even greater than rainforests and provide greater opportunity for discovering novel drug molecules. The first marine natural product to become a successful drug was cytarabine (Ara-C, cytosine arabinoside) isolated from *Cryptotheca crypta*, which is currently used for treatment of leukemia and lymphoma. Plitidepsin, a depsipeptide, was isolated from the Mediterranean Tunicate *Aplidium albicans*. It is effective against various types of cancers. Ecteinascidin 743 (terbactedin, yondelis) was isolated from the ascidian *Ecteinascidia turbinata* and is used in soft tissue sarcomas and ovarian cancer. Discovery of Halichondrin B obtained from sponge *Halichondria okadai* as anticancer agent led to the approval of eribulin for metastatic breast cancer in 2010 (Fig. 2.4).

2.3.2.4 Microbial

Microorganisms have already been proven as an excellent source of novel NPs primarily with antibiotic potential as well as several other therapeutic areas. The first antibiotic penicillin was discovered by Flemming from *Penicillium chrysogenum* (formerly *notatum*). Mitomycin is an antimitotic agent and rapamycin is an immunosuppressive agent. Chloramphenicol isolated from *Streptomyces venezuelae* is used in typhoid. Doxorubicin isolated from the fungus *Streptomyces*

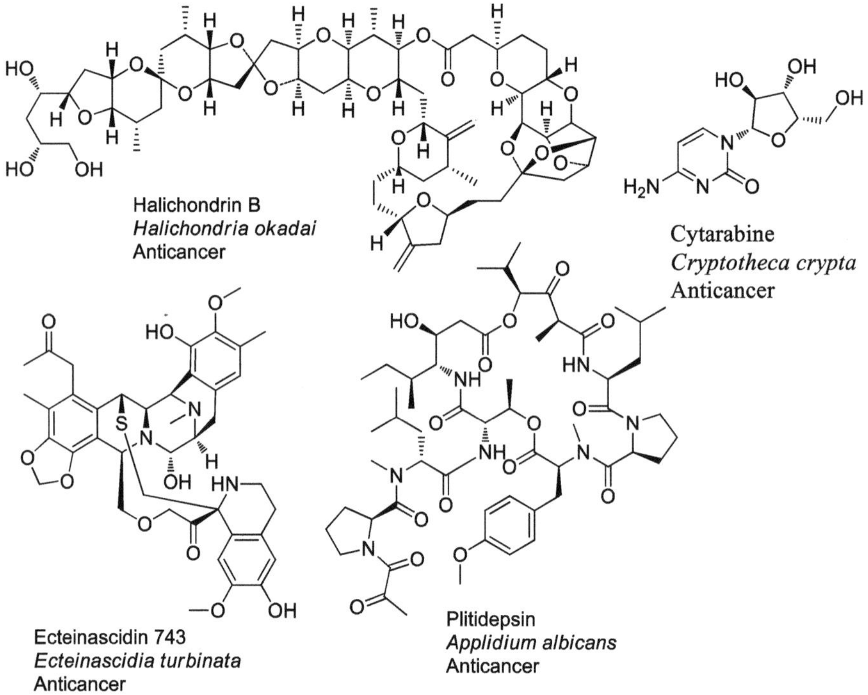

Fig. 2.4 Drugs from marine organisms

peucetius is used in the treatment of leukemia as well as Hodgkin and non-Hodgkin lymphomas [9] (Fig. 2.5).

2.3.3 *Extraction*

After careful selection of the natural resource for the discovery of known, new or novel drug molecules, the raw material has to be extracted with a suitable method/ solvent system so that the target class of molecules is extracted. Extraction is a process of removal of one or more component from liquid, semisolid, or solid raw material that may be a plant part, or microbial broth, or animal tissue, etc. It is the first step in the analysis of plant material and isolation and characterization of secondary metabolites. This is an intensive and time-consuming exercise, and together with the fact that amounts of active ingredients in natural raw materials are fairly low, the extraction and isolation of NPs is usually considered a bottleneck in the application of NPs in drug development. Solvent extraction is the most widely used method; the solvent penetrates into the solid matrix solubilizing the solute that diffuses out of the matrix. Any factor that would increase the solubility or diffusivity of the solute will facilitate the extraction. Therefore the properties of the extraction

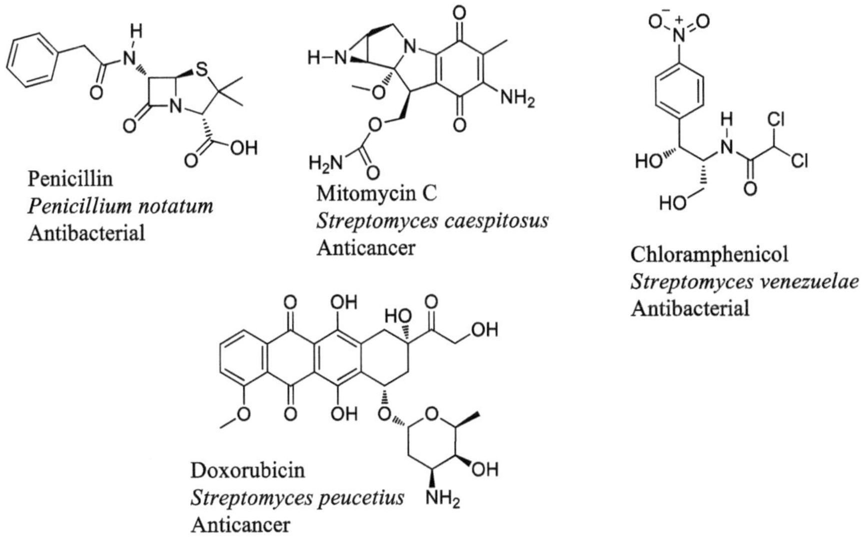

Fig. 2.5 Drugs from microbial sources

solvent, the particle size of the raw material, solvent-to-solid ratio, the extraction temperature, and the extraction duration considerably affect the extraction efficiency. The selection of solvent is crucial for solvent extraction; solvents with a polarity value close to the polarity of solute to be extracted are likely to give better extraction efficiency [10].

Several extraction methods have been employed for obtaining crude extracts from raw materials. **Maceration** is a very simple extraction method in which plant material is soaked in a suitable solvent with stoppered container kept at room temperature for a period of a minimum of 3 days with intermittent agitation followed by filtration. The disadvantage of this process is long extraction time, lesser efficiency, and huge solvent consumption. The major advantage is its suitability for the extraction of thermolabile components. **Percolator** is a conical vessel with an open top and adjustable bottom and has a closure to allow the passage of the fluid at a definite rate. The plant material is moistened with a sufficient amount of solvent prior to placing in the percolator. The material is placed in a percolator vessel so as to allow passage of fluid and complete contact with the plant material. The percolator must be filled with liquid and bottom outlet is opened for definite flow rate. The wet mass is pressed to extract the maximum residual fluid retained. Finally, an extract is obtained by filtration. This process is less time consuming and gives more complete extraction. In hot continuous extraction (**Soxhlet extraction**), the finely divided crude drug is placed in porous bag or thimble, the extracting solvent in the flask is heated and its vapors condensed into the condenser. The condensed extractant drops down into the thimble containing the crude drug. The liquid contents containing the extract siphon into the flask when the level of liquid in the chamber rises up to the top of the siphon tube. This process is continuous and carried out until a drop of solvent from the siphon tube does not leave a residue when evaporated. In Soxhlet extraction, larger amount of drug can be extracted with a small quantity of solvent. The major disadvantage is the raised temperature which may degrade thermolabile compounds. **Countercurrent extraction** (CCE) is a liquid–liquid extraction process in which two immiscible liquids move in the opposite direction in continuous contact with each other. In this process, crude material is crushed and prepared in the form of fine slurry. The material to be extracted (fine slurry) moves in one direction within a cylindrical extractor where it comes in contact with extraction solvent. Finally, concentrated extract comes out at one end of extraction assembly while the marc (residue left after extracting the desired constituents) comes out from the other end [11]. The **microwave-assisted extraction** (MAE) uses microwave as a source of energy to heat solvent in contact with the sample. MAE is very fast and uses very small amount of solvents and gives improved yield. In this process, microwave radiation interacts with dipoles of polarizable material (solvent and sample) making it a selective method for polar compounds. **Accelerated solvent extraction** (ASE) employs high temperature and pressure for extracting various constituents from the crude drug. The crude drug is packed with sand in an extraction cell forming layers of sand and sample. The automated extraction technology is able to control temperature and pressure and is a very fast technique usually complete within an hour. ASE is very efficient technique, despite using high temperature and pressure; it still can be used for thermally unstable compounds as the extraction time is very short. **Enzyme assisted extraction (EAE)** is a very safe, green, and novel approach for the extraction of bioactive compounds. The pressed material is treated with an enzyme such as cellulase, pectinase, and α-amylase. The use of this hydrolytic enzyme gives better yield due to the action of enzymes on the cell wall and membrane inside the cell that facilitates the release of the natural products [12]. **Ultrasound extraction (sonication)** uses ultrasound ranging from 20 to 2000 kHz, the ultrasound waves increase the surface contact between solvent and sample, the ultrasound energy breaks the cell wall that facilitates the release of compounds. It is a more effective extraction method for thermolabile compounds as moderate temperature is used and is very fast technique. The use of ultrasound more than 20 kHz may affect active compounds through the formation of free radicals [11]. **Super-critical fluid extraction (SCFE)** uses supercritical fluid as an extracting solvent. Carbon dioxide is the most commonly employed supercritical fluid (SCF). A supercritical fluid is more likely a gas but with solvation characteristics of a liquid. Supercritical carbon dioxide is good for extraction of non-polar solutes; for more polar compounds, small amount of methanol or ethanol is added as an auxiliary solvent. Major advantage of SCFE is that the gas is easily recycled, chemically inert, non-flammable, a natural substance, easily available, and easily removed from the product [10]. **Decoction** is the process in which a crude drug is boiled in water for certain time period, then it is cooled and the mixture is strained or filtered. This method is suitable for extracting water-soluble and heat-stable compounds. Generally roots, leaves, and flowers are boiled in water for about 15 min, while branches and other hard parts can require up to an hour [11]. **Aqueous alcoholic fermentation** is commonly employed in traditional medicine preparation such as Ashavas and

Arishtas in Ayurveda. The drug either in powder form or decoction is kept with jaggery and flowers of *Woodfordia fruticosa* at defined temperature for a certain period of time. Fermentation generates alcohol in situ, which serves as a preservative [11].

Once the material is extracted, two different strategies may be adopted for further research. In the olden times, where the main focus was on delineating the chemistry of different plants, the approach adopted was straight forward isolation and identification of compounds followed by biological evaluation. Most of the NP research before 1970s followed the above approach. The result was isolation and characterization of a huge number of natural products from different plant sources, which may or may not have shown or evaluated for any activity. For the isolation of a specific class of compounds, the individual fraction may be assayed through physico-chemical data such as analyzing each fraction by nuclear magnetic resonance (NMR) or mass spectrometry. The selection of plants for chemical investigations was indeed based on their ethnopharmacological information, traditional uses, or folkloric reputations. The other approach is the bioassay-guided (mainly in vitro) isolation of NPs as lead compounds. The crude extracts are subjected for preliminary screening against various biological or pharmacological assays in order to find the active extracts for a given assay relevant to the disease condition. A bioassay is an analytical procedure for qualitatively or quantitatively measuring the concentration or potency of target molecules/extracts by evaluating effect on living cells or tissues. A variety of preliminary bioassays to detect properties such as antifungal, anticancer, anti-HIV, antibacterial, and antidiabetic are available for various disease conditions and can be used to prioritize the extracts in terms of activity. These bioassays should be simple, fast, and sensitive since the amounts of active constituents in extracts may be very small. Bioassays could involve the use of in vivo systems (whole animal experiments), or in vitro systems (e.g., cultured cells). In vivo studies are more relevant to clinical conditions and can also provide toxicity data simultaneously. However, in vitro bioassays are microplate-based and require a small amount of extract, fraction, or compound for the assessment of activity [3].

2.3.4 Isolation and Purification

Each crude extract from a natural source usually contains a huge number of secondary and primary metabolites in a percentage usually ranging broadly between 1.0 and 0.0001 percent of the total biomass. The activity shown by the crude extract may be due to a single compound or to a mixture of compounds. Therefore, it becomes important to isolate and purify the NPs from the crude extract by chromatographic techniques. It is relatively easier to purify a compound which may be 1% of the biomass, but extremely difficult to purify a compound which is 0.0001%. Isolation is a process in which separation of a single pure compound occurs from many related molecules in the mixture. The design of isolation protocol is very important for successful accomplishment of this objective. First, it should be determined whether the objective of isolation is an unknown compound responsible for a particular biological activity, a known compound, a class of NPs within an organism, or structurally related compounds such as isomers or all the secondary metabolites produced by organisms. The isolation strategy will differ depending on what is to be isolated. For example, if an unknown compound responsible for a particular biological activity is to be isolated, bioassay-guided isolation should be tried where bioactivity at each stage of isolation is determined and only active fractions are taken forward for further isolation steps. For a known compound, it would be best to follow the literature methods used for isolation of given compound. For diastereoisomers, separation on normal phase or reverse phase chromatography can be tried, whereas for enantiomers, chiral chromatography is the answer. Ionic compounds are best separated using reverse phase chromatography, ion-pair chromatography, or ion-exchange chromatography. Biomolecules are separated using size exclusion chromatography. The general features of molecules that are helpful for the isolation process are its solubility, acid-base properties, charge, stability, and molecular size. For an unknown molecule, qualitative analytical assays such as TLC or the above properties would suggest the most suitable methods for isolation. Hydrophobicity or hydrophilicity of extract and compounds present in the extract is determined by trying to dissolve compound in a range of polarity solvents such as water, methanol, acetonitrile, and ethyl acetate. Acid-base properties are determined by dissolving the compound/extract in different pH solvents [10].

Another important question is the purpose of isolation, whether it is to obtain sufficient amount for chemical characterization or to generate a sufficient amount for biological evaluation. A pure compound can be characterized even with a few mg of the sample, with high-resolution NMR, it is possible to run all 1D and 2D NMR experiments with as small as 1–2 mg of the sample. In vitro experiments can also be conducted with few mgs of the sample. However, animal experiments require higher amounts of sample usually in gram quantities. Isolation strategies would vary depending on the above requirements. Further, 90–95% purity is good enough for chemical characterization, whereas more than 99% purity is required for biological evaluation. There are examples where even 0.1–0.2% impurity in a molecule has led to wrong assignment of biological activity to a major compound.

2.3.4.1 Chromatographic Techniques for Isolation

A variety of chromatographic techniques are available to a separation scientist for the isolation and purification of NPs. It is seldom possible to purify a NP using a single technique; rather it is generally a mix of techniques used one after the other to achieve final purification of a compound. A crude extract is generally first divided into different polarity fractions by liquid–liquid extraction using solvents in a wide polarity range, starting from the nonpolar hexane to chloroform to ethyl acetate to butanol. This would generally provide five fractions, the hexane fraction containing nonpolar compounds such as terpenoids, ethyl acetate fraction containing medium polar compounds such a flavonoids, and butanol fraction containing alkaloids, glycosides, or other polar compounds. These fractions are then subjected to different chromatographic techniques to isolate a pure natural product. A brief introduction to various chromatographic techniques is given below, for details on chromatography the reader is referred to several excellent books on chromatography.

2.3.4.2 Thin-Layer Chromatography (TLC)

It is a preferred method for qualitatively analyzing natural mixtures before proceeding with other chromatographic techniques. A crude extract dissolved in a suitable solvent is spotted on TLC plate using microcapillary and allowed to run using different strength mobile phases. Various compounds migrate to different distances on TLC plate depending on their adsorption on the stationary phase. Less polar compounds move a greater distance on silica based and show high R_f value, whereas more polar compounds move a lesser distance and show low R_f value. The developed plate is visualized under UV or by derivatizing with suitable reagents, which produce distinct colors to indicate the presence of certain classes of natural products. For example, spraying with ninhydrin reagent shows amino acids as brightly colored spots on TLC plate. The adsorption of solute on silica stationary phase may involve one or several mechanisms including dipole-dipole interaction, hydrogen bond interaction, π-complex formation, or steric factors. The preliminary TLC experiments indicate suitable stationary and mobile phases for further preparative chromatographic procedures.

2.3.4.3 Preparative Thin-Layer Chromatography (PTLC)

It is a thin-layer chromatography technique aimed at isolating compounds in smaller amounts (10–50 mg) for structure elucidation. Silica, alumina, cellulose, C_8, and C_{18} reverse-phase precoated plates are available; however, silica is most commonly used stationary phase for preparative separations. Stationary phases with particle size distribution range (5–40 μm) are used in layer thickness 0.5–2 mm plates. The sample dissolved in the solvent at a concentration of 10–20 mg/mL is applied. The band of desired compound is scrapped off and extracted with a suitable solvent. The purity of isolated compound can be assessed by analytical TLC. PTLC is simple and cost effective compared with instrumental techniques; however, separation efficiency is poorer compared with HPLC or other instrumental techniques. PTLC is preferred if ΔR_f is high between the two compounds to be separated [10].

2.3.4.4 Open Column Chromatography (CC)/Vacuum Liquid Chromatography (VLC)/Flash Chromatography (FC)

Once the TLC analysis has given indications of a suitable stationary phase and mobile phase, the crude extract is subjected to further fractionation using CC/FC/

VLC. A suitable stationary phase and mobile phase is selected based on literature, polarity of the compounds to be separated, or the results from analytical TLC showing good resolution of the compounds to be separated. A column is packed with stationary phase to sample ratio ranging from 1:30 to 1:300 based on the complexity of the mixture. The sample is loaded either dissolved in a solvent, preferably the mobile phase, or as a solid adsorbed on the stationary phase support. The column is developed and fractions are collected with mobile phase of increasing strength. The eluted fractions are monitored by TLC. Compared with CC which is very time consuming, FC and VLC are very fast techniques and can be easily performed with available equipment/glassware. **Flash Chromatography (FC)** is an air-pressure-driven column chromatography optimized for rapid separation. Smaller silica gel particles (40–63 μ) and small positive pressure (10–15 psi) are used to run the column. A column of 5 inch bed is prepared with the stationary phase and a flow rate of 2 + 0.1 inches/min is maintained in order to achieve good resolution, which is achievable in reasonable period of time if the recommended conditions are followed. A mobile phase that shows an R_f value of about 0.35 on analytical TLC for the compound of interest is selected for isocratic elution. Under these conditions, chromatography is usually complete in 15–30 min and requires very small amount of solvents. On the other hand, **Vacuum liquid chromatography (VLC)** is vacuum-driven chromatography (20–70 mmHg). It was originally proposed to reproduce the results of analytical TLC on a preparative scale using column. A Buchner funnel is used as a column and the stationary phase, usually TLC grade silica or alumina, is filled up to a height of about 5 cm, the same as height of a TLC plate. Gradient elution of column is mostly preferred but isocratic can also be used, 10–15 mL fractions are collected at each polarity stage for a sample load up to 1 g. On normal phase silica gel, petroleum ether/hexane with increasing amounts of more polar solvents such as methylene chloride, diethyl ether, ethyl acetate, or acetone followed by increasing percentage of methanol are used as mobile phase for gradient elution. Usually 20–25 fraction will elute all the compounds from the column. The column is dried after each fraction is collected in a manner similar to repetitive runs of a TLC plate. A packed column can be reused for similar separation by washing the column with methanol. The advantage is reduced time for separation, reduced solvent consumption, simple and less expensive, and universally available apparatus. All these chromatographies can be performed using normal phase silica gel, reverse phase silica gel, or bonded phases.

The fractions so obtained are then subjected to final purification using instrumental techniques such as MPLC, HPLC, and countercurrent chromatography.

2.3.4.5 Low-Pressure and Medium-Pressure Liquid Chromatography (LPLC and MPLC)

LPLC columns use approximately 40–60 micron particle size enabling high flow rates and pressures up to 10 bars. The pressure in MPLC columns can go up to 40 bars. These low pressures also allow the use of refillable glass columns. The packing material is filled into columns manually and held in place with porous glass frits. Prepacked columns are also available for use. These systems can be used to separate samples from milligrams to gram scale.

2.3.4.6 Countercurrent Chromatography (CCC)

It is a form of support-free liquid–liquid chromatography in which two immiscible liquids prepared by mixing two or more solvents serve as the stationary and mobile phase, an instrument keeps one phase stationary and the other is pumped through it as mobile phase. The principle of separation is the partitioning of solute between two immiscible liquid phases. The relative proportion of solute in each phase is determined by the partition coefficient. In practice, the phase which contains more of the solute is kept stationary. The major advantage of this chromatography is that no solid support is used in the column and there is no irreversible adsorption of solute, no tailing, no loss of material and minimal solvent consumption. Two variants, droplet countercurrent (DCCC) and centrifugal partition chromatography (CPC) rely on gravitational field and centrifugal force for the retention of liquid stationary phase, respectively. In DCCC, droplets of mobile phase pass through an immiscible stationary liquid phase for continuous partitioning of solute between two phases. These droplets can move either in ascending or descending mode, depending on whether the stationary phase has higher or lower density than the mobile phase. A typical DCCC instrument contains 200–600 interconnected glass columns of approximately 2–3 mm diameter. The choice of two liquid phases is crucial for separation, binary solutions are impractical, ternary or quaternary mixture forming two phases are generally used in CCC. CPC uses centrifugal force to speed up separation and achieves higher flow rate than DCCC. Choice of mobile phase and stationary phase can be guided by TLC.

2.3.4.7 High-Performance (or High Pressure) Liquid Chromatography (HPLC)

HPLC can be used for final separation of partially purified fractions to obtain pure natural products. HPLC can be done in analytical, semi-preparative, or preparative mode. Samples from microgram range to gram to kilogram range can be processed on HPLC. High resolution achieved in HPLC is due to very fine particle size of the stationary phases used in HPLC columns. Most commonly, a particle size of about 5 micron is used in HPLC columns. The correct operating conditions (mobile phase and stationary phase) can be selected on the basis of initial TLC runs using normal phase silica gel or reverse phase silica TLC plates. The analytical HPLC conditions can be optimized by changing certain parameters such as retention factor (k'), selectivity (α), or plate number (N). An increment in any of these parameters will generally lead to a better separation or increased resolution between the compounds. These parameters generally depend on strength and type of mobile phase, type of stationary phase, length of column, particle size of stationary phase, temperature, etc. The selection of suitable stationary phase is very important to achieve the desired results; huge variety stationary phases are available in HPLC columns. However, most of the applications on HPLC are done using nonpolar reverse phase C18 columns and polar mobile phases (methanol/acetonitrile/water). Analytical HPLC is normally used for obtaining information (qualitative and quantitative) on sample mixtures, the amounts injected are generally very small, usually 5–10 micrograms, and cannot be used to isolate pure NPs. In order to obtain sufficient amounts of pure NPs, semi-preparative or preparative HPLC is used. The particle size in preparative or semi-preparative HPLC is smaller than that used for MPLC, generally 5–30 μm; mostly about 5–10 μm; in semi-preparative HPLC. The analytical column is usually 0.46 cm ID while semi-preparative columns are about 1 cm internal diameter [12].

2.3.5 *Structure Elucidation of NPs*

Once a pure NP is isolated, it is characterized for physicochemical properties such as color, nature of compound, melting point, and optical rotation. The structure elucidation of a NP is a challenge as the natural compounds exhibit tremendous structural diversity. The advancements in spectral techniques have made structure elucidation simpler and faster. The chemical structure is elucidated by interplay of different spectral techniques such as ultraviolet spectroscopy (UV), infrared spectroscopy (IR), mass spectrometry (MS), nuclear magnetic resonance (NMR), optical rotatory dispersion (ORD), and circular dichroism (CD). In spectral techniques, the sample is subjected to excitation by photons from different regions of electromagnetic spectrum. The first step in structure elucidation is determining the molecular weight and molecular formula of the compound using mass spectrometry or elemental analysis.

2.3.5.1 UV-Visible Spectroscopy

UV spectroscopy is highly sensitive with detectability up to 10^{-9} M and provides information on UV absorbing chromophores present in the molecule. The absorption of radiation in the UV/visible region by a molecule results in electron transitions from lower energy levels to higher energy levels; all organic molecules absorb UV/visible light in a wavelength up to 800 nm. The only electronic transitions possible in alkanes are from low-energy σ orbital to high-energy σ^* antibonding orbital, which require very high energy and occur at very short wavelength. Consequently no useful structural information about saturated organic molecules can be extracted. However, low energy transition in molecules containing multiple unsaturations fall in the UV region of 200–400 nm and can be detected normally by the UV spectrometer and therefore provide structural information. For example, UV spectroscopy can easily distinguish among conjugated diene, triene, tetraene, α,β-unsaturated carbonyl groups, etc. Empirical rules to correlate UV absorption with structural moieties were given by Woodward and Fieser, which can fairly predict the UV absorption for simple organic molecules [13].

2.3.5.2 Fourier-Transform Infrared Spectroscopy (FTIR)

Absorption of electromagnetic radiation in the infrared region effects changes in vibrational and rotational energy levels. Fourier-transform infrared spectroscopy is a powerful tool for the identification of functional groups present in the natural compounds. Specific functional groups show IR absorption bands in defined region of the spectrum; for example, a carbonyl group appears in a range from 1650 to 1850 cm^{-1} depending on other features in the vicinity of carbonyl group. The carbonyl group in different functionalities appears in the region:

Amide	Acid	Ketone	Aldehyde	Ester	Anhydride band I	Acid chloride	Anhydride band II
1650	1700	1715	1725	1735	1760	1800	1810 cm^{-1}

Apart from functional group identity, the other important structural information obtained from IR includes the presence and nature of H-bonding (intermolecular or intramolecular), ring size (for example, cyclohexanone, cyclopentanone, cyclobutanone), tautomerism (keto-enol), conformational isomerism, etc. Any factors that cause increase in bond strength would lead to IR absorption at higher wavenumber and vice versa (Fig. 2.6).

2.3.5.3 Mass Spectrometry

Mass spectrometry is a powerful analytical tool for the identification of unknown compounds, quantification of known compounds, and to elucidate the structures of molecules. It gives information about the molecular mass, molecular formula, and fragmentation pattern. The mass spectrometry analyzes ions in gas phase produced

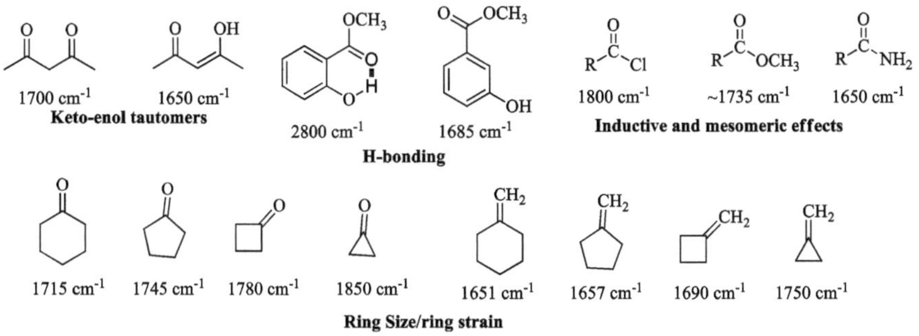

Fig. 2.6 Factors affecting absorption frequency in IR

by ionizing sample molecules by one of the several available ionizing techniques and measuring their mass-to-charge ratio (*m/z*) and relative abundance. Most commonly used ionization techniques are electron impact mass spectrometry (EIMS), chemical ionization mass spectrometry (CIMS), electrospray ionization mass spectrometry (ESIMS), and fast atom bombardment mass spectrometry (FABMS). In EI, the sample molecules in gas phase are bombarded by high-energy electrons (usually 70 eV) and are converted into high-energy positively charged ions (molecular ion or parent ion) by abstracting one electron to generate a radical cation. The molecular ions can fragment further into smaller ions (fragment ions or daughter ions). Mass spectrometry is highly sensitive and only picomolar concentration of the sample is required. MS has found wide applications in the screening of drug candidate, drug degradation analysis, and characterization of chemical compounds, drug metabolism, pharmacokinetics study and bioavailability studies. It is also frequently applied in dereplication process in natural products chemistry.

2.3.5.4 Nuclear Magnetic Resonance Spectroscopy (NMR)

Nuclear magnetic resonance (NMR) concerns magnetic properties of atomic nuclei, ^{1}H and ^{13}C or other magnetically active nuclei such as ^{19}F and ^{31}P or any other nuclei whose spin quantum number is greater than zero. The hydrogen nucleus possesses both the electric charge and mechanical spin and behaves like a tiny bar magnet and generates its own magnetic field. When this spinning nucleus with spin quantum number $I = 1/2$ is placed in a strong magnetic field B_0, it will orient itself in two directions (energy states), one aligned in the direction of the applied magnetic field (+1/2, lower energy) and the other aligned against the magnetic field (−1/2, higher energy), the energy difference between two states ($\Delta E = h\nu_0$) being proportional to the external applied field strength. The spinning nucleus will also precess around the axis of external applied field just like a spinning top does under the earth's gravitational field, at a precessional frequency (ν_0) that is proportional to the strength of external applied field. The irradiation of precessing nuclei with a radiofrequency matching ΔE will result in absorption of energy by nuclei in the lower energy state to jump to higher energy state. This absorption of energy is measured and recorded in the form of NMR spectrum. All the protons in a molecule do not experience the same applied magnetic field due to their different chemical environment and therefore absorb at slightly different radiofrequency, thus showing separate signals at different chemical shift values (δ) for different types of protons present in the molecule. The above is an oversimplification of the theory of NMR.

NMR provides information about the number, type, and electronic environment around the nucleus being studied. Advanced NMR experiments also give information of connectivities between the adjacent nuclei within the molecule as well as information on stereochemistry of the molecule. The number of signals in the ^{1}H NMR indicate the number of types of protons, the relative intensity of the signals gives the relative number of protons in that signal, splitting of a signal (a signal is split into $n + 1$ lines, n = no. of protons on adjacent carbons) indicates the number of neighboring hydrogens, the chemical shift indicates the chemical environment of the proton, the coupling constants suggest the stereochemistry of the molecule. The ^{1}H NMR spectrum of ethyl acetate shows three signals in ratio of 3:3:2 indicating there are three different types of protons. The signal at δ 1.3 has a relative intensity of 3, is split into three lines suggesting this signal is for three protons adjacent to a methylene group. The second signal for three protons appears as a singlet at δ 2.1 suggesting there are no protons on adjacent carbons. A signal at δ 4.2 for two protons appears as a quartet suggesting a methylene group adjacent to a methyl group. The ^{13}C NMR on the other hand does not show splitting and relative ratio of numbers. It provides information on number of carbon atoms and their chemical environment. More structural information is provided by advanced NMR experiments. Common one-dimensional (1D) NMR experiments include ^{1}H NMR, ^{13}C NMR, ^{19}F NMR, ^{31}P NMR, and DEPT; two-dimensional (2D) techniques include ^{1}H-^{1}H COSY,^{1}H-^{1}H DQF-COSY,[1] ^{1}H-^{1}H NOESY,^{1}H-^{1}H ROESY, TOCSY, ^{1}H-^{13}C HMBC,^{1}H-^{13}C HMQC,^{1}H-^{13}C HSQC, HSQC-TOCSY, etc. [14].

2.3.5.5 Hyphenated Techniques

It is a coupling of two different techniques with the help of an interface. It is an online combination of a chromatographic separation with a sensitive spectroscopic detector. GC-MS is a technique coupling of gas chromatography and mass spectrometry. Compounds that are volatile, small, and stable at high temperature can be analyzed by GC-MS. These two techniques are highly compatible with each other as sample is in the vapor phase in both the techniques. GC-MS has high resolving power and sensitivity and accomplishes rapid analysis of sample with good accuracy and precision. However, only volatile samples or those samples which can be volatilized after derivatization can be analyzed by GC-MS. In comparison, LC-MS is a combination of liquid chromatography (HPLC) and mass spectrometry. It is the most widely used method suitable for analysis of nonvolatile, thermally labile, and charged molecules. LC-NMR is least sensitive, however provides useful information towards the structure elucidation of compounds. LC-NMR is used for the analysis of complex mixtures of all types, but particularly for analysis of natural products and drug metabolites [10].

2.3.6 *Preclinical Studies and Clinical Trials*

Preclinical studies: These involve in vitro and in vivo studies to determine if the drug is safe for human testing. Wide range dosages for the study drug are given to the animal or to an in vitro substrate to obtain preliminary results such as efficacy, toxicity, and pharmacokinetic information and to assist decision-making to go-ahead for further testing of the drug. The mechanism of action, efficacy, and safety is determined during preclinical studies.

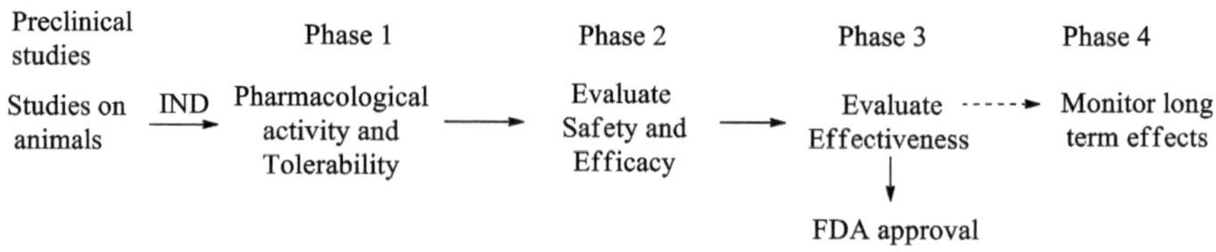

Fig. 2.7 Phases of clinical trials

Clinical trials: Clinical trials are research investigations in which people volunteer to test new treatments or tests to prevent, detect, treat, or manage various diseases or medical conditions. World Health Organization (WHO) defines a clinical trial as "*any research study that prospectively assigns human participants or groups of humans to one or more health-related interventions to evaluate the effects on health outcomes*." Clinical trials are essential for the development of new interventions to determine their safety and efficacy (Fig. 2.7).

2.3.6.1 Phase I

This is the first stage of testing on small group (20–100) healthy volunteers. The main objective is to establish the safety of the drug in humans. The trials are designed to assess the safety (pharmacovigilance), tolerability, pharmacokinetics, and pharmacodynamics of a drug; the absorption, metabolism, and elimination of the drug from the body. Any possible side effects are also investigated. These trials are conducted in an inpatient clinic, where the subject can be under the supervision of full staff. The subject is usually monitored until several half-lives of the drug have passed. Phase I trials also include dosing studies so that the appropriate dose for therapeutic use can be identified. Phase I trials mostly include healthy volunteers; however, there may be some exceptions such as patients who have an end-stage disease and lack other treatment options. This exception most often occurs in cancer and HIV drug trials.

2.3.6.2 Phase II

Once the initial safety of the study drug has been confirmed in Phase I trials, Phase II trials are performed on larger groups (100–500) of patients to evaluate the effectiveness of the drug and to examine the short-term side effects or adverse reaction. Phase II studies are divided into two phases such as Phase IIA and Phase IIB. Phase IIA is specifically designed to assess dosing requirements (how much drug should be given to the subject) and Phase IIB is specifically designed to study efficacy (how the drug works at the prescribed dose).

2.3.6.3 Phase III

Phase III studies are performed on large and diverse patient groups (1000–5000) and are designed to assess and generate significant data on the safety, effectiveness, and overall benefit-risk of the drug. The Phase III trials are the most expensive and time consuming. If the phase III clinical trials are satisfactory, then necessary documents are prepared for regulatory approval, which include description of the methods and results of human and animal studies. These documents also include all manufacturing procedures, details of formulation, and shelf life studies.

2.3.6.4 Phase IV

This phase of the clinical trial is also known as Post Marketing Surveillance. It involves the safety surveillance (pharmacovigilance) and ongoing technical support of a drug. The safety surveillance is designed to detect any rare or long-term adverse effects on a much larger patient population and longer time period than was possible during the Phase I-III clinical trials. Adverse effects reported in Phase IV trials may result in a restricted use or withdrawal of drug [15].

2.3.7 Discovery and Development of Eribulin

2.3.7.1 Extraction and Isolation of Halichondrin B

Halichondrin B, a macrocyclic lactone, was isolated from marine sponge *Halichondria okadai* collected from south of Tokyo in the mediolittoral zone of the Pacific Ocean. Approximately 600 kg specimens were collected and stored in a freezer; the frozen specimens were crushed in a blender with MeOH, kept for a period of 3 days, and filtered. The brown color filtrate was concentrated under reduced pressure at low temperature. The remaining aqueous solution was extracted with *n*-butanol saturated with water. The combined organic layers were concentrated under reduced pressure, the crude extract was further dissolved in 70% aqueous MeOH and washed with *n*-hexane. The 70% aqueous MeOH layer was concentrated under reduced pressure to give an oily material which was charged on TSK G3000S polystyrene gel column; the bioactive fractions (B-16 melanoma cells) were eluted with 50% ethanol and then 60% ethanol. Each fraction was further separated by repeated chromatography on LiChroprep RP-C8 column and YMC Pack A212 (C-8) column to yield eight active compounds (12.5 mg halichondrin B from 600 kg sponges) [16]. Halichondrin B was also isolated from other sponges with very less yield, 0.4 mg, $1.8 \times 10^{-8}\%$ from *Axinella* sp., 8 mg, $3.5 \times 10^{-9}\%$ from *Phakellia carter* [17, 18], and 2 mg, $4 \times 10^{-5}\%$ from *Lissodendoryx* sp. [19]. Further isolation was done for the most potent compound Halichondrin B from *Lissodendoryx* sp., which yielded 310 mg from 1 ton biomass [20] (Fig. 2.8).

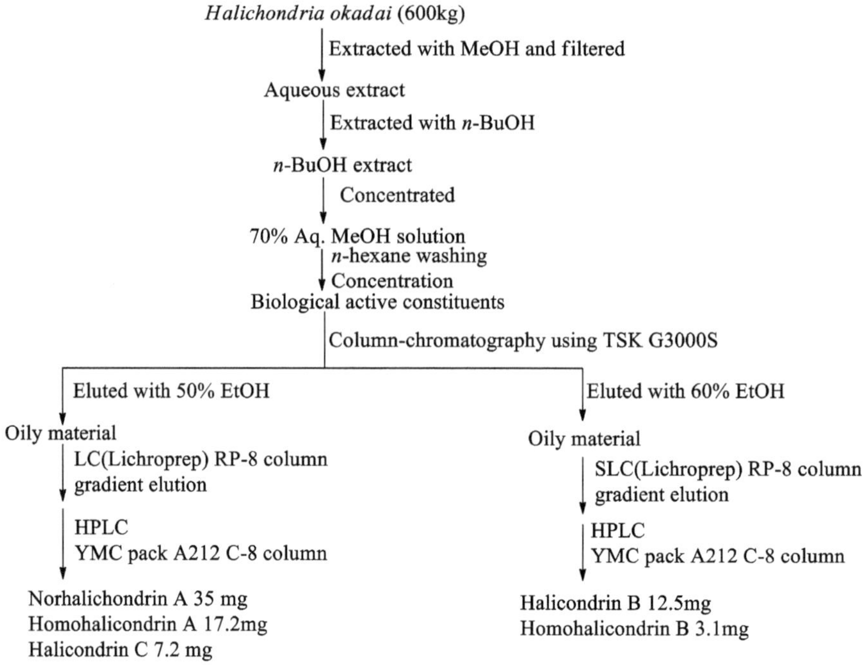

Fig. 2.8 Isolation of halichondrins

2.3.7.2 Characterization of Halichondrin B

The structure of norhalichondrin A (MW 1126) was established by single crystal X-ray analysis after preparing its ester with *p*-bromophenacyl bromide. The absolute configuration was determined by applying exciton chirality method. The structure of halichondrin B was determined by comparison of its spectral data with that of norhalichondrin A. The molecular weight (1110) of halichondrin B was obtained by FDMS *m/z*, 1133 $[M + Na]^+$ for the molecular formula $C_{60}H_{86}O_{19}Na$ suggesting one oxygen less in halichondrin B. Acetylation with acetic anhydride and pyridine gave a triacetate suggesting the presence of three hydroxyl groups. The structure was established by a comparison of COSY spectra of two compounds and assigning the partial structures from C1 to C14. Further the chemical shifts and coupling constants of each proton from C15 to C44 were assigned which were consistent with norhalichondrin A. A complete analysis of proton, carbon, ^1H-^1H COSY, and ^1H-^{13}C COSY led to the complete structure of halichondrin B [21] (Fig. 2.9).

2.3.7.3 Biological Activity of Halichondrin B

Halichondrin B showed in vitro cytotoxicity IC_{50} of 0.093 (ng/mL) against B-l6 melanoma cells. In vivo antitumor activity of halichondrin B was evaluated for B-16

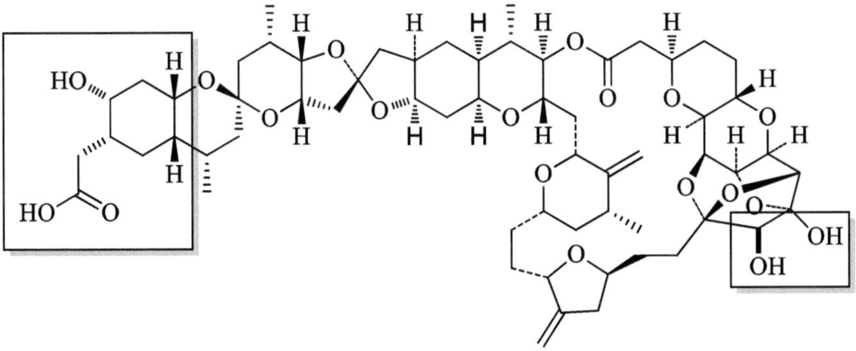

Norhalichondrin A

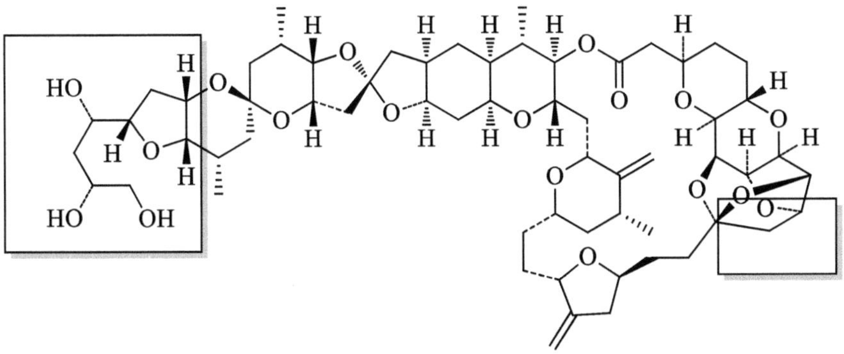

Halichondrin A

Fig. 2.9 Structures of halichondrins

melanoma, P-388 leukemia, and L-1210 leukemia in mice model. The mean survival time increased by almost 200% at a dose of 2.5–20 μg/kg in B-16 melanoma assay. Antitumor activity against P-388 leukemia and L-1210 leukemia also increased the mean survival rate of animals. Halichondrins inhibit microtubule dynamics by inducing nonproductive tubulin aggregates, leading to suppression of spindle microtubule. This mechanism differs from that of other tubulin-interactive drugs such as vinblastine or paclitaxel [22].

2.3.7.4 Optimization/Medicinal Chemistry

Halichondrin B was isolated in very small amount from sponges and showed very good activity at very low doses; it was necessary to obtain large amounts of this natural compound for further preclinical and clinical studies. The total synthesis of this big molecule with large number of chiral centers was a huge challenge. Kishi and coworkers accomplished the total syntheses of halichondrin B and norhalichondrin B [23]. However, the synthesis of halichondrin B was not cost

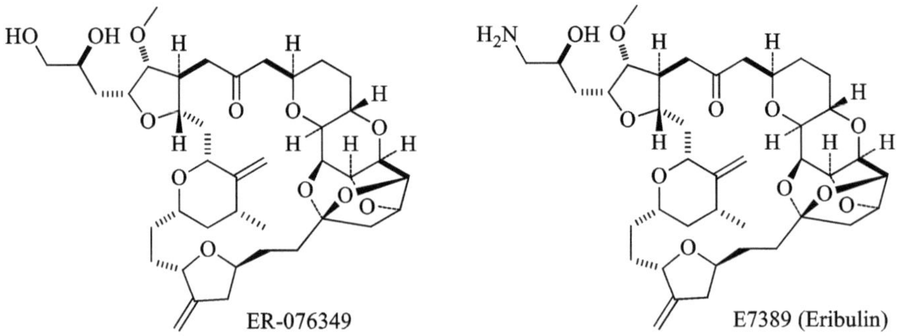

ER-076349 E7389 (Eribulin)

Fig. 2.10 Semisynthetic derivatives of halichondrin B

effective. An intermediate compound, a macrolactone diol, obtained during the total synthesis of halichondrin B showed IC_{50} 4.6 nM against human colon cancer cell lines that was comparable to parent compound. This promising result encouraged further studies to perform the structure activity relationship for this active intermediate. Two leads ER-076349 and E7389 were obtained after several modifications in the macrolactone diol (Fig. 2.10).

2.3.7.5 Preclinical Studies on Eribulin

Eribulin showed better antiproliferative activity than ER-076349, vincristine, and paclitaxel against MDA-MB-435 (IC_{50} 0.09 nm). It also showed high activity; at nanomolar or subnanomolar concentration against colon cancer COLO 205 and DLD-1, prostate cancer DU 145 and LNCaP, melansoma LOX, leukemia HL-60, lymphoma U-937 cells, this activity was better than vinblastine or paclitaxel [22]. Activity shown by eribulin in NCI 60 cell line screen was similar to halichondrin B. Eribulin also showed synergistic effects with several other conventional drugs in SK-BR-3 cell line. Eribulin showed tumor regression and increased life span in breast, lung, ovary, colon, melanoma, pancreatic, and fibrosarcoma in human tumor xenograft investigations in mice. Eribulin did not show any significant adverse effects in mice at maximum tolerated dose. Eribulin inhibits cancer cell growth through mitotic block at G_2-M phase disrupting the mitotic spindle formation [24].

2.3.7.6 Clinical Trials on Eribulin

Phase I

Several studies were conducted at different locations for the Phase I trials. The first study was done on 40 patients with advanced solid tumors. Eribulin was administered at a starting dose 0.125 mg/m^2/week in patients, dose escalation was done up to

2.0 mg/m^2/week. In the second study on 32 patients (1-h i.v. infusion, 1.4 mg/m^2) on days 1, 8, and 15 of a 28-day cycle suggested that 1.4 mg/m^2 dose level could not be administered in week 3 dose due to adverse effects. Therefore maximum tolerated dose was suggested to be 1.0 mg/m^2. The third study was done on 21 patients at doses ranging from 0.25 to 4 mg/m^2 with 1 h infusion on day 1. The MTD in this schedule was established as 2.0 mg/m^2. A dose of 1.4 mg/m^2 on days 1 and 8 every 3 weeks was recommended for phase II studies. The other studies in phase I were conducted on patients with renal dysfunction and advanced urothelial cancer, and combination with gemcitabine in patients with advanced solid tumors. The most common adverse events were neutropenia, alopecia, nausea, and fatigue among others; however, it was found to be fairly safe.

Phase II Trial

Phase II trial of eribulin was conducted in metastatic breast cancer patients (pretreated with either anthracycline or taxane) at 23 sites in the USA with bolus 1.4 mg/m^2/week on days 1 and 8 of a 21 day cycle. Eribulin achieved an overall response rate (ORR) of 11.5% and median overall survival was 275 days. Another phase II study sponsored by Eisai showed median overall survival of 10.4 months. Another study in Japan on MBC patients showed median overall survival of 331 days. Phase II trials were also conducted for non-small cell lung cancer, prostate cancer, ovarian cancer, sarcoma, pancreatic cancer, urothelial tract cancer (UC), and squamous cell carcinoma of the head and neck [24].

Phase III Trials

Total 762 patients were recruited and divided into eribulin treated groups (508) and other treatment of physician's choice (TPC) groups (254) with median age of 55.2 years. The median overall survival was 13.1 months for eribulin and 10.6 months for TPC. Median progression-free survival was (PFS) was 3.7 months for eribulin and 2.3 months for TPC. ORR was 12% (0.4% complete response, 11.5% partial response) for eribulin and 5% (0 CR; 5% PR) for TPC. About 10% of patients showed serious adverse effect—for eribulin arm 12% and for TPC arm 7%. Eribulin has been approved to treat the patients with metastatic breast cancer by the United State Food and Drug Administration (USFDA) in 2010. The discovery and development of eribulin is reviewed in detail by Swami et al. [24]. Complete discovery and development process of eribulin spanned 25 years, from isolation of Halichondrin B in 1985 to approval by USFDA in 2010 for treatment of metastatic breast cancer [24].

2.3.8 *Discovery and Development Process of Taxol*

2.3.8.1 Extraction and Isolation

Taxol was isolated for the first time from the bark of *Taxus brevifolia* Nutt. in 1966. The bark of *Taxus brevifolia* was extracted with alcohol and the concentrated extract was then portioned with water and chloroform. The isolation was performed by using bioassay-guided fractionation with 9KB cell and various leukemia systems. The chloroform fraction was first separated on florisil, followed by Sephadex LH-20 and silica gel. Final purification was done by crystallization from aqueous methanol and yielded 0.02% of taxol [25]. Other isolation methods have also been reported. In vivo bioguided isolation was performed with solid tumor in Walker-256 intramolecular rat carcinoma (5WM). The bark of *Taxus brevifolia* (12 kg) was extracted with 95% ethanol and the concentrated extract was partitioned with water and chloroform. Active chloroform fraction was applied to Craig Countercurrent Distribution (CCD) with a series of repetitions. The yield of pure taxol was 0.004% (0.5 g). This process was tedious and time consuming. Taxol has also been reported from different parts of other *Taxus* species in range of 0.001–0.016%. *Taxus brevifolia* is very slow growing plant and it takes about 200 years to reach maturity. Therefore, sufficient amount of taxol from *Taxus brevifolia* was practically not possible. It has been reported that baccatin III and 10 deacetylbaccatin III can be used for semisynthesis of taxol and its derivatives. Baccatin III and deacetylbaccatin III were isolated from *T. baccata*, *T. wallichina*, and *T. brevifolia* in considerable amount [26]. Elaborate studies were conducted in 1960s to find suitable species or varieties and parts of those trees which could generate maximal amount of taxol.

2.3.8.2 Structure Elucidation of Taxol

Taxol showed M^+ at *m/z* 853 corresponded to the composition $C_{47}H_{51}N0_{14}$. Hydroxy, ester, keto, and amide functionalities were established by IR spectroscopy. Characteristic moieties were also established through ^{1}H NMR. Wall and coworkers established the structure of taxol in 1971 by single crystal X-ray analysis after preparing its derivative *p*-bromobenzoate ester. Complete assignment of all proton and carbon signals using 1D and 2D NMR was reported in 1992 [27].

2.3.8.3 Optimization/Medicinal Chemistry

Semisynthesis of Taxol from Baccatin III and Deacetylbaccatin III

Adequate supplies of taxol were required for preclinical and clinical studies. Semisynthesis of taxol from readily available congeners ultimately provided sufficient supplies of taxol for commercialization. 10-Deacetylbaccatin (10-DAB) is

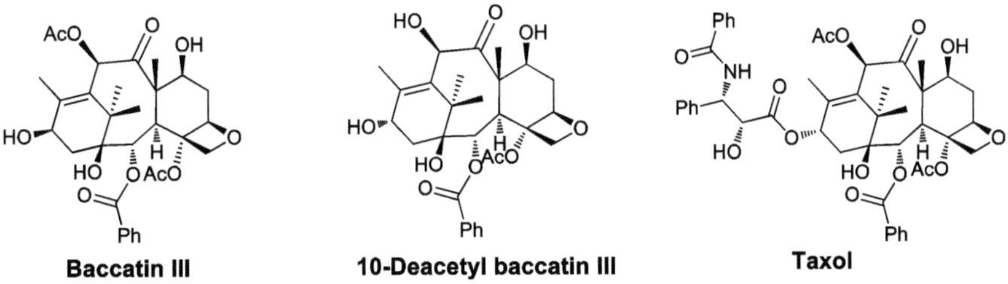

Fig. 2.11 Structures of taxol, Baccatin III, and 10-DAB

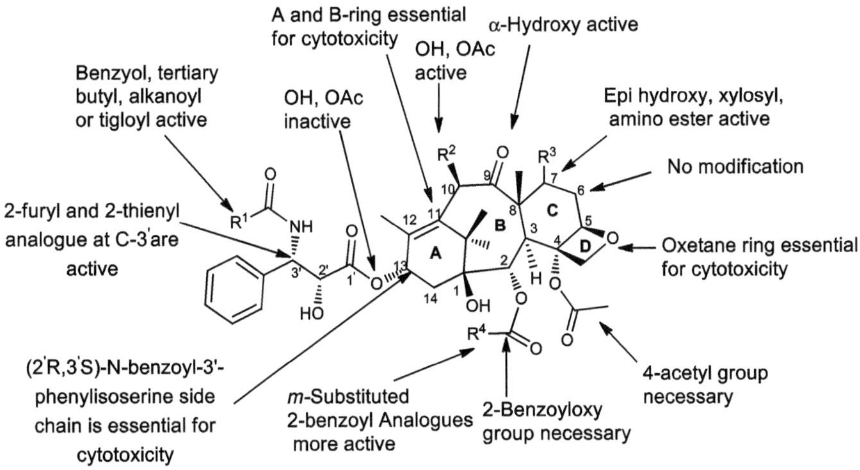

Fig. 2.12 Structure activity relationships in taxol

relatively more abundant taxane in the needles of English Yew *Taxus baccata* and its isolation is simpler than isolation of taxol. Therefore, a major breakthrough in the development of taxol came after semisynthesis of taxol from 10-deactylbaccatin III and baccatin III (Fig. 2.11) [28].

A huge number of analogs were synthesized for evaluation of anticancer activity along with several other naturally occurring taxanes. These included analogs/derivatives of 10-deacetyl taxol, 7-*epi*-taxol, 7-xylosyltaxol, modifications at C-7 and C-10 hydroxyl group, modifications at C-2, 4, 6, 9, and 19, and modifications in the C-13 side chain. A brief glimpse of medicinal chemistry of taxol is shown below and the structure activity relationships are summarized in Fig. 2.12 [28].

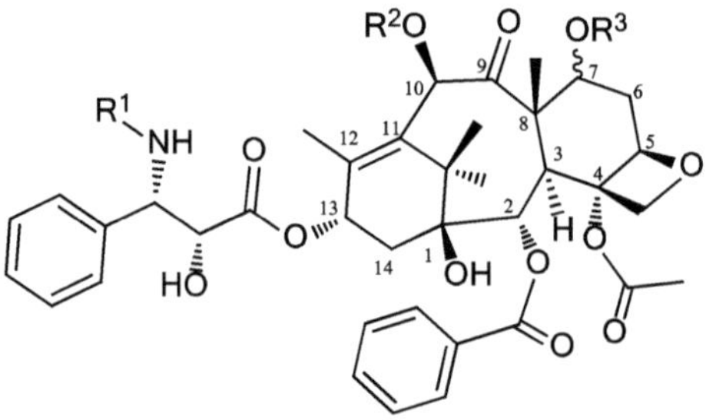

	Compound	Substitution			Biological activity				
		R^1	R^2	R^3	KB[a]	P388[b]	J774.2[c]	ID_{50}[d]/ ID_{50} (taxol)	ED_{50}[e]/ ED_{50} (taxol)
1	Taxol	Bz	Ac	β-OH	0.001	+	S	1.0	1.0
2	7-Xylosyltaxol	Bz	Ac	β-Xylosyl					0.4
3	Cephalomannine	Tigloyl	Ac	β-OH	0.004	+	M/S	1.5	
4	10-Deacetyltaxol	Bz	H	β-OH	0.003	+	M	1.3	
5	10-Deacetylcephalomannine	Tigloyl	H	β-OH	0.03	+	M	5.0	
6	7-Xylosylcephalomannine	Tigloyl	Ac	β-Xylosyl					0.5
7	10-Deacetyl-7-xylosyltaxol	Bz	H	β-Xylosyl					0.6
8	10-Deacetyl-7-epi-taxol	Bz	H	α-OH	0.03				
9	7-epi-taxol	Bz	Ac	α-OH	3×10^{-5}			3.0	
10	10-Deacetyl-7-epi-cephalomannine	Tigloyl	H	α-OH	0.05				

[a]ED_{50} in μg/mL for growth inhibition of KB cells
[b]Activity against P388 cells; + = Increase in life span >25%
[c]Growth inhibition of macrophages, i.e., J774.2 cells; *S* Strong, *M* Medium
[d]Microtubule disassembly
[e]Microtubule assembly activity

Modification at Diterpene Nucleus of Taxol and Taxotere Analogs

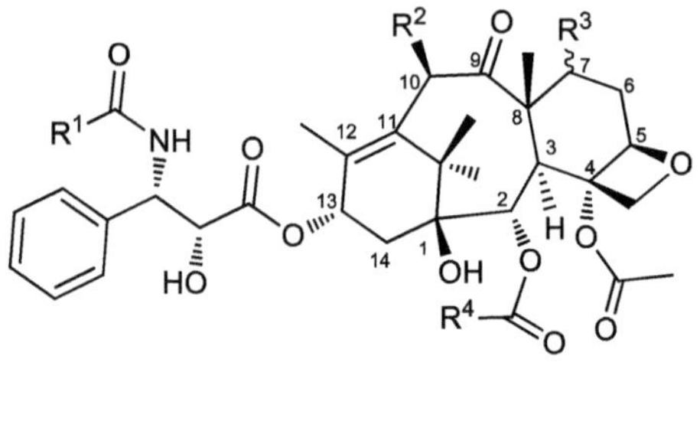

	Compound	Substitution				Biological activity
		R^1	R^2	R^3	R^4	ED_{50}[a]/ED_{50} (taxol)
1	Taxotere	OBu^t	OH	OH	Ph	0.5[b]
2	10-Deacetyltaxol	Ph	OH	OH	Ph	1.3[b]
3	10-Acetyltaxotere	OBu^t	OAC	OH	Ph	0.5[b]
4	10-Deacetoxytaxol	Ph	H	OH	Ph	0.5[c]
5	2-(3-Cyanobenzoyl)-2-debenzoyltaxol	Ph	OAc	OH	*m*-CNPh	0.2[d]
6	2-cyclohexylcarbonyl-2-debenzoyltaxotere	OBu^t	OAc	OH	Cyclohexyl	1.0[e]

[a]Microtubule assembly activity
[b]Microtubule disassembly
[c]Microtubule assembly
[d]P388 Murine leukemia cells
[e]$ED_{50}/ED_{50(Taxotere)}$ against P388 leukemia cells

Modification at C-13 Side Chain of Taxol

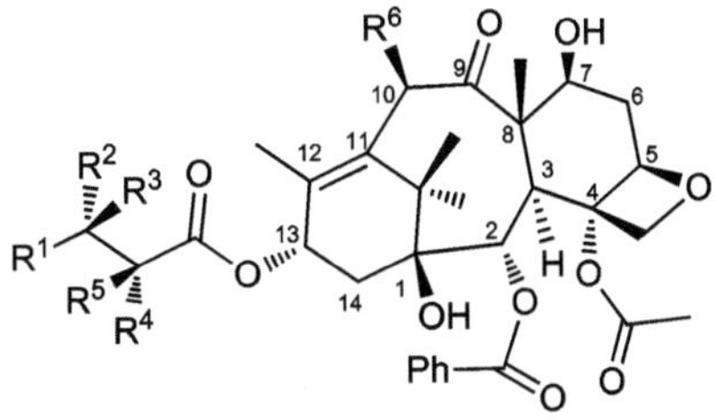

	Substitution						Biological activity				
	R^1	R^2	R^3	R^4	R^5	R^6	KB[a]	P388[b]	J774.2[c]	ID_{50}[d]/ $ID_{50(taxo)}$	ED_{50}[e]/ $ED_{50(taxo)}$
1	Ph	BzNH	H	OH	H	Ac	1.0	1.0	1.0	1.0	1.0
2	Ph	H	OH	H	OH	H				3.0	
3	Ph	H	H	H	OH	H				3.5	
4	Ph	H	H	H	OH	Ac			38.0		
5	Ph	BocNH	H	H	H	H				4.1	
6	Ph	H	H	OH	H	Ac					13.2
7	Ph	H	H	OH	H	H				4.5	
8	Ph	H	NH_2	H	OH	H				30.0	
9	Ph	NH_2	H	OH	H	Ac				44.0	
10	H	BzNH	H	H	OH	Ac			>500		>100

[a]ED_{50} in μg/mL for growth inhibition of KB cells
[b]Activity against P388 cells
[c]Growth inhibition of macrophages, i.e., J774.2 cells
[d]Microtubule disassembly
[e]Microtubule assembly activity [33]

2.3.8.4 Preclinical Studies on Taxol

In the initial studies, taxol was evaluated in vitro as well as in vivo. Taxol showed cytotoxicity to KB cells with ED_{50} 6.7 × 10^{-5} μg/mL. In vivo activity was performed against murine L1210, P388, and P1534 leukemias, the Walker 256 carcinosarcoma, sarcoma 180, and Lewis lung tumor. A number of in vitro studies against a variety of cell lines including human leukemias, melanomas, and carcinomas lung, CNS, breast, kidney, ovary, and colon showed taxol to be active with $IC_{50} < 2.5$ nM [28]. Taxol exhibited significant cytotoxic activity on human prostatic cancer cells at a concentration of 10 nM [29]. Taxol was evaluated against the human tumors transplanted into athymic mice (breast, endometrium, ovary, brain, lung tumor) at dose of 12.5 mg/kg subcutaneous injection every day for five consecutive days out of seven over a period of 3 weeks. Among all evaluated xenograft tumors, taxol was found most active against breast tumor, four out of five tumors disappeared in mice [30]. In mice, LD_{10} of taxol was found at 70 mg/m^2/d in 5-day schedule, whereas in dogs no toxicity was observed at dose of 22.5 mg/m^2 with the single dose and 7.5 mg/m^2/day in 5-day schedule [31]. It was shown by Horwitz and coworkers that taxol binds reversibly to microtubules and this binding site was different from the binding sites of GTP, colchicines, podophyllotoxin, and vinblastine. Taxol promotes microtubule assembly and stabilizes microtubules, which are stable even after treatment with calcium or low temperatures. Taxol induces formation of abnormal spindle asters during mitosis. Taxol has also been shown to prevent transition of cell from G0 phase to S phase.

2.3.8.5 Clinical Trials of Taxol

Phase I

Phase I clinical trial of taxol was initiated in 1983 under Division of Cancer Treatment (DCT). In one trial, a total of 20 patients (7 male and 13 female) with median age of 52 years (range 32–69 years) with metastatic cancer of colon, sarcoma, melanoma, lung, head and neck, ovary, and uterus were selected. Taxol was administered intravenously daily for 5 days at 3-week intervals and with starting dose 5 mg/m^2/day daily, and with highest dose 40 mg/m^2/day for 5 days [36]. A number of phase I trials were conducted by different research groups. The recommended dose for phase II trials was 150–250 mg/m^2/day. Adverse effects like leukopenia, neutropenia, hypersensitivity, neuropathy, and mucositis were noted [31–33].

Phase II

The phase II clinical trials of taxol were conducted at a number of institutes. In one trial, 47 patients were selected for treatment of advanced ovarian cancer of taxol. The patients were treated with varying dose 110–250 mg/m^2/day 22 days for 24-h infusion which showed 30% response rate [34]. Another study was conducted on 34 patients with metastatic ovarian cancer. Taxol was administered with dose ranging from 180 to 250 mg/m^2 for 24-h continuous infusion with 20% total response [35]. In another clinical study a group of 25 metastatic breast cancer patients were given 250 mg/m^2 of drug by 24-h infusion at 21 day intervals. This treatment showed total response rate of 56% (12% complete and 44% partial) [36]. Other research groups have reported the response rate of taxol 22–62 for treatment of metastatic breast cancer and 20–48 for ovary cancer. Phase II clinical trial was also evaluated for non-small cell lung, small cell lung, melanoma, renal, prostate, colon, cervix, gastric, pancreas, bladder, and head and neck lymphoma cancers. It has been observed that response rate also depends on the type of prior therapy [28].

Phase III

In a study on 391 patients, the response was obtained from 382 patients. Taxol was administered in dose ranging from 175 or 135 mg/m^2 for 24 or 3 h infusion. Response rate was found slightly higher at dose of 175 mg/m^2 (20%) than at 135 mg/m^2 (15%). Taxol administered as 24 h infusion showed high neutropenia. A longer time to progression was reported at a dose of 175 mg/m^2 [37]. In another trial on 471 patients with metastatic breast cancer patients, taxol was intravenously administered at a dose of 175 or 135 mg/m^2 every 3 weeks as a 3-h infusion. The higher dose 175 mg/m^2 showed overall response 29%, complete response 5%, median time to disease progression 4.2 months, and median survival time of 11.7 months [38]. Taxol was approved by United State Food and Drug Administration (USFDA) for the treatment of refractory ovarian cancer and refractory or anthracycline-resistant breast cancer in 1992 and 1994, respectively. Due to its unique mode of action, the extensive research has been done to develop more dugs, resulting in approval of docetaxel (Taxotere), cabazitaxel, and larotaxel by USFDA.

2.4 AYUSH (Indian Systems of Medicine)

Traditional systems of medicine have been used throughout the world since prehistoric periods. The World Health Organization (WHO) defines traditional systems of medicine as an aggregate of the knowledge, skills, and practices of different cultures used to maintain health and cure diseases. Awareness of plants and their healing properties have been found closely linked to human life and its social and cultural well-being. About 80% of the population of developing and underdeveloped countries relies on the traditional systems of medicine for the treatment of ailments. Several types of traditional systems of medicine are practiced worldwide such as the Indian Systems of Medicine (Ayurveda, Siddha, Unani), Traditional Chinese Medicine, Traditional Japanese Medicine, Traditional Korean Medicine, Traditional Aboriginal Medicine, Traditional African Medicine, and Russian Herbal Medicine [6, 39]. It is estimated that Ayurveda uses 1200 species of plants while Siddha and Unani include 900 and 700 species of plants, respectively, in their pharmaceutical preparations [40]. In 1995, the Government of India established the Department of Indian Systems of Medicine and Homoeopathy (ISM&H) under the Ministry of Health and Family Welfare for the growth and development of Ayurveda and other popular systems of Indian Medicines. In November 2003, Department of Indian Systems of Medicine and Homoeopathy was renamed as the Department of Ayurveda, Yoga and Naturopathy, Unani, Siddha and Homeopathy (AYUSH). Government of India upgraded the Department of AYUSH to a full Ministry of AYUSH in 2014. This ministry takes decisions for policy formulation and implementation of schemes and programs. The Ministry of Health and Family Welfare (AYUSH) has also recognized Sowa-Rigpa traditional system of medicine after the Indian Medicine Central Council Amendment Bill 2010 [41, 42].

2.4.1 Ayurveda

Ayurveda is the oldest and holistic system of Indian medicine that is popularly practiced in India and several other countries including Pakistan, Nepal, Bangladesh, and Sri-Lanka [42]. Ayurveda is a Sanskrit word (Ayur means life and Veda means the science or knowledge), so it is the science of life. It is believed that Ayurveda originated as a divine science of healing, transferred from the Hindu deity Lord Brahma to Dakshas and then further to Ashwini twins and to Lord Indra. It is assumed that medicinal information is recorded in the oldest scriptures Rig Veda and Atharva Veda [43].

The knowledge and experiences of practitioners and scholars have been documented in many texts since ancient times and have been translated into different languages including Greek (300 BC), Persian and Arabic (700 AD), Chinese (300 AD), and Tibetan [44]. The Ayurvedic literature originated from Indian Vedic times during 500–1500 BC, two main texts Charaka Samhita and the Sushruta Samhita (1000 BC) are the foundation of Ayurveda and these two were influential scriptures on traditional medicine during this era. The Charaka Samhita contains knowledge about foods, lifestyle, and drugs which give longevity, good health, and help disease prevention [43]. Sushruta Samhita is the oldest textbook of surgery and is divided into five parts, Rogaharas (physicians), Shaylyaharas (surgeons), Vishaharas (poison healers), Krityaharas (demon doctors), and Bhisagatharvans (magic doctors) [45]. Sushruta Samhita contains 184 chapters with description of 1120 illnesses, 700 medicinal plants, 64 preparations from mineral sources, and 57 preparations based on animal sources [46]. Charaka Samhita and Sushruta Samhita were later updated in the form of Astanga Sangraha and Astanga Hrdaya. Madhava Nidana (diagnosis of disease), Bhava Prakasa (information related to plant and diet), and Sarngadhara Samhita (formulation and dosage form) [47] also contributed towards Ayurveda. The eight major divisions of Ayurveda are Kayacikitsa (internal medicine), Salyatantra (surgery), Salakya (diseases of supra-clavicular origin), Kaumarabhrtya (pediatrics, obstetrics, and gynecology), Bhutavidya (psychiatry), Agadatantra (toxicology), Rasayanatantra (rejuvenation and geriatrics), and Vajikarana (aphrodisiology and eugenics) [45, 47]. The following concepts are basic principles of Ayurveda.

2.4.1.1 The Five Elements

According to Ayurveda, everything in the universe is composed of energy and this energy exists in five different states of density, giving rise to five factors or elements, namely Vayu (air), Jala (water), Aakash (space or ether), Prithvi (earth), and Teja (fire). Every person is made of these five basic elements called as Pancha Mahabhoota and helps understand physiology and pathology of the body. **Vayu (air)** exists in the gaseous state which is light, dry, clear, and dispersing. Air affects respiration, excretion, expansion, contraction, and voluntary and involuntary movements of body. **Jala (water)** is liquid, has no shape, and it holds all things together. It is present in body in the form of blood, stools, urine, saliva, and mucus. Water also controls taste buds and taste perception. **Aakash (space or ether)** has no physical shape and is associated with ears and throat. **Prithvi (Earth)** earth is a stable and solid matter. The characteristic feature is solidity, stability, and rigidity and is associated with nose and smell. **Teja (Fire)** is an energy and form without any matter. Fire has ability to convert a solid matter to liquid to gas and vice versa and is associated with eyes.

2.4.1.2 The Three Humors (Doshas)

The five elements are responsible individually or collectively to form the three basic humors of the human body which have different composition. The three humors are called as "Tridoshas," namely Vata dosha (wind/air), Pitta dosha (bile), and Kapha dosha (phlegm). The blood was also considered as a fourth dosha in some old schools of thought, but over periods the concept of three dosha has prevailed. These doshas play an important role in regulating the physiological and pathological functions of the human body [43]. **Vata dosha** is composed of air and ether which has great force, its action focuses on all movements in the body, it controls flow of blood, breathing, elimination of wastes. It is considered the most important of the three doshas. **Pitta dosha** contains fire and water. Pitta dosha controls metabolism and certain hormones linked to appetite. **Kapha dosha** is composed of water and earth and is concerned with stability in mind and body. It controls muscle growth, body strength, and immune system. It is believed that the body is composed of seven elements which is called "Saptadhatu," named as Chyle (rasa dhatu), blood (raktadhatu), fat (medhadhatu), flesh (mamsadhatu), bone (asthidhatu), marrow (majjadhatu), and semen or female reproductive tissue (shukradhatu) [42].

2.4.1.3 The Gunas (Quality)

According to Ayurveda, human beings possess three gunas, viz. sattva, rajas, and tamas. Sattva is the quality of purity and clarity of perception which is responsible for goodness and happiness. Rajas is the quality of all movements, and enjoyment, pleasure and pain, effort and restlessness. Tamas is the quality representing darkness, heaviness, inertia, and materialistic attitudes. These gunas are responsible for psychological constitution [42].

Ayurveda known as the Ashtanga-Ayurveda has eight distinct branches, *kayabalagrahaurdhwangashalyadaunshtrajaravrushan,* i.e., (i) general medicine, (ii) surgery, (iii) ear, nose, throat, eye, and mouth disease, (iv) psychiatry, (v) midwifery and pediatrics, (vi) toxicology, (vii) rejuvenation and tonics, and (viii) aphrodisiacs. Ayurvedic system of medicine promotes overall health rather than just treating the diseases.

2.4.2 Yoga and Naturopathy

Yoga originated in India and it has been associated with spiritual, self-care, and health practices and started thousands of years ago [48]. The term Yoga is derived from Sanskrit root "yuj" meaning "to join," "to balance," or "to unite." Yoga provides the ability to control the mind by Yoga Sutra. Yoga is a general practice for the growth of humans to divine heights which includes techniques useful for therapeutic applications in making life healthier. Several streams of yoga are reported but major four streams or schools are popular. **Karma Yoga** (the path of work) teaches to do service without any desire of fruit. **Gyana Yoga** (the path of knowledge) teaches humans to be rational, intellectual, and to acquire spiritual knowledge studying the scriptures and practices of meditation. **Bhakti Yoga** (worship or the control of emotions) is worship and surrender to the divinity. The bhakti path provides control over emotional instabilities. **Raja Yoga** (the path or Yoga for mind culture) is famous “Ashtanga Yoga” that provides all-round development of human beings. These are Yama, Niyama, Asana, Pranayama, Pratyahara, Dharana, Dhyana, and Samadhi [49].

2.4.2.1 Asana

It is a process to stay comfortable by adopting easy physical posture. Asana provides healthy body controlling power over mind. The Asanas are started gradually and then the postures are maintained for a longer time providing deep relaxation. By doing asana, the energy is channeled making the mind active and releasing the stress. The three stages in Asanas are Sthira, Cira, and Sukha. Sthira is making asana stable that requires much effort, full concentration, and will power in the beginning. The second stage is Cira that gives relaxation and third one is Sukha stage is the bliss. It starts from the body level moving towards muscular level to the breathing level to the emotional level, thereby managing the balance and calmness of mind. By doing so, one achieves Samatva, the ideal state of body and mind. Asana are classified as cultural, relaxational, and meditative postures. Relaxation postures are Savasana, the Makarasana (the crocodile postures), Sithila Dhanurasana and Sithila Taḍasana. The Padmasana, Siddhasana, the Vajrasana, and the Sukhasana are called the meditative postures. All the other postures are called cultural postures [50, 51].

2.4.3 Naturopathy

Naturopathy is a drugless alternative system of healing that was practiced in eighteenth and nineteenth centuries. The theory of naturopathy based on the sound philosophy, principles, techniques, and science. Naturopathy was first practiced by the Hippocratic School of Medicine in 400 BC [52]. Naturopathy was introduced to North America in 1895 and this medicinal system was at its peak in 1920–1930. In India, Naturopathy has been practiced since ancient times in the form of fasting, dugdha kalapa, and taking bath in holy rivers. Translations of Louis Kuhne’s famous book “The New Science of Healing” in Telugu by Shri Venkata Chelapati Sharma in 1894 and into Hindi and Urdu in 1904 by Shri Shroti Kishan Swaroop led to the revival of naturopathy in India. Naturopathy recognizes existence of vital curative force within the body. It is believed that naturopathy has no side effect. Naturopathy advocates that the cause of all diseases in body is accumulation of morbid matter in body and the treatment is the elimination of morbid matter from the body. Acute diseases are considered self-healing efforts of the body. The human body itself has the healing power and prevents itself from diseases. It treats all aspects including physical, mental, social, and spiritual at the same time. Naturopathy treats food as the only medicine. Various components of naturopathy include diet therapy, fasting therapy, mud therapy, hydrotherapy, massotherapy, acupressure, acupuncture, chromotherapy, air therapy, and magnet therapy. As per diet therapy, all foods must be taken in their natural form. The diets are classified into eliminative diet, soothing diet, and constructive diet. The diet should consist of 20% acidic and 80% alkaline foods. Fasting is considered as a therapy in various conditions. Fasting leads to increased insulin sensitivity resulting in reduced plasma glucose and improved glucose tolerance and reduced levels of oxidative stress. It also leads to enhanced immune function. Use of different forms of water such as solid, liquid, and vapor at different temperatures such as hot and cold baths, saunas, and wraps for treatment of disorders constitutes hydrotherapy. Massage with different lubricants such as mustard oil, olive oil, coconut oil, rose petals, and neem leaf powder constitutes massotherapy. Massage involves acting on the body applying pressure, tension, motion, vibration, etc. to target tissues such as muscles, tendons, ligaments, skin, and joints. The seven fundamental modes of massage are touch, effleurage, friction, petrissage, tapotement, vibration, and joint movement. Massage can affect vasodilation of arteries, stimulation of peristalsis, changing muscle tone, and stimulate heart and produce several other physiological effects. Medical research has shown that massage is highly beneficial in pain, anxiety, and depression. Acupressure is ancient healing that uses fingers or blunt objects to press key points "acu points" to stimulate self-curative ability of the body. Acu points release muscular tension and promote circulation of blood to help healing. Acupressure has been documented for treatment of over 3000 conditions. It is highly effective in headaches, eyestrain, neck pain, backache, arthritis, anxiety, and insomnia. Acupuncture uses pricking with fine needles into specific points on body to relieve pain. Acupuncture has been practiced in China since ancient times and the text documenting acupuncture dates back to 305–204 B.C. It has been shown to be effective in depression, headaches, nausea, vomiting, arthritis, sciatica, asthma, insomnia, etc. In magnet therapy, magnets are used on the body for the treatment. It helps in balancing the energy and improves circulation to the applied area [52–54].

2.4.4 Siddha

Siddha is one of the oldest systems of medicine that originated in South India (Tamil Nadu). All the literatures are available in the Tamil language. The Siddha medicinal system is popularly practiced even today for treatment of several diseases. It is largely therapeutic in nature. The principles and doctrine of Siddha is similar to Ayurveda. Siddha system believed that the human body is the replica of the universe and so are the food and drugs. This system also accepts the five-element theory and the tridosha theory as in Ayurveda. In Siddha medicinal system, the diseases are identified through pulse, urine, and different anatomical features such as the tongue, voice, complexion, eyes, touch (to find dry, warm, cold, sweating condition), and stools [45]. Metals and minerals are extensively used in this system. There are water-soluble inorganic alkalies and salts, water-insoluble mineral drugs that emit vapors when put in fire or when heated both natural and artificial. Metals like gold, silver, copper, tin, lead, and iron are regularly used after thorough processing through incineration. Even mercury and its different salts like mercuric chloride and red oxide of mercury are used in this system. Sulfur also finds usage for therapeutic applications in Sidha. Sidha system is used for the treatment of psoriasis, STD, urinary tract infections, gastric diseases, liver diseases, arthritis, and allergic disorders [55].

2.4.5 Unani

Unani system of medicine was started by Great philosopher and physician Hippocrates in Greece. Unani system of medicine was introduced and popularized in India by various Arab and Persian practitioners in the eleventh century. The famous books were written as Kitab-al-shifa (Book of Healing) and the Canon of Medicine. This system was known as UnaniTibb (Unani being the Arabic word for "Greek" and Tibb an Arabic word for "medicine") in 980 CE (WHO 2010). Unani system of Medicine was intensively propagated in nineteenth century by great Hakim Ajmal Khan. The Unani medicine is practiced in India, Pakistan, Bangladesh, Sri-Lanka, Nepal, China, Iraq, Iran, Malaysia, Indonesia, Central Asian and Middle Eastern countries, African and European countries. The Unani system of Medicine is sometimes called Unani-Tibbor Hikmat. According to Unani medicine system, the cause of disease is a natural process and symptoms are the responses to disease. The human body works on the basis of self-preservation mechanism. The fundamental theory of the Unani system is "humoral theory" of Hippocrates. According to this theory, four humors, namely Dam (blood), Balgham (phlegm), Safra (yellow bile), and Sauda (black bile), which maintain the body balance are present in the human body. Blood (Dam) is hot and moist, phlegm (Balgham) is cold and moist, yellow bile (Safra) is hot and dry, and black bile (Sauda) is cold and dry [56]. The human body is made up of seven components, which are Arkan (element), Mizaj (temperament), Akhlat (humors), Aaza (organs), Arwah (sprit), Qowa (faculties), and Afaal (functions). Unani system believes that every human body has its own temperament (mizaj) and this is expressed as sanguine, phlegmatic, choleric, and melancholic. Human body contains four elements, fire, water, earth, and air, and each of these has its own temperament. Disease diagnosis and treatment prescription is based on these components [57].

2.4.6 Homeopathy

Homeopathy is another form of alternative medicine introduced by German physician Samuel Hahnemann in 1796. The term homeopathy is derived from Greek words, homeo meaning "similar" and pathos meaning "suffering." Homeopathy is a relatively newer system of medicine compared with the other systems mentioned above. The Organon of Rationale Medicine written by Samuel Hahnemann is the

Bible of homeopathy. Homeopathic system of medicine is commonly based on the use of herbs. Homeopathic medicines cure the patients with very low concentrations and are supposed to cause effects similar to normal concentrations. Homeopathy system is based on the principle of Similia Similibus Curantur (like cured by like). Homeopathy believes that the vital force regulates the self-healing capabilities of good health, which equilibrate the mind and body. The weakening of the body causes illness. Samuel Hahnemann developed a method "potentization," a process in which a substance is diluted with alcohol or water and is shaken vigorously during preparation. According to this method, homeopathic agents are diluted until there is not a single molecule left in the final product. It is believed that higher potencies (i.e., more diluted remedies) are more effective than lower potencies [58, 59]. Homeopathic medicines originate from plants, animals, and minerals. Mahendra Lal Sircar, medical doctor who publically converted his practice into homeopathy in 1867, was the first Indian homeopathic physician [60, 61].

2.4.7 Sowa-Rigpa

Sowa-Rigpa is Tibetan system of medicine. It is believed that this is one of the oldest and well-documented medicinal systems. Sowa-Rigpa is a Tibetan word (Sowa meaning to nourish, heal, correct and Rigpa meaning science or knowledge, perception or erudition). Sowa-Rigpa is practiced in Tibet, India, Mongolia, Bhutan, China, and Nepal. Sowa-Rigpa is widely practiced in Himalayan states of India like Sikkim, Arunachal Pradesh, and regions of Darjeeling, Lahaul and Spiti, and Ladakh. The theory and practice is similar to Ayurveda. Sowa-Rigpa is also popularly known as Amchi system. Sowa-Rigpa is based on the principle of *Jung-wa-lan,* which means five elements (Panch-mahabhuta in Ayurveda), and Ngepa-Sum (Tridohsa in Ayurveda). It is believed that all activities in universe are composed of sa (earth), Chu (water), Mai (Fire), rlung (air), and Nam mkha (space), which in Sanskrit mean Dharti, Jal, Agni, Vayu, and Akash, respectively. The physiology, pathology, pharmacology, and materia-medica of Sowa-Rigpa was developed on the basis these theories. Sowa-Rigpa system of medicine was recognized in India in 2011 after the Indian Medicine Central Council Amendment Bill 2010 [40, 62].

2.4.8 Development of AYUSH Drugs

General guidelines for drug development of Ayurvedic formulations were recently published by Central Council for Research in Ayurvedic Sciences in 2018 in the form of three series, (1) drug development of Ayurvedic formulations, (2) safety toxicity evaluation of Ayurvedic formulations, and (3) clinical evaluation of Ayurvedic interventions. According to Rule 3a of Drugs and Cosmetics Act, 1940 "Ayurvedic, Siddha or Unani drug" includes all medicines for internal or external use for the diagnosis, treatment, mitigation, or prevention of disease or disorder in human beings or animals and manufactured exclusively in accordance with formulae described in authoritative books of Ayurveda, Siddha, and Unani Tibb systems of medicine. Ayurvedic drugs are majorly classified as classical drugs (raw/crude drugs, extracts, compound formulations, herbo mineral formulations), various dosage forms (asava and arista, arka, avaleha/leha/paka, kvatha, chuma, guggulu, chuma, ghrita/taila, lavanaksara, lepa, vati and gutika/pills, netrabindu and anjana, parpati, pisti, mandura, rasayoga, lauha, dhupa, bhasma), and patented and proprietary drugs (syrup, ointment, capsule, granules, confectioneries, dusting powder, tablet, suppositories) [63]. The raw drugs are generally procured from wild/cultivated sources. Different plant parts are used in preparation of drugs and these may include roots, fruits, bulbs, seeds, flowers, etc. Forty different parts have been listed in the general guidelines. The chemical composition and the content of active constituents vary between different parts. Good agriculture and collection practices guidelines published by WHO are an important step in ensuring the quality of herbal medicines. The protocol for development of AYUSH drugs is compiled in these guidelines and is divided into eight phases as listed in Table 2.1.

2.4.8.1 Prevalence Survey

Literature search is the key point for the development of drugs and vast literature on traditional systems of medicine in practice is available in the public domain. It is reported that globally about 35,000 plants are used for medicinal purposes [64]. The herbs/plants contain biologically active compounds that impart medicinal values to these plants. For the purpose, the medicinal plants can be divided into major categories, (1) herbs which occur in databases that provide detail and coherent description of the history and theory of use from Ayurvedic, Unani, Siddha, and Homeopathy medicine and (2) herbs used as folk medicine but lack literature information on their history and theory of use. The number of medicinal plants listed in Ayurveda (1200 species), Siddha (900 species), and Unani (700 species) [40] and thousands of formulations documented in the Ayurvedic Formulary of India, National Formulary of Unani Medicine, and Siddha Formulary of India can tremendously increase the success rate for developing medicinal formulations with sound scientific evidence.

2.4.8.2 Collection of Plant Materials

Good Agricultural Practices (GAP) of medicinal plants is defined as a cultivation program designed to ensure optimal yield in terms of both quality and quantity of any crop intended for health purposes. The supply and demand of plant materials has increased worldwide due to increased popularity of herbal medicines. Several reports have demonstrated adverse effects of many herbals that might be due to use of wrong plant species, adulteration with undeclared other medicines and/or potent

Table 2.1 General research guidelines and methodologies for drug development for Ayurveda, Siddha, and Unani medicine

Preparatory phase		
Phases	Activity	Considerations
1	Prevalence survey and formulation of drug/combination for specific targeted indication/activity	Appropriate basis of literary survey, previous clinical data of ingredients/any other data of folklore claims, classical evidences
Drug developments phases		
2	Collection of raw drugs	Current good agricultural practices, good field collection practices and Ayurvedic textual methods
3	Botanical identification/pharmacogonostic /chemical studies of ingredients	Based on available guidelines and classical methodology
4	Formulation of SOPs and standardization, stability studies, quality assurance	Considering the classical methods and currents available physical/chemical, biological parameters for standardization
5	Preclinical safety studies (acute/subacute/chronic studies as per the clinical use of the drug)	With appropriate animal ethical clearances as per available guidelines
6	Animal studies for biological activity/ efficacy	Specific/mechanism of action activity for clinical correlation
7	Design of study and formulation of clinical protocols (bulk preparation of quality assured drug for clinical trial, packing labelling, etc. as per need at appropriates time)	As per current guidelines and adopting classical methodology
8	Execution of clinical trial	Approval of IEC/IRB and CTRI Registration Trial conduct Trial monitoring Trial coordination Data analysis Publication

substances, contamination with toxic and/or hazardous substances, overdosage, inappropriate use by healthcare providers or consumers, and interaction with other medicines. Intrinsic (genetic) and extrinsic (environment, collection methods, cultivation, harvest, post-harvest processing, transport, and storage practices) factors are responsible for the safety and quality of raw materials of medicinal plant and final products. The environmental contaminations in medicinal plants are heavy metals, pesticides, herbicides, mycotoxins, and pathogenic microorganisms, all related to plant growth in polluted soils, cultivation requiring the use of pesticides, harvest, and storage of the raw materials in unclean warehouses. World Health Organization (WHO) has published guidelines for Good Agriculture and Collection Practices (GACPs) to ensure the affordable and sustainable supply with good quality of medicinal plant materials. WHO also addressed the impact of cultivation and collection of medicinal plants on the environment and ecological processes and the welfare of local communities [63, 65]. According to GACPs, proper documentation is required to minimize the contaminations of plant material.

2.4.8.3 Botanical Identification/Pharmacogonostic/Chemical Studies of Ingredients

Starting plant materials are the foundation of the any finished products, therefore determination of authenticity of plant material is essential to maintain the consistency of efficacy and safety of any finished products. Following parameters are recommended to ensure the quality of raw material.

Passport Data of Plant Material (Place and Date of Collection)

The cultivation and harvest/collection time, the place of cultivation and collection, the name of the part collected, and the chemicals or other substances used during cultivation or storage such as fertilizers, pesticides, herbicides, and fumigants should be documented [65, 66].

Foreign Matter

The raw material should be free from any foreign matter such as molds or insects, and other animal contamination including animal excreta. Plant material should be free from any poisonous, dangerous, or harmful foreign matter. Take about 100 g plant material and spread in a thin layer in an appropriate dish or tray. Examine the sample in daylight with eye or magnifying glass to separate the foreign matter. Weigh the foreign matter and calculate the excluded foreign matter in percent with reference to drug sample.

Macroscopic and Microscopic Characters

The identity and the degree of purity of plant materials can be observed through their characteristic macroscopic and microscopic features and should match the Pharmacopoeial standard as a reference. The macroscopic identity of plant materials is based on shape, size, color, surface characteristics, texture, fracture characteristics, and appearance of the cut surface. Microscopic examination is performed with chemical reagent for identification of plant materials. Microscopic characteristics features include the presence or absence of hairs (trichomes), canals, oil glands, particular cell types, pollen or seed morphology, and vascular traces.

Loss on Drying (Moisture Content)

It is a method to determine water and volatile matter in the raw material. The sample (10 g) is prepared by chopping, grounding, or shredding of the unground or unpowdered material to a thickness of not more than 3 mm. Care should be taken to avoid moisture loss during sample preparation. A suitable amount of sample is placed in evaporating dish for drying at 105 °C for 5 h and weighed. This is repeated after 1 h interval till the difference between two successive weighing corresponds to not more than 0.25%.

pH Value

Determination of pH value gives information about acidity or alkalinity of a solution. The pH value can be determined with a pH meter.

Ash Values

It is a method to measure the total amount of material remaining after ignition. Two types of ash are determined, physiological ash (derived from the plant tissue itself) and non-physiological ash (the residue of the extraneous matter, e.g., sand and soil). Ground sample is accurately weighed in a tared platinum or silica dish and heated at a temperature not exceeding 450 °C till free from the carbon, coal and reweighed. The obtained total ash is boiled with dilute HCl and ignited to determine the remaining acid-insoluble ash. Water-soluble ash is the difference between the total ash and the residue after treatment of the total ash with water.

Extractive Values

The extractive values indicate the amount of chemical constituents extracted with solvents from a given amount of sample. The air-dried and coarsely powdered sample (5 g) is macerated with 100 mL of water in a closed flask for 24 h, with frequent shaking during first 6 h and kept for 18 h. The percentage of water-soluble extractive value is calculated and compared with reference. Alcohol soluble extractive value is calculated after maceration with alcohol.

Volatile Oil (If Oil-Bearing Plants)

Volatile oils possess specific characteristics like odor, oil-like appearance, and ability to volatilize at room temperature. The volatile oil in sample is extracted by distilling with a mixture of water and glycerin using clevenger apparatus.

Test for Heavy/Toxic Metals

Analytical techniques like atomic absorption spectrophotometry or inductively coupled plasma mass spectrometry techniques are used for the determination of heavy metal elements and some nonmetal elements in the atomic state. The permissible limits of lead (10 ppm), arsenic (3 ppm), cadmium (0.3 ppm), and mercury (1 ppm) are given in API (API IX).

Pesticide Residue

A pesticide is any chemical or mixture of chemicals which are used for preventing, destroying, or controlling any pest, unwanted species of plants or animals causing harm during or otherwise interfering with the production, processing, storage, transport, or marketing of plant drugs. The plant material and products should be free from pesticide chemicals. Mostly chlorinated hydrocarbons and related pesticides like aldrin, chlordane, DDT, dieldrin, HCH, and a few organophosphorus pesticides like carbophenothion have a long residual action. Therefore these pesticides should be tested in the medicinal plant materials [73]. Several methods are employed for extraction, purification, and identification of pesticides.

Microbial Contamination

Microbial tests should be applied to raw plant materials for Enterobacteria and also for fungal count as per Ayurvedic Pharmacopoeia of India. Besides, microbial tests for specific pathogens like *Escherichia coli*, *Salmonella* spp., *Staphylococcus aureus*, and *Pseudomonas aeruginosa* should also be performed.

Aflatoxins

Aflatoxins are hazardous to health even in very small amounts. Aflatoxins (B1, B2, G1, and G2) are highly dangerous and precaution should be taken during aflatoxin test. Qualitative and quantitative analysis of aflatoxins can also be performed by analytical techniques like TLC and HPLC [67].

Chemical Standardization

Quality control and authentication of raw materials are the main concern for the acceptability of herbal medicines. Several assays are reported for determination of different classes of bioactive compounds such as total tannins, sugar contents, total phenolics, and total flavonoids. Many of the biological activities shown by plant drugs are ascribed to the phenolics contents, which are assayed

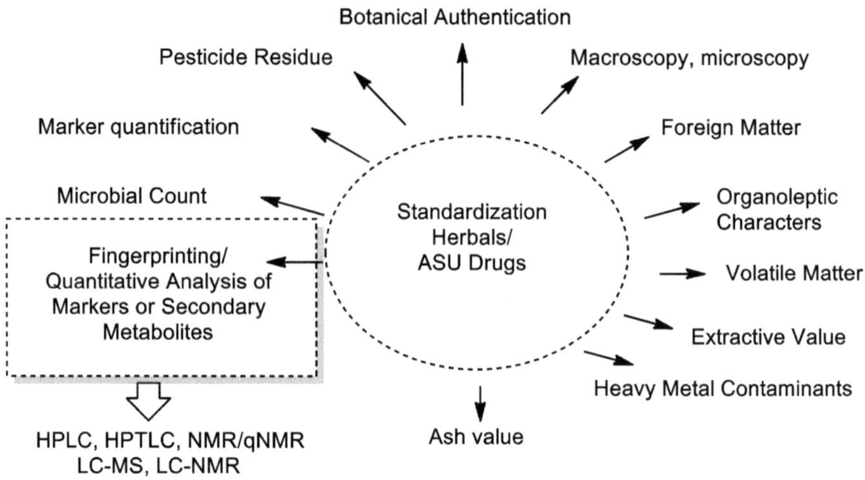

Fig. 2.13 Standardization of ASU drugs

spectrophotometrically using Folin Ciocalteu reagent. Total sugars, total reducing sugars, and total nonreducing sugars are estimated by Nelson-Somogyi photometric method.

In plants, the secondary metabolites are generally present inside the cells. Therefore, grinding of the raw material and breaking tissue are required to enhance the extractive yields. Standardization of plant extracts is very important as quality of plants material plays significant role in ensuring quality of the finished products. Traditional drugs can be single herbs or polyherbal. The identification and quantitation of one or two markers or medicinally active constituents in extracts are used for the quality and authenticity of herbal medicines. The extracts of single herb or polyherbal drugs may contain hundreds of compounds belonging to different chemical classes and some of these may be present in very low concentration, but may be highly active. Further, different compounds may be acting on different therapeutic targets leading to the overall observed benefits of these drugs. Therefore, standardization with respect to one or two markers may not provide true whole picture for quality control in terms of therapeutic benefits of extracts. This may also not account for the synergistic effects within these drugs. On other hand, employing a single analytical technique may also be inadequate for the effective analysis of complex herbal medicine, chromatographic and spectral fingerprints using TLC, HPTLC, HPLC, LC-MS, and NMR may be generated. Therefore, standardization of extract with identification and quantification of maximum chemical constituents is required to assure the reproducibility and repeatability of biological and clinical efficacy. Standardization parameters for quality control are summarized in Fig. 2.13.

Complete metabolite profiling of *Eugenia jambolana* fruit pulp has been reported. Sixty-eight chemically diverse metabolites were identified in the fruit pulp using qNMR, HPLC, LC-MS, GC-MS, and MALDI-TOF. The identified compounds included anthocyanins, anthocyanidins, amino acids, sugars, phenolics, and volatile

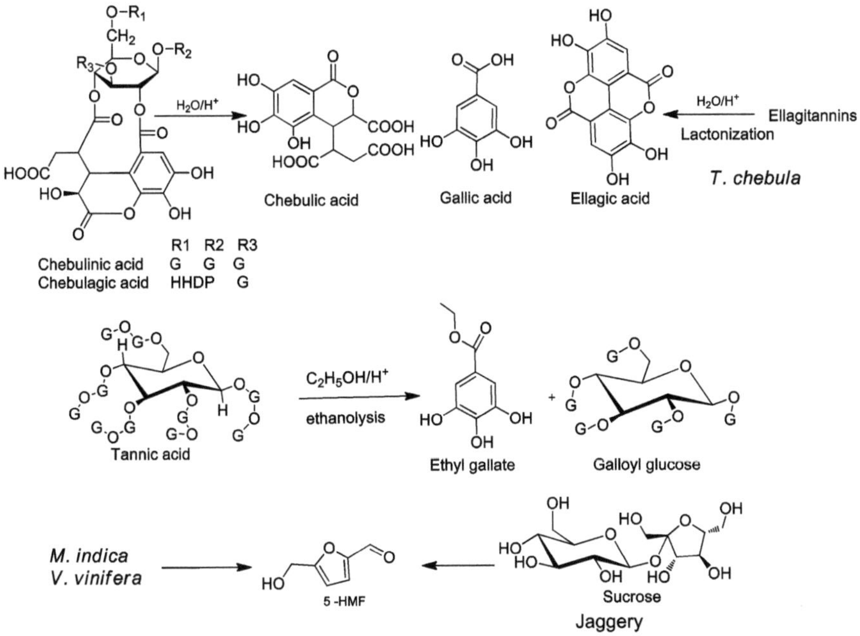

Fig. 2.14 Chemical changes during fermentation in Abhayarishta

compounds. Twenty-five metabolites were identified in the *n*-butanol and aqueous-methanolic extracts by qNMR. The main advantage of using qNMR is that reference standards of natural products being quantified are not required. Any commercially available compound of known purity can be used as an internal standard provided its NMR signals do not overlap with the NMR signal of the analyte being quantified. Further qNMR does not require the separation of analytes. HPLC and HPTLC require authentic reference standards for the preparation of calibration curves [68].

Arishtas are important Ayurvedic formulations that are prepared by anaerobic fermentation of decoctions of plant materials with jaggery. Fermentation generates alcohol which helps preserve these formulations. It has been shown that fermentation alters the chemical composition of the finished products. The amounts of chemical constituents in fermented product differed from that observed in the raw materials, which was explained in terms of chemical transformations occurring during fermentation (Fig. 2.14). Abhayarishta contains *Terminalia chebula* (pericarp), *Vitis vinifera* (fruits), *Embelia ribes* (fruits), and *Madhuca indica* (flowers). It was shown that chebulinic acid and chebulagic acid, two major constituents and marker compounds of *Terminalia chebula*, completely disappeared in the formulation. There was an increase in amounts of chebulic acid, ellagic acid, and gallic acid. This suggested that gallotannins and ellagitannins hydrolyzed during fermentation and were converted into monomeric constituents. Similarly, ethyl gallate, which is not present in any of these plants, appeared in the finished product. This was considered to form through ethanolysis of tannic acid and galloyl glucose. This study underlines the importance of standardization of raw materials as well as the finished products to ensure quality control [69].

2.4.8.4 ASU Formulations

ASU drugs can be formulated into a variety of dosage forms depending on the nature of chemical constituents and the intended uses. Pre-formulation studies are essential for developing a suitable herbal formulation (Fig. 2.15).

2.4.8.5 Stability Studies/Shelf Life

Stability depends on several factors like type of compounds present in the drug, packing material used, storage, etc. Stability studies of drugs give information about shelf life and storage conditions. This is important for providing evidence for the quality of a drug substance over a period of time under the influence of variety of environmental factors such as temperature, humidity, and light. Stability test of drugs can be studied by storing the sample at standard storage and accelerated storage conditions and test them at defined time intervals. Alternately, drugs can be evaluated by selecting samples manufactured over a period of 5 years and crossing 6 months. The stability test should be performed with suitable parameters like the physical, chemical, biological, and microbiological attributes. Stability for chemical parameters may include color reaction, pH value, weight variation, disintegration, bulk density, extractive values, and estimation of biomarker or marker compounds by suitable analytical methods. If possible, a suitable bioassay can be used. The date of expiry should be displayed on the label. The guidelines for stability studies are prescribed in the Ayurvedic Pharmacopoeia of India, Part-I,

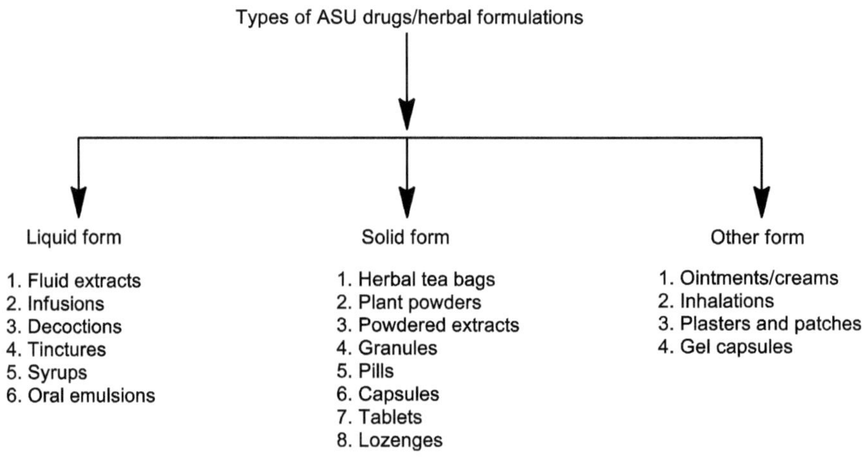

Fig. 2.15 Types of herbal formulations

Volume-VIII. In general, the shelf life of AYUSH drugs varies from 1 year (arka) to 10 years (asava and arista) for different dosage forms.

2.4.8.6 Preclinical Safety/Toxicity and Efficacy

The importance of safety and toxicity of food and drugs is known since ancient times and there are evidences that preclinical testing Vishaana/Virudhanna/Pareeksha on animals was well known in Ayurveda. Sushruta mentions testing of food /drugs on birds and animals to establish safety. The toxicity of some plants, metals, and minerals was well known to ancient Ayurveda practitioners as processing (shodhana, marana) of these drugs to make them safe is well documented in Ayurveda. It is important to do comprehensive literature search on plant ingredient of drug preparations before starting the preclinical safety/efficacy studies. The literature search should then be extended to gather information on closely related plant species for chemotaxonomic correlation. Generally, preclinical studies are performed to characterize toxicity with respect to target organs, dose dependence, and relationship to exposure. These data help for clinical experiment in the human trial for establishing safe starting dose and identify potential adverse effects. Ministry of AYUSH issued guidelines for safety requirements for different categories of ASU drugs. Recently, CCRAS has given detailed protocols for conducting safety/toxicity of ASU drugs; these include single-dose toxicity studies (Acute Toxicity), repeated-dose oral toxicity study, reproduction and developmental toxicity studies, and special toxicity tests. Several parameters are listed for toxicity studies in CCRAS manual for preclinical studies for different category of drugs (CCRAS Guidelines Series II) [70].

For any ASU drugs given in Section 3(a) of Drugs and Cosmetics Act 1940, which are manufactured and prescribed as per ASU texts, there is no requirement of any preclinical safety and efficacy data. Even if there are any changes in the dosage form, but the drugs ingredients are as per texts, still there is no requirement of preclinical safety and efficacy data. If these drugs are to be used for a new indication, preclinical efficacy data is required. For patented and proprietary drugs defined under Section 3(h) of the act, containing crude drugs/aqueous extracts or hydroalcoholic extracts as per text, not preclinical data are required. However, for patented and proprietary drugs containing other than aqueous/hydroalcoholic extract of any other solvent extract, detailed preclinical safety and efficacy data is required. These include, for oral preparations, acute and long-term toxicity in rats, reproductive and developmental toxicity, genotoxicity, and carcinogenicity. Dermal toxicity, photo toxicity, and allergenicity tests are required for topical formulations. Similar detailed toxicity and efficacy studies are required for patented or proprietary drugs defined under Section 3(h) containing any of the ingredients of Schedule E(1). Complete details required for licensing are given in CCRAS guidelines Series II. The efficacy study is required for proof and documentation of indicated claims [70].

2.4.8.7 Clinical Study

Ayurvedic texts have mentioned two-way approach of experimentation and drug trials, feasibility of test interventions, observational design and system validation are mentioned in Sushruta Samhita. Clinical studies for ASU drugs and patent and proprietary medicines are divided into Phase I, Phase II, Phase III, and Phase IV. Phase I determines the maximum tolerated dose (both single and multiple dose) and early measurement of drug activity. It also determines the nature of expected adverse reactions. Phase II explores the efficacy of drug for the indications and to determine any short-term side effects. The study should be able to determine the dose, therapeutic regimen, and target population (mild versus severe disease) for Phase III trials. The primary objective of Phase III trials is the confirmation of therapeutic benefits. This study should be designed to confirm the preliminary evidences on safety and efficacy in intended indication. It may explore dose–response relationships, usage in wider population in different stages of the disease, safety and efficacy in combination with other drugs. Phase IV trials are post-marketing surveillance and include additional drug–drug interactions, dose response or safety studies, any mortality/morbidity studies, epidemiological studies, etc. Trials for classical Ayurvedic drugs prepared as per Ayurvedic description with same textual indications can start directly from phase III/IV trial, whereas for drugs with new indications, trials can start directly from Phase II. Clinical trials should start from Phase I for patented and proprietary medicines containing Schedule E-1 ingredients (poisonous substances). These clinical studies are mainly of two types, observational study and experimental study. Observational study explores cause and effect relationships where the investigator observes the participants by asking questions, taking measurements, or studying clinical records. In experimental study, participants receive medical products/drugs as per the approved protocols and a comparison of placebo to a standard drug is done. Trial design is very important for any successful clinical trial. Complete details and requirements to conduct clinical trials are provided in CCRAS guidelines Series III. These guidelines also give detailed requirements for issuing of license for ASU drugs [71].

Several new drugs have been developed by the research councils (CCRAS, CCRH, CCRUM, and CCRS) and several others are being developed. AYUSH 82 as an antidiabetic and AYUSH SG as an anti-rheumatoid drug have been developed by CCRAS. AYUSH PJ-7 for Dengue fever, AYUSH Manas for mental retardation, and several other drugs for various conditions are currently being developed.

2.4.8.8 Development of AYUSH 82 for Diabetes

Ayush 82 consists of four plants: Amer Bija (*Mangifera indica* Linn seed), Jambu Bija (*Eugenia jambolana* seed), Karvellaka Bija (*Momordica charantia* Linn seed), and Gudmar (*Gymnema sylvestre* leaf). This polyherbal ASU formulation was developed based on the reported antidiabetic potential of these four plants.

Mangifera indica belongs to the family Anacardiaceae. It is an important medicinal plant of the Ayurvedic and indigenous medicinal systems. The fruit, bark, leaves, and seeds possess antidiabetic, antioxidant, antiviral, anti-inflammatory properties. A number of studies have proved the effect of *M. indica* for antidiabetic activity. Various classes of bioactive compounds, terpenoids, flavonoids, alkaloids, coumarins, terpenoidal saponins, polyphenolics, tannins, etc. have been reported from different parts of the *M. indica* [72]. The ethanol and aqueous extracts of leaves and stem bark of *M. indica* exhibited significant hypoglycemic effect in type 2 diabetes rat model at dose of 250 mg/kg body weight [73]. The ethanolic extract showed decreased blood glucose level and also restored the levels of glycated hemoglobin in streptozotocin-induced diabetic rats at dose of 300 mg/kg b.w./day for 21 days [74]. The methanol and aqueous extracts of seed kernel showed significant reduction in blood glucose level in diabetic rats. Purified mangiferin from *Mangifera indica* leaves showed significant antidiabetic, antihyperlipidemic, and antiatherogenic activity at a dose of 10 and 20 mg/kg b.w. [75]. The ethanolic extract of mango fruit peel showed significant reduction in blood glucose level in streptozotocin-induced diabetic rats [76].

Momordica charantia L., commonly known as bitter melon or bitter gourd, belongs to family Cucurbitaceae. It is traditionally used for the management of diabetes since ancient times [77]. Studies have been reported for antidiabetic effect of *M. charantia* [78]. The methanol extract of *M. charantia* fruit orally administered to alloxan monohydrate-induced diabetic rats for 30 days showed a significant decrease in triglyceride, low-density lipoprotein and a significant increase in high-density lipoprotein level. A significant effect was also observed for oral glucose tolerance [79]. *M. charantia* fruit juice showed the increment of β cell in the pancreas of streptozotocin (STZ)-induced diabetic rats as compared to untreated diabetics rats [80]. Several phenolic acids, flavonoids, phytosterols, and terpenoids have been reported [81].

Gymnema sylvestre belongs to the family Asclepiadaceae and is popularly known as gurmar. It is mentioned in Ayurveda for the treatment of several ailments, especially diabetes management. First experimental pharmacological study was reported in 1930 [82]. It possesses a variety of biological activities including antidiabetic, antihyperlipidemic, antiobesity, antioxidant, immunomodulatory, anti-inflammatory, anticancer, and wound healing [83]. The water-soluble extract of *Gymnema sylvestre* leaves was administered (400 mg/day) to 27 patients with insulin-dependent diabetes mellitus on insulin therapy. The water-soluble extract enhanced endogenous insulin, possibly by regeneration/revitalization of the residual *beta* cells in insulin-dependent diabetes mellitus [84]. Daily oral administration of methanolic extract of *Gymnema sylvestre* leaves (100, 200 and 400 mg/kg b.w.) and glibenclamide (5 mg/kg) in normal and streptozotocin-induced diabetic rats for 28 days showed significant reduction in blood glucose level with no histopathological change [85]. Gymnemic acid has been reported as an active compound with antidiabetic properties and inhibited glucose absorption [86].

Eugenia jambolana Lam. (Syn. *Syzygium cumini* Skeels or *Syzygium jambolana* or *Eugenia cuminii* Druce.) belongs to the family Myrtaceae. It is commonly known as Indian blackberry, jamun, java plum, Portuguese plum, black plum, Indian blackberry, jambu, jambul, and jambool. In traditional system of Indian medicine the fruit pulp and seed extracts are being used for the treatment of several diseases including diabetes. Jamun has been used for the treatment of diabetes alone or in combination with other antidiabetic plants in Europe [87]. A number of pharmacological activities like antioxidant, hepatoprotective, anticancer, anti-inflammatory antibacterial, antifungal, and gastroprotective activities have been reported [88]. Metabolite profiling of jamun fruit and pulp has shown the presence of phenolic acids, anthocyanins, flavonoids, carotenes, and phytosteroids [68, 89]. The anti-hyperglycemic effect of aqueous and alcoholic extracts was evaluated in diabetic animals. The different doses 32 mg/kg and 120 mg/kg of alloxan monohydrate for mild (plasma sugar >180 mg/dL, duration 21 days) and moderate (plasma sugar >280 mg/dL, duration 120 days) diabetes, respectively, and 150 mg/kg of streptozotocin for sever diabetes (plasma sugar >400 mg/dL, duration 60 days) were administered. The reduction of glucose level in moderate and severe diabetic rats was 55.62 and 17.72%, respectively, [90]. The hypoglycemic activity of ethanolic extract of seeds of *E. jambolana* was evaluated in alloxan-induced diabetic rabbits. The induced diabetic rabbits were treated with ethanolic extract with a dose of 100 mg/kg for (1 day), (7 days), and (15 days) for sub-diabetic, mild diabetic, and sever diabetic rabbits, respectively. The glucose level significantly reduced in sub-diabetic and mild diabetic rabbits [91].

Acute Toxicity

Administration of Ayush 82 orally in Swiss Albino (I. B) mice revealed no pre-terminal deaths, toxic signs, or abnormal behavior in the animals at 10 times of intended therapeutic dose.

Subacute Toxicity

Subacute toxicity studies of Ayush 82 was performed in Wistar rats which showed no significant effect in the blood biochemistry, hematology, and weight of the vital organs in comparison to the control suggestive of its safety.

Clinical Efficacy

Total 886 patients was selected in which 497 patients completed cases and 389 patient dropouts at Council's peripheral Central Research Institutes. The Ayush 82 was administered thrice daily. The results showed significant reduction in fasting and post prandial blood sugar level along with clinical improvement. No adverse effects were reported in the patients during the treatment period. The recommended dose is 15 g/day, which is divided into three doses along with 500 mg Shuddha Shilajita twice daily [92].

2.5 Summary

Nature is the richest source of medicines for the prevention and treatment of various diseases. Most of the blockbuster drugs available in the market are derived from natural sources. The process of natural product-derived drug discovery involves taxonomical identification of targeted plant followed by extraction, fractionation, isolation, and characterization. However, lesser yield of hit molecules from natural sources drives the focus mainly to semisynthetic derivatives of lead molecules derived from nature. The other major challenges faced by natural-product-based drug discovery and indigenous systems of medicines include poor bioavailability of most natural compounds, time-consuming process for discovery of novel skeletons, lack of strict regulations for the standardization of traditional formulations, etc. Despite the supremacy of synthetic drugs, there is a re-emergence of plant-based drugs in recent years. Molecules from traditional sources have great ethnopharmacological value. Natural-product-based drug discovery will continue to be a promising area in the field of research and development.

References

1. Gad SC. Drug discovery handbook. Hoboken, NJ: Wiley-Interscience; 2005.
2. Lipinski CA. Lead-and drug-like compounds: the rule-of-five revolution. Drug Discov Today. 2004;1(4):337–41.
3. Cragg GM, Newman DJ, Snader KM. Natural products in drug discovery and development. J Nat Prod. 1997;60:52–60.
4. Government of India, Ministry of Health and Family Welfare (http://www.fda.gov), Drug and Cosmetics Act 1940 & Rules; 1945.
5. Botanical Drug Development Guidance for Industry (http://www.cdsco.nic.in) Center for Drug Evaluation and Research (CDER).
6. Yuan H, Ma Q, Ye L, Piao G. The traditional medicine and modern medicine from natural products. Molecules. 2016;21(5):559–77.
7. Newman DJ, Cragg GM. Natural products as sources of new drugs from 1981 to 2014. J NatProd. 2016;79:629–61.
8. Castro MG, Zimmermann S, Sankar MG, Kumar K. Scaffold diversity synthesis and its application in probe and drug discovery. Angew Chem Int Ed. 2016;55:7586–605.
9. Cragg GM, Newman DJ. Biodiversity: a continuing source of novel drug leads. Pure Appl Chem. 2005;77(1):7–24.
10. Sarkar SD, Latif Z, Gray AL. Methods in biotechnology: natural products isolation. Totowa, New Jersey: Humana Press; 2006.

11. Handa S, Khanuja SP, Longo G, Rakesh DD. Extraction technologies for medicinal and aromatic plants. Italy: United Nations Industrial Development Organization and the International Centre for Science and High Technology; 2008.
12. Sticher O. Natural product isolation. Nat Prod Rep. 2008;25:517–54.
13. Kemp W. Energy and electromagnetic spectrum. In: Kemp W, editor. Organic spectroscopy. London: Macmillan Press; 1991.
14. Pavia DL, Lampman GM, Kriz GS. Introduction to spectroscopy, a guide for students of organic chemistry. Australia: Brooks Cole Thomson Learning; 1979.
15. Friedman LM, Furberg CD, Demets DL. Fundamentals of clinical trials. New York: Springer; 1942.
16. Hirata Y, Uemura D. Halichondrins-antitumor polyether macrolides from a marine sponge. Pure Appl Chem. 1986;58(5):701–10.
17. Pettit GR, Herald CL, Boyd MR, Leet JE, Dufresne C, Doubek DL, Schmidt JM, Cerny RL, Hooper JNA, Rutzler KC. Isolation and structure of the cell growth inhibitory constituents from the Western Pacific marine sponge *Axinella sp.* J Med Chem. 1991;34:3339–40.
18. Pettit GR, Tan R, Gao F, Williams MD, Doubek DL, Boyd MR, Schmidt JM, Chapuis JC, Hamel E, Bai R, Hooper JNA, Tackett LP. Isolation and structure of halistatin 1 from the eastern Indian Ocean marine sponge *Phakellia carter*. J Org Chem. 1993;58:2538–43.
19. Litaudon M, Hart JB, Blunt JW, Lake RJ, Munro HG. Isohomohalichondrin B, a new antitumour polyether macrolide from the New Zealand deep-water sponge *Lissodendoryr sp.* Tetrahedron Lett. 1994;35(50):9435–8.
20. Munro MHG, Blunt JW, Dumdei EJ, Hickford SJH, Lill RE, Li S, Battershill CN, Duckworth AR. The discovery and development of marine compounds with pharmaceutical potential. J Biotechnol. 1999;70:15–25.
21. Yu MJ, Kishi Y, Littlefield BA. Anticancer agents from natural products: discovery of E7389, a fully synthetic macrocyclic ketone analog of halichondrin B. London: CRC press; 2005.
22. Towle MJ, Salvato KA, Budrow J, Wels BF, Kuznetsov G, Aalfs KK, Welsh S, Zheng W, Seletsky BM, Palme MH, Habgood GH, Singer LA, DiPietro LV, Wang Y, Chen JJ, Quincy DA, Davis A, Yoshimatsu K, Kishi Y, Yu MJ, Littlefield BA. *In vitro* and *in vivo* anticancer activities of synthetic macrocyclic ketone analogues of halichondrin B. Cancer Res. 2001;61:1013–21.
23. Aicher TD, Buszek KR, Fang FG, Forsyth CJ, Jung SH, Kishi Y, Matelich MC, Scola PM, Spero DM, Yoon SK. Total synthesis of halichondrin B and norhalichondrin B. J Am Chem Soc. 1992;114(8):3162–4.
24. Swami U, Chaudhary I, Ghalib MH, Goel S. Eribulin-a review of preclinical and clinical studies. Crit Rev Oncol Hematol. 2012;81:163–84.
25. Wani MC, Taylor HL, Wall ME, Coggon P, Mcphail AT. Plant antitumor agents. VI. The isolation and structure of taxol, a novel antileukemic and antitumor agent from *Taxus brevifolia*. J Am Chem Soc. 1971;93(9):2325–7.
26. Kingston DGI. The chemistry of taxol. Pharmacol Ther. 1991;52:1–34.
27. Chmurny GN, Hilton BD, Brobst S, Look SA, Witherup KM, Beutler JA. ^{1}H and ^{13}C-NMR assignments for Taxol, 7-epi-Taxol, and cephalomannine. J Nat Prod. 1992;55(4):414–23.
28. Suffness M. Taxol: science and applications. London: CRC Press; 1995.
29. Roytta M, Laine KM, Harkonen P. Morphological studies on the effect of Taxol on cultured human prostatic cancer cells. Prostate. 1987;11:95–106.
30. Riondel J, Jacrot M, Picot F, Beriel H, Mouriquand C, Potier P. Therapeutic response to Taxol of six human tumors xenografted in to nude mice. Cancer Chemother Pharmacol. 1986;17:137–42.
31. Legha SS, Tenney DM, Krakoff IR. Phase I study of Taxol using a 5-day intermittent schedule. Int J Clin Oncol. 1986;4(5):762–6.
32. Wiernik PH, Schwartz EL, Einzig A, Strauman JJ, Lipton RB, Dutcher JP. Phase I trial of Taxol given as a 24-hour infusion every 21 days: responses observed in metastatic melanoma. J Clin Oncol. 1987;5(8):1232–9.

33. Brown T, Havlin K, Weiss G, Cagnola J, Koeller J, Kuhn J, Rizzo J, Craig J, Phillips J, VonHoff D. A phase I trial of Taxol given by a 6-hour intravenous infusion. J Clin Oncol. 1991;9(7):1261–7.
34. Mcguire WP, Rowinsky EK, Rosenshein NB, Grumbine FC, Ettinger DS, Armstrong DK, Donehower RC. Taxol: a unique antineoplastic agent with significant activity in advanced ovarian epithelial neoplasms. Ann Intern Med. 1989;111:273–9.
35. Einzig AI, Wiernik PH, Sasloff J, Runowicz CD, Goldberg GL. Phase II study and long-term follow-up of patients treated with Taxol for advanced ovarian adenocarcinoma. J Clin Oncol. 1992;10(11):1748–53.
36. Holmes AN, Walters RS, Theriault RL, Forman AD, Newton LK, Raber MN, Buzdar AU, Frye DK, Hortobagyi GN. Phase II trial of Taxol, an active drug in the treatment of metastatic breast cancer. J Natl Cancer Inst. 1991;83(24):1797–805.
37. Eisenhauer EA, Huinink WWTB, Swenerton KD, Gianni L, Myles J, Vander Burg MEL, Kerr L, Vermorken JB, Buser K, Colombo N, Bacon M, Santabarbara P, Onetto N, Winograd B, Canetta R. European-Canadian randomized trial of paclitaxel in ovarian cancer: high-dose versus low-dose and long versus short infusion. J Clin Oncol. 1994;12(12):2654–66.
38. Nabholtz JM, Gelmon K, Bontenbal M, Spielmann M, Catimel G, Conte P, Klaassen U, Namer M, Bonneterre J, Fumoleau P, Winograd B. Multicenter, randomized comparative study of two doses of paclitaxel in patients with metastatic breast cancer. J Clin Oncol. 1996;14(6):1858–67.
39. Misra LN. Traditional phytomedicinal systems, scientific validations and current popularity as nutraceuticals. Int J Tradit Nat Med. 2013;2:27–75.
40. Sen S, Chakraborty R. Towards the integration and advancement of herbal medicine: a focus on traditional Indian medicine. Bot Targets Ther. 2015;5:33–44.
41. Angmo K, Adhikari BS, Rawat GS. Changing aspects of traditional healthcare system in Western Ladakh. India J Ethnopharmacol. 2012;143:621–30.
42. Katoch D, Sharma JS, Banerjee S, Biswas R, Das B, Goswami D, Harwansh RK, Katiyar CK, Mukherjee PK. Government policies and initiatives for development of ayurveda. J Ethnopharmacol. 2017;197:25–31.
43. Chopra A, Doiphode V. Ayurvedic medicine: core concept, therapeutic principles, and current relevance. Med Clin North Am. 2002;86:75–89.
44. Jaiswal Y, Liang Z, Zhao Z. Botanical drugs in ayurveda and traditional Chinese medicine. J Ethnopharmacol. 2017;194:245–9.
45. Katiyar CK, Mukherjee PK. Some excerpts from Charaka Samhita-an ancient ayurvedic treatise on ayurveda & healthy living. J Ethnopharmacol. 2017;197:3–9.
46. Loukas M, Lanteri A, Ferrauiola J, Tubbs RS, Maharaja G, Shoja MM, Yadav A, Rao VC. Anatomy in ancient India: a focus on the Susruta Samhita. J Anat. 2010;217:646–50.
47. Dwivedi G, Dwivedi S. Sushruta—the clinician-teacher par excellence. Indian J Chest Dis Allied Sci. 2007;49:243–4.
48. Sen S, Chakraborty R. Revival, modernization and integration of Indian traditional herbal medicine in clinical practice: importance, challenges and future. J Tradit Complement Med. 2017;7:234–44.
49. Ostermann T, Vogel H, Boehm K, Cramer H. Effects of yoga on eating disorders—a systematic review. Complement Ther Med. 2019;46:73–80.
50. About the System, Yoga: http://ayush.gov.in/about-the-systems/yoga; last accessed 29 Aug 2019.
51. Iyengar BKS. The illustrated light on yoga: Herper Collins India; 2005.
52. WHO. WHO Benchmarks for Training in Naturopathy. Geneva, Switzerland: World Health Organization; 2010.
53. Rastogi R. History and development of naturopathy in India. J Ind Inst Hist Med. 2008;38:71–8.
54. About the system, naturopathy. http://ayush.gov.in/about-the-systems/naturopathy. Last accessed 29 Aug 2019.

55. About the system, siddha. http://ayush.gov.in/about-the-systems/siddha. Last accessed 29 Aug 2019.
56. Unani system of medicine publisher: Department of AYUSH, Ministry of Health and Family Welfare, Government of India, New Delhi; 2013.
57. WHO Benchmarks for Training in Unani Medicine. Geneva, Switzerland: World Health Organization; 2010.
58. Mukherjee PK, Wahile A. Integrated approaches towards drug development from Ayurveda and other Indian system of medicine. J Ethnopharmacol. 2006;103:25–35.
59. Frank R. Integrating homeopathy and biomedicine: medical practice and knowledge production among German homeopathic physicians. Sociol Health Illn. 2002;24(6):796–819.
60. Andrew L. Encyclopedia of homeopathy. New York: D. Kindersley Pub.; 2000.
61. Ghosh AK. A short history of the development of homeopathy in India. Homeopathy. 2010;99:130–6.
62. Gurmet P. "Sowa-Rigpa": Himalayan art of healing. Indian J Tradit Know. 2004;3:212–8.
63. CCRAS General guidelines for drug development of ayurvedic formulations (first edition Series-II); 2018.
64. Yirga G, Teferi M, Kasaye M. Survey of medicinal plants used to treat human ailments in Hawzen district, Northern Ethiopia. Int J Biodivers Conserv. 2011;3:709–14.
65. WHO. WHO guidelines on good agricultural and collection practices (GACP) for medicinal plants. Geneva, Switzerland: World Health Organization; 2003.
66. WHO. WHO quality control methods for medicinal plant materials. Geneva, Switzerland: World Health Organization; 1998.
67. Khan IA, Singh A, Mindala DP, Meena S, Vij B, Yadav AK, Roy S, Nandi U, Katare AK, Jaglan S, Singh D, Gupta AP, Gupta P, Singh S, Tikko MK, Singh G, Vishwakarma RA. Preclinical development of gastro-protective botanical candidate from *Woodfordia fruticosa* (Linn.) Kurz: chemical standardization, efficacy, pharmacokinetics and safety pharmacology. J Ethnopharmacol. 2019;241:112023.
68. Sharma RJ, Gupta RC, Bansal AK, Singh IP. Metabolite fingerprinting of *Eugenia jambolana* fruit pulp extracts using NMR, HPLC-PDA-MS, GC-MS, MALDI-TOF-MS and ESI-MS/MS spectrometry. Nat Prod Commun. 2015;10:969–76.
69. Lal UR, Tripathi SM, Jachak SM, Bhutani KK, Singh IP. Chemical changes during fermentation of Abhayarishta and its standaridization by HPLC-DAD. Nat Prod Comm. 2010;5:575–9.
70. CCRAS. General guidelines for safety/toxicity evaluation of ayurvedic formulations (first edition series-II); 2018.
71. CCRAS. General guidelines for clinical evaluation of ayurvedic interventions (first edition series-III); 2018.
72. Ediriweera MK, Tennekoon KH, Samarakoon SR. A review on ethnopharmacological applications, pharmacological activities, and bioactive compounds of *Mangifera indica* (mango). J Evidence-Based Complement Altern Med. 2017;2017:1–24.
73. Bhowmik A, Khan LA, Akhter M, Rokeya B. Studies on the antidiabetic effects of *Mangifera indica* stem-barks and leaves on nondiabetic, type 1 and type 2 diabetic model rats. Bangladesh J Pharmacol. 2009;4:110–4.
74. Gupta R, Gupta RS. Antidiabetic efficacy of *Mangifera indica* seed kernels in rats: a comparative study with glibenclamide evaluate the antidiabetic as well as antioxidant efficacy of *Mangifera indica* ethanolic extract in streptozotocin induced diabetic rats. Diabetol Croatica. 2011;40:107–12.
75. Muruganandan S, Srinivasan K, Gupta S, Gupta PK, Lal J. Effect of mangiferin on hyperglycemia and atherogenicity in streptozotocin diabetic rats. J Ethnopharmacol. 2005;97:497–501.
76. Gondi M, Rao UJSP. Ethanol extract of mango (*Mangifera indica* L.) peel inhibits α-amylase and α-glucosidase activities, and ameliorates diabetes related biochemical parameters in streptozotocin (STZ)-induced diabetic rats. J Food Sci Technol. 2015;52:7883–93.
77. Virdi J, Sivakami S, Shahani S, Suthar AC, Banavalikar MM, Biyani MK. Antihyperglycemic effects of three extracts from *Momordica charantia*. J Ethnopharmacol. 2003;88:107–11.

78. Raman A, Lau C. Anti-diabetic properties and phytochemistry of *Momordica charantia* L. (Cucurbitaceae). Phytomedicine. 1996;2(4):349–62.
79. Chaturvedi P, George S, Milinganyo M, Tripathim YB. Effect of *Momordica charantia* on lipid profile and oral glucose tolerance in diabetic rats. Phytother Res. 2004;18:954–6.
80. Ahmed I, Adeghate E, Sharma AK, Pallot DJ, Singh J. Effects of *Momordica charantia* fruit juice on islet morphology in the pancreas of the streptozotocin-diabetic rat. Diabetes Res Clin Pract. 1998;40:145–51.
81. Wang S, Li Z, Yang G, Ho CT, Li S. *Momordica charantia*: a popular health-promoting vegetable with multifunctionality. Food Funct. 2017;8:1749–62.
82. Porchezhian E, Dobriyal RM. An overview on the advances of *Gymnema sylvestre*: chemistry. Pharmacol Patents Pharmazie. 2003;58:5–12.
83. Yadav D, Kwak M, Jin J-O. Clinical applications of *Gymnema sylvestre* against type 2 diabetes mellitus and its associated abnormalities. Prog Nutr. 2019;21:258–69.
84. Shanmugasundaram ER, Rajeswari G, Baskaran K, Rajesh Kumar BR, Radha SK, Kizar AB. Use of *Gymnema sylvestre* leaf extract in the control of blood glucose in insulin-dependent diabetes mellitus. J Ethnopharmacol. 1990;30:281–94.
85. Prabhu S, Vijayakumar S. Antidiabetic, hypolipidemic and histopathological analysis of *Gymnema sylvestre* (R. Br) leaves extract on streptozotocin induced diabetic rats. Biomed Prev Nutr. 2014;4:425–30.
86. Pothuraja R, Sharma RK, Chagalamarri J, Jangra S, Kavadi PK. A systematic review of *Gymnema sylvestre* in obesity and diabetes management. J Sci Food Agric. 2014;94:834–40.
87. Helmstädter A. Antidiabetic drugs used in Europe prior to the discovery of insulin. Pharmazie. 2007;62:717–20.
88. Xu J, Liu T, Li Y, Liu W, Ding Z, Ma H, Seeram NP, Mu Y, Huang X, Li L. Jamun (*Eugenia jambolana* Lam.) fruit extract prevents obesity by modulating the gut microbiome in high-fat diet-fed mice. Mol Nutr Food Res. 2019;63:1801307.
89. Baliga MS, Bhat HP, Baliga BRV, Wilson R, Palatty PL. Phytochemistry, traditional uses and pharmacology of *Eugenia jambolana* Lam. (black plum): a review. Food Res Int. 2011;44:1776–89.
90. Grover JK, Vats V, Rathi SS. Antihyperglycemic effect of *Eugenia jambolana* and *Tinospora cordifolia* in experimental diabetes and their effects on key metabolic enzymes involved in carbohydrate metabolism. J Ethnopharmacol. 2000;73:461–70.
91. Sharma SB, Nasir A, Prabhu KM, Dev G, Murthy PS. Hypoglycemic and hypolipidemic effect of ethanolic extracts of seeds of *E. jambolana* in alloxan-induced diabetic model of rabbits. J Ethnopharmacol. 2003;85:201–6.
92. CCRAS. Drug development for select diseases evidence based approach based on CCRAS R&D contributions. New Delhi: Central Council for Research in Ayurvedic Sciences (CCRAS); 2016. p. 7–9.

Chapter 3
The Concept of Receptor and Molecule Interaction in Drug Discovery and Development

Ramarao Poduri and Gowraganahalli Jagadeesh

3.1 Introduction

Pharmacodynamics is the study of the physiological and biochemical effects of drugs and their mechanisms of drug action, including the correlation of actions and effects of drugs with their chemical structure. Structure–activity relationships form the basis for both the rational therapeutic use of drugs and the design of new molecular entities with improved efficacy, specificity, selectivity, and fewer side effects. Drugs are used to prevent, treat, or cure diseases and medical conditions (although it should be noted that, except for antimicrobials (where any interaction with the host is adverse) and anti-inflammatory drugs, they infrequently cure diseases). While doing so, they must perturb the physiological or cellular system by binding to a "pharmacological target." They mainly alleviate symptoms or control the subsequent development of chronic diseases.

A drug refers to an active pharmaceutical ingredient in a drug product and includes all the small molecules and biologics that are approved by the regulatory health agencies around the world to treat diseases. Disease processes are complex and involve a sequence of events. Drugs are targeted to intervene in the disease

The opinions expressed herein are those of GJ and do not necessarily reflect those of the US Food and Drug Administration.

R. Poduri
Department of Pharmaceutical Sciences and Natural Products, Central University of Punjab, Bathinda, Punjab, India
e-mail: ramaraop@cup.edu.in; Ramaraop@yahoo.com

G. Jagadeesh (✉)
Division of Pharmacology & Toxicology, Office of Cardiology, Hematology, Endocrinology, and Nephrology, Center for Drug Evaluation and Research, US Food and Drug Administration, Silver Spring, MD, USA
e-mail: Gowra.Jagadeesh@fda.hhs.gov; GJagadeesh2000@gmail.com

R. Poduri (ed.), *Drug Discovery and Development*,
https://doi.org/10.1007/978-981-15-5534-3_3

process. Most of the drugs are palliative and do not directly perturb the proteins that cause the disease. Thus, with genomic data and understanding of the genetic basis of disease, drug targets should become more related to disease-gene products. The advent of genomic science has advanced our knowledge in the development of new pharmaceuticals as we have increased our understanding of how genes are linked to diseases and code for proteins ([1–4], also see Chap. 9 of this book).

For the most effective treatment, the new molecules are identified by "targets," which are related to a disease and to a disease-associated gene or nonhuman target genes such as invading bacteria and viruses [1]. Determination of the optimal molecular targets for drug intervention provides the basis for the discovery of new medicines. The mechanisms by which drugs act are different and unique to each class as they involve interactions with pharmacological receptors, enzymes, ion channels, and cellular transport processes (Box 3.1) [1, 4, 5]. Hormones, neurotransmitters, and autocoids are endogenous chemical messengers that have a role in physiological and pathological processes. The interpretation of their action is determined by the localization and functional capacity of the specific receptor with which the "first messenger" interacts and the concentration of the chemical to which the receptor is exposed. Most biological processes in higher eukaryotes are controlled by membrane-bound receptor-based mechanisms. While chemicals acting on intracellular receptors should also be acknowledged, they are not discussed in this chapter.

Box 3.1 Cellular Sites of Drug Action

1. G protein-coupled receptors:

   The Rhodopsin family (e.g., Adrenergic, angiotensin, muscarinic, serotonergic), Secretin and Adhesion, Glutamate, Frizzled, Taste.

2. Enzyme-Associated Receptors:

   (a) Receptors linked to tyrosine kinase: Insulin, EGF, PDGF, CSF, VEGF.
   (b) Receptor protein-tyrosine phosphatases: CD45.
   (c) Receptors linked to guanylyl cyclase: Atrial natriuretic peptide, speract.

3. Ion channels (Ionophore receptors) and Transporters:

   (a) Transmitter-gated: nicotinic acetylcholine, $GABA_A$, Glycine, $5\text{-}HT_3$.
   (b) Voltage-gated: no physiological ligand. Effectors are sodium, calcium, and potassium.
   (c) Ion transporters, ion pumps.

4. Transcriptional regulators and Nuclear hormone receptors:

   Steroids: glucocorticoids, mineralocorticoids; non-steroids: thyroid hormone, retinoids, PPARγ.

5. Drug action not mediated by specific receptors:

(continued)

Box 3.1 (continued)
(a) Inhibition of metabolic pathway.
(b) Effect mediated by interrupting normal cellular DNA, RNA, and protein synthesis/function.
(c) Effect mediated through enzymes.
(d) Perturbation of structure of excitable cells/colligative effects.

Drug target innovation analysis shows four major privileged protein classes as drug targets (Box 3.1). Among these, receptors are the largest at >45% [this includes G protein-coupled receptors (GPCRs) as the largest (>30%), and nuclear hormone receptors], followed by enzymes (kinases are the predominant subgroup), transporters (the largest being voltage-gated calcium channels), and others (including enzyme-interacting proteins, structural and adhesion proteins, and antirheumatic ligands) [5–7]. As of 2017, 134 GPCRs are targets for more than 700 approved drugs, suggesting that approximately 35% of approved drugs target GPCRs [8].

The family of GPCRs is functionally diverse and responds to a wide variety of ligands, including light, odor, taste, and smell that bind and modulate the activity of GPCRs. Underlying the extraordinary therapeutic value of this class of receptors is the vast variety of such receptors (>1% of the human genome: currently more than 300 human genes encode more than 1000 proteins) employed by cells to detect the levels of important substances in their environment. Thus, GPCRs provide for the sensations of light, odor, and taste, but, more importantly, for drug actions, they sense the prevailing levels of neurotransmitters, neuromodulators, and endocrine hormones [6]. By initiating transmembrane signaling pathways, the occupancy of these receptors shifts the activity and resources of cells in a manner appropriate to their homeostatic role, as dictated by the level of the external stimuli. This control is, therefore, well suited for the targeting of therapeutic agents.

This chapter ranges from discussions of the traditional receptor theory to changes that have advanced our knowledge in predicting the behavior of drugs under varying conditions, with pharmacologic measures in improving their choice in therapeutics.

3.2 Descriptive

The classical concept in the initiation of pharmacological response involves the formation of a complex between the drug and its site of action or target. Paul Ehrlich (1854–1915) stated that agents do not work unless they first bind. Langley (1852–1926) referred to this as a *receptive substance* and was the first to propose a receptor theory for the action of drugs and neurotransmitter substances in the body (see Box 3.2). Receptors are proteins present on the cell surface and within cells that mediate the effect of chemical messengers, including the neurotransmitters and hormones, and the actions of many exogenous drugs in the body. Thus, "drug" and "receptor" are two sides of a coin in mitigating a disease. Historically, the concept of drug–receptor interaction that began with Ehrlich and Langley was gradually developed over a period by A.J. Clark, E.J. Ariëns, P. Stephenson, and R. Furchgott [9] (Box 3.2).

Box 3.2 Doyens in the Evolution of the Receptor Concept

Paul Ehrlich (1854–1915), the founder of chemotherapy, was a bacteriologist and immunologist. He was a master in many fields including chemotherapy, immunology, histochemistry, and toxicology. He experimented with differential staining of dyes as he was focused on the selectivity of agents. The study revealed that selectivity in drug effect was the result of selectivity in action. He proclaimed, "*corpora non agunt nisi fixata*" (a drug will not work unless it is bound). However, he never used the term "receptor" [10–12]. He was awarded the Nobel Prize for Physiology or Medicine in recognition of his contribution to immunology. Drug–receptor interaction that started with Paul Ehrlich was gradually developed over a period of time by Langley, Clark, Paton, Ariëns, Stephenson, Furchgott, and others.

John Newport Langley (1852–1926). He studied the interaction of alkaloid pilocarpine and atropine with regard to salivary secretion and noted their mutual antagonism in effects. Further studies showed that the antagonism was dose-dependent and that very large doses can overcome each other's effect. In 1905, he postulated that drugs act directly on the effector cells and formulated the concept of "receptive substances." He located "receptive substances" *in* the cell rather than *on* the cell. It was substantiated by additional experiments with nicotine and curare [11].

Alfred Joseph Clark (1885–1941). He introduced the concept that a drug interacts with the substance on the cell and the response is proportional to the occupancy, i.e., that there is a linear relationship between occupancy and effect. He presented a more qualitative approach to the description of receptor selectivity and saturability [13].

John Henry Gaddum (1900–1965): He developed a mathematical treatise describing the analysis of the competitive antagonism between adrenaline and ergotamine in the rabbit uterus. According to the "Gaddum equation," fractional receptor occupancy by drug A in the presence of another drug B that competes for the binding of A at the receptor. This concept was further developed by Schild, who proposed a measure to quantify drug antagonism.

R. Paul Stephenson (1925–2004). He differed with the view expressed by AJ Clark that a more complex relationship exists between receptor occupancy and response so that the response elicited by partial agonists can be explained. He studied a series of trimethylammonium salts on the guinea pig ileum and

(continued)

Box 3.2 (continued)
noted that lower homologs behaved like acetylcholine and the higher homologs acted like atropine. This distinguished drugs with low and high efficacy. The drugs with low efficacy were termed partial agonists. He emphasized the non-proportionality between occupancy and response. He put forward the spare receptor theory and postulated that a maximum effect can be achieved by an agonist by occupying a fraction of available receptors, the response is not linear but a hyperbola, and different drugs may have varying abilities to initiate a response and consequently occupy different proportions of the receptors when producing equal responses. Stephenson called this the "efficacy" of the drug [11, 13].

In the drug development arena, two main pharmacological indices of drug activity are "affinity" and "efficacy." The potency is a mixed function of both affinity and efficacy. These two measurable basic properties of a drug can be separable and can vary independently with changes in chemical structure. In the drug discovery process, medicinal chemists can individually manipulate a drug or class of drugs to allow better targeting of candidate molecules toward therapies. This can be well exemplified with the development of saralasin, the first angiotensin II receptor antagonist that has no efficacy but has a high affinity for the AT-1 receptors. Similarly, elimination of efficacy of the natural agonist histamine produced cimetidine, a potent H-2 receptor antagonist in the treatment of gastric ulcers. In addition to affinity and efficacy, 2 ancillary properties predict the activity of a molecule in all physiological systems. They are an orthosteric or allosteric manner of binding and kinetics of interaction of the molecule with the target [14].

Similarly, the cellular response of a target protein is dependent on two system-based properties, target (receptor) *density* and target (receptor) *effector coupling efficiency* to cellular metabolism [9, 15]. The coupling efficiency, in turn, depends on the levels of expression of signaling molecules, signaling regulators, and the type of G protein involved [15]. Thus, these four properties give drugs the ability to have a spectrum of activities and potencies in different cellular systems. Additionally, four physicochemical properties present in a molecule dictate good drug activity, and these may have different structure–activity relationships. These are: primary pharmacodynamic activity at the therapeutic target; absence of secondary or off-target pharmacologic activity [16, 17], studies conducted in compliance with Good Laboratory Practice to identify undesirable pharmacodynamic properties of a drug, particularly on vital physiologic functions, such as the central nervous, cardiovascular, and respiratory systems (collectively known as safety pharmacology assessment) (ICH S7A guidance) [18]; and favorable ADME properties (solubilize in aqueous media, cross lipid membranes, and be present for sufficient time at the target to produce an effect) [16].

Based on the nature of interaction with a receptor and generation of stimulus-response in a system, candidate molecules are classified through pharmacologic

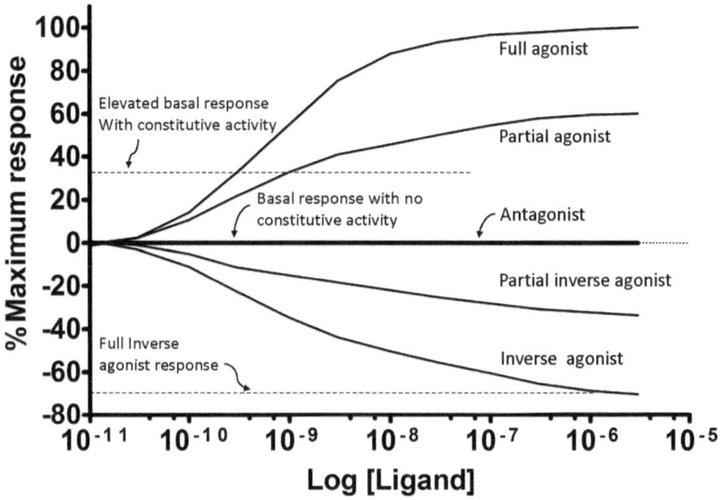

Fig. 3.1 The behavior of drugs in a system. Based on the characteristics of concentration (dose) response curves, the behaviors of drugs in a system or tissue are characterized. If it has the ability to bind to a receptor (*affinity*) and elicit a maximal response (*efficacy*), it is a full agonist. If it produces a submaximal response, it is a partial agonist. In some cases, a ligand of lower or no intrinsic efficacy binds to the receptor and reduces the response of a full agonist: this is an antagonist (efficacy is zero). According to traditional receptor theory, receptors are quiescent unless acted upon by an agonist. However, it is now known that most of the receptors can spontaneously adopt an "active conformation" capable of regulating a signal pathway in the absence of an agonist. The receptors are said to be in a constitutively active state. The magnitude of constitutive receptor activity is dependent on receptor isomerization from an inactive to an active conformation and the efficiency of receptor-coupling in the cell. The elevated baseline response can be decreased (below zero) by inverse (negative) agonists. These are also competitive orthosteric antagonists. The efficacy of an inverse agonist is negative. Partial inverse agonists reduce elevated basal responses less than full inverse agonists

procedures as agonist, antagonist, partial agonist, inverse agonist (or reverse antagonist), partial inverse agonist, or biased agonist. The behaviors (overt actions) of drugs characterized by the dose–response curve are illustrated in Fig. 3.1. The perception of agonists and antagonists was recognized hundreds of years ago with the dawn of the concept of receptors. Most of the drugs that are developed and approved for use in medicine and in research are in either of these two classes.

The performance of a drug is not limited to a boundary of agonist or antagonist: it may change in different systems or tissues of varying density (levels of sensitivity of a system), basal activity (constitutive receptor activity), receptor coupling efficiency, and the type of G protein. For example, labetolol, a nonselective α- and β-adrenoceptor blocking drug, also functions as an inverse agonist and as a partial agonist, while prenalterol, a cardioselective ß-adrenoceptor agonist, functions as a full agonist, partial agonist, or neutral antagonist.

3.3 Agonist

3.3.1 Drug Receptor Interaction

An agonist produces an observable change or response in a physiological system. The cellular response is directly related to its "efficacy" or "intrinsic efficacy." Paul Stephenson, a British pharmacologist, first used the term efficacy and E. J. Ariëns, a Dutch pharmacologist, called this intrinsic activity (see Box 3.2). It quantifies the extent of change produced as a result of the drug–receptor interaction-mediated stimulus-response system. The latter term describes the production of a second messenger (e.g., inositol trisphosphate, diacyl glycerol, and cAMP) to the end-organ response (e.g., vasoconstriction, increase in the myocardial force of contraction, and secretion).

The stimulus generated by the agonist–receptor complex may not be the same for all the agonists in different systems. Similarly, the same agonist binding to the receptor in different tissues may generate a stimulus to different degrees. The receptor occupation–response relationship of a ligand depends on the availability of spare receptors for that ligand in a given system [19]. Several factors, such as disease, age, and chronic administration of a drug, can modify these relationships.

Efficacy is different for each drug (or varies with the drug) and the system. A drug with high efficacy produces a maximal response with the fraction of receptor population occupied in a system that has high efficiency of receptor–effector coupling. The magnitude of response produced by an agonist displayed in the form of a dose–response curve is a function of agonist concentration used to engage the target. This ability of a molecule to bind the receptor is called "affinity." Affinity is a constant and unique for each drug–receptor pair, and is dependent on the chemical structure of the drug and the receptor [15].

In summary, the magnitude of cellular response in all physiological systems is governed by two agonist-dependent properties (efficacy and affinity) and two system-dependent properties (receptor density and efficiency of receptor–effector coupling).

The drug–receptor interaction is characterized by the binding of an agonist (drug) to a receptor (attributed to affinity) followed by the generation of a response (attributed to efficacy). This is illustrated by a simple model of ligand binding based on the law of mass action (Eq. (3.1)).

$$A + R \underset{k_{\text{off}}}{\overset{k_{\text{on}}}{\rightleftarrows}} AR \text{ complex } \underset{k_{\text{off}}}{\overset{k_{\text{on}}}{\rightleftarrows}} AR* \rightarrow \text{Stimulus-Response} \tag{3.1}$$

A is the agonist
R, the inactive state of the receptor
R^*, the active state of the receptor
$[AR]$ is the fraction of the receptor population bound by the drug.

The first reaction is the reversible formation of the agonist–receptor complex [*AR*]. This is formed as a result of the chemical property of affinity possessed by the drug. This forward reaction is dependent upon the "rate of association," denoted by k_{on}, and the reverse or dissociation rate constant, denoted by k_{off}. At equilibrium:

$$[A].[R].k_{on} = [AR].k_{off}$$

$$\frac{[A].[R]}{[AR]} = k_{off}/k_{on} = K_D(\text{mole/liter}) \tag{3.2}$$

K_D is the equilibrium dissociation constant ($1/K_D = K_A$ = affinity), defined as the concentration of the drug that measures half-maximal receptor occupancy.

The higher the K_D value, the lower the affinity of the receptor for the drug, and vice versa. Receptors are either free (*R*) or bound (*AR*). Free concentration of the receptor cannot be estimated, thus:

$$R = R_{total}(or\ B_{max}) - AR$$

As noted earlier, an agonist can elicit a response even by a fractional occupancy of receptors. This is given by:

$$\frac{[AR]}{[R] + [AR]}$$

The second reaction in Eq. (3.1) is the reversible formation of the active drug–receptor complex, *AR**. [*AR**] is generated in proportion to [*AR*] and leads to a stimulus–response. Thus, [*AR**] is the ability to interact with transmembrane signaling proteins (e.g., various G proteins, ion channels) to alter cellular function to produce an effect. This reflects efficacy, which again varies widely among drugs. A drug with high efficacy, called an agonist, produces a maximal response in contrast to a drug with lower efficacy, a partial agonist, at the same receptor. Antagonists lack efficacy whereas inverse agonists have negative efficacy.

GPCRs relay the critical extracellular information to the intracellular environment via activation of heterotrimeric GTP-binding proteins (G proteins) that modulate the activity of specific effector molecules such as adenylyl cyclase, phospholipase C, phospholipase A2, and phospholipase D to release second messengers [20]. The interaction between the receptor and the G-protein occurs at the inner surface of the plasma membrane of the cell where the G proteins are abundantly distributed. In the ternary complex model, activated receptors (*R**) activate heterotrimeric G proteins (composed of α.GDP, β, and γ subunits), whereby GDP is released and GTP binds to Gα subunit. Both free Gα and Gβγ activate downstream effectors.

$$A + R \underset{k_{\mathrm{off}}}{\overset{k_{\mathrm{on}}}{\rightleftarrows}} AR \text{ complex } \underset{k_{\mathrm{off}}}{\overset{k_{\mathrm{on}}}{\rightleftarrows}} AR * \mathrm{G} \rightarrow \text{Response} \tag{3.3}$$

The two-state model of receptor function demonstrates that, in the absence of a drug, receptors exist in equilibrium between two conformation states: an inactive conformation R that does not signal, and an active conformation R^* that effectively transduces a signal to its related G protein or effector system [10, 15]. Different agonists evoke different receptor conformations that are unique by nature revealing a single receptor's activation of multiple signaling pathways (see biased agonism) [21]. Additionally, conformations induced by partial agonists are distinct from those induced by full agonists [22–24]. The extended model is simplified below.

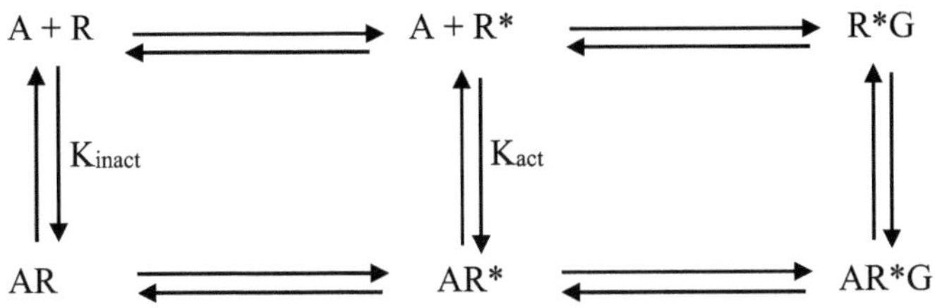

The extended ternary complex model. The receptor is in flux between R (inactive state) and R^* (active state) and coupled to a G protein in the absence of a ligand. An intramolecular constraint is proposed to control the equilibrium between the R and R^* states. In the case of the AT-1 receptor, two intrahelical hydrogen bonds exist: first between Val 108 and Tyr292, and second between Asn111 and Tyr295 [25–27], which are in close spatial proximity. This interaction is considered a major determinant of the inactive conformation of the AT-1 receptor [28]. R^* is generated following a loss of the constraint, e.g., the Asn111 mutation produces a partial and constitutively activated receptor [29]. Even at a low level of constitutive activity in the absence of any ligand, a small proportion of the receptor is in the active state ([R^*]) [30]. Agonist binding causes a transition to the [R^*] state (fully active state). Constitutively active mutants promote the spontaneous association of R^* with a G protein. Inverse agonists inhibit basal constitutive activity.

3.3.2 *Quantifying Agonist Activity: A Theoretical Consideration*

As noted previously, in a given class of drugs, the structure determines the affinity and efficacy of analogs compared with each other. The potency is a mixed function of both affinity and efficacy [31]. In high target densities (high expression) and efficient target-coupling mechanisms to cellular response, the EC_{50} of a full agonist is a complex function of both the affinity and efficacy of the agonist as well as the effect of stimulus-response amplification by the tissue [9]. Thus, EC_{50} does not equate to the concentration that occupies 50% of the binding sites and is not an estimate for K_{eq} for binding (1/Affinity) [9]. In contrast, a partial agonist must occupy all the available receptors to produce a maximal response, which is still lower than the system (tissue) maximal response. The concentration of the partial agonist producing the maximal response approaches the saturation binding concentration of the agonist. Thus, for a partial agonist, the EC_{50} approaches the K_{eq} for binding [9].

Relative potencies of full agonists are measured as the equiactive ratio of EC_{50} values. For this, all agonist analogs must produce a response by a similar mechanism of activation of the target, e.g., angiotensin II at the AT-1 receptor. In such cases, the measured potency ratios are system independent, wherein a similar rank order of potency for these analogs is observable in all tissues or cell lines. It should be noted that the relative potencies of the agonists are considerably different when the receptor is expressed in COS vs. CHO cells expressing AT-1 receptors. The agonist activity of octapeptide (Asp^1-Arg^2-Val^3-Tyr^4-Ile^5-His^6-Pro^7-Phe^8) Ang II at the AT-1 receptor is defined by two amino acids, Phe^8 and Tyr^4, the first and second agonist switches, respectively [29].

The phenylalanine at position 8 is critical for agonist activity, as des-Phe^8-heptapeptide is devoid of AT-1 receptor agonist activity. Substitution at position 8 with isoleucine weakens agonist activity and produces an agonist to antagonist transition (e.g., [Sar^1,Ile^8]Ang II and [Sar^1,Ile^4,Ile^8]Ang II are partial agonists, while [Sar^1,Ala^8]Ang II (saralasin) was the first high-affinity antagonist) [29, 32] although it has substantial affinity for the receptor with little or zero efficacy. The first amino acid, Asp^1, is important for binding affinity and duration of action but is not essential for biological activity [33]. In fact, its substitution with sarcosine (N-methylglycine) enhances binding affinity to the receptor with full-agonist activity. Deletion (des-Asp^1-Ang II/Ang III) still maintains full agonism but binds with a reduced affinity [29]. Thus, binding of an agonist to a receptor occurs through two types of functional interactions: those that contribute primarily to binding affinity, and those that contribute primarily to efficacy [34].

The binding of various angiotensin analogs to the receptor induces conformational changes to direct the functional groups or the receptor to enhance the signal transduction [22]. It is possible that distinct active conformations may be generated by different ligand–receptor complexes [22]. Ang II analogs reach different conformation states as AT-1 receptors activate multiple signaling pathways (PLCβ–PKC, adenylyl cyclase–cAMP, ERK1/2, transactivation of receptor tyrosine kinase, JAK-STAT, and nuclear transcription factors) with variable responses [20, 35].

Using a photoaffinity labeling approach, Fillion and colleagues [36] studied a series of Ang II analogs with substitutions at all eight positions of Ang II to characterize its interaction with the hAT-1 receptor. The binding affinities (equilibrium dissociation constant, K_i) in COS-7 cells showed wide variation in their affinities (a shift in the dose–response curve to the right) for the AT-1 receptor ranging from 0.9 nM (Ang II) to $\geq$500 nM. Such substitution also results in a large variation in efficacy (a decrease in maximal response), shifting full agonist to partial agonist and antagonist.

In agonist-based drug design, the receptor subtype selectivity of agonists is ranked based on EC_{50} values (potency). For example, cognitive dysfunction with degenerating neurons contributes to the development of Alzheimer's disease. Although restoration of acetylcholine deficiency with the acetylcholinesterase inhibitors is one of the mainstays of the treatments in Alzheimer's disease [37], current emphasis focuses on improving the efficiency of the cholinergic receptor signal. Among the various subtypes of muscarinic cholinergic receptors, M1 muscarinic cholinergic agonists emerge as the most sought-after potential candidates over M2, M3, M4, and M5 agonists in improving the cognition in Alzheimer's disease [38]. The receptor selectivity for M1 is evaluated based on EC_{50} values for a series of agonists.

3.3.3 *Assessment of Molecules for Agonist Activity*

Agonists are evaluated for two properties, affinity and efficacy, using in situ perfusion, animal models, and in vitro techniques using isolated tissues and/or organs, biochemical assays, and radioligand binding studies.

(a) *Animal models:* The pathophysiology of the disease being targeted by a new chemical entity should always be tested in whole animals or animal models [39]. These studies provide a valuable measure of the bioavailability of the drugs.

(b) *Isolated tissues:* Testing of drug activity using isolated intact tissue, isolated organs, and perfused organs with intact blood vessels began nearly 150 years ago with John Newport Langley (1852–1926) at Cambridge, UK, and Rudolf Magnus (1873–1972) at the University of Utrecht, Netherlands. This technique records tissue responses in a controlled environment without systemic interference of the intact animal. These models can be used to quantitate the physiological impact of the altered genetic sequences at the tissue or organ level. Measurement of responses (affinity and efficacy) include contraction, relaxation, changes in blood vessel diameter, blood flow, pressure, force, muscular and neural electrical activity, and myocardial activities such as rate, force, conduction, contractility, and derived functions. Radioisotopes can be used to monitor ion fluxes, drug metabolism, tissue uptake, and release of chemicals.

    Isolated tissues are also used in a number of biochemical and molecular biological techniques. For example, relative affinity, potency, and efficacy of several structurally related α-1 adrenergic receptor ligands were studied by Tiwari [40] on the functional response (contraction) in three different tissues that differed in levels of sensitivity based on receptor density. In isolated rabbit thoracic aorta, a highly sensitive tissue with a large receptor reserve, three full agonists—phenylephrine, amidephrine, and cirazoline—had similar efficacy and

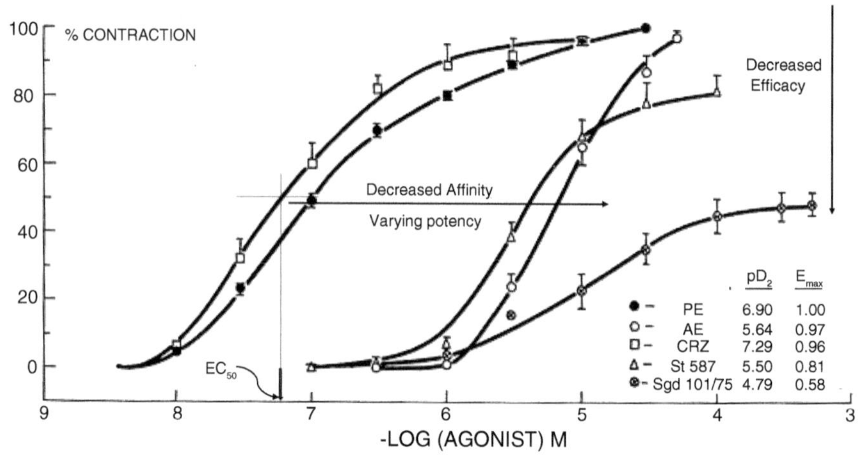

Fig. 3.2 Relative potency of alpha-adrenergic agonists in high-sensitivity tissue—isolated rabbit thoracic aorta. The respective potencies of agonists depend on two drug factors, "affinity" and "efficacy," and two tissue factors, "receptor density" and "efficiency of receptor coupling." All three drugs except two partial agonists (St587, Sgd101/75) produced tissue maximal response (similar efficacy, no suppression of maxima for three full agonists). In such a case, the observed maximal response is a nonlinear function of their relative efficacies (E_{max} varied from 1 for the full agonist to 0.58 for the partial agonist). Partial agonists differ with agonists in their ability to induce a maximal response, a property described as "efficacy." They may not induce any response in the least sensitive tissues (see Fig. 3.4). The concentration of each agonist producing 50% of their maximal response is termed ED_{50} (or EC_{50}) and the negative log of EC_{50} is pD_2. The relative potencies of agonists are measured as the ratio of EC_{50} values. Although they differ in potency, all of them have similar efficacy (no suppression of maxima). For two full agonists, PE (reference) and AE, the potency ratio would be p6.9 − p5.64 = p1.26, = 18. CRZ and PE have similar efficacy (=1) but differ in potency, where CRZ is more potent by 2.5 times in this high-sensitivity tissue. The dose–response curve indicates that the structural difference in the molecules has different effects on efficacy and affinity. *PE* Phenylephrine, *AE* Amidephrine, *CRZ* Cirazoline. Vertical lines on the symbol show standard error of the mean (SEM). The figure is reproduced with permission from reference [40]

showed maximal responses but with differing affinities and potencies. As per expectations, the two partial agonists St587, Sgd101/75 did not produce the full system maximal response (Fig. 3.2). In normally sensitive tissue such as rat vas deferens, the affinity of all agonists was less than that noted in the receptor-dense tissue, aorta. Maximal response to partial agonists was further reduced compared to that in the aorta (Fig. 3.3). Finally, in the rabbit vas deferens, a low sensitive system that operates with no receptor reserve, both partial agonists were unable to elicit any response, thereby indicating a lack of efficacy in tissue with low-receptor expression. In such cases, the partial agonist acts as an antagonist (hybrid). Although all full agonists retained maximal response by occupying all 100% receptors, they exhibited poor affinity since their EC_{50} increased from nanomolar to micromolar (Fig. 3.4). Thus, structural differences in the molecules and differences in receptor density contribute to variation in affinity and efficacy.

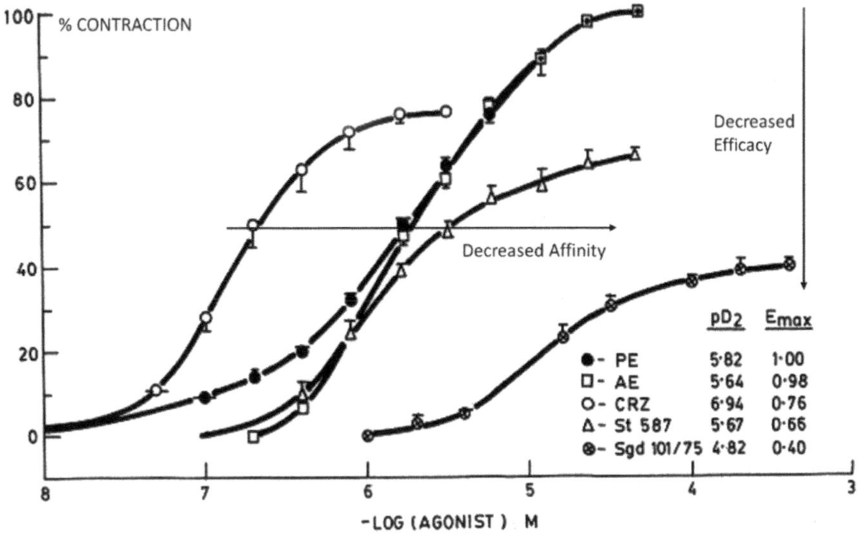

Fig. 3.3 Relative potency of alpha-adrenergic agonists in natural or normal sensitivity tissue—isolated rat vas deferens. In this tissue of normal sensitivity, only PE and AE are full agonists with a potency ratio of 1.5 (compared with 18.2 in high-sensitivity tissue). The EC_{50} of PE was 12 times lower (1.5μM) than that in the aorta (0.125μM). Furthermore, CRZ, a full agonist in high-sensitivity tissue (see Fig. 3.2), is a partial agonist (E_{max} 0.76) in this tissue. Additionally, CRZ still has a high affinity as demonstrated by a high potency ratio of 12 relative to PE. This suggests that, although affinity gets a drug to a receptor, the functional consequences resulting from its interaction with the receptor is independent of it [15]. The EC_{50} of an agonist is a complex function of both the affinity and efficacy of the agonist. It does not equate to the concentration that occupies 50% of the available receptors (see Fig. 3.5). Both partial agonists (St587, Sgd101/75) were less efficacious in normal sensitivity relative to high-sensitivity (see Fig. 3.2) tissue as receptor reserve is dwindling. *PE* Phenylephrine, *AE* Amidephrine, *CRZ* Cirazoline. Vertical lines on the symbol show standard error of the mean (SEM). The figure is reproduced with permission from reference [40]

The impact of receptor reserve on the affinity and efficacy in the dose–response curve is further illustrated in Fig. 3.5.

(c) *Biochemical assays:* These are cell-based assays that fall into three broad categories: second messenger assays, reporter gene assays, and cell proliferation assays [41]. Second messenger (e.g., cAMP, IP_3, intracellular calcium) assays monitor signal transduction following activation of membrane receptors. Cell-based accumulation of second messengers such as cAMP is measured using fluorescence- or luminescence-based assays [42]. Reporter gene assays are based on the principle of receptor signaling or second messengers eventually inducing gene transcription by various response elements. The synthesis of a reporter gene is followed by monitoring of the reporter protein expression by its enzymatic activities assessed with a variety of colorimetric or luminescent read-outs [43, 44]. Proliferation assays involve ligands that stimulate or inhibit overall growth (proliferation) of the cells that express the receptor- and ligand-induced effects read by marker-assays [45]. Specific pharmacologic responses as a

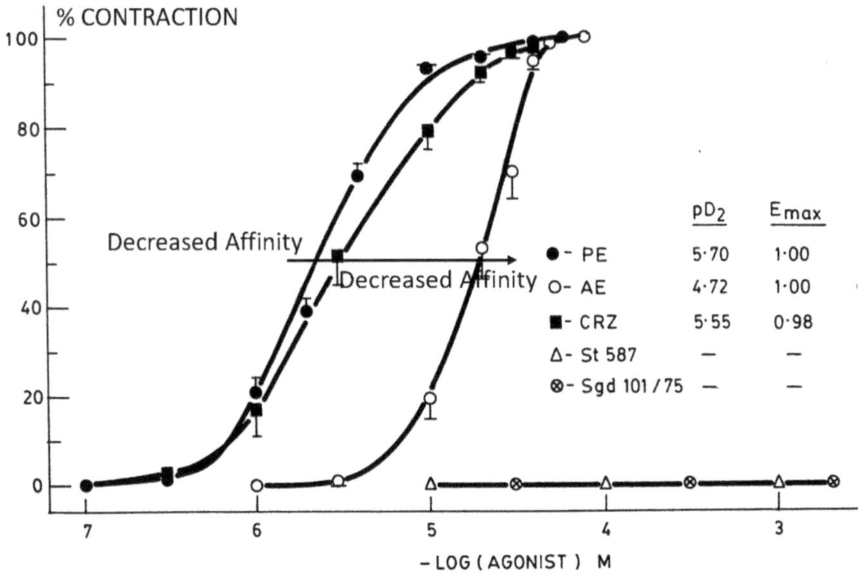

Fig. 3.4 Relative potency of alpha-adrenergic agonists in low-sensitivity/receptor tissue—isolated rabbit vas deferens. In this tissue of low sensitivity (compared with rabbit thoracic aorta (Fig. 3.2) and rat vas deferens (Fig. 3.3), PE, AE, and CRZ are still full agonists but with far reduced EC_{50}s. Both partial agonists (St587, Sgd101/75) did not produce any response, suggesting no receptor reserve or low receptor density and less than efficient receptor–effector coupling. With the decreasing sensitivity of the tissue, the dose–response curve to the full agonist shifted to the right along with the concentration range. Thus, both PE and AE were less potent in this tissue relative to rabbit thoracic aorta and rat vas deferens. *PE* Phenylephrine, *AE* Amidephrine, *CRZ* Cirazoline. Vertical lines on the symbol show standard error of the mean (SEM). The figure is reproduced with permission from reference [40]

function of changes in cell proliferation rates are then quantified. This receptor selection and amplification technique can quantitatively differentiate ligands based on various intrinsic efficacies [41].

(d) *Radioligand binding studies:* Radiolabeled ligands are widely used as molecular probes since they can associate with a multitude of biological receptors, enzymes, or transporters. Measuring the interaction between a radioligand and receptor provides a wealth of information on the number of receptor binding sites (B_{max}), their affinity (K_I or K_D), and accessibility to other drugs. This provides a measure of the relative affinities of agonists acting on the receptor. Binding studies lack a functional response as seen with isolated tissues or in vivo studies. However, the agonist interaction with binding sites and its sensitivity to various ions and guanine nucleotides can provide a valuable measure of efficacy (see below). Radioligand binding experiments present a pivotal tool in the processes of drug discovery and receptor classification. The purified membrane preparations for the assay help determine receptor number in health and disease

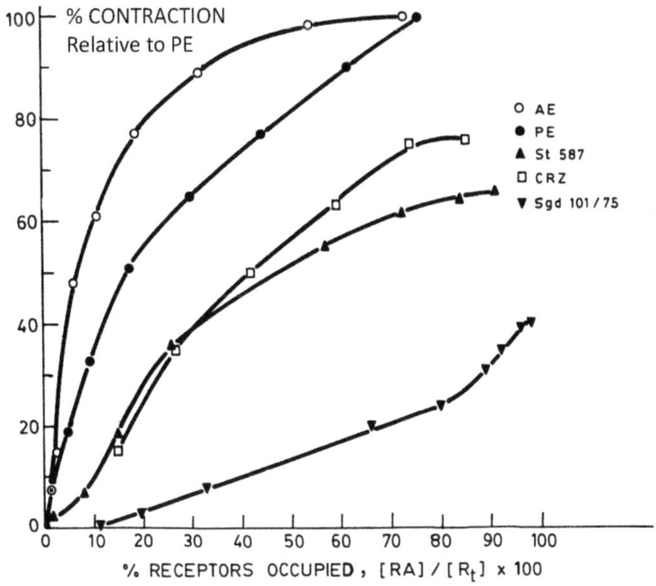

Fig. 3.5 Receptor occupancy and response in rat vas deferens. In many tissues, the relationship between receptor occupancy and tissue response is not linear. Full agonists need only a fraction of the existing receptor population to produce the maximal response. The extent of response depends on the sensitivity of the tissue (as described in Figs. 3.2, 3.3, and 3.4) and the efficacy of the agonist. Thus, the magnitude of receptor reserve is different for different agonists: an agonist with a high efficacy will have a greater receptor reserve than one with a lower efficacy. Partial agonists may not produce any response in tissues with no receptor reserve (as shown in Fig. 3.4 for rabbit vas deferens). This phenomenon is studied by irreversible inactivation of a fraction of the receptor by phenoxybenzamine. The response for full agonists AE and PE in this tissue (rat vas deferens, natural sensitivity) was hyperbola, a characteristic of full agonists or agonists with high intrinsic efficacy [19]. In this tissue, AE and PE produced a half-maximal response by occupying 6.72% and 17%, respectively. In contrast, Sgd101/75, a partial agonist, had to occupy 69% to produce half-maximal and all available receptors to produce a maximal response

with fewer diffusional barriers. With the automation of radioligand binding assays, novel compounds can be screened with lower microgram quantities for activities against a large number of receptors. Additionally, radioligand binding assays minimize the time needed to evaluate large numbers of compounds for potential therapeutic activity.

Analyses of radioligand binding experiments are based on a common model of equilibrium binding. Receptor–ligand interaction follows the law of mass action as described in Eq. (3.1). Radioligand binding experiments can be divided into two types: (i) equilibrium binding studies and (ii) kinetic binding experiments, of which the equilibrium binding experiments can be further divided into saturation binding and competitive binding.

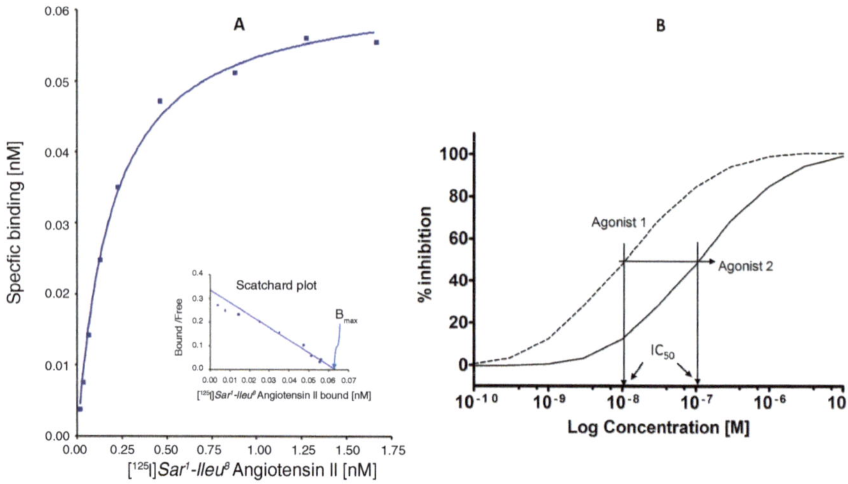

Fig. 3.6 Radioligand binding studies. (**a**) Saturation binding. Specific binding of increasing concentrations of radiolabeled AT-1 receptor antagonist Sar[1] –Ileu[8] Angiotensin II. The data are plotted on a linear scale showing a hyperbola. Scatchard plot (inset) linearizes the curve by plotting specific bound vs. bound/free. In this example, Bmax is 0.06351 nM/mg protein, K_D 0.1897 nM. (**b**) Dose–response curves for the inhibition of [^{3}H] prazosin binding to alpha-1 adrenergic receptors by an investigative drug (solid line) relative to the reference drug (dotted line). The competition curves were analyzed by a nonlinear regression curve fitting program for a single class of binding sites. In this example, an investigative drug (IC_{50}, 100 nM) is approximately 10 times less potent than the reference drug (IC_{50}, 10 nM). The respective K_I values are 7.60 and 8.69

Saturation binding: Assays are performed at equilibrium using a series of radioligand concentrations (>10), up to a concentration at which the ligand virtually occupies all of the receptors. The relationship between binding (*Y*-axis) and ligand concentration (*X*-axis) shows a rectangular hyperbola on a linear scale (Fig. 3.6a) and a sigmoid on a logarithmic scale. However, the curves can be linearized using methods such as the Scatchard (Rosenthal) plot, Lineweaver–Burke, Eadie–Hofstee, and Hill plots.

In a Scatchard plot, one of the most popular graphical presentations of ligand–receptor binding, the specific bound is plotted against the ratio of specific binding to the concentration of free radioligand (bound/free). The *X*-intercept is the B_{max} and the negative reciprocal of the slope is K_D (equilibrium dissociation constant) [46–48] (Fig. 3.6a, inset). Linear Scatchard plots imply ligand binding to a single homogenous population of binding sites. However, deviations from linearity, which may indicate multiple receptor classes, are also observed in these plots, as in the case of a ligand binding to two or more sites simultaneously, leading to a curvilinear Scatchard plot [46].

Competitive binding: Competition binding experiments are used to compare drug potencies (Fig. 3.6b), investigate low affinity drug–receptor interactions, multiple affinity forms such as receptors linked to G proteins, and in the screening of receptor binding of drugs. In these assays, increasing concentrations (10–20×, spanning over at least six orders of magnitude) of an unlabeled drug competes with a fixed concentration of a radioligand for receptor binding. For example, varying concentrations of unlabeled agonist phenylephrine competes with [^{3}H]-prazosin in rabbit aorta. In this case, a Hill slope of 1 is indicative of a simple bimolecular interaction involving one ligand interacting with one receptor type such as an α_1-adrenergic receptor. Both ligands compete for a single class of binding sites resulting in the displacement of bound radioligand by the agonist, phenylephrine. Shallow competition binding curves, on the other hand, demonstrate a deviation from the simple bimolecular model with a slope of 1 [46]. In the absence of GTP, the receptors linked to G proteins bind the agonist with high affinity. On the other hand, in the presence of GTP, or in case of receptors that do not interact with G proteins, the receptors bind agonists with low affinity. Thus, in the absence of GTP, the competition binding curves tend to be shallow, demonstrating more than one class of binding sites. In the presence of GTP, the ternary complex (*AR*G*) is not stable, the G protein dissociates into α_{GTP} and ßγ subunits, the receptor gets uncoupled from G protein, and binds the agonist with low affinity [47].

The data are analyzed by a nonlinear regression curve fitting program to determine the IC_{50}, defined as the concentration of unlabeled drug that reduces half the specific binding. The IC_{50} determined from radioligand binding studies approximates to that determined from functional (in vitro tissue) studies. Since IC_{50} varies with the concentration of the radioligand, it is corrected mathematically using the Cheng–Prusoff equation [49] to obtain K_I, the equilibrium dissociation binding constant for the unlabeled ligand [50].

$$K_I = \frac{IC_{50}}{1 + \frac{[\text{radioligand}]}{K_D}}$$

where [radioligand] is the concentration of the radiolabeled ligand used in the study and K_D is the dissociation constant of the radioligand as obtained from the Scatchard plot. The Cheng–Prusoff relationship measures the affinity of a drug/ligand for the enzyme/receptor used primarily in the radioligand binding assays. Schild analysis (pA_2), on the other hand, is used for the isolated tissue studies (see next section, antagonist). However, the values obtained by the two methods are typically in good agreement [51].

Kinetic binding: Kinetic binding experiments are used to calculate both the association and dissociation rate constants for a radioligand. The radioligand binding is determined as a function of time. The rate of binding of radioligand to its binding site is quantitated from the rate of agonist A associated with the association rate constant, k_{on}, and dissociated from the dissociation rate constant, k_{off}, with the putative receptor *R* [47]. At equilibrium, the rate of formation of *AR* equals the rate of dissociation of the *AR* complex, since there is no net formation of *AR* (see Eqs. (3.1) and (3.2)) [52]. Typically, the K_D for the radioligand is determined via a Scatchard plot using steady-state incubations and is similar to the value calculated from kinetics (Eq. (3.2)).

Efficacy Determination from Radioligand Binding Studies

The functional activity of a ligand can be determined for GPCRs by taking advantage of "GTP-shift" as an indicator of ligand efficacy, whereby the agonist dose–response curve is constructed in both the absence and presence of excess GTP and the resulting rightward shift in the curve is an indirect biochemical estimator of agonist efficacy [53]. Thus, both the agonists and GTP exert reciprocal effects on the stability of the GPCR-G protein complex. The ability of agonists to increase GTP binding to G proteins is the basis of their efficacy. Because of complex agonist binding that is often resistant to modulation by GTP, the GTP-shift cannot be utilized for all GPCRs to detect agonist efficacy [54]. Three types of heterologous binding studies are utilized to detect the GTP-shift:

(a) Competition study for agonist against labeled antagonist in the presence of GTP. In this assay, the degree of a rightward shift in the presence of GTP and/or the ratio of receptors exhibiting high versus low agonist affinity are computed [47].
(b) Displacement of bound [^{35}S]GTPγS by increasing concentrations of unlabeled agonists [55]. This study is used to determine EC_{50} and relative maximal activity (equivalent to relative intrinsic efficacy).
(c) Displacement of a bound labeled agonist by GTP. The degree or size of reduction in agonist binding in the presence of GTP is a measure of the intrinsic efficacy of an agonist. Data from [^{35}S]GTPγS binding studies, when plotted against agonist K_A values reported from functional studies, show a high degree of correlation for both relative potency and relative efficacy [55].

3.4 Antagonists

Endogenous hormones or neurotransmitters and exogenous agonists interact with specific targets to produce responses. In Eq. (3.1), we considered a single molecule (A) interacting with a receptor (R) on the cell. Here, we will consider two molecules (A and B), interacting with the same receptor at the same binding site (orthosteric) or at a different site (allosteric) on the receptor. In this case, molecule A is an agonist that causes a change in the cellular function, while molecule B interferes with the interaction of A with the target. The net outcome—potentiation or inhibition—depends on the nature of the interaction. An increase in response is attributed to inhibition of transport or neuronal uptake, or metabolism by an enzyme. Inhibition or blockade of response, on the other hand, is characterized as "drug antagonism" and molecules that bind to the target and interfere with the agonist response are termed antagonists. Antagonism leading to a reduction or overactivation of endogenous hormones (e.g., angiotensin receptor blockers curtailing excessive action of Ang II), neurotransmitters (e.g., adrenergic blockers against norepinephrine/epinephrine), autocoids (e.g., antihistamines), inflammatory molecules (e.g., leukotrienes), etc., have a prominent place in drug discovery. Thus, receptor antagonists have therapeutic use in many diseases (sometimes life-long), where chronic endogenous agonism is present and is harmful if not treated in a timely manner. Most of the drugs approved by regulatory agencies fall into the category of drug antagonism.

3.4.1 *Quantifying Antagonism*

For a class of drugs, antagonism potency is a measure of the extent of the affinity of the antagonist for the receptor. The affinity of an antagonist is quantified as the equilibrium dissociation constant (K_D or K_I) of the antagonist–receptor complex in an in vitro system (isolated tissue or radioligand binding) or in a functional efficacy assay [9]. The dose–response curve of an agonist in the absence, followed by the presence, of an antagonist, produces two effects—the nature and extent of shift in the dose–response curve (calculated as a dose ratio, DR) to the right [56], and the level of change (suppression) in maximal response. These two effects depend on the affinity of the antagonist for the receptor and the range of concentration of the antagonist studied in the system.

The mechanism of antagonism is identified based on the type of agonist, the assay used, the selectivity of the antagonist for the receptor subtype, and the sensitivity of the functional system (receptor number and the efficiency of coupling). A selective and competitive antagonist acts at the same site as that of the agonist and precludes the binding of agonist to the receptor. However, a high concentration of agonist can overcome the blockade by binding to the still available receptors resulting in a normal response even in the presence of an antagonist. A parallel shift in the dose–response curve to the right by a factor of X with the retention of maximal effect for the agonist in the presence of increasing concentrations of an antagonist is described as competitive antagonism. The magnitude of the shift, calculated as DR, denotes a decrease in the potency of the agonist and is proportional to the degree of antagonism since it depends on the concentration of the antagonist and its affinity for the receptor. A mathematical approach for studying the interaction of agonist and competitive antagonist was initially developed by Gaddum (1926). Schild ([57]: for a review on this subject, see reference [56]) devised a method to measure the affinity of competitive antagonists in a variety of isolated tissues which was then extended to in vivo systems [58].

The extent of the shift in the dose–response curve of an agonist can be used for the determination of the affinity of an antagonist. The logarithm of the affinity of the antagonist is related to the logarithm of its molar concentration which causes a shift by a factor 2. This negative logarithm has been referred to by Arunlakshana and Schild (1959) [59] as pA_2.

$$pA_2 = pA_x + \log\ (DR - 1) \tag{3.4}$$

A plot of log (DR-1) *versus* $-\log\ [B]$ producing a straight line with a slope of approximately 1 is indicative of competitive antagonism in which the *X*-axis intercept equals pA_2. The determination of pA_2 values is used to classify drugs (e.g., adrenergic drugs) targeting the same receptor (e.g., alpha-1 receptor) in a given tissue (e.g., thoracic aorta), or different tissues with the same type of receptor (e.g., vas deferens).

The apparent dissociation constant, K_B, for the receptor–antagonist complex in isolated tissues such as thoracic aorta was calculated by Furchgott (1966) [60].

$$K_B = \frac{[B]}{(DR - 1)} \tag{3.5}$$

$$pA_2 = -\log K_B$$

where $[B]$ is the concentration of the antagonist and DR is the dose ratio, defined as the ratio of the concentration of the agonist in the presence of the antagonist and that in the absence of the antagonist.

Competitive Antagonism: Competitive antagonists can be (i) surmountable (reversible) or (ii) insurmountable (irreversible).

Competitive surmountable antagonism: In high-sensitivity tissues, a potent agonist can produce a maximal response with the occupation of a small fraction of receptors leaving a large receptor reserve [61]. An orthosteric antagonist with fast dissociation kinetics weakens the sensitivity of the system to the agonist with a dextral displacement of the agonist dose–response curve. However, higher concentrations of the agonist can restore the response depending on the sensitivity of the tissue and the potency of agonist.

In a high- to moderate-sensitivity system, the agonist dose–response curve is also shifted to the right with no depression of maximal response. The magnitude of the shift in the agonist dose–response curve is antagonist concentration-dependent. Such antagonism is termed competitive and surmountable (Fig. 3.7a). For partial agonists that operate with no receptor reserve, the antagonist produces a depression of the maximal response at all concentrations.

The intrinsic efficacy of partial agonists lies between 0 and 1. The low level of efficacy in partial agonists produces direct agonism in a more sensitive tissue (Fig. 3.2) which behaves as a competitive antagonist in an insensitive tissue (Fig. 3.7b). Such dual behavior (Fig. 3.7b) of agonist was noted with dichloroisoproterenol (DCI) in guinea pig trachea by James Black [11]. Stimulatory activity of the partial agonist is noted at low concentrations in tissues of a low basal level of activity whereas antagonism dominates at higher concentrations (Fig. 3.7b). DCI is also a protean agonist as it produces positive agonism in normal systems and negative effects (inverse agonism) in constitutively active systems [9].

Competitive insurmountable antagonism: Repeated and increasing concentrations of antagonist that binds tightly to and dissociates slowly from the receptor may deplete all available receptors for an agonist causing saturability of the system. Furthermore, such antagonists reduce the ratio of R:R*, thereby reducing the coupling efficiency of the receptor. In this case, the dose–response curve shifts to the right with the depression of maxima (Fig. 3.8a). This is termed insurmountable

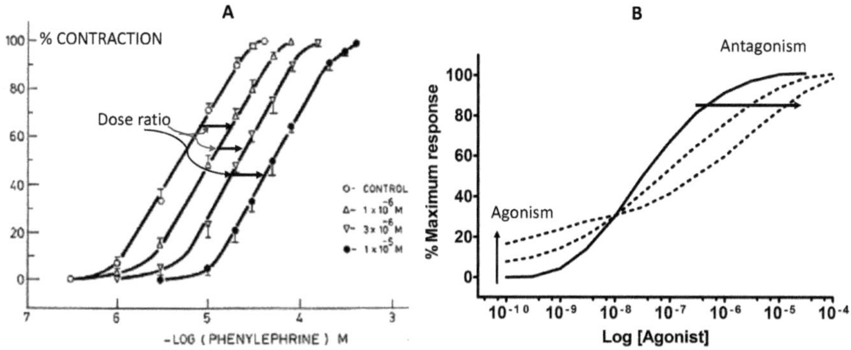

Fig. 3.7 Drug antagonism, competitive. (**a**) Surmountable. In receptor antagonism, two molecules, an agonist and an antagonist, compete for the same binding site on the receptor. The relative proportions of the receptor bound to agonist and antagonist depend on the kinetics of binding of both the agonist and antagonist and are governed by the law of mass action [9]. Cumulative dose–response curves to phenylephrine (PE) were constructed (intersecting points are not shown) in the absence and presence of three different concentrations of an orthosteric antagonist (equilibrated for 30 min) in rabbit vas deferens. The PE dose–response curve was shifted to the right (at higher concentrations of agonist) in a parallel fashion without suppressing maxima at each of the concentrations of the antagonist, indicating competitive, surmountable nature of antagonism. The magnitude of the shift in the agonist dose–response curve is proportional to the degree of antagonism as it relates to the concentration of antagonist added and its affinity. The extent of shift measured at ED_{50} at various concentrations of the antagonist (known as dose ratio) when plotted against the molar concentrations of the antagonist should yield a straight line with a slope of unity for a competitive antagonist. The procedure is termed a "Schild plot" [59]. The X-intercept is pA_2, an empirical measure of antagonist potency. In this example, the pA_2 of the antagonist is 5.87. (**b**) Dual nature of partial agonists. In a quiescent system or at low basal activity, partial agonists can cause activation (agonism, elevated baseline). This effect is weakened or absent at high basal activity and demonstrates antagonism of response (shifting the dose–response curve to the right) in a system activated by full agonists. The figure shows the effect of varying concentrations of a partial agonist (dotted line) on the theoretical dose–response curve to an agonist (normal line)

and is also evident in systems of low sensitivity. A slowly dissociating orthosteric antagonist (that also binds to the agonist receptor site) is irreversible in action (e.g., candesartan, an AT-1 receptor antagonist) [29].

Noncompetitive antagonism: An antagonist may bind to the receptor covalently and continue to inactivate the receptors until a dose–response relationship for the agonist is no longer possible (Fig. 3.8b). Examples include alpha-adrenergic antagonist β-haloalkylamine phenoxybenzamine and organic phosphorus compounds such as the acetylcholinesterase inhibitor diisopropylfluorophosphate. This is similar to noncompetitive antagonism as produced by an allosteric (allotopic) antagonist. In this case, the antagonist binds to the receptor on a site different from that of the agonist (that is, outside the agonist receptor binding site). This causes a change in conformation of the protein altering the agonist interaction and affecting affinity and/or efficacy. Since the antagonist cannot dissociate from the receptor, a steady-state cannot be attained, and thus K_B cannot be measured [9]. In the natural system, negative allosteric antagonist induces signaling bias on the endogenous agonist

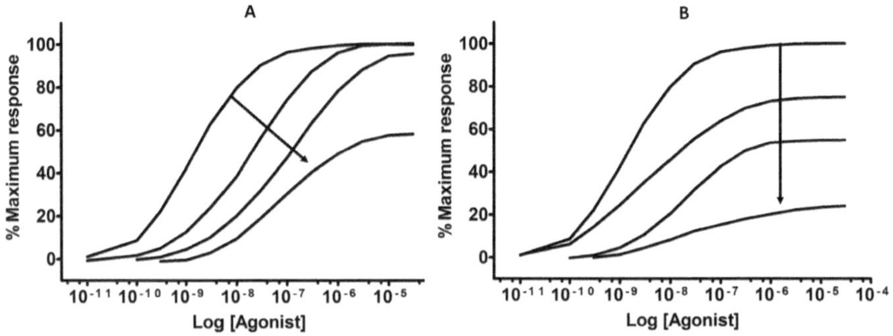

Fig. 3.8 Drug antagonism, insurmountable. (**a**) Antagonist binds to the same site as the agonist but does so irreversibly or pseudo-irreversibly. The antagonist has slow dissociation kinetics (e.g., calcium channel blocker, amlodipine; opiate partial agonist, buprenorphine) but does not bind covalently. At low concentrations, an insurmountable antagonist may produce reversible antagonism in tissues in which activation of only a fraction of available receptors is sufficient to produce a maximal response to an agonist. When blockade exceeds the available receptors (in the absence of receptor reserve), in addition to shifting the curve to the right the maximal response is depressed. (**b**) Noncompetitive antagonism. Irreversible interaction of the antagonist as both the agonist and the antagonist bind at different sites. Such binding causes a conformational change in the receptor that inhibits the response to the bound agonist. A noncompetitive antagonist does not shift the agonist dose–response curve to the right but will depress the maximal response to the agonist at all concentrations. For example, photoaffinity labels and alkylators (e.g., phenoxybenzamine, an alpha-adrenergic receptor antagonist) binds covalently with the receptor

signaling, an important difference between orthosteric and allosteric antagonists [14] (see biased agonism below, and Chap. 5).

3.4.2 Allosteric Modulators

GPCRs are intrinsically allosteric proteins [62]. Positive allosteric modulators (PAMs) enhance the endogenous agonist (affinity (cooperativity) and/or efficacy (modulation)) of the natural system [63]. It is proposed that these modulators increase agonist stabilization of the receptor active state. In preclinical models of Parkinson's disease, valiglurax (VU295) elicited activation of the metabotropic glutamate receptor subtype 4 (mGlu4) via an allosteric site on the receptor to improve the endogenous glutamate signal [64]. PAMs prefer one specific GPCR subtype over others. PAMs such as MK-7622 and GSK1034702 act primarily to potentiate the action of endogenous orthosteric agonist acetylcholine at muscarinic acetylcholine M_1 receptors but not at the other four receptor subtypes, and are being developed as potential therapeutic agents for the symptomatic treatment of Alzheimer's disease [65, 110]. Similarly, PAM LY2033298 activity is predictive of clinical antipsychotic drug efficacy and potentiates acetylcholine-mediated calcium responses at the M4 mAChR but not at the other four receptor subtypes in animal models [66]. Thus, positive cooperativity among binding sites enhances the response.

Positive cooperativity: The binding of a molecule to one subunit promotes the binding of the next molecule to the second subunit, resulting in the binding curve to be steeper (with a slope greater than unity for a Hill coefficient) than for a monomer. This is known as positive cooperativity between binding sites. An example is the binding of four molecules of oxygen to each of the four heme moieties of the hemoglobin, a tetrameric protein. The binding is allosterically cooperative wherein the binding of each oxygen molecule increases the affinity of the remaining subunits for oxygen [13]. Both positive and negative allosteric modulations have been demonstrated for sodium ion for the α-2 adrenergic receptor. Similarly, both negative (NAM) and positive allosteric modulations have been demonstrated with amiloride and its analogs for a number of GPCRs such as adenosine, adrenergic, dopamine, chemokine, muscarinic, serotonin, gonadotropin-releasing hormone, $GABA_B$, and taste receptors [48, 67].

An allosteric ligand can alter the affinity of the co-binding ligand, modify the efficacy of the co-binding ligand, or itself produce direct agonism as in the case of positive allosteric modulator agonists. Another unique feature of the modulator is a variation of its activity as it modifies the protein interaction with various probes (e.g., orthosteric agonist). The magnitude and direction of the allosteric effect can change depending on the orthosteric ligand [9, 68]. This is known as Probe dependence. Thus, the effects of allosteric ligands can be linked to the nature of the co-binding ligand and the receptor subtype. For example, eburnamonine (muscarinic ACh receptor M3 > M2 receptor allosteric agonist) enhanced the affinity of pilocarpine, acetylcholine and carbachol, reduced the affinity of arecoline and had no effect on the affinity of arecaidine propyl ester to the M1-M4 subtypes of muscarinic receptors in the membranes of CHO cells [112]. Besides PAM and NAM, there is another class of ligand that does not alter the action of an orthosteric agonist, activator, or antagonist and is termed neutral allosteric ligand (NAL). It can prevent the binding of other allosteric ligands to the same allosteric site [68].

Upon binding, a positive allosteric modulator increases the affinity of the receptor for full and partial agonists and shifts the dose–response curve to the left (Fig. 3.9a, c). In tissues with large receptor reserve, an increase in the efficacy by PAM shifts the agonist curve to the left (since the system cannot show further increased maximal response), while for a partial agonist, it increases the maximal response (Fig. 3.9d). In contrast, for full agonists, a negative allosteric modulator upon binding decreases the affinity resulting in a shift in dose–response curve to the right with or without the suppression of maximal response (decreased efficacy), and is dependent on the sensitivity of the system or receptor reserve (Fig. 3.9a, b). For partial agonists, an allosteric ligand may increase or decrease affinity (Fig. 3.9c) and/or efficacy (Fig. 3.9d) based on the receptor density. Additionally, an allosteric modulator can move agonist affinity in one direction and efficacy in another direction (Fig. 3.9e, f). The direction (positive or negative) of the allosteric interaction and its magnitude vary depending on the receptor subtype and the nature of the agonist and allosteric ligand [112]. Some of these states are illustrated in Fig. 3.9.

Many effective drugs (e.g., astemizole, cisapride, dofetilide) have been withdrawn from the market or are restricted in their use as a result of QT interval

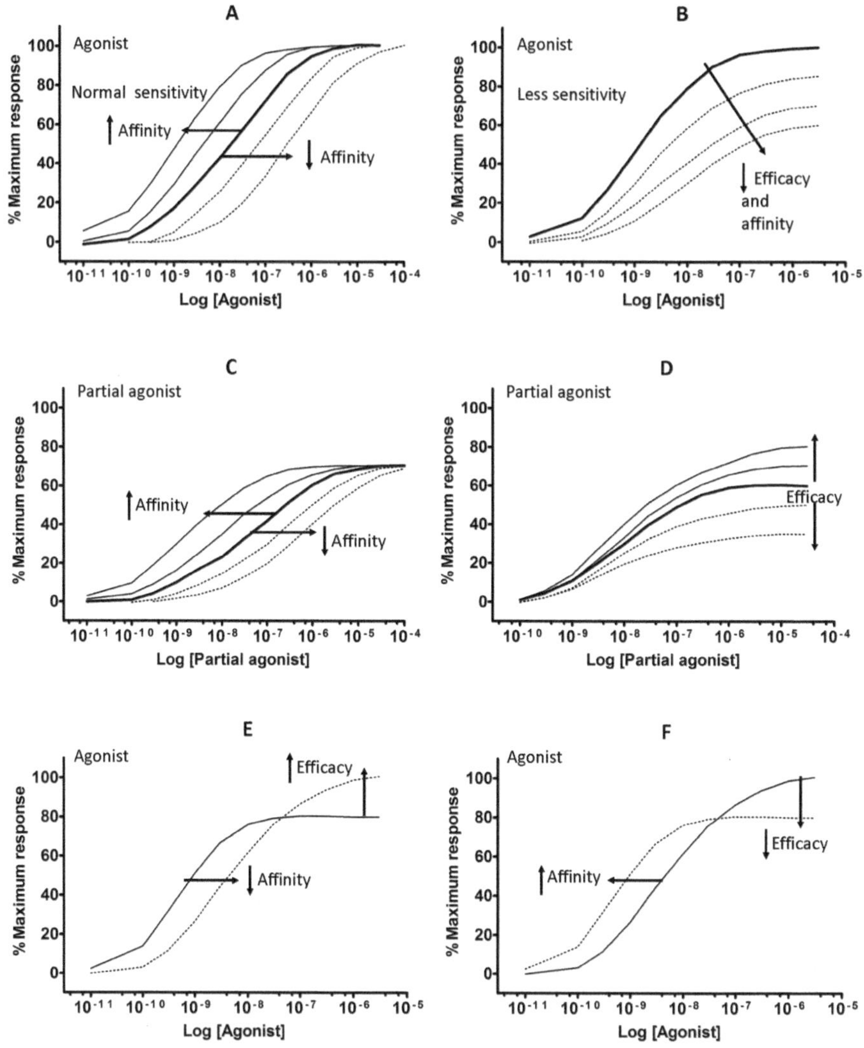

Fig. 3.9 Allosteric effects, antagonism, or potentiation. An allosteric modulator (positive or negative) produces its effect by binding a site on the receptor distinct from that of the agonist. This results in changing the affinity of the receptor for the agonist. Additionally, it might also affect the efficacy of the agonist. (**a**) In normal sensitivity tissue, upon binding, a positive allosteric modulator (PAM) increases the affinity of a full agonist by shifting the curve to the left with no effect on the efficacy as the system maximum has already been reached by the agonist. In contrast, a negative allosteric modulator (NAM) decreases the affinity of the agonist/partial agonist by shifting the curve to the right. Both sodium and Gpp(NH)p increase the affinity (by an increase in the association rate) of rauwolscine, an α-2 adrenergic receptor antagonist, binding to the receptor in the bovine aorta. In contrast, both decrease the affinity of the agonist epinephrine at the same receptor site. Another example is amiloride, a NAM and Na^+ /H^+ exchange inhibitor, which reduces the affinity of rauwolscine for α-2 adrenergic receptors by decreasing the association rate constant [48]. TAK-071, a selective muscarinic cholinergic M_1 receptor PAM, is shown to reverse cognitive impairment in models of Alzheimer's disease by potentiating the action of endogenous

prolongation and the induction of torsades de pointes (TdP). All of them block I_{kr} (rapid delayed rectifier potassium current). In a series of preclinical studies, Qile et al. [70] reported that LUF7244, a negative allosteric modulator/activator of Kv11.1 channel, prevented cardiac proarrhythmias of dofetilide by inducing a change in receptor conformation, thereby decreasing its affinity for binding to Kv11.1 channel. Thus, such a combination approach would prevent arrhythmias resulting from drug-induced blockade of I_{kr}.

3.5 Constitutive Activity and Inverse Agonism

According to classical receptor theory, receptors are quiescent (R) and get activated (R*) upon binding to an agonist. The studies of Cerione et al. [71] on β-2 adrenergic receptors and Costa and Herz [72] on delta-opioid receptors have revolutionized that concept to show that most receptors can signal in the absence of an agonist. In a high-sensitive system that has highly expressed receptor density, the functional system is constitutively active [73]. A number of receptors spontaneously transition to the active state and produce an elevated baseline response in the absence of endogenous agonists [14] (Fig. 3.1). The degree of constitutive receptor activity is dependent on: (a) conformational flexibility—ease of changing from inactive (constrained) state *R* or *Ri* to active state (*R**), (b) receptor density (sensitivity of the system), and (c) receptor–effector coupling efficiency in the cell [15]. In the case of GPCRs, several amino acid residues located in transmembrane domains and the second or third cytoplasmic loops are implicated in constitutive activation [29, 74].

Many ligands previously classified as antagonists reduce the constitutive receptor activity and have been now termed "inverse agonists" since their effects are opposite to that of agonists. In addition, inverse agonists have different degrees of negative intrinsic efficacy leading to the classification as full or partial inverse agonists. Inverse agonists have a higher affinity for *R* than for *R**. This shifts the equilibrium toward *R*, stabilizing the inactive conformation of the receptor ("off" state) [10], enriching the receptors in the inactive conformation, and depleting the receptors in active conformation, thereby reducing G protein signaling. This results in a decrease in elevated basal activity (Fig. 3.1) as their intrinsic efficacy is below zero or

Fig. 3.9 (continued) acetylcholine [38]. (**b**) In a less sensitive tissue, NAM decreases both affinity and efficacy of an agonist. (**c, d**) For partial agonists, an allosteric ligand may increase or decrease affinity and/or efficacy. (**e, f**) Dual role: Allosteric modulators may produce opposite effects on agonist affinity and efficacy. LP1805, a conformation-specific allosteric antagonist, is a partial and noncompetitive inhibitor of neurokinin A binding to NK2 receptors. It inhibits the NKA-induced cAMP response while mildly enhancing the NKA-induced calcium response [69]. (**e**) Depending on the sensitivity of the system, the allosteric modulator reduces the agonist affinity and at the same time increases its efficacy. (**f**) Allosteric modulator increases the affinity of the agonist with a reduction in its efficacy negative. In the absence of constitutive receptor activity, inverse agonists behave as simple antagonists shifting the dose–response curves of the full agonist to the right. Thus, inverse agonists are competitive orthosteric antagonists [9]. On the other hand, conventional competitive antagonists are not inverse agonists (although vice versa is true) as their intrinsic efficacy is zero and does not decrease the baseline. Thus, there is a distinction between a reduction in constitutive activity by inverse agonists and reduction in agonist-induced receptor activity by simple antagonists.

Inverse agonism is detected more readily by biochemical (e.g., adenylyl cyclase or GTPase activity) and receptor binding assays than by functional systems (e.g., in vitro atrial tension, in vitro smooth muscle contraction or relaxation, in vivo left ventricular dp/dt measurement). For example, enhanced basal adenylyl cyclase activity in membranes from the transgenic mouse heart is significantly depressed by the ß-adrenergic receptor inverse agonist, ICI-118,551. This inverse agonism is reversed by alprenolol, a ß-adrenergic receptor antagonist [75]. It is useful to classify clinically available receptor antagonists based on their ability to block basal activity into three subtypes: neutral antagonists, partially inverse agonists, and inverse agonists [76]. In this context, of the 8 clinically available angiotensin II receptor (AT-1) antagonists, valsartan, irbesartan, candesartan, olmesartan, and azilsartan are both inverse and neutral antagonists, whereas losartan, eprosartan, and telmisartan are pure neutral antagonists [29].

Constitutively active receptor mutants: Increase in constitutive activity due to mutations in receptors [75] is linked to an increasing number of cancer and noncancer diseases. In a high-sensitivity system, increased constitutive activity may play a role in cancer progression, tumor growth, and metastasis [77–79]. The constitutively active mutations in GPCRs are widespread and are present in a variety of diseases involving tumor growth, angiogenesis, and metastasis [80–82]. Table 3.1 lists some constitutively activating receptor mutants in disease.

Inverse agonists find their way in therapeutics due to the physiological relevance and the role of constitutive receptor activity in pathophysiologic functions. Although the regulation of mutant receptors by inverse agonists suggests the potential therapeutic value of inverse agonists [74, 75], their clinical utility has not yet been explored. Therapeutic efficacy attributed to inverse agonism is explicitly found in the labeling of pimavanserin, approved by the FDA for the treatment of hallucinations and delusions associated with Parkinson's disease psychosis. The mechanism of action of pimavanserin in the treatment of hallucinations and delusions associated with Parkinson's disease psychosis is unclear. However, the effect of pimavanserin could be mediated through a combination of inverse agonist and antagonist activity at serotonin 5-HT2A receptors and to a lesser extent at serotonin 5-HT2C receptors (statement based on the FDA label). Prolonged use of inverse agonists can result in the development of apparent tolerance, perhaps due to the upregulation of receptor density and the consequential increase in signaling efficiency [98].

Table 3.1 Constitutively activating G protein-coupled receptor mutants in pathology

GPCR	Disease	Reference
Adenosine A_1	Genetic obesity	[83]
Angiotensin AT-1	Prostate cancer, cholangiocarcinoma	[84, 85]
$ß_1$- and $ß_2$-adrenergic	Breast cancer cardiomyopathy	[86]
CXCR2	Ovarian cancer	[87]
CXCR4	Small-cell lung cancer, head and neck cancer; also implicated in breast cancer	[88, 89]
Endothelin-1, ET	Colorectal cancer, prostate cancer	[90]
Frizzleds -Wnts	Colorectal cancer, prostate cancer, hepatocellular carcinoma	[91]
Gastrin releasing Peptide receptor	Pancreatic cancer	[92]
LH receptor	Testicular cancer	[93]
Rhodopsin	Color blindness, night blindness, retinal degeneration	[94, 95, 111]
Thyrotropin, TSH	Hyperfunctioning thyroid adenoma and congenital hyperthyroidism, thyroid cancer	[96]
Vasopressin V_2	Nephrogenic diabetes insipidus	[97]

3.6 Biased Agonism or Functional Selectivity (See Also Chap. 5)

Biased agonism: Drugs differentially activate receptor subtypes and the ensuing signaling pathways. Thus, drug selectivity is based on receptor subtypes and signal selection. Many receptors are pleiotropically linked to multiple signaling pathways. If an agonist activates multiple signaling pathways and is biased toward some pathways, then it is termed biased signaling. Receptor selectivity is defined as activating only the target receptors. The availability and knowledge of an array of drugs selective for different receptor subtypes indicate that drugs can be functionally selective for a cellular signaling pathway. The active state of a receptor most stabilized by the agonist determines activation of the specific signaling pathway. This is known as "selective stabilization of different receptor active states." If two drugs stabilize different receptor states of the same receptor subtype, they may differentially activate multiple signaling pathways. Each one of them may produce favorably biased signaling with respect to the other [9]. Thus, receptors can adopt many more conformations and each active conformation could regulate one or more responses. Biased signaling is a function of the ternary complex of ligand, receptor, and signaling molecule rather than being a ligand–receptor effect [99].

GPCRs regulate cellular function via transducing molecules such as G proteins or non-G proteins such as β-arrestin. This led to the search for drugs that are biased toward downstream signaling over the other pathways, e.g., an agonist can be a strong inverse agonist for the phospholipase C (PLC) pathway while at the same time a weak agonist for phospholipase A2 pathway. Norbinaltorphimine, a κ-opioid receptor antagonist, is an antagonist for Gi-mediated responses but is an agonist for the activation of c-Jun-N terminal kinase. Carvedilol, a nonselective β-adrenergic blocking agent with α1-blocking activity for the treatment of heart failure, is an inverse agonist for cAMP production and has exhibited partial (biased) agonism for β-arrestin-mediated extracellular signal-regulated kinase (ERK) 1/2 [63, 100]. This suggests that a drug acting at the same receptor can be an agonist, antagonist, inverse agonist, and/or bias agonist for different signaling pathways stimulated by the receptor. This is known as functional selectivity or biased agonism based on multiple response-dependent efficacies. Efficacy is a system-dependent drug constant and influences potency. A drug can have multiple efficacies depending on the number of effectors regulated by a receptor. These efficacies can vary for different responses such that a drug that acts on a single receptor subtype can have selectivity for the disparate cellular signaling pathways regulated by the receptor (Fig. 3.10). Furthermore, the efficacy depends on the magnitude of constitutive activity (highest in the highly sensitive system), expression of receptors, type of signaling protein, the efficiency of receptor coupling which is unique for each signaling pathway, and the affinity of a ligand for the receptor conformation for each response [103].

Functional selectivity: Functional selectivity varies with the structure of the molecule. Changes in molecular structure can affect the efficiency of receptor coupling and consequently of different downstream effectors. Drugs selective for receptor subtypes have been developed over the past several decades. Ligand-pathway-selective agonists or mixed agonist-antagonist activating specific cellular signaling pathways, while at the same time inhibiting endogenous agonist activation of other pathways (see below, Biased antagonism), are being developed. For example, G protein and arrestin signaling pathways mediate distinct physiologic processes (discussed in depth in Chap. 5). Responses elicited by biased agonists cannot be attained with conventional agonists or antagonists [103]. For example, structure–activity relationships of the angiotensin II analogs with the AT-1 receptor have shown bias toward β-arrestin versus $G_{\alpha q}$ signaling. The aromatic phenylalanine^8 and tyrosine^4 side chains of Ang II are critical for full agonist activation and for G_q signaling [29]. Replacement of the phenylalanine^8 aromatic side-chain abolishes G protein-dependent signaling but makes it more selective for engaging β-arrestin-dependent signaling [104–106]. These findings have important implications in the development of AT-1 receptor-targeted drug development.

Biased drug development is believed to improve the treatment of diseases with fewer adverse effects as therapeutic and adverse effects of a drug are mediated by distinct pathways involving G proteins and β-arrestins [107]. This is best illustrated with TRV120027, a β-arrestin-biased agonist for the AT-1 receptor, for the treatment of heart failure. In preclinical studies, TRV120027 was shown to bind to the AT-1 receptor, did not activate Gq protein, but selectively stimulated β-arrestin signaling. This was shown to improve cardiac myocyte contractility and thus deemed beneficial in heart failure [108]. However, animal studies did not translate in humans and the drug failed in phase II clinical trials. Recently (August 2020), for the first time the US FDA has approved μ- opioid agonist oliceridine (Olinvyk™) for moderate to

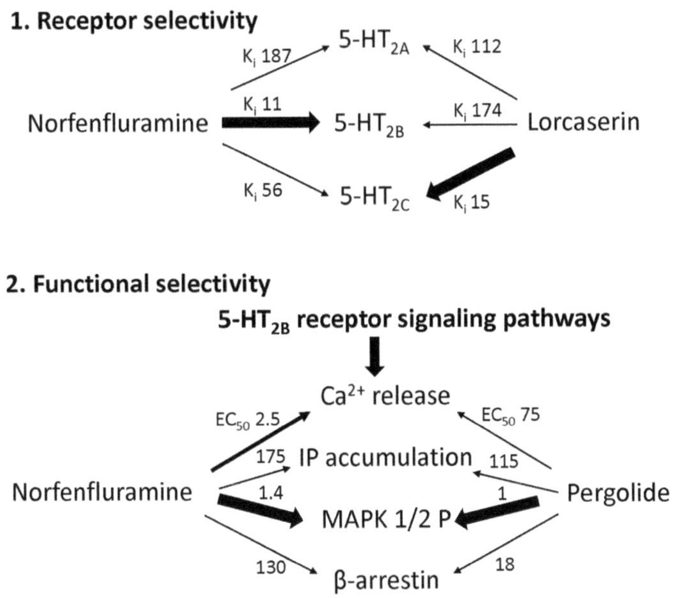

Fig. 3.10 Drug selectivity for receptor subtypes and signaling pathways. (1) Receptor selectivity is based on the relative affinity of norfenfluramine and lorcaserin for human 5-HT_{2A}, 5-HT_{2B}, and 5-HT_{2C} receptors, a subfamily of 5-HT receptors implicated in the cardiovascular, feeding, and mitogenic effects of 5-HT. Norfenfluramine showed the high affinity of 5-HT_{2B} receptors (K_i 11 nM) consistent with the hypothesis that valvular lesions are the result of activation of 5-HT_{2B} receptors. In contrast, lorcaserin exhibited selectivity for the 5-HT_{2C} receptor (K_i 15 nM). However, the current label for lorcaserin contains a warning about possible valvulopathy based on radioligand binding affinity data [17]. (2) Functional selectivity is based on the efficacy of agonists (EC_{50} in nM) in activating various signaling pathways (e.g., calcium release, inositol phosphate accumulation, MAP kinase 1/2 phosphorylation, β-arrestin translocation). Thus, both the affinity and efficacy of the ligand for the receptor are molecular determinants of bias as they interact with these pathways. Both norfenfluramine and pergolide are known valvulopathic compounds that activate more than one cellular signaling pathway. Their selectivity profile is similar to MAPK phosphorylation signaling (EC_{50} 1–1.4 nM) as the prime suspect in contributing to valvular lesions. In contrast, the non-valvulopathic drugs lorcaserin and ropinirole exhibited an EC_{50} of 1800 and 2500 nM, respectively, for MAPK signaling (not shown in the figure) (Data were taken from references 17, 101, 102]). Thus, norfenfluramine and pergolide are not only selective for specific receptors (5-HT_{2B}), but also selective for regulation of specific cellular signaling pathways

severe acute pain in adults (https://www.accessdata.fda.gov/scripts/cder/daf/index.cfm?event=BasicSearch.process). Oliceridine is a Gi-biased agonist for analgesic activity with minimal or reduced β-arrestin-2 association (relative to morphine) that purported to mediate adverse events (respiratory depression, gastrointestinal complications). This dissociation in analgesia and respiratory depression reflects differential recruitment of β-arrestin-2 signalling so as to increase therapeutic window compared to conventional opioids such as morphine, hydromorphone, and fentanyl.

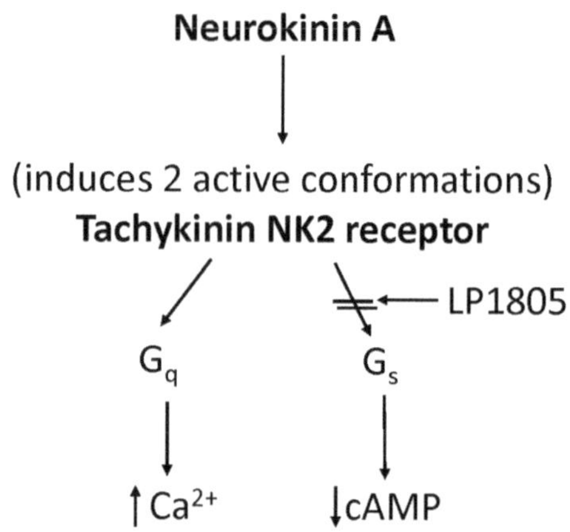

Fig. 3.11 Biased antagonism. A single receptor interacting with multiple G proteins. Neurokinin A (an orthosteric agonist) stabilizes the tachykinin NK2 receptor in two different active conformations, each with different dissociation kinetics: one generates intracellular calcium elevation, and the other generates cAMP production via activation of G_q and G_s proteins, respectively. LP1805, a conformation-specific allosteric antagonist, destabilizes one of the NKA-NK2 receptor complexes, resulting in selectively blocking G_s signaling (cAMP response) while leaving the other (G_q) conformation unchanged (Ca release) [69]

3.7 Biased Antagonism

Biased antagonism is similar to negative allosteric modulation. The allosteric antagonist has a different affinity for different GPCR signaling protein complexes. An allosteric ligand can cause biased modulation by promoting more than one type of active state. Depending on the signaling pathway linked to each receptor conformation, the allosteric ligand can increase the potency of an endogenous agonist for pathway 1 and at the same time can decrease the potency of the agonist for pathway 2 (Fig. 3.11). This may result in normal signaling of a GPCR for a natural agonist leading to a biased signal and enhancing one G protein (e.g., Gq) while diminishing a second G protein (e.g., Gs) [109]. Thus, allosteric ligands provide an opportunity for engaging an endogenous agonist to exhibit biased signaling so that it can selectively target desired pathways.

For example, neurokinin A stimulates tachykinin NK2 receptors that regulate two different signaling pathways; a G_q pathway coupled to phospholipase C generating inositol phosphate and calcium release, and a Gs pathway coupled to adenylyl cyclase generating cAMP. Similarly, LP1805, a conformation-specific allosteric antagonist, alters the equilibrium between the two active conformations of NKA-NK2 receptor complexes resulting in selectively inhibiting NK2 receptor-mediated cAMP signaling while slightly potentiating NK2 receptor-mediated calcium release (Fig. 3.11) [69].

3.8 Conclusions

The concept of receptor theory that evolved in the late nineteenth century was the guiding principle in the development of drugs for a variety of diseases. That was based on the understanding that drugs behaved as either agonists or antagonists. Generation of new molecules with combinatorial chemistry, recombinant DNA strategies, and structure–activity relationship have forged a new era in drug development where a drug can be simultaneously a biased agonist, an antagonist, an allosteric modulator, and an inverse agonist at the same receptor in treating specific diseases (e.g., carvedilol). Allosteric modulators may have great potential as therapeutic drugs by blocking the adverse effects of effective drugs when given in combination therapy. Understanding the role of constitutive receptor activity and multiple signaling pathways in physiological functions and disease might provide new opportunities for the discovery of functionally selective ligands as drugs with increased specificity and fewer adverse effects. Recent advances in GPCR-biased signaling suggest that the biased ligands may have an important impact in treating diseases that include cardiovascular diseases, neuropsychiatric disorders, and cancers as several of them are entering clinical studies.

Acknowledgments The authors are thankful to Dr. John Turner of the US FDA for his editorial assistance.

References

1. Hopkins AL, Groom CR. The druggable genome. Nat Rev Drug Discov. 2002;1:727–30.
2. Goh K-II, Cusick ME, Valle D, Childs B, Vidal M, Barabási A-L. The human disease network. PNAS. 2007;104:8685–90.
3. Swinney, Anthony J. How were new medicines discovered? Nat Rev Drug Discov. 2017;10:507–829. https://doi.org/10.1038/nrd3480.
4. Yıldırım MA, Goh K, Cusick ME, Barabási A-L, Vidal M. Drug–target network. Nat Biotechnol. 2007;25(10):1119–26. https://doi.org/10.1038/nbt1338.
5. Rask-Andersen M, Sällman Almén M, Schiöth HB. Trends in the exploitation of novel drug targets. Nat Rev Drug Discov. 2011;10:579–90. https://doi.org/10.1038/nrd3478.
6. Hauser AS, Attwood MM, Rask-Andersen M, Schiöth HB, Gloriam DE. Trends in GPCR drug discovery: new agents, targets and indications. Nat Rev Drug Discov. 2017;16:829. https://doi.org/10.1038/nrd.2017.178.
7. Santos R, Ursu O, Gaulton A, et al. A comprehensive map of molecular drug targets. Nat Rev Drug Disc. 2017;16:19–34.
8. Sriram K, Insel PA. G protein-coupled receptors as targets for approved drugs: how many targets and how many drugs? Mol Pharmacol. 2018;93:251–8.
9. Kenakin TP. Pharmacology in drug discovery and development, understanding drug response. 2nd ed. New York: Academic Press; 2017.
10. Limbird LE. Evolution of the receptor concept. Mol Interv. 2004;4:326–35.
11. Prüll C-R, Maehle A-H, Halliwell RF. A short history of the drug receptor concept. New York: Palgrave Macmillan; 2009. https://doi.org/10.1057/9780230583740.
12. Kaufmann SHE. Paul Ehrlich: founder of chemotherapy. Nat Rev Drug Discov. 2008;7:373.

13. Limbird LL. Cell surface receptors. A short course on theory and methods. 3rd ed. New York: Springer; 2005.
14. Kenakin TP. Prescient indices of activity: the application of functional system sensitivity to measurement of drug effect. Trends Pharmacol Sci. 2019;40:529–39.
15. Berg KA, Clarke WP. Making sense of pharmacology: inverse agonism and functional selectivity. Int J Neuropsychopharmacol. 2018;21:962–77. https://doi.org/10.1093/ijnp/pyy071.
16. Papoian T, Chiu HJ, Elayan I, Jagadeesh G, et al. Secondary pharmacology data to assess potential off-target activity of new drugs: a regulatory perspective. Nat Rev Drug Discov. 2015;14:294–6. https://doi.org/10.1038/nrd3845-c1.
17. Papoian T, Jagadeesh G, Saulnier M, et al. Regulatory forum review: utility of in vitro secondary pharmacology data to assess risk of drug-induced valvular heart disease in humans: regulatory considerations. Toxicol Pathol. 2017;45:381–8. https://doi.org/10.1177/0192623317690609.
18. ICH Expert Working Group. Safety pharmacology studies for human pharmaceuticals S7A; 2011. http://www.ich.org/fileadmin/Public_Web_Site/ICH_Products/Guidelines/Safety/S7A/Step4/S7A_Guideline.pdf
19. Ruffolo RR Jr. Important concepts of receptor occupation theory. J Auton Pharmacol. 1982;2:277–95.
20. Balakumar P, Jagadeesh G. A century old renin-angiotensin system still grows with endless possibilities: AT1 receptor signaling cascades in cardiovascular physiopathology. Cell Signal. 2014;26:2147–60.
21. Perez DM, Hwa J, Gaivin R, et al. Constitutive activation of a single effector pathway: evidence for multiple activation sites of a G protein-coupled receptor. Mol Pharm. 1996;49:112–22.
22. Karnik SS. Conformational theories on receptor activation from studies of the AT1 receptor. In: Husain A, Graham RM, editors. Drugs, enzymes and receptors of the renin–angiotensin system: celebrating a century of discovery. The Netherlands: Harwood Academic Publishers; 2000. p. 117–30.
23. Ghanouni P, Bryzcynski Z, Steenhuis JJ, et al. Functionally different agonists induce distinct conformations in the G protein-coupling domain of the β2-adrenergic receptor. J Biol Chem. 2001;276:24,433–6.
24. Seifert R, Wenzel-Seifert K, Gether U, Kobilka BK. Functional differences between full and partial agonists: evidence for ligand-specific receptor conformations. J Pharm Exp Ther. 2001;297:1218–26.
25. Groblewski T, Maigret B, Larguier R, et al. Mutation of Asn111 in the third transmembrane domain of the AT1A angiotensin II receptor induces its constitutive activation. J Biol Chem. 1997;272:1822–6. https://doi.org/10.1074/jbc.272.3.1822.
26. Balmforth AJ, Lee AJ, Warburton P, Donnelly D, Ball SG. The conformational change responsible for the AT1 receptor activation is dependent upon two juxtaposed asparagine residues on transmembrane helices III and VII. J Biol Chem. 1997;272:4245–51. https://doi.org/10.1074/jbc.272.7.4245.
27. Singh KD, Unal H, Desnoyer R, Karnik SS. Mechanism of hormone peptide activation of a GPCR: angiotensin II activated state of AT_1R initiated by van der Waals attraction. J Chem Inf Model. 2019;59:373–85.
28. Hunyady L, Ji H, Jagadeesh G, et al. Dependence of AT1 angiotensin receptor function on adjacent asparagine residues in the seventh transmembrane helix. Mol Pharmacol. 1998;54:427–34.
29. Balakumar P, Jagadeesh G. Structural determinants for binding, activation, and functional selectivity of the angiotensin AT1 receptor. J Mol Endocrinol. 2014;53:R71–92.
30. Miura S, Kiya Y, Hanzawa H, et al. Small molecules with similar structures exhibit agonist, neutral antagonist or inverse agonist activity toward angiotensin II type 1 receptor. PLoS One. 2012;7:e37974. https://doi.org/10.1371/journal.pone.0037974.

31. Ross EM, Kenakin TP. Pharmacodynamics. Mechanisms of drug action and the relationship between drug concentration and effect, Chapter 2. In: Hardman JG, et al., editors. Goodman and Gilman's the pharmacological basis of therapeutics. 10th ed. New York: McGraw-Hill; 2001.
32. Wilkes BC, Masaro L, Schiller PW, Carpenter KA. Angiotensin II vs its type I antagonists: conformational requirements for receptor binding assessed from NMR spectroscopic and receptor docking experiments. J Med Chem. 2002;45:4410–8. https://doi.org/10.1021/jm0103155.
33. Fillion D, Lemieux G, Basambombo LL, et al. The amino-terminus of angiotensin II contacts several ectodomains of the angiotensin II receptor AT1. J Med Chem. 2010;53:2063–75. https://doi.org/10.1021/jm9015747.
34. Kobilka BK. Commentary: conformational theories on receptor activation from studies of the AT1 receptor. In: Husain A, Graham RM, editors. Drugs, enzymes and receptors of the renin–angiotensin system: celebrating a century of discovery. The Netherlands: Harwood Academic Publishers; 2000. p. 131–5.
35. Hunyady L, Catt KJ. Pleiotropic AT1 receptor signaling pathways mediating physiological and pathogenic actions of angiotensin II. Mol Endocrinol. 2006;20:953–70. https://doi.org/10.1210/me.2004-0536.
36. Fillion D, Cabana J, Guillemette G, et al. Structure of the human angiotensin II type 1 (AT1) receptor bound to angiotensin II from multiple chemoselective photoprobe contacts reveals a unique peptide binding mode. J Med Chem. 2013;288:8187–97. https://doi.org/10.1074/jbc.M112.442053.
37. Verma S, Kumar A, Tripathi T, Kumar A. Muscarinic and nicotinic acetylcholine receptor agonists: current scenario in Alzheimer's disease therapy. J Pharm Pharmacol. 2018;70:985–93. https://doi.org/10.1111/jphp.12919.
38. Mandai T, Kasahara M, Kurimoto E, et al. In vivo pharmacological comparison of TAK-071, a positive allosteric modulator of muscarinic M1 receptor, and xanomeline, an agonist of muscarinic M1/M4 receptor, in rodents. Neuroscience. 2019;414:60–76.
39. Morgan SJ, Elangbam CS, Berens S, et al. Use of animal models of human disease for nonclinical safety assessment of novel pharmaceuticals. Toxicol Pathol. 2013;41:508–18.
40. Tiwari P. Characterization of postsynaptic alpha-adrenoceptors in rabbit thoracic aorta, vas deferens and rat vas deferens. A thesis submitted for the degree of master of pharmacy in pharmacology of Banaras Hindu University, Varanasi, India; 1986.
41. Sundberg SA. High-throughput and ultra-high-throughput screening: solution- and cell-based approaches. Curr Opinion Biotechnol. 2000;11:47–53.
42. Thomsen W, Frazer J, Unett D. Functional assays for screening GPCR targets. Curr Opin Biotechnol. 2005;16:655–65. https://doi.org/10.1016/j.copbio.2005.10.008.
43. Cheng Z, Garvin D, Paguio A, et al. Luciferase reporter assay system for deciphering GPCR pathways. Curr Chem Genomics. 2010;4:84–91.
44. Paguio A, Stecha P, Wood KV, Fan F. Improved dual-luciferase reporter assays for nuclear receptors. Curr Chem Genomics. 2010;4:43–9.
45. Messier TL, Dorman CM, Braiiner-Osborne H, Eubanks D, Brann MR. High throughput assays of cloned adrenergic, muscarinic, neurokinin, and neurotrophin receptors in living mammalian cells. Pharmacol Toxicol. 1995;16:308–11.
46. Munson PJ, Rodbard D. Ligand: a versatile computerized approach for characterization of ligand-binding system. Anal Biochem. 1980;107:220–39.
47. Jagadeesh G, Deth RC. Different affinity states of alpha-1 adrenergic receptors defined by agonists and antagonists in bovine aorta plasma membranes. J Pharmacol Exp Ther. 1987;243:430–6.
48. Jagadeesh G, Cragoe EJ, Deth RC. Modulation of bovine aortic alpha-2 receptors by Na+, 5′-guanylylimidodiphosphate, amiloride and ethylisopropylamiloride: evidence for receptor G-protein precoupling. J Pharmacol Exp Ther. 1990;252:1184–96.

49. Cheng Y-C, Prusoff WH. Relationship between the inhibition constant (K_i) and the concentration of inhibitor which causes 50 percent inhibition (I_{50}) of an enzymatic reaction. Biochem Pharmacol. 1973;22:3099–108.
50. McKinney M. Practical aspects of radioligand binding. In: Enna SJ, Williams M, Ferkany JW, Kenakin T, Porsolt RD, Sullivan JP, editors. Current protocols in pharmacology, vol. 1. New York: Wiley; 2002. p. 1.3.1–1.3.33.
51. Eglen RM, Whiting RL. Problems associated with the application of the Cheng-Prusoff relationship to estimate atropine affinity constants using functional tissue responses. Life Sci. 1989;44:81–94.
52. Schwartz KR. The principles of receptor binding studies. In: Fozzard HA, et al., editors. The heart and cardiovascular system. New York: Raven Press; 1986. p. 169–88.
53. Christopoulos A, El-Fakahany EE. Qualitative and quantitative assessment of relative agonist efficacy. Biochem Pharmacol. 1999;58:735–48.
54. Christopoulos A, Grant MKO, El-Fakahany EE. Transducer abstraction. A novel approach to the detection of partial agonist efficacy in radioligand binding studies. J Pharmacol Toxicol Methods. 2000;43:55–67.
55. Tian W-N, Miller DD, Deth RC. Bidirectional allosteric effects of agonists and GTP at αa2A/D-adrenoceptors. J Pharmacol Exp Ther. 2000;292:664–71.
56. Schild HO. Drug antagonism and pA_x. Pharmacol Rev. 1957;9:242–6.
57. Schild HO. pA_2 and competitive drug antagonism. Br J Pharmacol. 1949;4:277–80.
58. Tallarida RJ, Murray RB. Manual of pharmacological calculations with computer programs. 2nd ed. New York: Springer; 1987.
59. Arunlakshana O, Schild HO. Some quantitative uses of drug antagonists. Br J Pharmacol. 1959;14:48–58.
60. Furchgott RF. The use of β-haloalkylamines in the differentiation of receptors and in the determination of dissociation constants of receptor-agonist complexes. In: Harper NJ, Simmonds AB, editors. Advances in drug research, vol. 3. London: Academic Press; 1966. p. 21–55.
61. Stephenson RP. A modification of receptor theory. Br J Pharmacol. 1956;11:379–93.
62. Thal DM, Glukhova A, Sexton PM, Christopoulos A. Structural insights into G-protein coupled receptor allostery. Nature. 2018;559:45–53. https://doi.org/10.1038/s41586-018-0259-z.
63. Wootten D, Christopoulos A, Sexton PM. Emerging paradigms in GPCR allostery: implications for drug discovery. Nat Rev Drug Discov. 2013;12:630–44. https://doi.org/10.1038/nrd4052.
64. Panarese JD, Engers DW, Wu YJ, et al. Discovery of VU2957 (valiglurax): an mGlu4 positive allosteric modulator evaluated as a preclinical candidate for the treatment of Parkinson's disease. ACS Med Chem Lett. 2018;10:255–60.
65. Broad LM, Sanger HE, Mogg AJ, et al. Identification and pharmacological profile of SPP1, a potent, functionally selective and brain penetrant agonist at muscarinic M1 receptors. Brit J Pharmacol. 2019;176:110–26.
66. Chan WY, McKinzie DL, Bose S, et al. Allosteric modulation of the muscarinic M4 receptor as an approach to treating schizophrenia. Proc Natl Acad Sci USA. 2008;105:10978–83.
67. Massink A, Amelia T, Karamychev A, IJzerman AP. Allosteric modulation of G protein-coupled receptors by amiloride and its derivatives. Perspectives for drug discovery? Med Res Rev. 2019;40:1–26. https://doi.org/10.1002/med.21633.
68. Christopoulos A, Changeux J-P, Catterall WA, Fabbro D, Burris TP, Cidlowski JA, et al. International union of basic and clinical pharmacology. XC. Multisite pharmacology: recommendations for the nomenclature of receptor allosterism and allosteric ligands. Pharmacol Rev. 2014;66:918–47.
69. Maillet EL, Pellegrini N, Valant C, et al. A novel, conformation-specific allosteric inhibitor of the tachykinin NK2 receptor (NK2R) with functionally selective properties. FASEB J. 2007;21:2124–34.

70. Qile M, Beekman HDM, Sprenkeler DJ, Houtman MJC, van Ham WB, Stary-Weinzinger A, et al. LUF7244, an allosteric modulator/activator of Kv11.1 channels, counteracts dofetilide-induced torsades de pointes arrhythmia in the chronic atrioventricular block dog model. Br J Pharmacol. 2019;176:3871–85.
71. Cerione RA, Codina J, Benovic JL, et al. The mammalian beta 2-adrenergic receptor: reconstitution of functional interactions between pure receptor and pure stimulatory nucleotide binding protein of the adenylate cyclase system. Biochemistry. 1984;23:4519–25.
72. Costa T, Herz A. Antagonists with negative intrinsic activity at delta opioid receptors coupled to GTP-binding proteins. Proc Natl Acad Sci U S A. 1989;86:7321–5.
73. Samama P, Cotecchia S, Costa T, Lefkowitz RJ. A mutation-induced activated state of the β2-adrenergic receptor: extending the ternary complex model. J Biol Chem. 1993;268:4625–36.
74. Pauwels P, Wurch T. Review: amino acid domains involved in constitutive activation of G-protein-coupled receptors. Mol Neurobiol. 1998;17:109–35.
75. Kenakin T. The classification of seven transmembrane receptors in recombinant expression systems. Pharmacol Rev. 1996;48:413–63.
76. Miura S, Saku K, Karnik SS. Molecular analysis of the structure and function of the angiotensin II type 1 receptor. Hypertens Res. 2003;26:937–43.
77. Liu G, Duranteau L, Carel JC, et al. Leydig-cell tumors caused by an activating mutation of the gene encoding the luteinizing hormone receptor. N Engl J Med. 1999;341:1731–6.
78. Zhao H, Guo L, Zhao H, et al. CXCR4 over-expression and survival in cancer: a system review and meta-analysis. Oncotarget. 2015;6:5022–40.
79. Xu C, Zheng L, Li D, et al. CXCR4 overexpression is correlated with poor prognosis in colorectal cancer. Life Sci. 2018;208:333–40.
80. Dorsam RT, Gutkind JS. G-protein-coupled receptors and cancer. Nat Rev Drug Discov. 2007;7:79–94.
81. Insel PA, Sriram K, Gorr MW, et al. GPCRomics: an approach to discover GPCR drug targets. Trends Pharmacol Sci. 2019;40:378–87.
82. Nieto Gutierrez A, McDonald PH. GPCRs: emerging anti-cancer drug targets. Cell Signal. 2018;41:65–74.
83. Lanoue KF, Martin LF. Abnormal A1 adenosine receptor function in genetic obesity. FASEB J. 1994;8:72–80.
84. Uemur H, Hasumi H, Ishiguro H, et al. Renin-angiotensin system is an important factor in hormone refractory prostate cancer. Prostate. 2006;66:822–30.
85. Saikawa S, Kaji K, Nishimura N, et al. Angiotensin receptor blockade attenuates cholangiocarcinoma cell growth by inhibiting the oncogenic activity of yes-associated protein. Cancer Lett. 2018;434:120–9.
86. Dethlefsen C, Hansen LS, Lillelund C, et al. Exercise-induced catecholamines activate the hippo tumor suppressor pathway to reduce risks of breast cancer development. Cancer Res. 2017;77:4894–904.
87. Lee Z, Swaby RF, Liang Y, et al. Lysophosphatidic acid is a major regulator of growth-regulated oncogene alpha in ovarian cancer. Cancer Res. 2006;66:2740–8.
88. Kijima T, Maulik G, Ma PC, et al. Regulation of cellular proliferation, cytoskeletal function, and signal transduction through CXCR4 and c-Kit in small cell lung cancer cells. Cancer Res. 2002;62:6304–11.
89. Almofti A, Uchida D, Begum NM, et al. The clinicopathological significance of the expression of CXCR4 protein in oral squamous cell carcinoma. Int J Oncol. 2004;25:65–71.
90. Wang Z, Liu P, Zhou X, et al. Endothelin promotes colorectal tumorigenesis by activating YAP/TAZ. Cancer Res. 2017;77:2413–23.
91. Katoh M, Katoh M. Molecular genetics and targeted therapy of WNT-related human diseases. Int J Mol Med. 2017;40:587–606.
92. Szepeshazi K, Schally AV, Nagy A, Halmos G. Inhibition of growth of experimental human and hamster pancreatic cancers in vivo by a targeted cytotoxic bombesin analog. Pancreas. 2005;31:275–82.

93. Liu G, Duranteau L, Caarel JC, et al. Leydig-cell tumors caused by an activating mutation of the gene encoding the luteinizing hormone receptor. N Engl J Med. 1999;341:1731–6.
94. Hwa J, Klein-Seetharaman J, Khorana HG. Structure and function in rhodopsin: mass spectrometric identification of the abnormal intradiscal disulfide bond in misfolded retinitis pigmentosa. Proc Natl Acad Sci U S A. 2001;98:4872–6.
95. Spiegel AM. Defects in G protein-coupled signal transduction in human disease. Annu Rev Physiol. 1996;58:143–70.
96. Parma J, Duprez L, Van Sande J, et al. Somatic mutations in the thyrotropin receptor gene cause hyperfunctioning thyroid adenomas. Nature. 1993;365:649–51.
97. Rosenthal W, Antaramian A, Gilbert S, Birnbaumer M. Nephrogenic diabetes insipidus: a V2 vasopressin receptor unable to stimulate adenylyl cyclase. J Biol Chem. 1993;268:13,030–3.
98. Smit MJ, Leurs R, Alewijnse AE, et al. Inverse agonism of histamine H2 antagonist accounts for upregulation of spontaneously active histamine H2 receptors. PNAS USA. 1996;93:6802–7.
99. Onaran HO, Costa T. Where have all the active receptor states gone? Nat Chem Biol. 2012;8:674–7.
100. Wisler JW, et al. A unique mechanism of β-blocker action: carvedilol stimulates β-arrestin signaling. Proc Natl Acad Sci U S A. 2007;104:16,657–62.
101. Fitzgerald LW, Burn TC, Brown BS, et al. Possible role of valvular serotonin 5-HT(2B) receptors in the cardiopathy associated with fenfluramine. Mol Pharmacol. 2000;57:75–81.
102. Huang XP, Setola V, Yadav PN, et al. Parallel functional activity profiling reveals valvulopathogens are potent 5-hydroxytryptamine(2B) receptor agonists: implications for drug safety assessment. Mol Pharmacol. 2009;76:710–22.
103. Luttrell LM, Maudsley S, Bohn LM. Fulfilling the promise of "biased" G protein-coupled receptor agonism. Mol Pharmacol. 2015;88:579–88.
104. Nikiforovich GV, Zhang M, Yang Q, et al. Interactions between conserved residues in transmembrane helices 2 and 7 during angiotensin AT1 receptor activation. Chem Biol Drug Des. 2006;68:239–49.
105. Zimmerman B, Beautrait A, Aguila B, Charles R, Escher E, Claing A, Bouvier M, Laporte SA. Differential b-arrestin-dependent conformational signaling and cellular responses revealed by angiotensin analogs. Sci Signal. 2012;5:ra33. https://doi.org/10.1126/scisignal.2002522.
106. Valero TR, Sturchler E, Jafferjee M, et al. Structure–activity relationship study of angiotensin II analogs in terms of β-arrestin-dependent signaling to aldosterone production. Pharmacol Res Per. 2016;4(2):e00226, pp 1–10. https://doi.org/10.1002/prp2.226.
107. Ehlert FJ. Analysis of biased agonism. Prog Mol Biol Transl Sci. 2018;160:63–104. https://doi.org/10.1016/bs.pmbts.2018.08.001.
108. Lymperopoulos A. Arrestins in the cardiovascular system: an update. Prog Mol Biol Transl Sci. 2018;159:27–57. https://doi.org/10.1016/bs.pmbts.2018.07.003.
109. Kenakin T. What is pharmacological 'affinity'? Relevance to biased agonism and antagonism. Trends Pharmacol Sci. 2014;35:434–41.
110. Silva GM, Barcelos MP, JGC P, et al. Allosteric modulators of potential targets related to Alzheimer's disease: a review. ChemMedChem. 2019;14:1467–83. https://doi.org/10.1002/cmdc.201900299.
111. Hamm HE. How activated receptors couple to G proteins. Proc Natl Acad Sci USA. 2001;98:4819–21.
112. Jakubík J, Bačáková L, El-fakahanY EE, and Tuček S. Positive cooperativity of acetylcholine and other agonists with allosteric ligands on muscarinic acetylcholine receptors. Mol Pharmacol. 1997; 52:172–179

Chapter 4
Dynamic Axial Chirality in Drug Design and Discovery: Introduction to Atropisomerism, Classification, Significance, Recent Trends and Challenges

Gaurav Joshi, Manpreet Kaur, and Raj Kumar

4.1 Introduction

Nature has always fascinated the living world with its chirality and stereochemistry [1, 2]. The discoveries and development of drugs cannot be kept aside from the impact of chirality [3]. As the knowledge was gained in advanced drug discovery, it became pretty clear that problems raised by the stereochemistry of racemic drugs caused multiple cases of drug–drug interactions, fatal adverse reactions leading to their withdrawal from the market [4]. This aspect has immensely affected drug discovery industries and research [5]. It was further found significantly important to develop stereochemically pure drugs that could overcome the drawbacks and possess better pharmacokinetics and pharmacodynamics profiles [5, 6]. It has always been claimed that losing switch over a molecule's configuration can be disastrous in drug discovery. Recently, the impact of dynamic axial chirality was foreseen, where chirality is induced via bond rotation between hindered biaryl systems leading to the development of enantiomers called atropisomers by the phenomenon atropisomerism [7, 8]. Although atropisomers were revealed a century ago in 1922 [9], by two researchers James Kenner and George Hallatt Christie from University of Sheffield, the actual term "Atropisomerism" was coined by Richard Kuhn in the year 1933 [10]. The term was Greek derived meaning "a—not" and "tropos—to turn". The impact of atropisomerism is prevalent in drug discovery and development, which has introduced many challenges that need to be addressed [11, 12]. The chapter is therefore put forth to discuss the impact of atropisomerism in drug discovery considering some important drug examples, classification system,

G. Joshi · M. Kaur · R. Kumar (✉)
Laboratory for Drug Design and Synthesis, Department of Pharmaceutical Sciences and Natural Products, Central University of Punjab, Bathinda, Punjab, India
e-mail: raj.khunger@cup.edu.in

R. Poduri (ed.), *Drug Discovery and Development*,
https://doi.org/10.1007/978-981-15-5534-3_4

techniques to detect atropisomerism, application in drug discovery and beyond, the methodology involved for atropselective conversion along with a brief discussion on atropisomers found in naturally occurring drug entities.

4.2 Nomenclature and Classification of Atropisomers Based Drug Candidates

The axial stereochemistry of atropisomers is primarily designated by Cahn–Ingold–Prelog priority rules, where the priority is made in terms of Newman projection. The configurations are termed S_a and R_a based upon the locked configuration orientation and in close analogy to the traditional R/S nomenclature for tetrahedral stereocenter (Fig. 4.1). The molecules in the chiral axes may also be observed as spirals or helices. In order to put forth this nomenclature, the designation of ligands with the highest priority from 1 to 3 position in the front and back of frameworks are considered. If priority from ligand 1 to 3 is in clockwise, this is designated by 'plus' (*P*), and for counterclockwise, it is 'minus' (*M*) [13–15].

The important requirements to exhibit atropisomerism by molecules having axis includes (a) length and rigidity between axial groups, (b) steric requirement of substituents to produce hindered rotation and the mechanism to induce hindered rotation that may vary between photochemically or chemical induced processes rather than mere physical rotation along the axis. The activation energy to enable

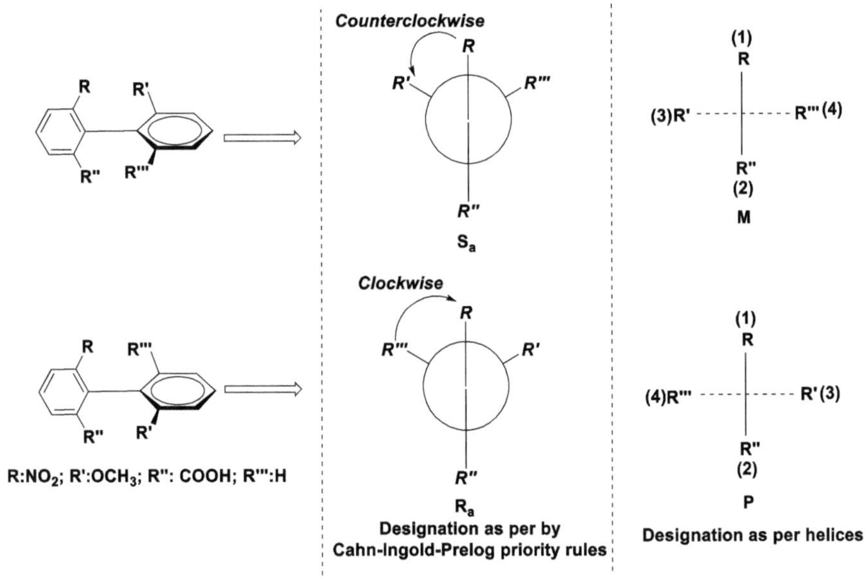

Fig. 4.1 General nomenclature and descriptors for atropisomers compounds

this rotation is convoyed by the electron distribution system of the scaffold, which is further complemented by steric hindrance, temperature and solvent conditions. Molecules possessing low energy barrier to rotation (proatropisomers) are more prone towards developing into atropisomers under physiological conditions. The atropisomers are further classified by their ease to racemize by the bond rotation and based on the biaryl ring system present [11, 13–15].

Based on the ease to racemize by the bond rotation solely involving energy barrier [16], atropisomers are broadly classified into three classes. Class 1 type atropisomers can racemize instantly within few seconds at room temperature with a mere requirement of 20 kcal/mol energy barriers for rotation around the chiral axis; class 2 atropisomers can racemize between few hours to month scale and possesses energy barrier of around 20–28 kcal/mol for rotation between the axis, and class 3 atropisomers are considered to be the most stable and get racemized on the year or greater timescale at room temperature with an energy barrier of 28 kcal/mol [12, 15]. Atropisomers can be separated physically to individual species only when they possess half-life of minimum 1000 s or 16.7 min at given temperature conditions [15]. Considering this statement class 3 drugs are considered stable, class 1 are unstable whereas class 2 drugs maintain a variable state as they may racemize on an intermediate time scale and may lead to different pharmacokinetic and pharmacodynamic parameters. Some examples of Classes 1 to 3 (FDA approved drugs) along with potential proatropisomers are collected in Fig. 4.2 [12, 17–19].

Next, the classification of atropisomers for biaryl linkage falls into two broad classes: bridged and non-bridged atropisomers [11, 12, 15–18]. Bridged biaryl includes lignans [20], lactotones [21], cyclopeptides [22] and multicyclic or monocyclic carbon bridged biaryls [23]; the second category includes isocyclic biaryl, fused heterocycles, heterobiaryls and multiple coupled biaryls [16]. The relevant examples of both categories are compiled in Fig. 4.3. The atropisomerism on the stereogenic axis could emerge as a result of hindered rotation around sp^2 hybridized carbon atom (Fig. 4.4) either to form C–C bond (e.g. atropisomeric styrenes, amides imines); C–N bond (e.g. anilides, carbamates, urea, lactams, imides); C–P bond (e.g. phosphine oxides); C–O bond (e.g. diaryl ethers); or C–S bond (e.g. sulphones, sulphoxides, sulphides) [11, 14–16].

The half-lives of racemization of atropisomers having sp^2–sp^2 single bond of biphenyls have been determined and extensively covered in some books [24, 25]. The general consideration is as under:

(a) Biphenyls with tetra-*ortho* substitution are resolvable and stable to racemization unless the two substituents out of four are fluorine or methoxy groups.
(b) In the case of tri-*ortho* substitution in biphenyls, they get easily racemized or undergo slow racemization when one of the substituent groups is small.
(c) In the case of di-*ortho* substitution on biphenyls, they undergo racemization and are separable only when the substituents are large, restricting the rate of racemization. The size of substituents is based on their van der Waals radii that affect the racemization. The general order is $I > Br > CH_3 > Cl > NO_2 > COOH > OCH_3 > F > H$.

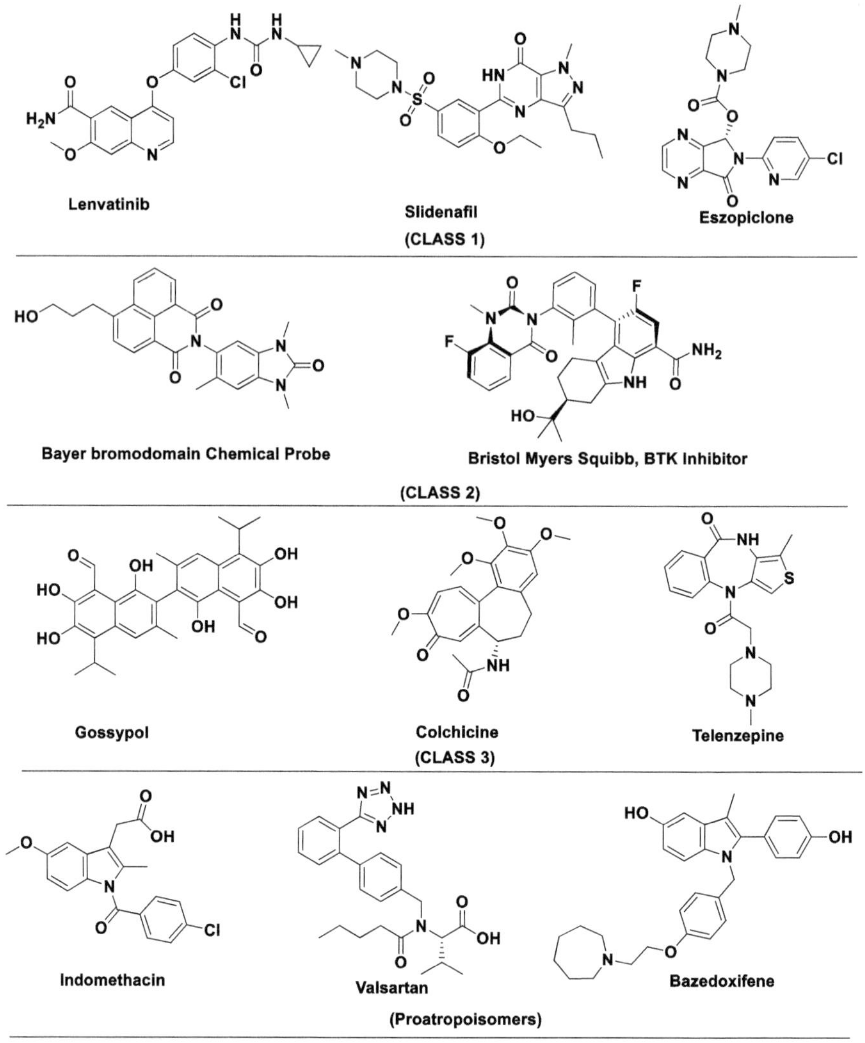

Fig. 4.2 Classes of atropisomers and proatropisomers with relevant examples

(d) In the case of mono-*ortho* substitution on biphenyls, they are not separable.
(e) *Meta* substituted biphenyls enhance the rate of racemization via buttressing

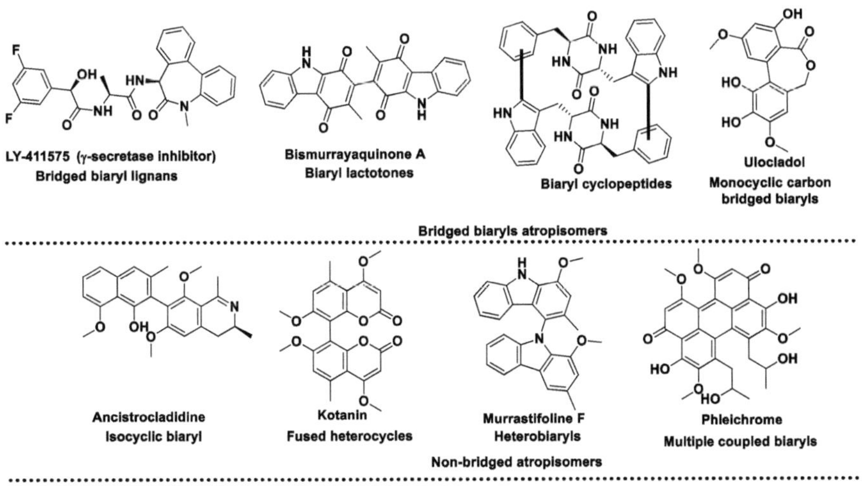

Fig. 4.3 Reported examples and structures of atropisomers classified based on biaryl linkage

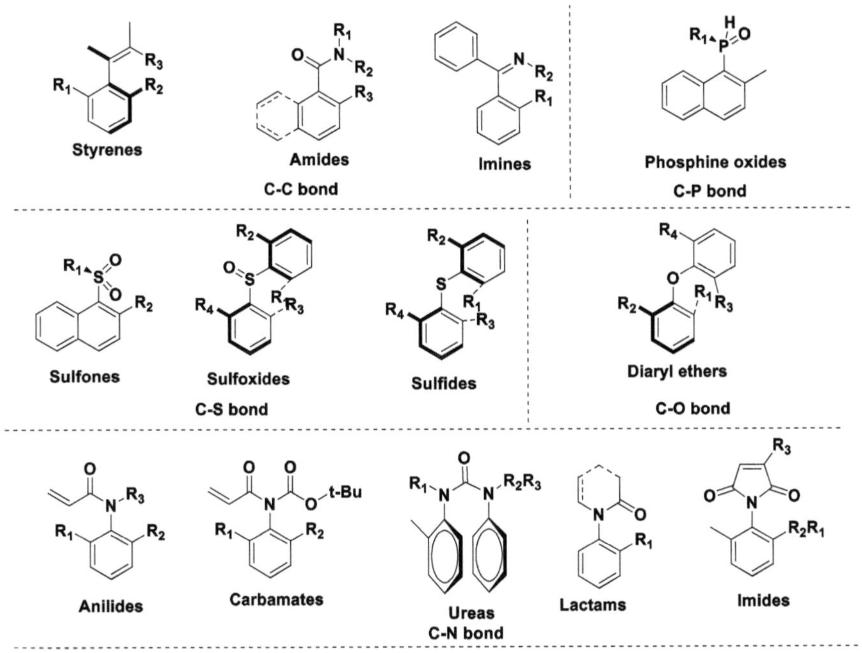

Fig. 4.4 Examples of various chemical classes where atropisomerism developed as a result of hindered rotation around sp^2 hybridized carbon atom

effect, which allows ortho substituents to rotate around each other more rapidly involving energy minimization.

4.3 Applications of Atropisomers: Drug Discovery and Beyond

Atropisomerism is found to arise in many common scaffolds used in drug discovery; this includes lactones, anilides, biaryls, diaryl ethers, diaryl amines, benzamides, etc. A thorough investigation on 1900 USFDA approved drugs revealed that approximately 285 drugs contain one atropisomeric axis that can render chirality by restricted rotation in a biological system. Further, the investigation also revealed around 190 drugs can be considered as proatropisomeric, which may get converted to atropisomers by a mere break in symmetry along the axis [12, 18]. A big surprise comes from the field of kinase inhibitors in which approximately 80% FDA approved drugs are considered prone to atropisomerism based on the presence of rapidly interconverting axis. Currently, as the industry standard approach is concerned, stable atropisomers are avoided, and rapidly interconverting atropisomers are designated as achiral [15, 18].

The active site of protein is always folded into a specific conformation, to which only similar conformation of a ligand bind. The three-dimensional ligand structure, therefore, possesses prime importance in such interactions. Atropisomeric molecules developed as a result of hindered rotation may present different stereospecificity conformation towards the receptor and may lead to alteration in pharmacokinetics and pharmacodynamics of the said molecule [11–13, 26].

Chirality developed via bond rotation usually arises challenges and intricacies in the drug discovery and development processes [14]. Some examples (Fig. 4.5) include research by Leivers et al. that disclosed the discovery of phosphatidylinositol 4-kinase alpha inhibitors. The molecule was found to be class 3 type atropisomer where R_a—form (**1**) was found to be more active (25 times) than its counterpart isomers (**2**), with improved kinase selectivity profile. The atropisomerism was generated by restricted interconversion of *ortho*-trifluoromethyl present on 3-phenylquinazolin-4(3*H*)-one moiety [27]. Chen et al. explored the atropisomeric property of compound 1-(2,6-difluorobenzyl)-3-[(2*R*)-amino-2-phenethyl]-5-(2-fluoro-3-methoxyphenyl)-6-methyl uracil (**3**) as an oral antagonist of the human GnRH hormone receptor. The atropisomers were found to possess short half-lives at room temperature (35 and 114 min in chloroform and water, respectively) that led their failure to be developed as GnRH hormone receptor inhibitors. The group provided plausible explanation for dynamic interconversion which was due to electronegativity by fluorine atom and repulsion by an oxygen atom of the carbonyl group that hindered the rotation beyond electronegative groups [28]. Kirshenbaum and group investigated N-aryl peptoids oligomers (**4**) which exhibited atropisomerism that hindered their specific biological activities. The study revealed stereochemical features of folded oligomers due to atropisomerism can induce

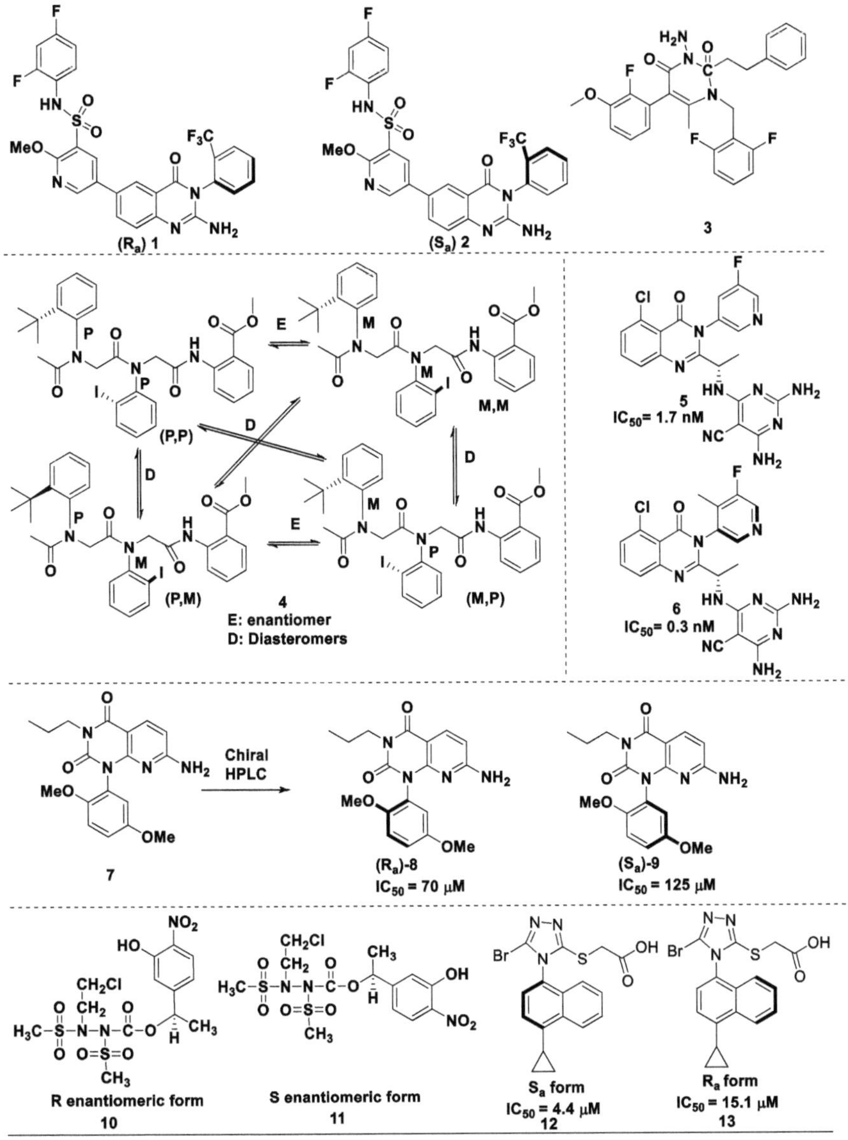

Fig. 4.5 Example of some reported atropisomers in literature

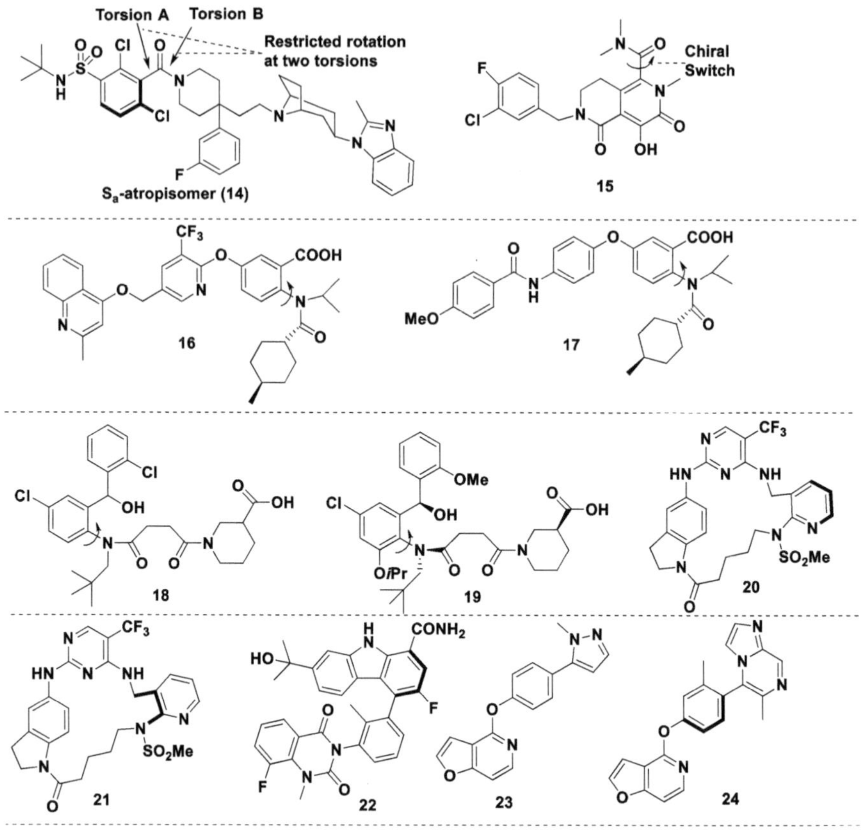

Fig. 4.5 (continued)

dynamicity and in contrast with the static nature of peptides and proteins stereochemistry [29]. Research by Patel and team led to the development of phosphoinositide 3-kinase (PI3K) inhibitor derived from 3-(pyridin-3yl) quinazolin-4(3*H*)-one as the active nucleus (**5**). The group, however, also observed acidic instability in series leading to quinazolinone ring opening. An *ortho*-methyl (**6**) group was introduced to overcome the instability, that further removed quinazolinone conjugation with pyridine ring conferring the stability, but at the same time led to the development of atropisomers (class 3), that on the resolution were found that activity was retained by only single atropisomer [30]. Research conducted by Hasegawa et al. highlighted the development of 1-aryluracil (**7**) with ortho substitution as PDE4 inhibitor. The thorough investigation suggested molecules to be atropisomers (class 3) due to hindered rotation around the axis. The atropisomers were separated via chromatographic techniques, where **8** (R_a) was found to be more potent and stable towards microsomal liver enzymes than **9** (S_a;) form [31]. Sartorelli and group developed a prodrug (1,2-bis(methylsulphonyl)-1-(2-chloroethyl)-2-[[1-(4-nitrophenyl)ethoxy]carbonyl]hydrazine) having cytotoxic potential possessing induced chirality due to hindered rotation around the central axis leading to the development of atropisomers. The atropisomers were further separated based on their stability where **10** (R_a) was found to have better $t_{1/2}$ than **11** (S_a), **11** was found to possess much better anticancer potential and distribution in hypoxic cancer masses [32]. Wang and research group studied lesinurad (hURAT1 inhibitor) and found that compound exists as atropisomers due to hindered rotation of C–N bond due to substitution of thioacetic acid and bromine on axial triazole ring. Both the atropisomers were separated and individually tested for hURAT-1 inhibitory profile, where S_a (**12**) form was found to be more potent (3.5 times) as compared to R_a (**13**) form [33]. Kazmierski and group identified and characterized GSK214096 as CCR5 inhibitor that was found to exist as a separable mixture of four atropisomer based on hindrance due to two different types of torsional strain (**A** & **B**). Torsional **B** led to the development of class 1 atropisomers that interconverted rapidly with $t_{1/2}$ of half-hour. However, torsion **A** led to the development of class 3 atropisomers due to hindrance assisted by the presence of 2,6-dichlorobenzene substitution. Further, the stable atropisomers around torsion **A** were resolved using circular dichromatism, which further revealed S_a-atropisomer (**14**) to be a more potent CCR5 inhibitor than the R_a form [66]. Egbertson and group serendipitously discovered pH-dependent development of atropisomer from 6-(3-chloro-4-fluoro-benzyl)-4-hydroxy-2-methyl-3,5-dioxo-2,3,5,6,7,8-hexahydro-[2,6]naphthyridine-1-carboxylic acid dimethylamide (**15**), intended for pharmaceutical development. The atropisomer was developed because of strained dimethylamide group in molecule **15** in aqueous buffer. The racemization processes were essentially affected by pH. High pH led to fastest racemization in few minutes (class 1), whereas lower pH led to slow conversion and formation of atropisomers (class 3) revealing the development of chiral molecular switches [34]. LaPlante and co-workers analysed the atropisomers based on anthranilic acid (HCV polymerase inhibitors; **16** and **17**). The X-ray co-crystallized structure revealed R_a enantiomer (**17**) binds to HCV Polymerase active site, and S_a enantiomer (**17**) was found to bind on HIV matrix, which further provided strong insight of R_a enantiomer as a selective HCV Polymerase inhibitor [35]. Similar evidence based on preferential selectivity of co-crystallized ligand was provided by Ichikawa and group, where they developed squalene synthase inhibitors reported atropisomerism due to hindrance in rotation by anilide group leading to the generation of conformers in a solution having different bioactive conformation at protein's active site. The C–N bond of **18** was explored to overcome the atropisomerism, and an isopropoxy group was introduced at *ortho*-position that increased rotational barrier giving a major form of R_a atropisomer (**19**) as potent squalene synthase inhibitors [36]. In a quest to develop proline-rich tyrosine kinase 2 inhibitors, Farand and group developed some macrocyclics which were found to possess atropisomerism depending upon the projection of pyridine ring (**20** & **21**). Although both atropisomers expressed rapid interconversion, atropisomer **21** exhibited better potency [37]. Giants in pharmaceutical drug development have also come across technicalities of atropisomerism in their products. Some important examples include Bruton's kinase inhibitor (BMS-986142; **22**) developed by Bristol Myers-Squibb for rheumatoid arthritis. They identified temperature-dependent atropisomer during Phase II development of a drug candidate. The investigation revealed that **22** consists of a chiral centre along with two atropisomeric chiral axes, thereby generating eight possible isomers. Considering the temperature dependency, the team developed various synthetic routes with precise monitoring of temperature. The atropisomers were also resolved using column chromatography at chilling conditions to obtain a single diastereomer. The further investigation identified different solubility profiles of enantiomers and diastereomers that led to the simple purification and isolation using solubility gradient and crystallization [38]. Davoren and research group discovered and lead optimized atropisomer as D_1 agonists with reduced desensitization through high-throughput screening. During the course of their research they identified PF-4211 (**23**) as lead that was optimized further to provide locked biaryl ring system that resulted in development of two atropisomers (barrier to rotation of >15 kcal/mol). The atropisomers were separated by the use of chiral supercritical fluid chromatography technique and evaluated further for their individual biological effect. Enantiomer **24** was found to be more potent and possessed superior affinity as orthosteric agonist of the D_1 [39].

4.3.1 Atropisomeric Natural Products

Natural products have always inspired drug discovery. The biological profile of these natural compounds is often determined with various pharmacophoric features that complement specifically to a particular protein target. With the advent in atropisomers detection technologies, many naturally derived biologically active compounds were found to possess hindrance in rotation, and thus more prone to develop atropisomerism. The important natural products that have been identified as atropisomers are classified into two groups, one those possess stereogenic centre along with hindered rotation, and other groups consist of enantiomers developed solely by hindered rotation (non-traditional atropisomers). A list of reported class of atropisomers occurring in nature is compiled in Fig. 4.6. The list covers vancomycin, a glycopeptide used as antibiotic and produced by bacteria (Amycolatopsis orientalis) residing in soil. Gossypol is produced by Gossypium cotton genus and is known to possess anticancer, antiapoptotic and antifertility properties. Napthylisoquinoline alkaloids include some important bioactive molecules, including mbandakamines-A, hibarimicinone, possessing antibiotic and anticancer properties, and flavomannins, taloaromannins, viriditoxin, and rugulotrosin with antibiotic potential. Some non-traditional examples of atropisomers that include diazepine moieties, abyssomicin C, marinopyrroles, dixiamycins, streptorubin-B, ustiloxins and haouamine-A are included [16].

Apart from their impact on drug discovery, atropisomers find their use in allied fields too [14]. They have been explored in the development of organocatalysts, kinetic resolution of secondary alcohols, and atropselective reactions that include cyclization reactions [14], [4 + 2]-cycloaddition, [3 + 2]-cycloaddition, radical cyclization, Heck cyclization, anionic cyclization and iodo lactonization. In the

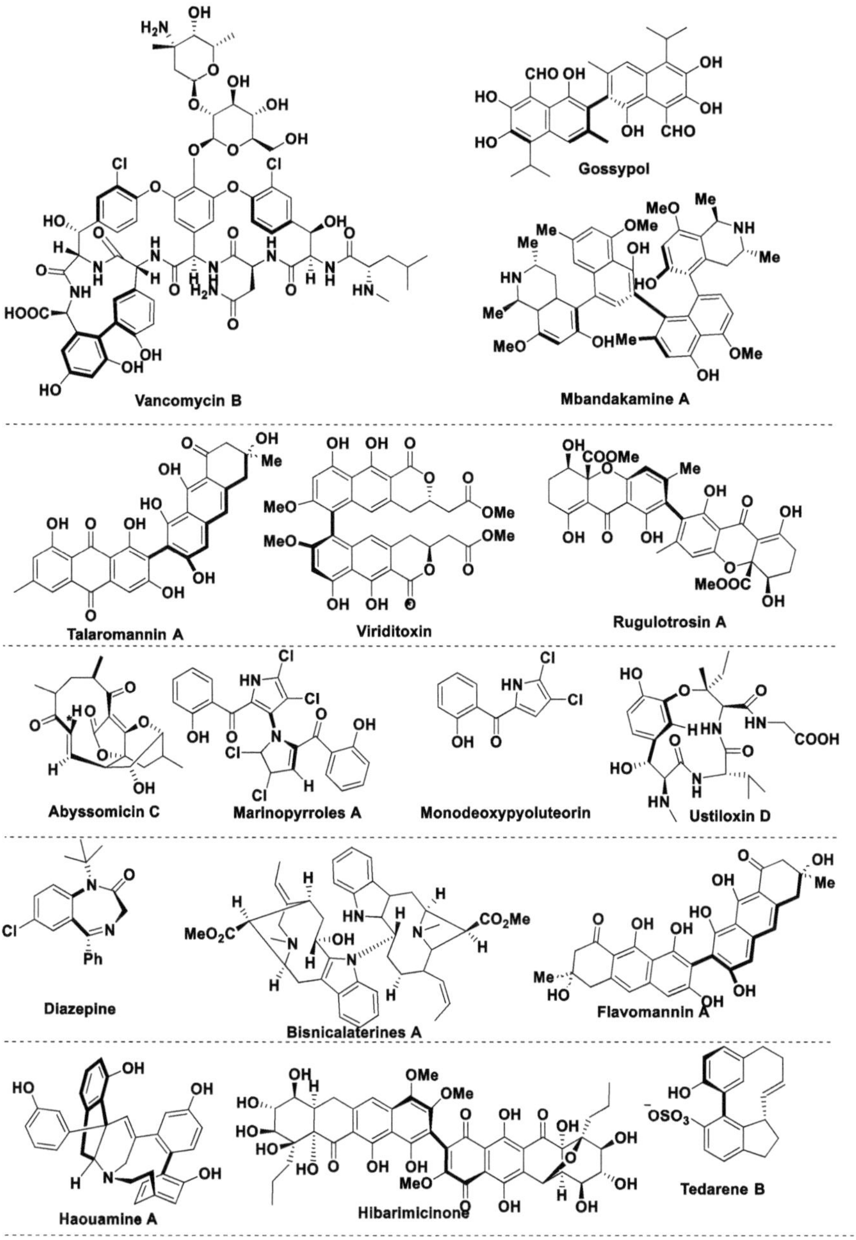

Fig. 4.6 Natural products based atropisomers as drug candidates

field of Catalysis [14], e.g. palladium catalysed asymmetric synthesis using phosphoramidites as chiral ligands with benzylic electrophiles via Mizoroki–Heck Reaction [40], in multicomponent reactions [14] like Domino reaction as hydrogen bond-forming catalyst, in the total synthesis of the calphostins [41], desymmetrization, Grignard coupling, asymmetric Heck reaction, Suzuki–Miyaura cross-coupling, benzoin and Stetter reaction, and Michael addition. In atropselective photoreactions [14], e.g. 6π-photocyclization, 4π-photocyclization, [2 + 2]-photocycloaddition, hydrogen abstraction photochemistry of chiral crystals and photochemistry involving atropisomeric compounds under elevated pressures. Besides, they also find their applications in the development of molecular machines [42], gears, and in conformation control via stereochemical relay reactions [43].

4.4 Methods for Measurement of Atropisomers Racemization

Overlooking atropisomerism might lead to some of the lethal drawbacks in drug discovery. With the advent of newer chemical methodologies, the drug development has resulted in multiple incidences of atropisomeric products, that may possess possible differences in their pharmacokinetic and pharmacodynamics depending upon the level of racemization. Thorough knowledge regarding activation of a particular atropisomer not only assists in making important decisions about handling, storage or for the formulation development or clinical development. Numerous reports on measurements of atropisomers have facilitated the detection and are considered now as an important tool in drug discovery [15, 44]. Various methods classified are based upon speed, accuracy, sample requirement, automation, safety and reliability. Some of the reliable methods reported for detection of atropisomers are as follows.

4.4.1 Segmented Flow Technology

This method allows the measurement of thermal interconversion kinetics of atropisomers, by regulating the reaction temperature and allowing the instantaneous heating and cooling resulting in more precise results. The instrument importantly covers separate segments for the solvent system and the substrate or investigational drug that is present in solubilized form. Both substrate and solvent are delivered alternatively to a small tube reactor, which further passes to UV detection unit, that analyses the residence time of a particular form of atropisomers. Davoren et al. reported kinetic measurement of AMPA antagonist CP-465021 and its atropenantiomer CP465022 using segmented flow technique and corroborated the finding with traditional thermal methods (Fig. 4.7a and b) [44].

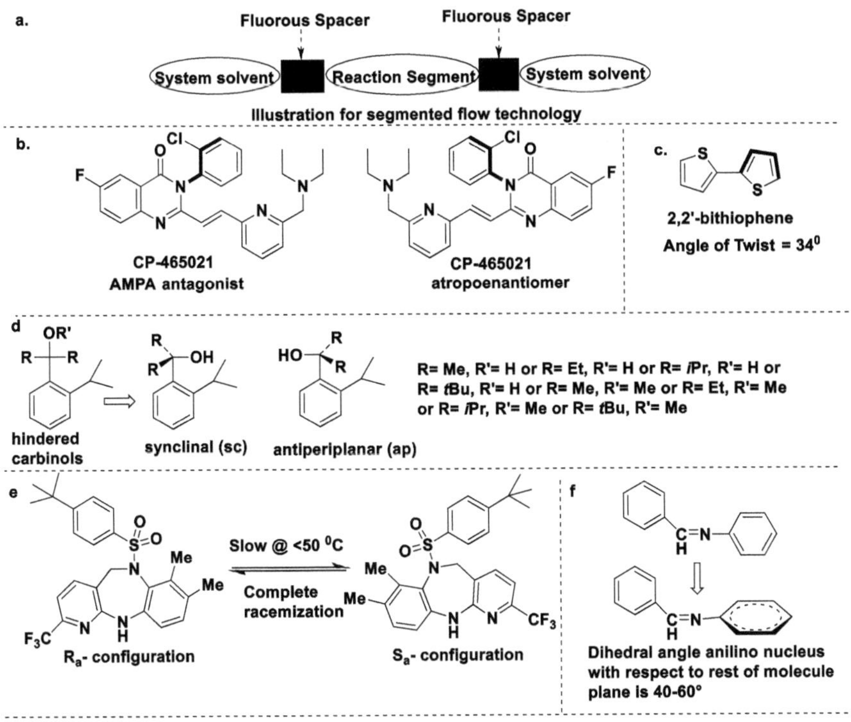

Fig. 4.7 Examples of various methodologies reported for measurement of atropisomers racemization

4.4.2 Electron Diffraction

The methodology is employed to determine the chiral crystal symmetries by measuring the orbital angular momentum [45]. The methodology was used by Svendsas et al. to determine the angular angle (34°) of 2,2′-dithienyl vapours (Fig. 4.7c) [46].

4.4.3 Dynamic Nuclear Magnetic Resonance

Temperature variable NMR spectroscopy, also called as dynamic NMR, is used for investigating stereodynamic processes. This method may assist in detecting energy barriers (4.5–23 kcal/mol) due to internal motion of atropisomers. Atropisomers are efficiently detected if they possess half-life of more than 10^{-2} s. For example, hindered carbinols gave rise to two compounds, i.e. synclinal (sc) and antiperiplanar (ap) as conformational atropisomers (Fig. 4.7d) [15, 47, 48].

4.4.4 X-Ray Crystallography

X-ray crystallography is a reliable technique used as high-throughput means of determining absolute stereochemistry during the discovery process [49]. For example, unusual planer chirality developed in conformationally restrained diazepine ring system was detected using X-ray in Benzodiazepine Sulphonamide-Based Bombesin Receptor Subtype 3 Agonists (Fig. 4.7e) [50].

4.4.5 Electronic Spectroscopy Techniques

Steric hindrance due to rotation alters the resonance patterns in the biaryl systems leading to the alteration in the absorption or dipole movement. This difference is exploited in the detection of atropisomers [51]. For example, the theoretical studies predicted azomethine having stable non-planar conformation in which dihedral angle was found to be 40–60° between the amino group to rest of molecule plane. However, azomethines exhibit a peculiar difference in physical and chemical properties due to different electronic transitions suggesting the possibility of finding atropisomers (Fig. 4.7f) [52].

4.4.6 pKa Measurements

Steric relaxation leads to a decrease in basicity, whereas inhibition of resonance increases the basicity. The difference in basicity is, therefore, an important parameter used for pKa measurement of sterically hindered conformers [53].

4.4.7 Measurement of Hammett and Taft Constants

The sensitivity of Hammett and Taft values in determining steric hindrance is often used to assess the degree of non-planarity induced by the aryl groups, thus providing an insight about hindrance in axial rotation, which is a peculiar feature of atropisomerism [54, 55].

4.4.8 Computational Methods

Many computational methodologies including density field theory and surface potential energy determination techniques have been utilized for detection of atropisomers [56].

4.5 The Methodology Involved in Atropselective Conversion

Nonracemic axially chiral molecules usually find their varied applications in the field of drug discovery, development of chiral ligands, as organocatalysts and much more. Among these, biaryl atropisomers are the most common ones, and many synthetic methodologies are developed for their enantioselective development [57, 58]. Pharmaceutical industries put much emphasis on the development of single enantiomers as a therapeutically active entity. Therefore, thorough investigation of active conformer of atropisomers is needed to be performed beforehand precisely at early stages of drug design and discovery. The atropisomerism is importantly introduced in biaryl compounds via coupling reaction that includes Suzuki–Miyaura reaction, Ullmann coupling, etc. The enantioselective coupling is achieved by various methodologies that fix the axial configuration to a most stable one. The important considerations include "barrier to atropisomerization" [11]. If the barrier is high, development of pure and stereochemically stable isomer of the drug is given priority (e.g. atropisomeric enantiomers of telenzepine possessing slow interconversion; Fig. 4.8a), whereas in case, the barrier is low, rapid interconverting isomers (e.g. fast interconversion between the atropisomeric enantiomers of lipoxygenase inhibitor (Sch40120); Fig. 4.8b) are developed as drug candidates. In order to

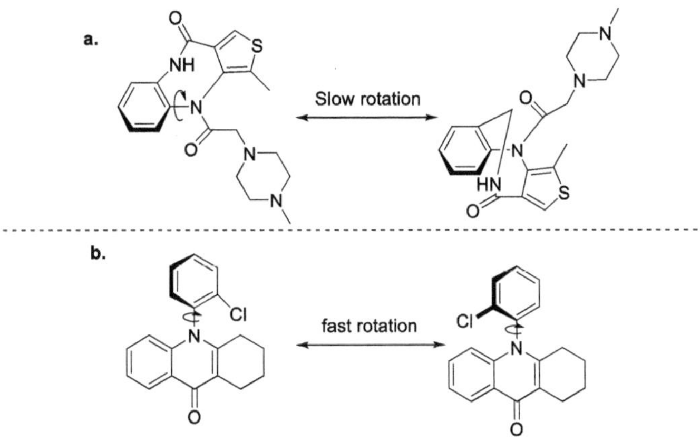

Fig. 4.8 Examples of the development of pure and stereochemically stable isomers of the drugs

achieve this, four common methodologies are involved with enantioselective synthesis [11, 14]: (a) elimination of labile chirality by inducing internal symmetrization; (b) modification to allow the facile rotation around biaryl-axis by decreasing barrier to rotation; (c) making hindered rotation by increasing the energy barrier to rotation; and (d) introduction of an artificial chiral bridge for atropdiastereoselective coupling of arenes. These are explained below.

4.5.1 Elimination of Labile Chirality by Inducing Internal Symmetrization

Labile chiral compounds are those having low energy barriers and easily undergo racemization to form atropisomers. The energy barrier is so less that it is impossible to separate atropisomers and establish their exact profile [59]. The classical example includes Tachykinin 243, which is a neuropeptide used as NK_1 receptor antagonist in depression. The presence of multi diastereomers necessitated the redesign of tachykinin 243 (**25**) to a newer molecule (**26**), thus overcoming the labile chirality and retaining the biological potential of **26** (Fig. 4.9a).

4.5.2 Modification to Allow the Facile Rotation Around Biaryl-Axis by Decreasing Barrier to Rotation

This method is employed to develop enantioselective atropisomers whereby compounds possess low barrier to the rotation, leading to the development of multiple diastereomers (each having small $t_{1/2}$) possessing different pharmacokinetics and pharmacodynamics properties. The energy barrier to the rotation, in this case, might be overcome by simple substitution around a rotating axis, allowing the facile rotation. Example in this category includes enolate chemistry with racemic atropisomeric amides (Fig. 4.9b) [60]. The work involved alkylation reactions on anilide substrate (**27**) by treatment with LDA. The reaction yielded **28** but with very modest diastereoselective property (around 2–4:1). The modest diastereoselectivity was due to rotation around C–N bond; this was further improved by introducing an *N*-MEM (methoxyethoxymethyl) derivative (**29**) that ensured the axial stability and the reaction advanced with exceptional diastereoselective property ($\geq$25:1).

4.5.3 Increasing the Energy Barrier to Rotation via Hindered Rotation

This method encompasses those atropisomers that possess high energy of rotation and as such are inseparable under normal conditions. The example includes research

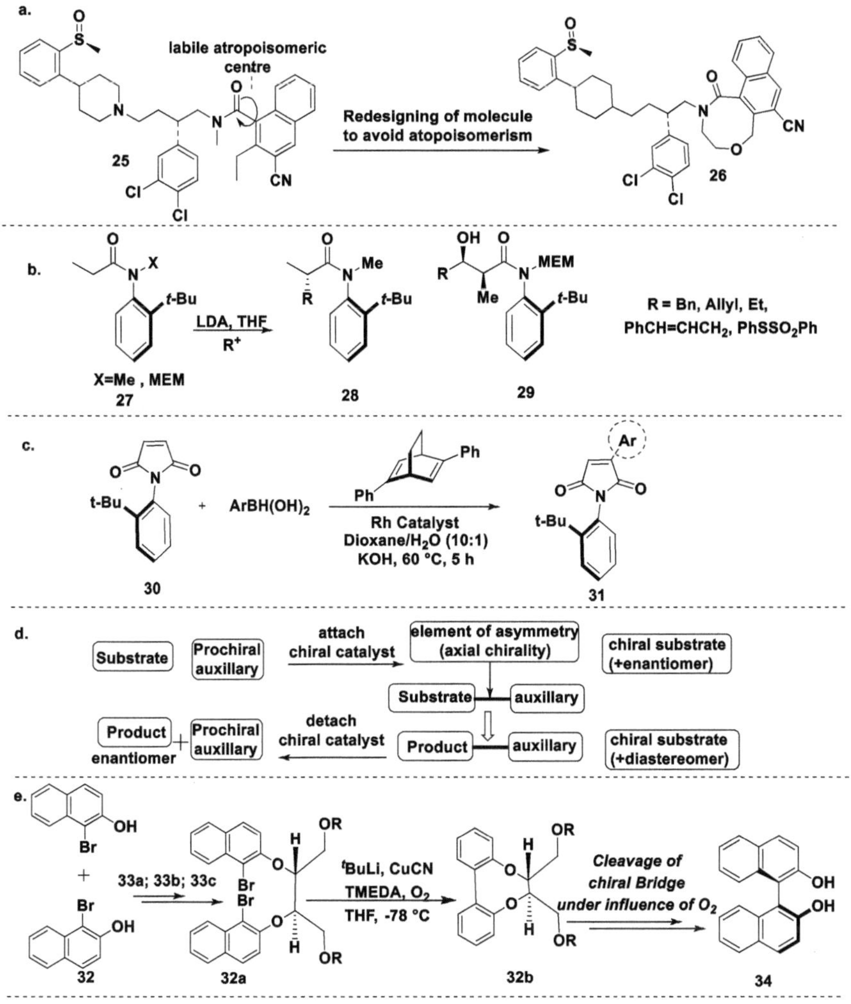

Fig. 4.9 The various methodologies involved in atropselective conversion

work done by Hayashi and group [61] (Fig. 4.9c), whereby they reported involvement of rhodium catalysed asymmetric 1,4-addition reaction for the asymmetric synthesis of N-aryl imides (**30**). The reaction yielded a hindered enantiomer (**31**) with excellent enantio- and diastereoselectivities. The plausible explanation to the selectivity was owed to the approach of Rh complex from the opposite side of bulkier substituent *tert*-butyl, thus hindering the rotation.

4.5.4 *Introduction of an Artificial Chiral Bridge for Atropdiastereoselective Coupling of Arenes*

This method includes the introduction of an optically pure chiral auxiliary to access atropisomeric-selective compounds. In this methodology, an optically pure chiral auxiliary is incorporated adjacent to an evolving axial chiral centre, thus influencing the equilibrium of conformation which is finally detached yielding pure enantiomeric product [62]. The example includes research performed by Lipshutz and group. The group disclosed the development of chemical methods by inducing chiral auxiliary that yielded nonracemic biaryls. In their approach, 1-bromo-2-naphthol (**32**) was converted to dibromide compound (**32a**) using lactic acid ester (prochiral auxiliary) (**33a**). Intermediate **32a** was then transformed into **32b** via sequential reaction of ^{t}BuLi with the addition of CuCN. **32b** further produced binaphthyl **34** consequently upon reaction with aerial oxygen with 66% of diastereomeric excess. The replacement of tartaric acid with mandelic acid (**33b**) increased the diastereomeric excess to 80%. The stable gauche interaction was further ensured by converting tartaric acid to diol form (**33c**) yielding 90% diastereomeric excess (Fig. 4.9d, e) [63].

4.6 Regulatory Guidelines for the Development of Atropisomers

Currently, no formal regulatory policy has been developed to define drug candidates possessing axial rotation and leading to the development of atropisomers. Therefore, atropisomers are considered under the same guidelines and treated similarly as stereochemical molecules with classical chiral centres, whereby their rate of isomerization into different conformers is taken into consideration. Numerous countries started issuing regulatory guidelines for drugs possessing chiral centre(s) since mid of the 1980s. These agencies essentially investigated the influence of each enantiomer on the pharmacokinetic and pharmacodynamic pattern. Japanese regulatory was the first to take a stand on ever-rising problems of chirality in drug development, although the official document related to chiral drug development was not released [64]. The FDA policy statement (1992) laid enough emphasis on understanding the important therapeutic efficiency of the individual isomers that are in the drug development phase. The knowledge thus attained via these individual studies on isomers helps in providing market preferences for such drugs. The worldwide share of chiral drug candidates as single enantiomer has grown at a sustained rate. Fifteen single-enantiomers have been approved so far by USFDA [65]. Apart from USFDA guidelines on "Development of new stereoisomeric drugs", other regulatory sources and guidance which include, Health Canada's "Guidance for Industry: Stereochemical Issues in Chiral Drug Development" along with the European Medicine agency guidelines on "Investigation of Chiral Active Substances" are also considered for the expansion of stereoisomeric drugs or medicines. Irrespective of the nature of regulatory agencies, whether National or International, all emphasize on the vital importance of chirality that lies within active ingredients during testing of bulk drugs, assessment of final products developed with the use of chiral ingredients, development of protocols for stability testing of chiral drugs along with the proper labelling of such drug candidates. The agencies also urge strongly to pharmaceutical industries to comply with the terms laid therein [64].

4.7 Conclusion

Atropisomers importantly consist of two or more molecules, when we are expecting only one. This unique isomerism is generated by the restricted rotation along the biaryl axis. Atropisomers can dampen the thought and goal of designing and discovering stable pharmaceutical substances. The prevalence of atropisomerism in drug discovery has been increasing from the last decades. Atropisomers as such are overlooked for a long time, although they may behave and interact differently in biological systems like chiral molecules and may exert differential pharmacokinetics and pharmacodynamics effect, including toxicities too. Recent USFDA report revealed 1900 small molecules are having at least one atropisomeric centre and approximately 10% of the drugs from the data bank are proatropisomeric. As each atropisomer is unique, it is therefore advised to approach each drug candidate case wise. The recent literature also insights that atropisomeric drugs have not been explored so far for their impact on drug resistance and toxicity. The USFDA indicates the preference for developing single enantiomer; however, atropisomers have failed to find their place in the regulatory guidelines as a separate entity.

References

1. Nicolaou K. Organic synthesis: the art and science of replicating the molecules of living nature and creating others like them in the laboratory. Proc Roy Soc A Math Phys Eng Sci. 2014;470 (2163):20130690.
2. Feringa BL. The art of building small: from molecular switches to motors (Nobel lecture). Angew Chem Int Ed. 2017;56(37):11,060–78.
3. Agranat I, Caner H, Caldwell J. Putting chirality to work: the strategy of chiral switches. Nat Rev Drug Discov. 2002;1(10):753.
4. Nguyen LA, He H, Pham-Huy C. Chiral drugs: an overview. Int J Biomed Sci. 2006;2(2):85.
5. Wozniak TJ, Bopp RJ, Jensen EC. Chiral drugs: an industrial analytical perspective. J Pharm Biomed Anal. 1991;9(5):363–82.
6. Williams K, Lee E. Importance of drug enantiomers in clinical pharmacology. Drugs. 1985;30 (4):333–54.
7. Alkorta I, Elguero J, Roussel C, Vanthuyne N, Piras P. Atropisomerism and axial chirality in heteroaromatic compounds. Adv Heterocyclic Chem. 2012;105:1–188.
8. Lassaletta JM. Atropisomerism and axial chirality. An Quím. 2019;115(5):451–2.

9. Christie GH, Kenner J. LXXI.—The molecular configurations of polynuclear aromatic compounds. Part I. The resolution of γ-6: 6′-dinitro-and 4: 6: 4′: 6′-tetranitro-diphenic acids into optically active components. J Chem Soc Trans. 1922;121:614–20.
10. Freudenberg K. Stereochemie, Leipzig u. Wien, Franz Deutliche. 1933:695.
11. Clayden J, Moran WJ, Edwards PJ, LaPlante SR. The challenge of atropisomerism in drug discovery. Angew Chem Int Ed. 2009;48(35):6398–401.
12. Toenjes ST, Gustafson JL. Atropisomerism in medicinal chemistry: challenges and opportunities. Future Med Chem. 2018;10(4):409–22.
13. Bringmann G, Price Mortimer AJ, Keller PA, Gresser MJ, Garner J, Breuning M. Atroposelective synthesis of axially chiral biaryl compounds. Angew Chem Int Ed. 2005;44(34):5384–427.
14. Kumarasamy E, Raghunathan R, Sibi MP, Sivaguru J. Nonbiaryl and heterobiaryl atropisomers: molecular templates with promise for atropselective chemical transformations. Chem Rev. 2015;115(20):11239–300.
15. LaPlante SR, Fader LD, Fandrick KR, Fandrick DR, Hucke O, Kemper R, Miller SP, Edwards PJ. Assessing atropisomer axial chirality in drug discovery and development. J Med Chem. 2011;54(20):7005–22.
16. Bringmann G, Gulder T, Gulder TA, Breuning M. Atroposelective total synthesis of axially chiral biaryl natural products. Chem Rev. 2010;111(2):563–639.
17. Glunz PW. Recent encounters with atropisomerism in drug discovery. Bioorg Med Chem Lett. 2018;28(2):53–60.
18. LaPlante SR, Edwards PJ, Fader LD, Jakalian A, Hucke O. Revealing atropisomer axial chirality in drug discovery. ChemMedChem. 2011;6(3):505–13.
19. Mann A. Conformational restriction and/or steric hindrance in medicinal chemistry. In: Wermuth CG, editor. The practice of medicinal chemistry: Elsevier; 2008. p. 363–79.
20. Durairaj K. Modern concepts and strategies in synthesis of biaryl compounds. Curr Sci. 1994:833–8.
21. Bringmann G, Menche D. Stereoselective total synthesis of axially chiral natural products via biaryl lactones. Acc Chem Res. 2001;34(8):615–24.
22. Mendive-Tapia L, Bertran A, García J, Acosta G, Albericio F, Lavilla R. Constrained cyclopeptides: biaryl formation through Pd-catalyzed C−H activation in peptides—structural control of the cyclization vs cyclodimerization outcome. Chem Eur J. 2016;22(37):13,114–9.
23. Xue X, Gu Z. Synthesis of bridged biaryl atropisomers via sequential cu-and Pd-catalyzed asymmetric ring opening and cyclization. Org Lett. 2019;21:3942.
24. Finar I. Organic chemistry, volume 2, Stereochemistry and the chemistry of natural products. Edinburgh: English Language Book Society. Longman Group Limited; 1970.
25. Eliel EL, Wilen SH. Stereochemistry of organic compounds. New York: Wiley; 2008.
26. Smyth JE, Butler NM, Keller PA. A twist of nature–the significance of atropisomers in biological systems. Nat Prod Rep. 2015;32(11):1562–83.
27. Leivers AL, Tallant M, Shotwell JB, Dickerson S, Leivers MR, McDonald OB, Gobel J, Creech KL, Strum SL, Mathis A. Discovery of selective small molecule type III phosphatidylinositol 4-kinase alpha (PI4KIIIα) inhibitors as anti hepatitis C (HCV) agents. J Med Chem. 2013;57 (5):2091–106.
28. Tucci FC, Hu T, Mesleh MF, Bokser A, Allsopp E, Gross TD, Guo Z, Zhu YF, Struthers RS, Ling N. Atropisomeric property of 1-(2, 6-difluorobenzyl)-3-[(2R)-amino-2-phenethyl]-5-(2-fluoro-3-methoxyphenyl)-6-methyluracil. Chirality. 2005;17(9):559–64.
29. Paul B, Butterfoss GL, Boswell MG, Renfrew PD, Yeung FG, Shah NH, Wolf C, Bonneau R, Kirshenbaum K. Peptoid atropisomers. J Am Chem Soc. 2011;133(28):10,910–9.
30. Patel L, Chandrasekhar J, Evarts J, Forseth K, Haran AC, Ip C, Kashishian A, Kim M, Koditek D, Koppenol S. Discovery of orally efficacious phosphoinositide 3-kinase δ inhibitors with improved metabolic stability. J Med Chem. 2016;59(19):9228–42.

31. Hasegawa F, Kawamura K, Tsuchikawa H, Murata M. Stable C–N axial chirality in 1-Aryluracil scaffold and differences in in vitro metabolic clearance between atropisomers of PDE4 inhibitor. Bioorg Med Chem. 2017;25(16):4506–11.
32. Penketh PG, Baumann RP, Shyam K, Williamson HS, Ishiguro K, Zhu R, Eriksson ES, Eriksson LA, Sartorelli AC. 1, 2-Bis (methylsulfonyl)-1-(2-chloroethyl)-2-[[1-(4-nitrophenyl) ethoxy] carbonyl] hydrazine (KS119): a cytotoxic prodrug with two stable conformations differing in biological and physical properties. Chem Biol Drug Des. 2011;78(4):513–26.
33. Wang J, Zeng W, Li S, Shen L, Gu Z, Zhang Y, Li J, Chen S, Jia X. Discovery and assessment of atropisomers of (±)-lesinurad. ACS Med Chem Lett. 2017;8(3):299–303.
34. Welch CJ, Biba M, Pye P, Angelaud R, Egbertson M. Serendipitous discovery of a pH-dependant atropisomer bond rotation: toward a write-protectable chiral molecular switch? J Chromatogr B. 2008;875(1):118–21.
35. LaPlante SR, Forgione P, Boucher C, Coulombe R, Gillard J, Hucke O, Jakalian A, Joly M-A, Kukolj G, Lemke C. Enantiomeric atropisomers inhibit HCV polymerase and/or HIV matrix: characterizing hindered bond rotations and target selectivity. J Med Chem. 2013;57 (5):1944–51.
36. Ichikawa M, Yokomizo A, Itoh M, Haginoya N, Sugita K, Usui H, Terayama K, Kanda A. Discovery of atrop fixed alkoxy-aminobenzhydrol derivatives: novel, highly potent and orally efficacious squalene synthase inhibitors. Bioorg Med Chem. 2011;19(17):5207–24.
37. Farand J, Mai N, Chandrasekhar J, Newby ZE, Van Veldhuizen J, Loyer-Drew J, Venkataramani C, Guerrero J, Kwok A, Li N. Selectivity switch between FAK and Pyk2: macrocyclization of FAK inhibitors improves Pyk2 potency. Bioorg Med Chem Lett. 2016;26 (24):5926–30.
38. Beutner G, Carrasquillo R, Geng P, Hsiao Y, Huang EC, Janey J, Katipally K, Kolotuchin S, La Porte T, Lee A. Adventures in atropisomerism: total synthesis of a complex active pharmaceutical ingredient with two chirality axes. Org Lett. 2018;20(13):3736–40.
39. Davoren JE, Nason D, Coe J, Dlugolenski K, Helal C, Harris AR, LaChapelle E, Liang S, Liu Y, O'Connor R, Orozco CC, Rai B, Salafia M, Samas B, Xu W, Kozak R, Gray D. Discovery and lead optimization of atropisomer D1 agonists with reduced desensitization. J Med Chem. 2018;61(24):11,384–97.
40. Yang Z, Zhou J. Palladium-catalyzed, asymmetric Mizoroki–Heck reaction of benzylic electrophiles using phosphoramidites as chiral ligands. J Am Chem Soc. 2012;134(29):11,833–5.
41. Merlic CA, Aldrich CC, Albaneze-Walker J, Saghatelian A, Mammen J. Total synthesis of the calphostins: application of fischer carbene complexes and thermodynamic control of atropisomers. J Org Chem. 2001;66(4):1297–309.
42. Zhang Y, Calupitan JP, Rojas T, Tumbleson R, Erbland G, Kammerer C, Ajayi TM, Wang S, Curtiss LA, Ngo AT. A chiral molecular propeller designed for unidirectional rotations on a surface. Nat Commun. 2019;10(1):1–9.
43. Petit M, Lapierre AJ, Curran DP. Relaying asymmetry of transient atropisomers of o-iodoanilides by radical cyclizations. J Am Chem Soc. 2005;127(43):14,994–5.
44. Davoren JE, Bundesmann MW, Yan QT, Collantes EM, Mente S, Nason DM, Gray DL. Measurement of atropisomer racemization kinetics using segmented flow technology. ACS Med Chem Lett. 2012;3(5):433–5.
45. Juchtmans R, Verbeeck J. Orbital angular momentum in electron diffraction and its use to determine chiral crystal symmetries. Phys Rev B. 2015;92(13):134108.
46. Almenningen A, Bastiansen O, Svendsas P. Electron diffraction studies of 2, 2′-dithienyl vapour. Acta Chem Scand. 1958;12:1671–4.
47. Casarini D, Lunazzi L, Mazzanti A. Recent advances in stereodynamics and conformational analysis by dynamic nmr and theoretical calculations. Eur J Org Chem. 2010;2010 (11):2035–56.
48. Kessler H. Detection of hindered rotation and inversion by NMR spectroscopy. Angew Chem Int Ed Engl. 1970;9(3):219–35.
49. Flack H, Bernardinelli G. The use of X-ray crystallography to determine absolute configuration. Chirality. 2008;20(5):681–90.

50. Liu P, Lanza TJ Jr, Chioda M, Jones C, Chobanian HR, Guo Y, Chang L, Kelly TM, Kan Y, Palyha O. Discovery of benzodiazepine sulfonamide-based bombesin receptor subtype 3 agonists and their unusual chirality. ACS Med Chem Lett. 2011;2(12):933–7.
51. Katritzky AR. Advances in heterocyclic chemistry, vol. 99. Oxford: Academic; 2009.
52. Minkin V, Zhdanov YA, Medyantzeva E, Ostroumov YA. The problem of acoplanarity of aromatic azomethines. Tetrahedron. 1967;23(9):3651–66.
53. Gilli P, Pretto L, Bertolasi V, Gilli G. Predicting hydrogen-bond strengths from acid−base molecular properties. The p K a slide rule: toward the solution of a long-lasting problem. Acc Chem Res. 2008;42(1):33–44.
54. Hansch C, Leo A, Taft R. A survey of Hammett substituent constants and resonance and field parameters. Chem Rev. 1991;91(2):165–95.
55. Ten Thije P, Janssen M. Hammett and Taft-Ingold relations in heterocyclic compounds. Recueil des Travaux Chimiques des Pays-Bas. 1965;84(9):1169–76.
56. Kapłon K, Demchuk O, Pietrusiewicz K. The DFT study on racemisation of atropisomeric biaryls. Curr Chem Lett. 2015;4(4):145–52.
57. Raut VS, Jean M, Vanthuyne N, Roussel C, Constantieux T, Bressy C, Bugaut X, Bonne D, Rodriguez J. Enantioselective syntheses of furan atropisomers by an oxidative central-to-axial chirality conversion strategy. J Am Chem Soc. 2017;139(6):2140–3.
58. Hu Y-L, Wang Z, Yang H, Chen J, Wu Z-B, Lei Y, Zhou L. Conversion of two stereocenters to one or two chiral axes: atroposelective synthesis of 2, 3-diarylbenzoindoles. Chem Sci. 2019;10:6777.
59. Trapp O. Interconversion of stereochemically labile enantiomers (enantiomerization). In: Schurig V, editor. Differentiation of enantiomers II. Berlin Heidelberg: Springer; 2013. p. 231–69.
60. Hughes A, Price D, Simpkins N. Atropisomeric amides: stereoselective enolate chemistry and enantioselective synthesis via a new SmI 2-mediated reduction. J Chem Soc Perkin Trans 1. 1999;10:1295–304.
61. Duan W-L, Imazaki Y, Shintani R, Hayashi T. Asymmetric construction of chiral C–N axes through rhodium-catalyzed 1, 4-addition. Tetrahedron. 2007;63(35):8529–36.
62. Curran DP, Hale GR, Geib SJ, Balog A, Cass QB, Degani ALG, Hernandes MZ, Freitas LCG. Rotational features of carbon-nitrogen bonds in axially chiral o-tert-butyl anilides and related molecules. Potential substrates for the 'prochiral auxiliary'approach to asymmetric synthesis. Tetrahedron Asymmetry. 1997;8(23):3955–75.
63. Lipshutz BH, Kayser F, Liu ZP. Asymmetric synthesis of biaryls by intramolecular oxidative couplings of cyanocuprate intermediates. Angew Chem Int Ed Engl. 1994;33(18):1842–4.
64. Lin GQ, Zhang JG, Cheng JF. Overview of chirality and chiral drugs. Chiral Drugs Chem Biol Action. 2011:14–8.
65. Calcaterra A, D'Acquarica I. The market of chiral drugs: chiral switches versus de novo enantiomerically pure compounds. J Pharm Biomed Anal. 2018;147:323–40.
66. Kazmierski WM, Danehower S, Duan M, Ferris RG, Elitzin V, Minick D, Sharp M, Stewart E, Villeneuve M. Biological and structural characterization of rotamers of C–C chemokine receptor type 5 (CCR5) inhibitor GSK214096. ACS Med Chem Lett. 2014;5(12):1296–9.

Chapter 5
Biased Agonism: Renewing GPCR's Targetability for the Drug Discovery

Ravinder Reddy Gaddam and Ajit Vikram

5.1 Introduction

GPCRs, also known as seven transmembrane domain receptors, make up the largest family of transmembrane receptors that transmit extracellular cues into intracellular responses. Over 30% of the Food and Drug Administration (FDA) approved drugs target at least one member of the GPCR gene family, and they still represent an attractive molecular target for drug development. Classically, the ligands for GPCRs are grouped as agonists, partial agonists, and antagonists depending on their ability to stabilize conformational changes in the GPCR that results in heterotrimeric G proteins coupling and activation of second messenger systems [1]. In the last couple of decades, several ligands were found to induce distinct effects through the same GPCRs [2–4]. The discovery that upon binding to GPCRs the β-arrestins (βarrs) mediates downstream signaling and different G-proteins can interact to a single GPCR and mediate distinct signaling originated the idea of biased agonism. A single GPCR can mediate multiple signaling pathways and a preferred activation of one specific signaling pathway over another by ligand can elicit a distinct response and such ligands are known as "biased agonists." In this chapter, we aim to systematically understand the conceptual framework of biased agonism, the experimental approach in determining it, and its potential in discovering new drugs (Fig. 5.1).

R. R. Gaddam · A. Vikram (✉)
Department of Internal Medicine, Carver College of Medicine University of Iowa, Iowa City, IA, USA
e-mail: ajit-vikram@uiowa.edu

R. Poduri (ed.), *Drug Discovery and Development*,
https://doi.org/10.1007/978-981-15-5534-3_5

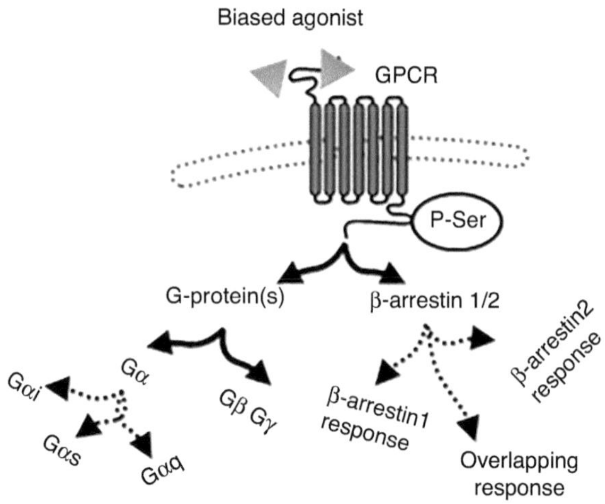

Fig. 5.1 A single GPCR can mediate multiple responses, and a preferred activation of one specific signaling response over others is known as "biased agonism"

5.2 The Classical Model of the GPCRs Function

The binding of agonist initiates coupling of GPCRs to heterotrimeric G proteins, which are formed by Gα, Gβ, and Gγ, and is followed by the generation of second messengers and downstream signaling. In humans, there are 16 Gα, 5 Gβ, and 13 Gγ subunits that can combine to form a wide range of heterotrimeric G proteins. Each Gα subunit can signal independently, whereas the Gβ subunits and Gγ subunits are obligate heterodimers that function as a single unit (Gβγ). The GPCRs undergo phosphorylation in the intracellular loop by the GRKs, which allows βarrs to bind to the GPCRs. The binding of βarrs sterically hinders further binding of G proteins and leads to receptor desensitization. The βarrs serve as an adaptor for the components of clathrin machinery to mediate receptor endocytosis. The C-tail of βarrs binds with the clathrin and adaptor protein 2 (AP2) to initiate internalization machinery. In the unbound/basal conformation of βarrs the C-tail is anchored to the N-domain and does not compete for the clathrin and AP2 binding. The binding of βarrs to GPCRs releases C-tail, and thereby it triggers the receptor internalization. The expression of separated arrestin C-tail carrying these sites inhibits GPCR internalization, apparently by winning the competition with the arrestin–receptor complexes for clathrin and AP2. [Please see Chap. 3 for more detailed information about the ligand-GPCR interaction and downstream signaling.]. In addition to desensitization and endocytosis, the βarrs also initiate distinct signaling pathways, and differential activation of G-protein and βarrs signaling by a ligand through a single GPCR forms the basis for the biased agonism.

5.3 Biased Agonism

The term biased agonism, originally introduced by Jarpe et al., describes the phenomenon that a ligand preferentially engages with a subset of the classical signaling pathways over others mediated by the same receptor [5]. This phenomenon is also referred to as functional selectivity, stimulus trafficking, biased inhibition, differential engagement, and ligand directed signaling. The molecular basis of biased agonism includes stabilization of a specific conformation of the GPCR by the ligand. Since different receptor conformations are likely to exhibit different affinities for G proteins and βarrs, the GPCRs can mediate distinct signaling [6]. Biased agonism is not only restricted to GPCRs but could also be applied to other receptors where a ligand-activated receptor can bind to at least two different signaling molecules such as steroid hormone receptors [7] and protease-activated receptors [8]. The biased agonism helped in explaining the paradoxical effects observed with some antagonists [e.g., losartan inhibits both maladaptive G protein response and beneficial βarrs signaling of angiotensin II type I receptor (AT1R) unlike biased agonists TRV027 that only inhibits maladaptive G protein response] [9]. The conceptual framework of biased agonism also renewed the opportunity of developing biased agonists as drug molecules aiming at increasing the benefit-to-adverse-effects ratio through already known GPCRs. The dedicated development of biased agonists has led to drug candidates, such as the G protein-biased μ opioid receptor agonist oliceridine and βarrs biased AT1R antagonist TRV027. A distinct function through a single GPCR can arise at multiple levels, such as differential binding of GPCRs to (**a**) G-proteins and (**b**) subtypes of βarrs. Interestingly, for some receptors, one signaling pathway is identified to mediate beneficial effects, whereas the other pathway appears to mediate the undesirable outcomes. Thus, based on the receptor system, it might be possible to develop biased agonists for one of these two pathways. For example, the G-protein response of apelin-receptor (APJ) is anticipated to rescue cardiac dysfunction, a concept that is being tested in human studies [10, 11], the βarrs signaling is believed to mediate maladaptive response and a decline in the efficacy of the APJ agonists. Another example is TRV027, a βarrs-biased agonist of AT1aR, which improves cardiac performance in animal models in addition to reducing the blood pressure similar to the receptor antagonists [12].

5.3.1 Coupling of GPCRs to Different G-Proteins

Historically, it was believed that the GPCRs couple to a specific type of G-protein to mediate its signaling, such as AT1R and α1-adrenergic receptors couple to Gαi while the β-adrenergic receptors couple to Gαs. The exceptions to this rule were reported and it was soon accepted that the binding of GPCRs to different G-proteins is the rule and not the exception. Receptors typically coupling to Gαq also couples to Gαi proteins or Gαs proteins. Similarly, the Gαi-coupled receptors to Gαq and typically Gαs-coupled receptors to the Gαi and Gαq. Some of the conventional GPCRs which are known to bind to different G proteins are summarized in Table 5.1.

Table 5.1 The GPCRs interaction with different G-proteins

S.No.	GPCRs	Gαs	Gαi	Gαq	References
1	AT1R		Y	Y	[37]
2	α1b	Y		Y	[38]
3	M2 muscarinic acetylcholine receptor		Y	Y	[39]
4	β2 and β3 adrenergic receptor	Y	Y	Y	[40, 41]

The G-proteins which typically couple with GPCRs are marked with underlined "Y" while those G-proteins to which these receptors are known to bind are marked with "Y". Y denotes "yes"

5.3.2 *Binding of GPCRs to Different Subtypes of βarrs*

The two isoforms of βarrs (βarrs1 and βarrs2) have high sequence homology and similar overall structure [13]. The βarrs preferentially bind to active (phosphorylated) GPCRs, and it was initially considered that the primary function of βarrs is to desensitize GPCRs [14]. However, later it was identified that it also binds to clathrin, clathrin adaptor protein 2(AP2), and small G protein ADP-ribosylation factor 6 (ARF6) proteins and regulates GPCR endocytosis [15, 16]. The discovery that the arrestins are ubiquitinated upon receptor binding and regulate ubiquitination of GPCRs revealed another mechanism [17], whereby arrestins regulate receptor trafficking indirectly. Both βarrs are recruited for most of the GPCRs and exhibit functional differences. For example, the absence of βarrs2, but not βarrs1, leads to an increase in the agonist-induced cAMP production by β2AR. As βarrs are not known to directly affect the cAMP level in cells, this suggests that the absence of βarrs2 delays β2AR internalization [18]. The divergence of βarrs specific response is evident in in vivo conditions as well, as the βarrs1 absence improves while the absence of βarrs2 impairs cardiac function in the mouse model of myocardial ischemia [19, 20]. Another example is the biased agonist for the parathyroid hormone 1 receptor (PTH1R) which promotes bone formation independent of the G protein activation [21]. (For more examples please see ref. [22]).

5.4 Measurement of Biased Agonism

The determination of biased agonism at GPCRs is difficult as it requires to separate the ligand-bias from the system bias. The general approach in determining the bias agonism involves identification of the physiologically relevant cell type, selection of assay, which generates comparable signal amplification for the different signaling pathways, and assessment of response in a time-dependent manner to ascertain that the bias persists across valid biological time scale [23]. At present, the pharmacological parameters such as E_{max}, EC_{50}, and reversal of rank order for agonists in signal specific assays (e.g., cAMP, Ca^{2+}, βarrs recruitment) are used to recognize the signaling bias. However, these methods do not account for the difference in the receptor reserve and signal amplification of the assays. A simple model for the determination of signaling bias would be to plot the βarrs activity on the *x*-axis and

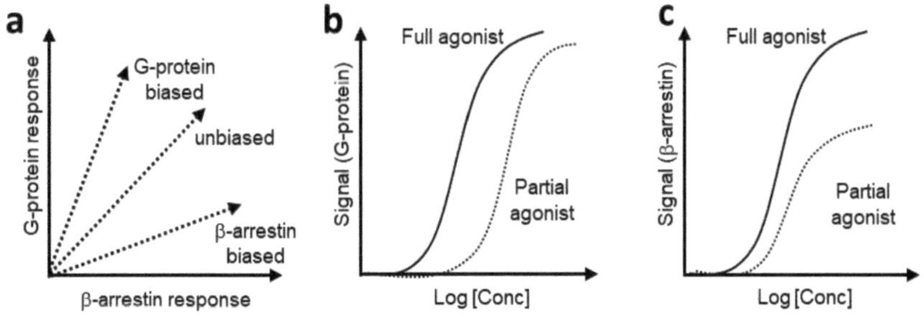

Equation 1

$$\text{Bias factor} = \text{Log}\left[\left\{\frac{E_{max\text{-}path1} \times EC_{50path2}}{E_{max\text{-}path2} \times EC_{50path1}}\right\}_{test} \Big/ \left\{\frac{E_{max\text{-}path1} \times EC_{50path2}}{E_{max\text{-}path2} \times EC_{50path1}}\right\}_{Reference}\right]$$

Equation 2

$$\text{Response} = \frac{E_{max} \times \tau^n \times [A]^n}{\tau^n \times [A]^n + ([A] + K_A)^n}$$

Equation 3

$$\text{Bias factor} = \Delta\Delta \text{ Log } (\tau/K_A)$$
$$= [\{(\log(\tau/K_A)_{test} - (\log(\tau/K_A)_{reference}\}_{pathway1}] - [\{(\log(\tau/K_A)_{test} - (\log(\tau/K_A)_{reference}\}_{pathway2}]$$

Fig. 5.2 Determining the signaling bias of the agonist. (**a**) A schematic showing that an unbiased agonist will have equal efficacy for pathway 1 (G-protein) as well as for the pathway 2 whereas the biased agonist's response will shift either in favor of G protein response or βarrs response. (**b and c**) In a system with high receptor reserve or assay which involves signal amplification, the full agonist as well as the partial agonist will reach maximal response (**b**), while in the system with no receptor reserve or assay which does not involve signal amplification, the partial agonist will not reach maximal response (**c**), and partial agonist may get classified as biased agonist

G-protein response on the *y*-axis. The unbiased agonist will have equal efficacy for βarrs and G-protein mediated responses and appear on a straight line (Fig. 5.2a). However, the G-protein biased agonist will appear above the midline while the βarrs biased agonists will appear below the midline (Fig. 5.2a). In assays that involve signal amplification, such as measurement of cAMP formation, both full and partial agonists can reach the maximal response. However, in assays with little amplification such as recruitment of βarrs to a GPCR by enzyme complementation, the partial agonists will have significantly lower maximal responses than the full agonists. In such cases, a partial agonist that reaches a maximal effect in one assay while reaches a half-maximal effect in another assay would be incorrectly classified as a biased agonist (Fig. 5.2b, c).

The bias factor of test ligand can be mathematically determined by comparing the E_{max} and EC_{50} derived from the Hill equation for the test ligand and that of the reference ligand (Eq. 1) [24–26]. Another method uses the Black and Leff operational model to quantify functional selectivity. As shown in the Eq. (2), the operational model defines the "τ" (coupling efficiency) and K_A (conditional affinity) as the ligand's (A) intrinsic efficacy and dissociation constant of agonist–receptor–signal transducer complex, respectively (n = transducer slope) [27]. The determination of the τ/K_A ratios (Eq. 3) quantify pharmacological responses regardless of signal amplification and receptor reserve, which is the key advantage of the ligand bias analyses through this method [27–29].

5.5 Possible Therapeutic Applications or Clinical Studies

The opioid receptor ligands are primarily used for their analgesic effects while they also have unwanted effects such as tolerance, respiratory depression, and constipation. The beneficial effect of the opioid receptor is mediated through the G-protein response while the unwanted effects are mediated through βarrs signaling. Thus, a biased agonist that favors G-protein response over βarrs signaling might offer a similar analgesic effect with much lower unwanted effects compared to the unbiased ligand. G protein-biased μ opioid receptor agonist oliceridine is in the clinical trials. In the Phase IIb trial oliceridine (TRV130) was found to be effective in producing rapid analgesia in patients with postoperative pain, with an acceptable safety profile [30, 31]. Similarly, the apelin-receptor's (APJ's) beneficial effect against nitric-oxide production and the cardiac contractile effect is mediated through the G-protein dependent response while the βarrs signaling mediates receptor internalization and cardiac hypertrophic effects. Thus, the G-protein biased agonist is believed to improve endothelial function and cardiac function with minimal unwanted effects, a concept that is being tested in human studies [10, 11]. In a study involving human volunteers, MM07, a G-protein biased agonist of APJ produced a dose-dependent increase in forearm blood flow and repeated doses of MM07 produced reproducible effects [32]. The carvedilol acts as an antagonist for the G-protein pathway and partial agonist for the βarrs pathway of β-adrenergic receptor and is potentially beneficial against heart failure [33]. The anandamide, an agonist for the cannabinoid type 1 receptor (CB1R), favors Gαi pathway over Gαs and βarrs pathway and it might be useful against Huntington's disease [3, 4]. The LY2828360 favors Gαi pathway over Gαs and βarrs pathway of cannabinoid type 2 receptor and might offer better analgesia with less tolerance [34]. The G-protein response of dopamine receptor is beneficial in the Parkinson's disease despite its decreased CNS availability and desensitization which are major challenges in the Parkinson's disease treatment. The PF-8294 and PF-6142 are the biased agonists for dopamine type 1 receptor and they favor G-protein signaling over βarrs signaling and might be useful in achieving antiparkinson's effect with less receptor desensitization [35]. The cholinergic muscarinic receptor M2 ligand 77-LH-28-1 is biased toward G-protein response over βarrs response and might have a pro-cognitive effects [36]. Many other examples of the potential clinical use of biased agonists are summarized in Table 5.2.

Table 5.2 The biased agonists for the selected GPCRs and their pharmacological effects (anticipated or experimentally determined)

Receptor	Subtype	Biased ligand	Signaling bias	Anticipated/experimentally determined pharmacological effects	References
Adenosine receptor	A1	VCP746 and Capadenson	Favors Gαi/o response over G-protein independent response.	Protection against ischemia without causing bradycardia	[2]
Adrenergic receptor	β_1/β_2	Bisoprolol and metoprolol	Favors Gαi/o over Gαs	Unknown	[36]
		Carvedilol	Antagonist for G-protein pathway but partial agonist for βarrs pathway	Potential beneficial effects against heart failure	[33]
		Salmeterol	Antagonist for βarrs signaling but full agonist for G-protein signaling	Maintenance of chronic obstructive pulmonary disease (COPD) symptoms	[42]
	β_3	CL316243 and SR59230A	Biased toward cAMP accumulation or p38MAPK activation	Cardioprotective effects	[43]
Apelin receptor (APJ)		MM07	G-protein biased over βarrs signaling	Improvement in cardiac contractility with reduced risk of hypertrophy	[32, 44]
Cannabinoid receptor	CB_1-R	Anandamide	Favors Gi pathway over Gαs and βarrs pathway	Gαi/o biased activity may be useful in Huntington's disease	[3, 4]
	CB_2-R	LY2828360	Favors Gαi pathway over Gs and βarrs pathway	Analgesia with less potential for tolerance	[34]
Cholinergic muscarinic receptor	M1	VCP794	Biased toward ERK1/2 pathway over Gαi/o activation	Pro-cognitive effects	[45, 46]
	M2	77-LH-28-1	Biased toward G-protein response over βarrs pathway	Pro-cognitive effects	[36]
	M3	Pilocarpine	Biased toward βarrs and pERK1/2, but do not induce Insulin secretion	Unknown	[47, 48]

(continued)

Table 5.2 (continued)

Receptor	Subtype	Biased ligand	Signaling bias	Anticipated/experimentally determined pharmacological effects	References
Chemokine receptor	CXCR3	FAUC1036	G-protein biased over βarrs pathway	Biased ligands may display different effects in chemotaxis and receptor internalization	[49, 50]
		FAUC1104	βarrs biased over G-protein pathway		
Dopamine receptor	D_1 R	PF-8294 and PF-6142	G-protein biased over βarrs signaling	G-protein biased agonists may show antiparkinsonian properties without desensitization	[35]
	D_2 R	BRD5814	Selective antagonistic activity in βarrs pathway	Selective antagonism of β-arrestin pathway may provide antipsychotic effect without extrapyramidal side effects	[51]
Glucagon like Peptide-1 receptor (GLP-1R)		EXENDIN-P5	G-protein biased over βarrs signaling	Weak insulin secretagogue but effective in long-term glycemic control	[36]
5-Hydroxytryptamine receptor (5-HT-R)	5-HT_1-R	Aryl piperazine derivative	G-protein biased over βarrs signaling	Unknown	[52]
	5-HT_2-R	LSD, bromocriptine and pergolide	Biased toward Gq/11over calcium release	Gαq/11 biased signaling may be associated with hallucinogenic activity	[53, 54]
Opioid receptor (OP-R)	μ-OP R	TRV130, and PTI-609	G-protein biased over βarrs signaling	The Gi biased agonists induced analgesia with less potential for tolerance, respiratory suppression and constipation	[55, 36]
	K-OP R	Nalfurafin and 6′-GNTI display	G-protein biased over βarrs signaling	Might have less dysphoric effects	[56, 57]
Parathyroid hormone receptor type 1 (PTHR1)		Backbone modified PTH and abaloparatide	Favors cAMP production from surface receptor instead of internalization	Unknown	[58]

5.6 Conclusions

A single GPCR elicits functionally distinct signalings, and the biased agonist preferentially activates one over another. There are multiple examples where one signaling pathway of a GPCR is identified to mediate physiological effects, whereas the other pathway appears to mediate the pathological effects. A biased agonist acting as an agonist for one pathway while not activating the other pathway promises to increase the benefit-to-risk ratio in the therapeutic management of health disorders. For example, TRV130, a G-protein biased agonist of μ-opioid receptor, is approved for the pain management and has comparable analgesic effects with wider therapeutic window compared to morphine. The discovery of biased agonist is an active area of research and holds great promise to maximize the benefit-to-risk ratio for different disease conditions.

References

1. Wisler JW, Rockman HA, Lefkowitz RJ. Biased G protein-coupled receptor signaling: changing the paradigm of drug discovery. Circulation. 2018;137(22):2315–7.
2. Valant C, May LT, Aurelio L, Chuo CH, White PJ, Baltos JA, et al. Separation of on-target efficacy from adverse effects through rational design of a bitopic adenosine receptor agonist. Proc Natl Acad Sci U S A. 2014;111(12):4614–9.
3. Laprairie RB, Bagher AM, Kelly ME, Denovan-Wright EM. Biased type 1 cannabinoid receptor signaling influences neuronal viability in a cell culture model of Huntington disease. Mol Pharmacol. 2016;89(3):364–75.
4. Bonhaus DW, Chang LK, Kwan J, Martin GR. Dual activation and inhibition of adenylyl cyclase by cannabinoid receptor agonists: evidence for agonist-specific trafficking of intracellular responses. J Pharmacol Exp Ther. 1998;287(3):884–8.
5. Jarpe MB, Knall C, Mitchell FM, Buhl AM, Duzic E, Johnson GL. [D-Arg1,D-Phe5,D-Trp7,9, Leu11]Substance P acts as a biased agonist toward neuropeptide and chemokine receptors. J Biol Chem. 1998;273(5):3097–104.
6. Ghosh E, Dwivedi H, Baidya M, Srivastava A, Kumari P, Stepniewski T, et al. Conformational sensors and domain swapping reveal structural and functional differences between beta-arrestin isoforms. Cell Rep. 2019;28(13):3287–99. e6
7. Keenan CR, Lew MJ, Stewart AG. Biased signalling from the glucocorticoid receptor: renewed opportunity for tailoring glucocorticoid activity. Biochem Pharmacol. 2016;112:6–12.
8. Russo A, Soh UJ, Trejo J. Proteases display biased agonism at protease-activated receptors: location matters! Mol Interv. 2009;9(2):87–96.
9. Ikeda Y, Kumagai H, Motozawa Y, Suzuki J, Komuro I. Biased agonism of the angiotensin II type I receptor. Int Heart J. 2015;56(5):485–8.
10. Japp AG, Cruden NL, Barnes G, van Gemeren N, Mathews J, Adamson J, et al. Acute cardiovascular effects of apelin in humans: potential role in patients with chronic heart failure. Circulation. 2010;121(16):1818–27.
11. Barnes GD, Alam S, Carter G, Pedersen CM, Lee KM, Hubbard TJ, et al. Sustained cardiovascular actions of APJ agonism during renin-angiotensin system activation and in patients with heart failure. Circ Heart Fail. 2013;6(3):482–91.

12. Violin JD, DeWire SM, Yamashita D, Rominger DH, Nguyen L, Schiller K, et al. Selectively engaging beta-arrestins at the angiotensin II type 1 receptor reduces blood pressure and increases cardiac performance. J Pharmacol Exp Ther. 2010;335(3):572–9.
13. Gurevich VV, Gurevich EV. The molecular acrobatics of arrestin activation. Trends Pharmacol Sci. 2004;25(2):105–11.
14. Kuhn H, Hall SW, Wilden U. Light-induced binding of 48-kDa protein to photoreceptor membranes is highly enhanced by phosphorylation of rhodopsin. FEBS Lett. 1984;176 (2):473–8.
15. Claing A, Chen W, Miller WE, Vitale N, Moss J, Premont RT, et al. Beta-arrestin-mediated ADP-ribosylation factor 6 activation and beta 2-adrenergic receptor endocytosis. J Biol Chem. 2001;276(45):42,509–13.
16. Goodman OB Jr, Krupnick JG, Santini F, Gurevich VV, Penn RB, Gagnon AW, et al. Beta-arrestin acts as a clathrin adaptor in endocytosis of the beta2-adrenergic receptor. Nature. 1996;383(6599):447–50.
17. Shenoy SK, McDonald PH, Kohout TA, Lefkowitz RJ. Regulation of receptor fate by ubiquitination of activated beta 2-adrenergic receptor and beta-arrestin. Science. 2001;294 (5545):1307–13.
18. Ahn S, Nelson CD, Garrison TR, Miller WE, Lefkowitz RJ. Desensitization, internalization, and signaling functions of beta-arrestins demonstrated by RNA interference. Proc Natl Acad Sci U S A. 2003;100(4):1740–4.
19. Bathgate-Siryk A, Dabul S, Pandya K, Walklett K, Rengo G, Cannavo A, et al. Negative impact of beta-arrestin-1 on post-myocardial infarction heart failure via cardiac and adrenal-dependent neurohormonal mechanisms. Hypertension. 2014;63(2):404–12.
20. Watari K, Nakaya M, Nishida M, Kim KM, Kurose H. Beta-arrestin2 in infiltrated macrophages inhibits excessive inflammation after myocardial infarction. PLoS One. 2013;8(7):e68351.
21. Gesty-Palmer D, Flannery P, Yuan L, Corsino L, Spurney R, Lefkowitz RJ, et al. A beta-arrestin-biased agonist of the parathyroid hormone receptor (PTH1R) promotes bone formation independent of G protein activation. Sci Transl Med. 2009;1(1):1ra.
22. Srivastava A, Gupta B, Gupta C, Shukla AK. Emerging functional divergence of beta-arrestin isoforms in GPCR function. Trends Endocrinol Metab. 2015;26(11):628–42.
23. Gundry J, Glenn R, Alagesan P, Rajagopal S. A practical guide to approaching biased agonism at G protein coupled receptors. Front Neurosci. 2017;11:17.
24. Ehlert FJ. On the analysis of ligand-directed signaling at G protein-coupled receptors. Naunyn Schmiedeberg's Arch Pharmacol. 2008;377(4–6):549–77.
25. Griffin MT, Figueroa KW, Liller S, Ehlert FJ. Estimation of agonist activity at G protein-coupled receptors: analysis of M2 muscarinic receptor signaling through Gi/o,Gs, and G15. J Pharmacol Exp Ther. 2007;321(3):1193–207.
26. Rankovic Z, Brust TF, Bohn LM. Biased agonism: an emerging paradigm in GPCR drug discovery. Bioorg Med Chem Lett. 2016;26(2):241–50.
27. Black JW, Leff P. Operational models of pharmacological agonism. Proc R Soc Lond B Biol Sci. 1983;220(1219):141–62.
28. Kenakin T, Christopoulos A. Signalling bias in new drug discovery: detection, quantification and therapeutic impact. Nat Rev Drug Discov. 2013;12(3):205–16.
29. Kenakin T, Watson C, Muniz-Medina V, Christopoulos A, Novick S. A simple method for quantifying functional selectivity and agonist bias. ACS Chem Neurosci. 2012;3(3):193–203.
30. Singla N, Minkowitz HS, Soergel DG, Burt DA, Subach RA, Salamea MY, et al. A randomized, phase IIb study investigating oliceridine (TRV130), a novel micro-receptor G-protein pathway selective (mu-GPS) modulator, for the management of moderate to severe acute pain following abdominoplasty. J Pain Res. 2017;10:2413–24.
31. Soergel DG, Subach RA, Burnham N, Lark MW, James IE, Sadler BM, et al. Biased agonism of the mu-opioid receptor by TRV130 increases analgesia and reduces on-target adverse effects versus morphine: a randomized, double-blind, placebo-controlled, crossover study in healthy volunteers. Pain. 2014;155(9):1829–35.

32. Brame AL, Maguire JJ, Yang P, Dyson A, Torella R, Cheriyan J, et al. Design, characterization, and first-in-human study of the vascular actions of a novel biased apelin receptor agonist. Hypertension. 2015;65(4):834–40.
33. Wisler JW, DeWire SM, Whalen EJ, Violin JD, Drake MT, Ahn S, et al. A unique mechanism of beta-blocker action: carvedilol stimulates beta-arrestin signaling. Proc Natl Acad Sci U S A. 2007;104(42):16,657–62.
34. Lin X, Dhopeshwarkar AS, Huibregtse M, Mackie K, Hohmann AG. Slowly signaling G protein-biased CB2 cannabinoid receptor agonist LY2828360 suppresses neuropathic pain with sustained efficacy and attenuates morphine tolerance and dependence. Mol Pharmacol. 2018;93(2):49–62.
35. Gray DL, Allen JA, Mente S, O'Connor RE, DeMarco GJ, Efremov I, et al. Impaired beta-arrestin recruitment and reduced desensitization by non-catechol agonists of the D1 dopamine receptor. Nat Commun. 2018;9(1):674.
36. Seyedabadi M, Ghahremani MH, Albert PR. Biased signaling of G protein coupled receptors (GPCRs): molecular determinants of GPCR/transducer selectivity and therapeutic potential. Pharmacol Ther. 2019;200:148–78.
37. Crawford KW, Frey EA, Cote TE. Angiotensin II receptor recognized by DuP753 regulates two distinct guanine nucleotide-binding protein signaling pathways. Mol Pharmacol. 1992;41 (1):154–62.
38. Horie K, Itoh H, Tsujimoto G. Hamster alpha 1B-adrenergic receptor directly activates Gs in the transfected Chinese hamster ovary cells. Mol Pharmacol. 1995;48(3):392–400.
39. Schmidt M, Bienek C, van Koppen CJ, Michel MC, Jakobs KH. Differential calcium signalling by m2 and m3 muscarinic acetylcholine receptors in a single cell type. Naunyn Schmiedeberg's Arch Pharmacol. 1995;352(5):469–76.
40. Cao W, Luttrell LM, Medvedev AV, Pierce KL, Daniel KW, Dixon TM, et al. Direct binding of activated c-Src to the beta 3-adrenergic receptor is required for MAP kinase activation. J Biol Chem. 2000;275(49):38131–4.
41. Wenzel-Seifert K, Seifert R. Molecular analysis of beta(2)-adrenoceptor coupling to G(s)-, G(i)-, and G(q)-proteins. Mol Pharmacol. 2000;58(5):954–66.
42. Carter AA, Hill SJ. Characterization of isoprenaline- and salmeterol-stimulated interactions between beta2-adrenoceptors and beta-arrestin 2 using beta-galactosidase complementation in C2C12 cells. J Pharmacol Exp Ther. 2005;315(2):839–48.
43. Sato M, Horinouchi T, Hutchinson DS, Evans BA, Summers RJ. Ligand-directed signaling at the beta3-adrenoceptor produced by 3-(2-ethylphenoxy)-1-[(1,S)-1,2,3,4-tetrahydronapth-1-ylamino]-2S-2-propanol oxalate (SR59230A) relative to receptor agonists. Mol Pharmacol. 2007;72(5):1359–68.
44. Read C, Fitzpatrick CM, Yang P, Kuc RE, Maguire JJ, Glen RC, et al. Cardiac action of the first G protein biased small molecule apelin agonist. Biochem Pharmacol. 2016;116:63–72.
45. Keov P, Lopez L, Devine SM, Valant C, Lane JR, Scammells PJ, et al. Molecular mechanisms of bitopic ligand engagement with the M1 muscarinic acetylcholine receptor. J Biol Chem. 2014;289(34):23,817–37.
46. Keov P, Valant C, Devine SM, Lane JR, Scammells PJ, Sexton PM, et al. Reverse engineering of the selective agonist TBPB unveils both orthosteric and allosteric modes of action at the M (1) muscarinic acetylcholine receptor. Mol Pharmacol. 2013;84(3):425–37.
47. Pronin AN, Wang Q, Slepak VZ. Teaching an old drug new tricks: agonism, antagonism, and biased signaling of pilocarpine through M3 muscarinic acetylcholine receptor. Mol Pharmacol. 2017;92(5):601–12.
48. Stewart GD, Sexton PM, Christopoulos A. Detection of novel functional selectivity at M3 muscarinic acetylcholine receptors using a Saccharomyces cerevisiae platform. ACS Chem Biol. 2010;5(4):365–75.
49. Milanos L, Brox R, Frank T, Poklukar G, Palmisano R, Waibel R, et al. Discovery and characterization of biased allosteric agonists of the chemokine receptor CXCR3. J Med Chem. 2016;59(5):2222–43.

50. Quoyer J, Janz JM, Luo J, Ren Y, Armando S, Lukashova V, et al. Pepducin targeting the C-X-C chemokine receptor type 4 acts as a biased agonist favoring activation of the inhibitory G protein. Proc Natl Acad Sci U S A. 2013;110(52):E5088–97.
51. Weiwer M, Xu Q, Gale JP, Lewis M, Campbell AJ, Schroeder FA, et al. Functionally biased D2R antagonists: targeting the beta-arrestin pathway to improve antipsychotic treatment. ACS Chem Biol. 2018;13(4):1038–47.
52. Stroth N, Niso M, Colabufo NA, Perrone R, Svenningsson P, Lacivita E, et al. Arylpiperazine agonists of the serotonin 5-HT1A receptor preferentially activate cAMP signaling versus recruitment of beta-arrestin-2. Bioorg Med Chem. 2015;23(15):4824–30.
53. Cussac D, Boutet-Robinet E, Ailhaud MC, Newman-Tancredi A, Martel JC, Danty N, et al. Agonist-directed trafficking of signalling at serotonin 5-HT2A, 5-HT2B and 5-HT2C-VSV receptors mediated Gq/11 activation and calcium mobilisation in CHO cells. Eur J Pharmacol. 2008;594(1–3):32–8.
54. Wacker D, Wang C, Katritch V, Han GW, Huang XP, Vardy E, et al. Structural features for functional selectivity at serotonin receptors. Science. 2013;340(6132):615–9.
55. Viscusi ER, Skobieranda F, Soergel DG, Cook E, Burt DA, Singla N. APOLLO-1: a randomized placebo and active-controlled phase III study investigating oliceridine (TRV130), a G protein-biased ligand at the micro-opioid receptor, for management of moderate-to-severe acute pain following bunionectomy. J Pain Res. 2019;12:927–43.
56. Schattauer SS, Kuhar JR, Song A, Chavkin C. Nalfurafine is a G-protein biased agonist having significantly greater bias at the human than rodent form of the kappa opioid receptor. Cell Signal. 2017;32:59–65.
57. Schmid CL, Streicher JM, Groer CE, Munro TA, Zhou L, Bohn LM. Functional selectivity of 6′-guanidinonaltrindole (6′-GNTI) at kappa-opioid receptors in striatal neurons. J Biol Chem. 2013;288(31):22,387–98.
58. Liu S, Jean-Alphonse FG, White AD, Wootten D, Sexton PM, Gardella TJ, et al. Use of backbone modification to enlarge the spatiotemporal diversity of parathyroid hormone receptor-1 signaling via biased agonism. J Am Chem Soc. 2019;141(37):14,486–90.

Chapter 6
Computer-Aided Drug Design

Prasad V. Bharatam

6.1 Introduction

Thinking in three dimensions is an important and essential requirement for a successful Drug Discovery Scientist [1]. This is a difficult proposition; however, it has been made easy by computational methods. From this chapter, we learn many techniques of Computer-Aided Drug Design (CADD) which help the drug discovery scientist to " think in three dimensions". There are several synonyms for this term CADD—for example, Computer-Aided Drug Discovery, in silico drug discovery, and pharmacoinformatics. Several overlapping terms are also available for CADD—for example, Structure-Based Drug Design (SBDD), Ligand-Based Drug Design (LBDD), and Fragment-Based Drug Design (FBDD)—which employ computers to analyze the macromolecular, small molecular structures and their interactions, to design drugs. In addition, topics like chemoinformatics and bioinformatics are being extensively employed for CADD purpose. Several subtopics which fall under CADD are already sufficiently large in scope, for example, QSAR (Quantitative Structure Activity Relationship). Molecular modelling is a computer-aided method of modelling chemicals/biochemical, which found applications in material science in addition to drug design. Similarly, artificial intelligence is a topic which found extensive applications in Drug Discovery but the scope of which is huge (beyond drug design). In this chapter, all the above topics are considered equivalent wherever they are found suitable for the discovery of drugs/leads/hits.

CADD is associated with many concept-based methods. It offers many clues to the drug discovery teams, such that they can take informed decisions. It should not be considered as a platform technology. CADD helps in increasing rationality in

P. V. Bharatam (✉)
Department of Medicinal Chemistry, National Institute of Pharmaceutical Education and Research (NIPER), Mohali, Punjab, India
e-mail: pvbharatam@niper.ac.in

R. Poduri (ed.), *Drug Discovery and Development*,
https://doi.org/10.1007/978-981-15-5534-3_6

drug discovery process. Science and technology are equally important in CADD. All the efforts during drug discovery which can be expressed in mathematical terminology and which can be analyzed using numerical methods can be subjected to CADD. Many atomic-level details about small and macromolecules cannot be obtained using experimental methods but can be acquired using CADD methods. These methods employ data and produce more data, in the process they offer many hints. It is the job of the scientist to interpret the information obtained from the CADD towards the required goal; thus CADD is an "art in science." CADD methods should be used as complimentary techniques to the experimental methods during drug discovery, to obtain viable results [2–10].

CADD method development involves extensive computer programming and use of mathematics as well as physics towards understanding the interplay of chemistry and biology of molecules. Quantitative Structure Activity Relationship (QSAR) methods employ statistical methods extensively. Quantum chemical methods which provide electronic level details of the drugs employ calculus and trigonometry to solve Schrodinger wave equation, which is the most fundamental topic in physics. Molecular mechanics (MM) methods are based on classical physics. Pharmacophore mapping is based on the mathematics of mapping. Molecular dynamics utilizes statistical thermodynamics as its background. Artificial intelligence employs many modern concepts of computer science, including machine learning. Homology modelling is based on comparative analysis. Virtual screening technology utilizes all the methods of CADD along with a set of empirical rules. Since all the CADD methods are designed to be user friendly, the scientists with broad-level knowledge in basic sciences can also efficiently use these methods. Each of these CADD methods may appear to be simple technologies, but it is important to appreciate the fundamental science behind the technology to be able to make best use of the same. A word of caution is warranted—CADD methods do not provide YES or NO answers, they only provide improved clarity towards the rational design of drugs.

This chapter is divided into many sections; each section deals with a specific CADD method. In each of these sections, (i) the scientific concept behind the method has been explained, (ii) the computational strategies utilized in converting the scientific concepts into useful tools have been provided, (iii) current scope and limits of the methods have been discussed, (iv) the best way of using the method has been included, and (v) a case study has been incorporated.

6.2 Three-Dimensional Structures of Drugs and Macromolecules

Structure is the most important characteristic feature of any molecule (small/large). Every chemical species is characterized by a 3D structure; this was proposed during 1850s by Kekule and others. Valency of the elements has been considered as a guiding principle to determine 3D structure. Generally, while writing, 2D structures are employed; however, 3D structures represent the reality. While representing in 3D, each atom in a drug molecule can be defined with the help of Cartesian coordinates. Thus, all atoms in a drug can be defined in terms of a coordinate matrix. Similarly, each macromolecule (peptides, proteins, enzymes, nucleic acids, antibodies, transporters, lipid bilayers defining the cell walls, etc.) contains many atoms and thus the 3D structures of the macromolecules can also be defined in terms of coordinate matrices.

This 3D definition of the drugs and macromolecules is convenient to display the structures on a computer screen. The coordinate matrix helps in many ways—(i) to define the interaction between drugs and macromolecules, (ii) to estimate the energies of the drugs, macromolecules, and their complexes, (iii) to visualize the hydrogen bonds, hydrophobic interactions, short contacts, α-helices, and β-sheets of proteins, (iv) to examine the protein–protein, protein–DNA, and receptor–cell wall contacts, (v) to explore the path traversed by drugs while entering the active site of receptors/while navigating through the transporters, or channels, (vi) to explore the changes in the position of atoms in molecules during a dynamical process, (vii) to explore the reaction pathways under biochemical conditions, (viii) to understand the conformational changes in the drugs as well as macromolecules which are responsible for the therapeutic action, (ix) to explore mutations in the macromolecules which are responsible for drug resistance, and (x) to rationally design new drugs/leads/hits.

The advantages of computational 3D models of molecules include—(i) represent in many ways (ball & stick/wireframe/CPK models, etc.) to get clarity, (ii) zoom to a comfortable size, (iii) rotate in three dimensions, (iv) spin (if necessary), (v) ribbon representation of macromolecules, (vi) position the small vs. large molecules in perspective to each other, (vii) convert the 3D structures into Cartesian matrices and then submit the data for CADD analysis, and many more [11].

Experimentally, the 3D structures of the drugs can be obtained using X-ray diffraction methods. The experimentally solved 3D structures of a few drugs are available from CSD (the Cambridge Crystallographic Structural Database). The 3D structures of macromolecules can also be obtained using X-ray diffraction, if the crystals of the macromolecules are available. Such structures are stored in PDB (Protein Data Bank). The 3D structures stored in CSD and PDB are in the form of Cartesian coordinates. The 3D structures of the macromolecules help in defining the position of each atom and help in realizing (i) the number and orientation of water molecules embedded inside, (ii) active sites in which the drugs can be conveniently accommodated, (iii) paths through which the drugs can reach the active site, (vi) grooves in which substrate enzymes or DNA can interact, (v) the location of cofactors or coenzymes, (vi) to explore the proximity between the various reaction centers (e.g., ATP-binding site vs. the phosphorylation site in kinases), and many more studies in CADD.

The 3D structure of the drug molecule helps in understanding its shape, volume, and surface properties. The active sites of macromolecules are also characterized by shape, volume, and surface properties which can be estimated with the help of the 3D structure of macromolecules. The complementarity between the active site and the

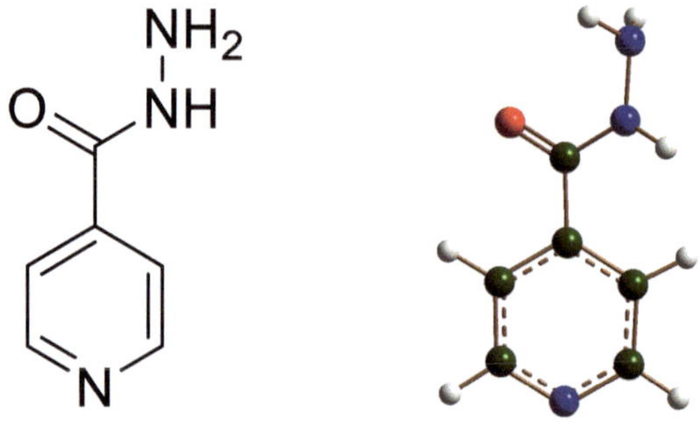

Fig. 6.1 2D and 3D structures of Isoniazid. The 3D structure provides a lot more information in comparison to 2D structure. For example, the pyramidalization at the nitrogen atoms of the –$CONHNH_2$ group are not clear from the 2D structure. The interatomic distance information can also be easily obtained from the 3D structure. This figure also indicates that 3D structures are little more complicated than the 2D structures. It is important to pay attention to all the extra facts available in 3D structures as these represent more realistic picture of drugs (or macromolecules). Every successful drug discovery scientist pays attention to 3D structure of small and macromolecules [1]

Table 6.1 List of Cartesian coordinates of the drug Isoniazid. Position of each and every atom can be defined using this Cartesian coordinate matrix. This forms the basic information behind any CADD software. To estimate energies also this information is required because energy of a molecule is a function its coordinates

	Coordinates (Angstroms)		
Atomic number	*X*	*Y*	*Z*
C	−0.2423	0.1005	0.0430
C	−0.8059	−1.1597	0.2038
C	−2.1851	−1.2861	0.1445
C	−2.4535	0.9646	−0.2085
C	−1.0877	1.1842	−0.1558
H	−0.2062	−2.0254	0.4015
H	−2.6577	−2.2379	0.2703
H	−3.1377	1.7715	−0.3675
H	−0.6792	2.1678	−0.2591
N	−2.9888	−0.2470	−0.0637
C	1.2271	0.3539	0.0990
O	1.6740	1.4539	0.4136
N	2.0404	−0.6759	−0.230
H	1.6936	−1.5197	−0.612
N	3.4263	−0.5645	−0.196
H	3.7618	0.1594	−0.7953
H	3.7737	−0.4789	0.7336

drug can be determined and visualized with the help of computationally defined structures of these species. This helps in exploring whether a small molecule shall interact with a macromolecule or not. Figure 6.1 shows the 2D and 3D structures of Isoniazid (an anti-TB drug) and Table 6.1 lists its Cartesian coordinates.

6.3 Energy of Drugs and Macromolecules

Every chemical/biological entity carries energy; this is another important characteristic feature of drugs as well as macromolecules. Molecular energy is the energy possessed by the molecules; thus, it is the potential energy of the molecules. It is a unique property of any molecule; hence, it is the absolute energy of molecules. This factor was extensively explored, in the form of chemical thermodynamics, during second half of nineteenth century, mainly by Willard Gibbs. Energies of two molecules become comparable (and relative energy can be estimated) only if the molecular formulae are same. Energy of a chemical reaction (Gibbs free energy) becomes relevant only when the chemical equation is balanced. Energy is released when a drug molecule is formed from its elements; it is known as the absolute energy, which is due to the formation of bonds.

Structure of a molecule (small or macro) determines its energy. Structural changes cause energy changes; alternatively, energy changes drive structural changes. When a drug and a receptor interact with each other, energy gets released; this is energy due to association, often referred as binding energy due to drug–receptor complex formation. As a drug molecule approaches the macromolecule, conformational changes happen in drugs as well as in macromolecules; this leads to a change in the energy (mostly increase in energy). When the two interact with each other, after undergoing induced fit (conformational changes to facilitate improved interactions), energy gets released. The interplay of structural changes and the energy changes determine the strength as well as the life of an association between drug and receptor. This also leads to additional conformational and energy changes, via cascade of events which drive the biochemical processes happening in human body. Exploring this manifestation in terms of structural changes and energy changes is practical using computational methods, and thus this became an essential component of CADD.

The absolute energy/relative energy/reaction energy/binding energy can be experimentally estimated using calorimetric methods. One such important calorimetric method useful in drug discovery process is Isothermal Titration Calorimetry (ITC). Such experimental methods can be applied only in a few cases, in practice. Alternatively, computational methods became available to estimate the energy and its changes as described in the next few sections. Absolute and relative energy values of four important molecules are given in Fig. 6.2.

6.4 Electronic Structure of Drugs

Any molecule, whether small or large, is a colony of electrons and nuclei. The electrons are distributed inside the molecule in various molecular orbitals (MOs). This internal distribution of electrons is responsible for the characteristics of the molecules. These details are known as electronic structure, which can be explored

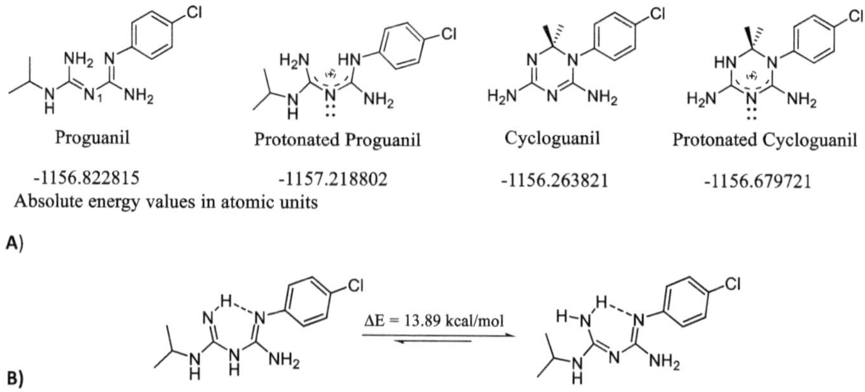

Fig. 6.2 (**a**) Structure of Proguanil, its protonated state, structure of Cycloguanil, its protonated state. Absolute energy values were estimated using HF/6-31G* level. (**b**) Relative energy (ΔE) value between two tautomers of Proguanil is the energy difference between them. Proton affinity value in Proguanil (~248 kcal/mol) is the energy difference between most stable tautomer and the protonated state of Proguanil. Intramolecular hydrogen bonds are also shown in (**b**)

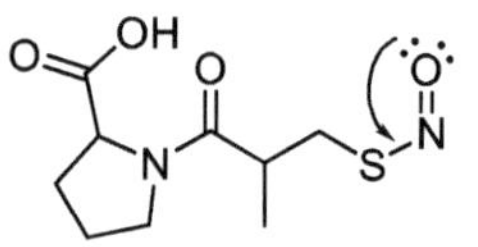

Fig. 6.3 Structure of nitrosocaptopril showing the negative hyperconjugation. The lone pair of electrons from oxygen atom donate electron density to the σ^*_{S-N} bond, making the S–N bond weaker, facilitating the release of NO gas

using quantum chemical methods. The nucleophilicity or electrophilicity of drugs is attributable to this electronic distribution. For example, the prodrug Proguanil is highly basic because the HOMO (highest occupied MO) of this species is the lone pair type MO at the N1 center (Fig. 6.2) [12]. The reactivity of any drug metabolite is also due to its internal electronic distribution. For example, the inhibitory properties of Cycloguanil are due to the presence of two lone pairs of electrons at the central nitrogen of the protonated-Cycloguanil (Fig. 6.2), which are pointed towards the NADPH cofactor in the active site of *pf*DHFR (*plasmodium falciparum* Dihydrofolate Reductase) and prevent the transfer of hydride ion from NADPH. The susceptibility of an atom/a bond in a drug towards reactions (like metabolic/toxic reaction) is due to the electronic characteristics of the drug. For example, in the molecule Nitrosocaptopril, the S-N bond requires only about 28 kcal/mol to break because it suffers from negative hyperconjugation due to the transfer of electrons from the oxygen atom to the adjacent S–N σ* MO, making it the weakest [13]. Because of this small S–N bond strength, Nitrosocaptopril can be used as a vasodilator as well as an antihypertensive agent (Fig. 6.3).

The catalytic reactivity of an enzyme towards the production of another species is due to its electronic structure. For example, the catalytic center of the metabolizing

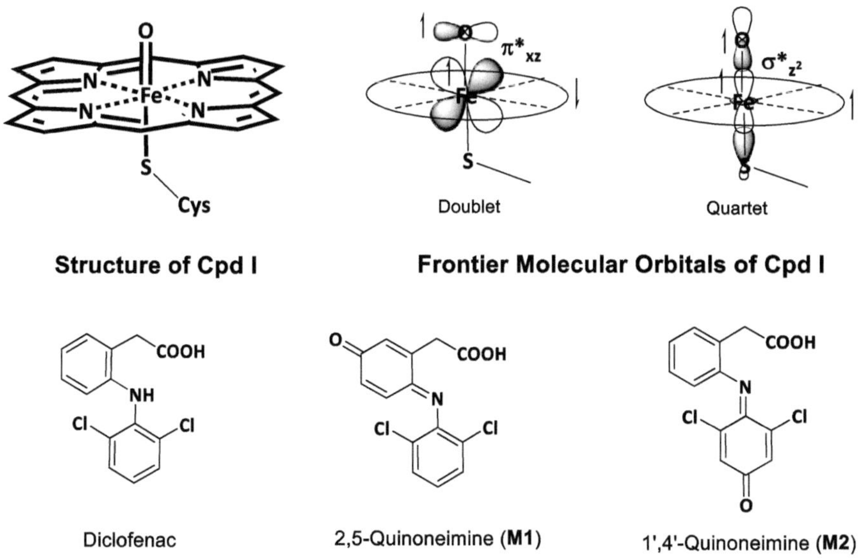

Fig. 6.4 The structure of the active center of cytochrome P450 (**Cpd-I**). It has iron atom in the oxidation state IV and the complex is a radical species. It can oxidize Diclofenac to give two quinonimine metabolites (**M1** and **M2**)

enzyme CYP3A4 (a cytochrome) is the heme-porphyrin center, which is bound to the enzyme by a covalent bond. The MO arrangement (Fig. 6.4) at the Fe atom in the heme-porphyrin center is suitable to capture an oxygen molecule as well as to accommodate electrons and subsequently break oxygen–oxygen bond. In this process, it produces a very reactive radical species known as **Cpd-I**. This reactive species contains an unpaired electron in the Fe–O bond, which is suitable to oxidize C–H bonds in drugs to produce metabolites.

To understand the electronic structure of a given drug, it is important to solve Schrodinger wave equation ($\hat{H}\Psi = E\Psi$; $\hat{H}$ is Hamiltonian operator (Eq. (6.1)), Ψ is wave function, E is energy) of a given molecule. A solution to Schrodinger wave equation of a drug provides the shapes and energies of every molecular orbital in the drug. A molecular Schrodinger equation is an analytical equation, which can be solved using calculus. Methods were developed to solve the same using numerical approaches and thus a numerical form of Schrodinger equation was developed (Eq. (6.2)), which employs matrices. Equation (6.2) can be solved using computers.

$$\widehat{H} = -\sum_{A} \frac{\hbar^2}{2M_A} \nabla_A^2 - \sum_{i} \frac{\hbar^2}{2} \nabla_i^2 + \sum_{A>B} \frac{Z_A Z_B}{R_{AB}} - \sum_{A_i} \frac{Z_A}{r_{A_i}} + \sum_{i>j} \frac{1}{r_{ij}} \qquad (6.1)$$

where M is the mass of nucleus, Z is the atomic number, i, j refer to electrons, and A, B refer to nuclei, $r_{ij} = |r_i - r_j|$ and $R_{AB} = |r_A - r_B|$.

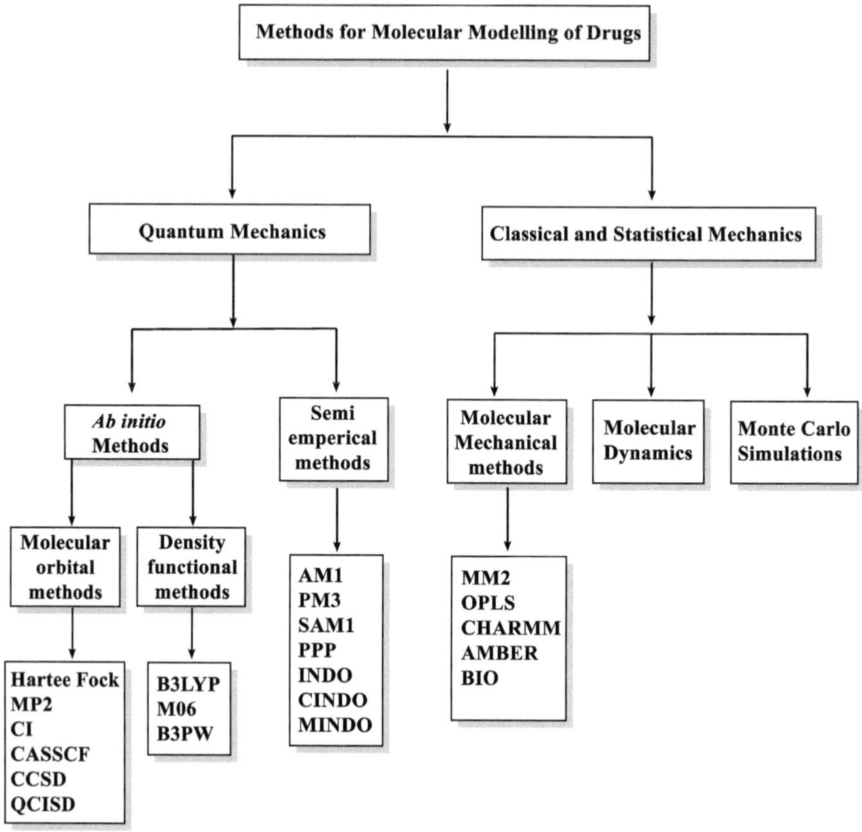

Chart 6.1 Showing a few molecular modelling methods useful to study the 3D structure, electronic structure, and dynamics of drugs (and macromolecules)

$$FC = SCE \tag{6.2}$$

where F is the Fock matrix, C is the coefficient matrix, S is the overlap matrix, and E is the energy matrix.

Many quantum chemical theories have been developed to understand the electronic structure of the drugs [14, 15]. Among them (Chart 6.1), the most useful one in CADD is the Density Functional Theory (DFT). This theory solves the electronic structure of the drugs in a cost-effective manner. The DFT method B3LYP/6-31+G (d) is being extensively used to explore the electronic structure of drugs. The theoretical formalism behind such methods is very robust and very extensive. It is suffice, for a beginner, to know that B3LYP is a quantum chemical density functional method which defines the Hamiltonian operator ($\hat{H}$) and 6-31+G(d) is a basis set of Gaussian type mathematical functions, which are employed to define the wave function (Ψ) while solving the Schrodinger equation $\hat{H}\Psi = E\Psi$ of a drug molecule.

Solving the Schrodinger equation with the help of quantum chemical methods provides MO diagrams and MO energies, which can be effectively used to estimate the internal characteristics of drugs. The absolute energy of the drugs (Fig. 6.2) can be estimated exactly by solving the Schrodinger equation (by supplying the coordinates of the atoms in the drugs). The 3D structure of drugs can be generated very accurately with these methods by performing geometry optimization and conformational analysis (vide infra). The partial atomic charges present on the drugs can be easily estimated using electronic structure methods. From the energy values of HOMO and LUMO, and their difference, many chemical properties of the drugs can be estimated. For example, the electrophilicity of the reactive metabolites can be easily estimated. All the drugs/metabolites whose electrophilicity index value is more than 3 are toxic, leading to Mechanism-Based Inhibition of cytochromes [16].

The 3D structures of many intermediates during the drug action inside the human body are not known because they cannot be isolated and subjected to X-ray diffraction. However, they can be generated using quantum chemical methods. For example, when Isoniazid is interacting with various cytochromes in the human body, it produces many intermediates/MICs (Mechanism-Based Inhibitory complexes) with the enzymes. Formation of such complexes can lead to the self-kill process of the cytochromes. The structures of the intermediates along the reaction path can be estimated using quantum chemical methods [16]. The structures of the transition states during any reaction can be reliably obtained using these methods. For example, the hydride ion transfer from the NADPH to dihydrofolate inside the cavity of the *pf*DHFR is due to the proximity of the two reactive centers. The structures of the transition state during this hydride ion transfer cannot be obtained using any experimental methods, but it can be easily and reliably obtained using quantum chemical methods by solving Schrodinger wave equation.

QM/MM methods are hybrid of quantum chemical and molecular mechanical methods (vide infra), which are useful to study the drug–macromolecule interactions [14]. The most important reactive center can be evaluated using the detailed quantum chemical methods and the rest of the molecular system can be modelled using molecular mechanics methods. Such hybrid methods are useful to consider the whole complex while paying special emphasis at the required center of action. The QM/MM methods became especially important to model the enzymes containing transition metals. These methods are being used, often, to explain the chemistry and the biology of drug action but they did not yet find applications in drug design, on a routine basis.

Quantum chemical methods become prohibitively expensive when the solutions are sought for many drugs/toxins/metabolites. To simplify, semi-empirical quantum chemical methods have been introduced. The semi-empirical methods do not solve the Schrodinger equation very accurately and hence they cannot provide the absolute energy values. But they can easily provide the heats of formation data of drugs. The 3D structures of the drugs obtained from the semi-empirical methods are sufficiently reliable. The time required to perform semi-empirical analysis on drugs is very small in comparison to that of the ab initio MO/DFT methods. The most popular semi-empirical method among the drug discovery scientists is the AM1 method. For

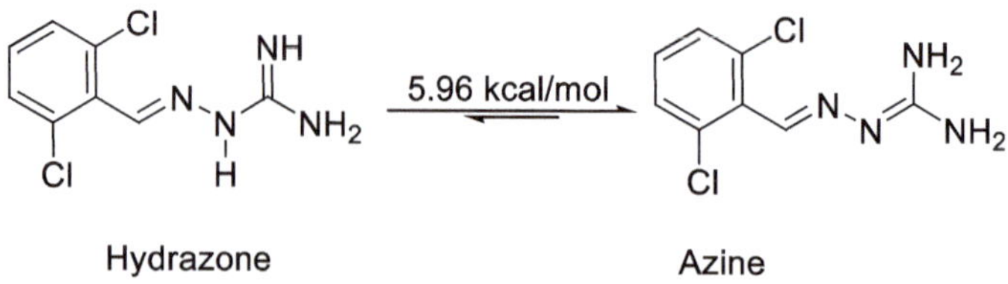

Fig. 6.5 Tautomeric preferences in Guanabenz free base. Hydrazone tautomer was always considered; however, electronic structure analysis confirmed that the real state of Guanabenz contains the azine tautomeric state

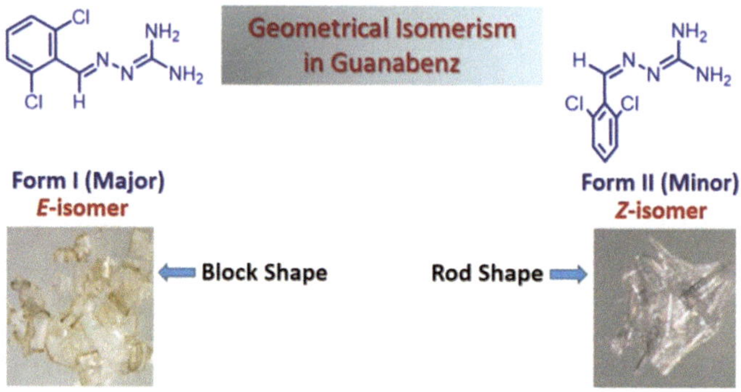

Fig. 6.6 Two crystalline forms of Guanabenz showing geometrical polymorphism. Reproduced with permission from American Chemical Society, "Ref. [18]"

example, when a database of the 3D structures of drugs and their reactive metabolites are required (about 6500 drugs and average 5 reactive metabolites of each drug making a total 39,000 species), experimental X-ray diffraction methods cannot be used because the corresponding crystals are not available. The ab initio quantum chemical methods also cannot be used because the time required to obtain such data is extremely high. Semi-empirical methods can easily be adopted because they can quickly provide the necessary reliable data.

Case Study 1

Electronic Structure studies on the Antihypertensive drug Guanabenz, leading to new molecule design, polymorphism. Quantum chemical studies using DFT method indicated that Guanabenz is more stable in the azine tautomeric state rather than in the hydrazone tautomeric state. This is contrary to the general belief that Guanabenz is a hydrazone derivative. The energy difference between the two tautomers is about 5.6 kcal/mol (in favor of azine tautomeric state, Fig. 6.5), X-ray diffraction studies on the neutral Guanabenz confirmed the expectation from quantum chemical studies [17]. Polymorphism is a phenomenon according to which a drug molecule may exist in two or more crystalline forms. In case of Guanabenz, water/methanol (3:1) system could produce two crystalline forms, block shape crystals (Form I) and the rod shape crystals (Form II) in a 95:5 ratio (Fig. 6.6). These details could be obtained after performing electronic structure analysis (on the *E* and *Z* isomers of Guanabenz) and subsequent experimental analysis. [18]

6.5 Surface Properties of Drugs and Macromolecules

Every drug is associated with a 3D structure and the corresponding electronic structure; in addition, it is also associated with surfaces. Most of the interactions between drugs and macromolecules are due to surface-to-surface contacts. The molecular recognition between drugs and receptors is due to surface properties. The molecular surface carries electrostatic characteristics. Based on the complementarity of the electrostatics on the surfaces of the drug and that of the macromolecules, the recognition between these two species takes place. The percentage polar surface area (PSA) of drugs is a very important property, which determines the hydrophobicity vs. hydrophilicity balance of a drug. The molecular volume of a drug is the entire volume present within the surface of the drug. The molecular surface provides the space within which the electrons and other subatomic particles reside; thus molecular surface may be considered as the boundary of the molecules. It can also be considered as the outer boundary of the molecular field.

The surface of any drug cannot be obtained from experimental methods. Only computational methods are suitable to generate and visualize the surface of drugs. Many innovative computational methods of defining the molecular surfaces have been developed, which provide sufficient and reliable information regarding molecular surfaces. van der Waals surface is one way of defining the molecular surface, in which the van der Waals radii of atoms of drug are considered in a given conformation. Excluding the overlapping molecular volumes and plotting the rest of the surface can provide the van der Waals surface. But this shall have several sharp points/edges. Hence, Connolly suggested that it is important to smoothen the surface, such a smoothened surface is known as Connolly surface.

When two molecules approach each other, without forming any permanent interaction, there is a limit, up to which, they can come close to each other—i.e., the surfaces of the two molecules cannot approach each other beyond a certain point. Based on this concept, another way of defining surface of a molecule was identified that is "solvent accessible surface". The centroid of the water molecule can reach only up to a certain point close to the surface of any drug. The path of the centroid traced by rolling a water molecule on every direction on the surface of the drug molecule can be plotted, such a 3D path defines the solvent accessible surface of the drug. The space encompassed by the Connolly surface is smaller than the space encompassed by the solvent accessible surface.

The atoms of a drug closer to its surface define the electrostatic surface. Functional groups like COO^- provide electronegative surface, which can be plotted in red color. The functional groups like NH_3^+ provide electropositive surface, which can be plotted in blue color. Thus, the entire surface of the drug molecule can be color coded. Similarly, the potential energy associated with the surface can also be

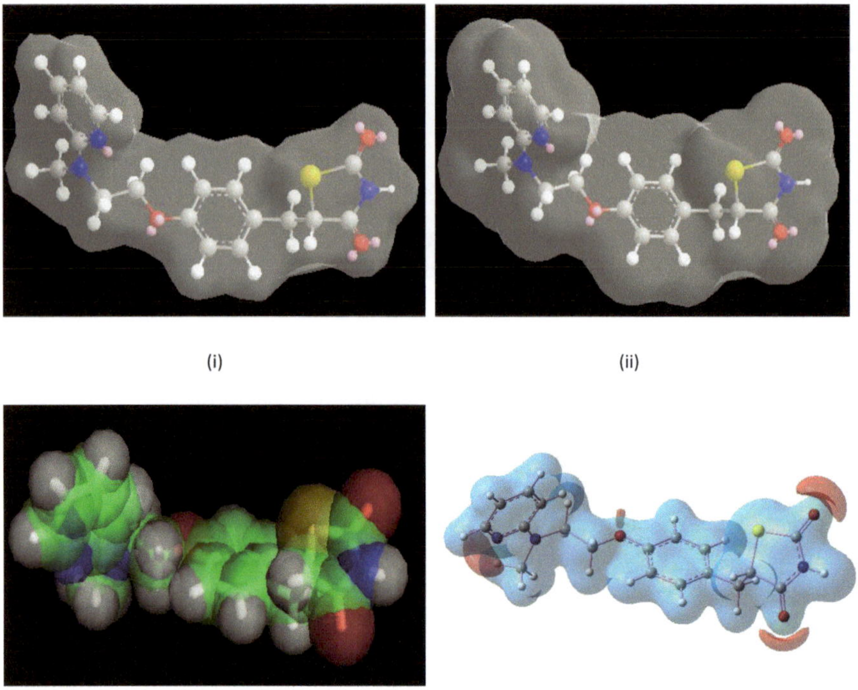

(i) (ii) (iii) (iv)

Fig. 6.7 The four different surfaces of the antidiabetic drug Rosiglitazone—(**i**) Connolly surface (**ii**) solvent accessible surface, (**iii**) van der Waals surface, and (**iv**) Molecular Electrostatic Potential Surface (MESP). The solvent accessible surface is larger than that of the Connolly surface. The van der Waals surface is more rugged in comparison to the Connolly surface. In (**iii**) and (**iv**) the lone pairs of electrons are shown as pink balls

color coded and plotted. For example, when a unit positive charge is rolled on the entire surface of the drug molecule, it experiences attractive or repulsive forces at every point. The potential energy due to these interactions can be plotted and a surface of the drug can be defined. Such a surface is known as MESP (Molecular ElectroStatic Potential) surface. Such a surface can be plotted using the charges or by estimating the energy accurately (quantum chemical estimation). The surface representations of the drug and macromolecule became an essential part of CADD. Figure 6.7 shows the various surfaces of the drug Rosiglitazone.

6.6 Computational Design of the 3D Structures of Drugs and the Estimation of Their Energies

3D structures of less than 10% of the drugs and less than 5% macromolecular druggable targets are solved using X-ray diffraction methods till now. The 3D structures of these drugs and macromolecules are required on many occasions during CADD process. The 3D structures of drugs and small molecules can be generated using computational methods (quantum chemistry or molecular mechanics, etc.). Similarly, the 3D structures of macromolecules can also be obtained using computational methods (e.g., Homology modelling). Once a 3D structure of any drug is created using structure building tools, it needs to be subjected to "cleaning process", which involves adjusting the geometrical parameters as per the standard values and using simple chemical rules as well as known structural knowledge. Such cleaned models carry physically acceptable structures. To make them chemically appropriate, they need to be subjected to energy minimization methods (which may be based on quantum chemical/molecular mechanical methods). They also may have to be subjected to computational conformational analysis. In this section, many essential steps of molecular modelling which need to be performed. Before "trusting the molecular structure available on computer screen", are described.

(i) 3D structure building
(ii) Cleaning the structure using knowledge-based methods
(iii) Performing geometry optimization
(iv) Performing conformational analysis
(v) Attributing electronic/force field/charge, etc. features to the molecular structure

When the structure of any drug is not available from X-ray diffraction analysis, its structure can be very reliably established with the help of molecular modelling techniques [14, 15]. The reliability increases after performing energy minimization (also geometry optimization). This is because the structure of any drug is optimal when its energy is lowest. Several computational techniques are available to carry out energy minimization of drugs. The energy minimization process leads to the optimization of the structure. There are many computational algorithms to carry out energy minimization, including simplex method, conjugate gradient method, steepest descent method, Newton–Raphson method, and Genetic Algorithms. In general, the energy of a given molecule is first estimated with the current geometry and subsequently the geometrical features are adjusted till the energy of the molecule becomes minimum on the potential energy surface. All these methods iteratively solve the dual problem, i.e., geometry optimization and energy minimization. The energy estimation during the energy minimization may be carried out using ab initio quantum chemical/semi-empirical quantum chemical/molecular mechanics methods (listed in Chart 6.1). The ab initio quantum chemical methods provide very accurate estimates of structure and absolute energy, but they are time consuming. The semi-empirical methods only provide heats-of-formation information, they need much less time. The molecular mechanical methods also provide only heats-of-formation but at faster rates. The choice of the method for geometry optimization depends on the scientific problem at hand. When electronic level information is required, quantum chemical methods are the best. When atomic-level information is sufficient, molecular mechanics methods are more appropriate. When fast geometry estimation is required on large number of molecules, knowledge-based methods are being employed.

Conformational analysis is another important step in molecular modelling. This will help in realizing which is the most appropriate structure of the drug (out of many possible alternatives). Manual methods of performing conformational analysis was in practice, which is highly dependent on the chemical intuition of the scientist. A few computational methods of conformational analysis have been developed. Systematic conformational search method is based on the systematic modification of torsional angles across all rotatable bonds in a drug molecule. This method produces large number of (physically acceptable) conformers of drugs, which can be rank ordered on the basis of the estimated energy. Random conformational analysis is associated with randomly changing the torsional angles of all rotatable bonds; after each change, geometry optimization needs to be taken up. This method produces fewer (chemically acceptable) conformers of the drug; based on the energy, it is possible to choose the best suited conformer. Monte Carlo methods, genetic algorithms, molecular dynamics, etc. also can be used to identify the appropriate conformers of the drugs [15]. One of the biggest limitations of these methods is that they do not help in identifying the bioactive conformation of the drug.

Quantum chemical methods (Sect. 6.4) are being extensively used to estimate the energy and structure of drug molecules, using Eq. (6.2). Molecular mechanical methods (force field methods) are also helpful in estimating the heats-of-formation. These methods are based on the force fields associated with molecules. The bond stretching, angle bending, torsional angle changing, van der Waals, electrostatic and miscellaneous forces can be easily estimated finally to obtain E_{FF} of molecules (Eq. (6.3)). Energies of macromolecules are generally estimated using molecular mechanical methods. A few empirical methods are available to estimate the drug–receptor, substrate–enzyme, inhibitor–enzyme, DNA–protein, and protein–protein binding energies.

$$E_{FF} = \sum E_b + \sum E_a + \sum E_t + \sum E_{es} + \sum E_{vdw} + \sum E_{nb} \tag{6.3}$$

where E_{FF} is energy of the molecule due to force field methods

E_b is the energy due to bond stretch motions

E_a is the energy due to angle bending motions

E_t is the energy due to torsional angle motions

E_{es} is the energy due to electrostatic interactions

E_{vdw} is the energy due to van der Waals contacts

E_{nb} is the energy due to nonbonded contacts

6.7 Computational Design of the 3D Structures of Macromolecules

Proteins contain many atoms and their structures are very complicated in comparison to that of drugs. Hence, the structures of proteins can be understood mainly in four levels—primary, secondary, tertiary, and quaternary. Whereas the primary structure of a protein is defined in terms of sequence of amino acids, the secondary structure of a protein is defined in terms of the α-helices, β-sheets, loops, folds, kinks, bends, hinges, etc. present in them. The spatial arrangement of these secondary structural elements determines the 3D structure of proteins. The tertiary structural information is the 3D structure of the proteins, which can be defined in terms of the Cartesian coordinates of each and every atom. The quaternary structure of proteins is due to the arrangement of several proteins in clusters—homodimers (e.g., nitric oxide synthase), heterodimers (Dihydrofolate Reductase and Thymidylate Synthase), or dodecamers (Phospodiesterases). Figure 6.8 shows the four different descriptions of the structure of *Plasmodium falciparum* Dihydrofolate Reductase (*Pf*DHFR).

One of the important components of CADD is Structure-Based Drug Design (SBDD), in which case, the term structure mainly implies the structure of the macromolecules [19–21]. The main source of 3D structural information of

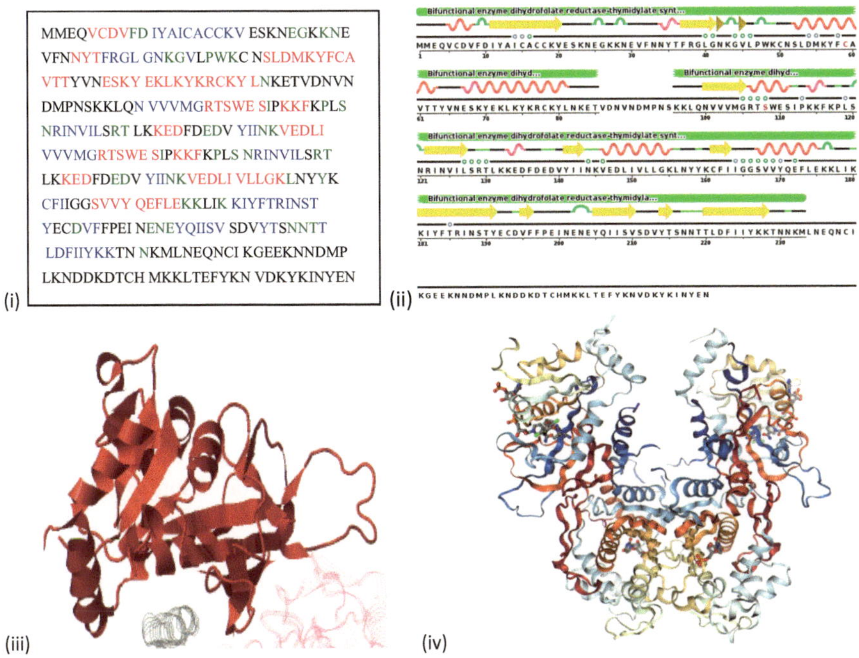

Fig. 6.8 The four types of representation of *Pf*DHFR enzyme—**(i)** primary structure, **(ii)** secondary structure, **(iii)** tertiary structure of DHFR (highlighted in red color), and **(iv)** quaternary structure of four chains of enzymes—two DHFR units along with two TS (Thymidyl synthase) domains

```
CLUSTAL O(1.2.4) multiple sequence alignment

sp|Q9UMR5|PPT2_HUMAN      MLGLCGQRLPAAWVLLLPF------LPLLLLAAPAPHRASYKPVIVVHGLFDSSY---S  51
sp|P45478|PPT1_BOVIN      -----MASSSCLWLLALAFLLGSCASLALGHLDPPA-----PLPLVIWHGMGDSCCNPLS  50
sp|P50897|PPT1_HUMAN      -----MASPGCLWLLAVALLPWTCASRALQHLDPPA-----PLPLVIWHGMGDSCCNPLS  50
                                    . *:* :  :        *  *  **       *::: **: **.    *

sp|Q9UMR5|PPT2_HUMAN      FRHLLEYINETHPGTVVTVLDLFDG-RE-SLRPLWEQVQGFREAVVPIMA---KAPQGVH  106
sp|P45478|PPT1_BOVIN      MGAIKKMVEKKIPGIHVLSLEIGKTLREDVENSFFLNVNSQVTTVCQILAKDPKLQQGYN  110
sp|P50897|PPT1_HUMAN      MGAIKKMVEKKIPGIYVLSLEIGKTLMEDVENSFFLNVNSQVTTVCQALAKDPKLQQGYN  110
                          :  : : ::::. **  *  *:: .   *   . :: :*:.   :*   :*   *  ** :

sp|Q9UMR5|PPT2_HUMAN      LICYSQGGLVCRALLSVMDDHNVDSFISLSSPQMGQYGDTDYL------KWLFPTSMRSN  160
sp|P45478|PPT1_BOVIN      AMGFSQGGQFLRAVAQRCPSPPMVNLISVGGQHQGVFGLPRCPGESSHICDFIRKTLNAG  170
sp|P50897|PPT1_HUMAN      AMGFSQGGQFLRAVAQRCPSPPMINLISVGGQHQGVFGLPRCPGESSHICDFIRKTLNAG  170
                           : :**** . **: .   .  : .:**:.. : * :*             :: .::.:.

sp|Q9UMR5|PPT2_HUMAN      LYRICYSPWGQEFSICNYWHDPHHDDLYLNASSFLALINGERDHPNATVWRKNFLRVGHL  220
sp|P45478|PPT1_BOVIN      AY---NKAIQERLVQAEYWHDPIREDIYRNHSIFLADINQERGVN--ESYKKNLMALKKF  225
sp|P50897|PPT1_HUMAN      AY---SKVVQERLVQAEYWHDPIKEDVYRNHSIFLADINQERGIN--ESYKKNLMALKKF  225
                           *    .   :.:  .:***** ::*:* * * *** ** **.      ::**:: : ::

sp|Q9UMR5|PPT2_HUMAN      VLIGGPDDGVITPWQSSFFGFYDANET--VLEMEEQLVYLRDSFGLKTLLARGAIVRCPM  278
sp|P45478|PPT1_BOVIN      VMVKFLNDTIVDPVDSEWFGFYRSGQAKETIPLQESTLYTQDRLGLKAMDKAGQLVFLAL  285
sp|P50897|PPT1_HUMAN      VMVKFLNDSIVDPVDSEWFGFYRSGQAKETIPLQETSLYTQDRLGLKEMDNAGQLVFLAT  285
                          *::   :* :: * :*.:**** :.:: .:  ::*  :* :* :*** :   * :*

sp|Q9UMR5|PPT2_HUMAN      AGISHTAWHSNRTLYETCIEPWLS      302
sp|P45478|PPT1_BOVIN      EGDHLQ---LSEEWFYAHIIPFLE      306
sp|P50897|PPT1_HUMAN      EGDHLQ---LSEEWFYAHIIPFLG      306
                           *        ..  : : * *:*
```

Fig. 6.9 The sequence comparison between the two isozymes PPT1 and PPT2 of the family of enzyme palmitoyl protein thioesterase (in Human and Bovine species). "*" symbol is used for perfect conservation (exact match) with a column of all the same residues; ":" symbol is used for high conservation where all residues are biochemically very similar; "." symbol is used to indicate weak conservation of residue type. Sequence identity and sequence similarity values between PPT1-H and PPT1-B are 88.5% and 94.5%, respectively. Sequence identity and sequence similarity values between PPT1-H and PPT2-H are 33.8% and 56.0%, respectively

macromolecules is Protein Data Bank (PDB). When the 3D structures of a protein are not available from PDB, the same can be generated using computational methods. The most important method to generate such structure is the homology modelling, which provides a reliable structure of the concerned macromolecule. The main hypothesis behind this method is that the sequence similarity between two proteins leads to their structural similarity, which may in turn lead to functional similarity. Though this hypothesis is only partially trustworthy, the 3D structures obtained from homology modelling provides sufficient information to confidently take up research based on SBDD (molecular docking, molecular dynamics, de novo design, etc.).

Homology between two proteins or enzymes in bioinformatics is associated with sequence or structural similarity between two or more proteins (which presumably originates from a common ancestor). Primary sequence homology between two proteins belonging to a family is generally very high. This is generally measured in terms of sequence identity and sequence similarity. Figure 6.9 provides an example of sequence identity and similarity between PPT1 and PPT2. Sequence homology gets reduced among proteins belonging to two different families. As a rule of thumb, minimum sequence similarity required between two proteins to take up homology modelling is ~30%.

The very first step in homology modelling is to know whether the primary sequence of the query protein matches with that of any of the proteins reported in PDB. Based on this analysis, a template structure needs to be chosen, which becomes a guide in computationally generating the 3D structure of the query protein. Generally, the 3D structure of the protein (in PDB), which possess maximum similarity with the query sequence, is chosen as a template. Next step is to employ any software which can generate the 3D structure of the query protein. This comparative modelling algorithm (for example in MODELLER software) adopts the principle "satisfaction of spatial restraints", which is inspired by the NMR-based structure generation method. The geometry optimization on the resultant structures can be carried out using the known structural features of the amino acids and associated geometry restraints. The generated model (3D structure of the macromolecule) needs to be validated in many ways and fine-tuned till a satisfactory model is obtained. For this purpose, Ramachandran plots, Prosa plots, DOPE (Discrete Optimized Protein Energy) scores, Z-score, verify-3D, ERRAT plots, ProCheck, etc. tools are being employed.

Apart from the homology modelling methods, a few alternative computational methods, namely "ab initio macromolecule structure prediction methods", "threading methods", and "De novo macromolecule structure prediction" are also available to generate the 3D structures of proteins. In fact, this problem "protein structure prediction" has been an unsolved problem for the past many years; it is often referred as "protein folding problem". Several software tools have been introduced for protein structure prediction, all with partial success. Some of them require a template but many are template-independent methods. PRIME, MOE, and Swiss-Model are some CADD software which are being widely used and they are template dependent. Protein threading (or fold recognition) is one of the methods which does not directly depend on the homology. It uses statistical knowledge gathered from the structures deposited in the PDB, since the number of folds recognized till now are not large (< 1500). DeepFR, AlphaFold, and SPMP (Selvita Protein Modelling Platform) are a few threading-based protein modelling software, among many, attempting to succeed in this field. "Ab initio protein structure prediction" and "De novo protein structure prediction" are quite synonymous with each other. QUARK, NovaFold, and Rosetta are three examples of tools which are useful in making protein structure prediction using de novo methods. Bhageerath is one example of the successful methods (ab initio structure prediction method) to obtain reliable 3D structures of macromolecules [22]. Initially, it produces numerous 3D structures of macromolecules based on the known conformational space. Subsequently, it adopts Monte Carlo sampling methods to remove steric clashes among various folds and employs energy minimization and score-based techniques to finally suggest plausible 3D structures of the proteins. A validated 3D structure of a protein obtained from computational methods carry information regarding the active site, allosteric site, catalytic site, substrate-binding domain, cofactor-binding domain, and many other details, which are useful for CADD.

The PDB contains structural information of proteins obtained from X-ray diffraction studies; however, the structures reported in PDB also need to be subjected to computational modification before using them for CADD purpose. This is because a few limitations have been noticed in the reported structures—for example, (i) missing position of hydrogen atoms, (ii) artificially introduced mutations in the protein to increase the crystallization of proteins, (iii) no clarity on the tautomeric/conformational states of the amino acids in the proteins, (iv) presence of many water molecules in the crystal structures, (v) missing residues in the reported structures, (vi) wrong representation of the structures of the small molecules embedded inside the structure of the macromolecules, and (vii) need for the modulation of the spaces in the active site to facilitate the induced fit of the ligands in the active site.

The 3D structures of the proteins generated from computational models or the experimentally generated structures which are suitably modulated using computational methods need to be used for CADD with some caution, because they all are model systems. The CADD scientist should be very watchful while employing these structures because they are not subjected to rigorous mathematical modelling (like quantum chemical modelling) and also because these static models do not account for the dynamical state of the proteins. Hence, SBDD becomes an art which is based on science. The comparative model building methods described in this section are not limited to proteins. With minor variations, these computational methods can be used to obtain the 3D structures of all macromolecules, including nucleic acids, protein–protein complexes, protein–DNA complexes, cell walls, chromosomes, and antibodies.

All the 3D structures of macromolecules are characterized by energies. Estimating the absolute energies of these macromolecules is not practical using ab initio or semi-empirical quantum chemical methods. Hence, the energies of the macromolecules are generally obtained using force field methods (or using knowledge-based methods). During geometry optimization (and energy minimization) of these structures, the energies obtained using force field methods are generally employed. For this purpose, AMBER, CHARMM, OPLS, TRIPOSE, UFF, etc. force fields have been generated and are being successfully used in various molecular modelling software. Most of them use equations similar to the one given in Eq. (6.3).

6.8 Computational Analysis of Drug–Macromolecule Interactions and the Estimation of Associated Energies

Drugs/cofactors/substrates/inhibitors, etc. small molecules interact with macromolecules. Protein–protein interactions are quite common in biological environment. Protein–DNA interactions are also very prominent. Receptors, transporters and pumps get embedded in the cell walls, which originate from interactions between

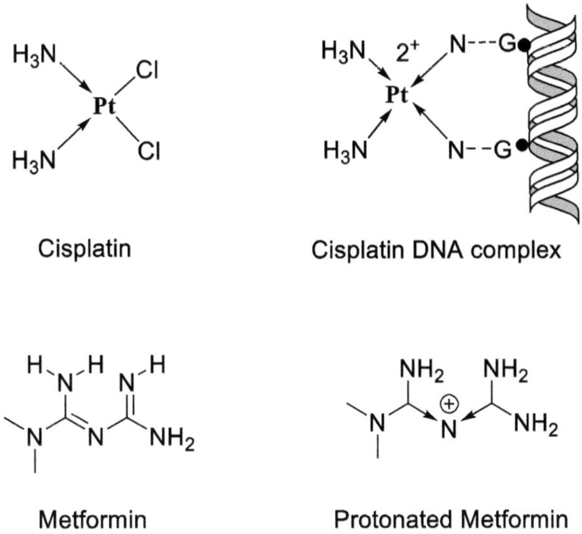

Fig. 6.10 Coordination bonds in drugs. N-Pt coordination bonds in Cisplatin before and after complexation with DNA. C–N coordination bonds in Metformin in its salt form

cell wall and the proteins. The above listed interactions are generally temporary and hence they lead to associations rather than bonds. All such association–dissociation processes are governed by small atomic-level interactions. Each of these atomic-level interactions is characterized by geometric as well as energy features. In CADD, it is very essential to analyze these interactions in terms of structural as well as energy factors. In this section, a brief introduction to the atomic-level interactions, their energies, and the computational methods of estimating the same are described.

Covalent bonds are generally less important in receptor–drug interactions; however, they become important in enzyme action. For example, the phosphorylation reaction by any kinase requires breaking a P–O coordination bond (from ATP) and formation of another P–O covalent bond with substrate. Every metabolic process is associated with covalent modification of the drugs. For example, the conversion of the antiviral prodrug Remdesivir into active metabolite involves many covalent bond modifications. The strength of a covalent bond is in the range of 60–75 kcal/mol for single bonds. These covalent bonds provide specific structures to reactants and products. Quantum chemical methods can be used to estimate the strength of any covalent bond, its formation, and dissociation.

Coordination bonds are also important in drug discovery. Whenever metalloenzymes or metal-based drugs are involved, the chances of coordination interactions become very prominent. For example, in Cisplatin (Fig. 6.10), there are two coordination bonds between Pt and NH_3 units and when Cisplatin binds to two adjacent guanine units in DNA, two covalent bonds (Pt-Cl) get broken and two coordination bonds (Pt $\leftarrow$ N7 (guanine)) get formed. Another interesting example is Metformin which carries all covalent bonds in the neutral form, but carries two important (C$\rightarrow$N) coordination bonds in the corresponding protonated state, thus introducing incredible structural and energy changes between Metformin and Metformin.hydrochloride [12, 23, 24]. The strength of a coordination interaction

(~25 kcal/mol) is generally weaker than that of a covalent bond. The coordination bonds also provide specific and stable structures wherever they are present.

Hydrogen bond is a bond as well as an electrostatic interaction. Bifurcated hydrogen bond, C–H ... O hydrogen bond, dihydrogen bond, and agostic interactions are very related interactions. These interactions are increasingly weak, in that order. The bond length of a regular hydrogen bond is in the range of 1.8–2.8 Å, the strength of the same varies between 4 and 7 kcal/mol. Bifurcated hydrogen bond is 2–4 kcal/mol strong and the C–H ... O hydrogen bond is 1–3 kcal/mol strong. Since hydrogen bonds are directional, they provide specific structural feature, which can be easily measured. Hydrogen bonds are ubiquitous in chemistry and biology and are very crucial in CADD.

Electrostatic interactions can be noticed between drug and receptor when two oppositely charged centers interact with each other (as in salt bridge or as in dipole–dipole interactions). They are not directional and hence do not contribute to specific structural feature but contribute to the specificity in the structural orientation of the drugs in the cavity of receptors. The strength of any electrostatic contact can be estimated using Coulomb's law.

Van der Waals interaction (vdw) is due to the instantaneous dipole moment generated when two electroneutral objects approach each other. This instantaneous dipole generates temporary attraction between the two objects, which may not break away until 1–2 kcal/mol energy is supplied. When drugs interact with macromolecules, many vdw contacts can be noticed. As a result, the total energy gain due to the collective effect of the vdw contacts, sometimes, becomes larger than that due to hydrogen bond and electrostatic interactions. The strength of a vdw interaction can be estimated using Lennard-Jones potential. π–π interactions, T-type π interactions, charge transfer interactions, and polarization interactions are also important between drugs and macromolecules.

The total strength of the interaction between drug and macromolecule can be estimated as a sum of the above-discussed fundamental interactions. In data intensive CADD efforts, these interactions are generally estimated using empirical formulae (scoring functions), but in science intensive research projects, the strength of these interactions are generally estimated using either quantum chemical (or molecular mechanical) methods. This energy accounts for the affinity between the drug and macromolecule. The structural preference of the drug in the active site of the macromolecule (the pose of the drug) is to maximize the complementarity between the drug and the macromolecule. The conformational change associated with this interaction leads to the cascade of events, which produce the observed (and desired) therapeutic action. Thus, the structure and energy originating from the atomic-level interactions lead to the complementarity between drugs and receptors, which is responsible for the observed therapeutic action.

6.9 Pharmacoinformatics

Information technology (IT) has proliferated into every walk-of-life. This proliferation into the pharmaceutical sciences lead to the growth of the topic called pharmacoinformatics. This field emerged as a result of integration of two powerful fields—pharmaceutical sciences and information technology. In this section, the data intensive component of pharmaceutical sciences being handled by information technology has been covered. Though the whole field of CADD, including molecular modelling, computational chemistry, and computational biology, are being considered under the umbrella of pharmacoinformatics, this section is limited to data management techniques in pharmaceutical sciences. Pharmacoinformatics can also be considered as an integration of chemoinformatics and bioinformatics with a common goal to apply IT in drug design, drug discovery, drug disposition, and drug development.

6.9.1 Chemoinformatics in CADD

Chemical information is vast. Collecting data about chemical species, chemical structures, chemical reactions, chemical concepts, chemical processes, chemistry patents, etc. started more than 120 years ago (w.r.t. year 2020). However, the application of computer-based methods to evaluate the chemical properties and analyzing them using statistical methods, in a data driven manner, started only about 50 years ago in the form of chemometrics. Visualization of chemical structures using molecular graphics became possible about 30 years ago. Alternatively, applying fundamental principles of physics to understand the theoretical concepts of chemistry using computers became practical with the advent of computational chemistry over 40 years ago. Integration of X-ray data with theoretical chemistry concepts lead to the development of molecular modelling tools about 25 years ago. With increasing data in chemical sciences and learning from the techniques of bioinformatics, the science of chemoinformatics became a standard topic about 20 years ago [25–27]. Currently, chemoinformatics topic came of age and it deals with large data of chemical systems. Given below are some of the important aspects of chemoinformatics which found application in CADD [28].

One-dimensional notation of drugs can be expressed in the form of SMILES (Simplified Molecular Input Line Entry System) notations or, more reliably, using InChI (IUPAC International Chemical Identifier) or InChIKeys. Such one-dimensional information carries structural information of the drugs. Connectivity between various atoms, stereochemistry, tautomerism, isomerism, isotope variation, charge information, etc. are also part of these notations. Such notations are very useful in chemical data search. Among many ways of representing chemical information using one-dimensional notation, InChI is very useful in chemical library searches because these are unique representations. Chart 6.2 shows the various

Name:	Rosiglitazone.
SMILES:	O=C1NC(=O)SC1Cc3ccc(OCCN(c2ncccc2)C)cc3
InChI:	1S/C18H19N3O3S/c1-21(16-4-2-3-9-19-16)10-11-24-14-7-5-13(6-8-14)12-15-17(22)20-18(23)25-15/h2-9,15H,10-12H2,1H3,(H,20,22,23)
InChiKey:	YASAKCUCGLMORW-UHFFFAOYSA-N
Bitmap:	1110100101100110011001110011
Hashcode:	E96667

Chart 6.2 Showing various 1D representations of the antidiabetic drug Rosiglitazone

1D representations of the antidiabetic drug molecule Rosiglitazone. In a particular databank containing all antidiabetic drugs, leads, and investigational compounds along with the reported biological activity data (in vitro, in vivo), if it is required to search for the drug molecule and retrieve all the information associated with it, the InChI of the query molecule can be supplied and the relevant information may be retrieved. To improve the speed in search, this one-dimensional information can be converted into bitmaps, or hashcodes. A lot of information is available with each drug—for example, molecular formula, trade name, IUPAC name, 2D structure, chemical graph, 3D structure, legal status, FDA approval information, metabolism, various identifiers (e.g., CAS number), all physicochemical parameters. All such information can be part of the chemical databases. Based on the information provided, many properties can be calculated on the fly (whenever required) easily using software—for example, the molar refractivity (MR) value may not be part of the database, but can be quickly calculated from SMILES notation. Many chemical graph-based descriptors can be quickly calculated.

One special advantage of chemoinformatics is the properties like logP, molar refractivity (MR) of a drug molecule can be reliably calculated using the one-dimensional information. For example, the MR of a drug is a summation of the MR values of various small chemical fragments present in the drug (additive property), which can be obtained from the database of fragments. In the case of Rosiglitazone, the MR values of C_6H_4 ring, CH_2 group, pyridine ring, NMe fragment, thiazolidinedione fragment, and divalent oxygen atom can be added to obtain the MR value of the drug.

Several chemoinformatics tools useful for CADD purpose have been developed. For example, the $MPDS^{TB}$ (Molecular Property Diagonistic Suite) is a specialized suite of chemoinformatics tools developed to help in anti-TB agent design [29]. This suite uses the chemoinformatics tools PaDel and CDK (Chemistry Developing Kit) to calculate the descriptors. It can perform many ways of data analyses including QSAR, docking, screening, and visualization. It uses a Galaxy platform to create a workflow linking databases and analysis tools. It uses a compound library of about >95 million compounds; the compound searches can be made using a unique set of molecular fingerprints.

6.9.2 Bioinformatics in CADD

Bioinformatics deals with the application of information technology tools on biological information. It was only a service-oriented platform in the beginning; however, several scientific improvements happened in this field, making it, a progressive scientific field today [30–35]. It utilizes information from molecular biology, structural biology, computational biology, and systems biology to provide very powerful tools to carry out modern biological analyses. It also integrates information from proteomics, genomics, and all other omics platforms with information technology. Broadly, there are two interdependent components of bioinformatics—sequence analysis and structural analysis.

Sequence analysis utilizes the primary sequence information of biological macromolecules. Each macromolecule is associated with a unique sequence of basic elements. For example, the receptor PPARγ is an antidiabetic target. It recognizes drugs like Rosiglitazone and produces the necessary drug action. The genes associated with this receptor express this protein, always producing the same sequence of 505 amino acids. This sequence can be compared with many other protein sequences. A comparison with the sequence of *Hs* $PPAR_{\alpha}$ produces 63% sequence identity and 78% sequence similarity. Such a comparison is the most important activity in bioinformatics [30]. Several algorithms have been developed for the purpose of sequence comparison and application of the same was discussed in Sect. 6.7 of this chapter.

There are several applications of bioinformatics in CADD [36–38]. A few of them are included in sections on molecular docking, de novo design, molecular dynamics and virtual screening (vide infra). One important application of bioinformatics in CADD is the target identification and target validation. Identifying a new biological target, so as to design chemical species which can better fit inside its active site and produce any therapeutic action, is a challenge. With the advancements in genomics and proteomics, available bioinformatics data is expanding on daily basis. However, how many of them are actually useful for CADD? Such information is only occasionally forthcoming. Bioinformatics (sequence and/or structural) can significantly contribute towards this goal. Also when a specific drug is interacting with many macromolecular targets, and produce interfering biochemical reactions leading to toxicity, often information is not available; to obtain this, bioinformatics methods are being efficiently used.

Functional Genomics is a sub-branch of genomics, which extensively uses bioinformatics tools. Many experimental paradigms are associated with functional genomics, including stem-cell research and microarray analysis. Several bioinformatics tools are being introduced to do this—ImtRDB [39], PhenPath [40], and Shiny-Seq [41] are a few examples which were introduced in 2019. DAVID (the Database for Annotation, Visualization and Integrated Discovery) and bioinformatics tools [42] are high-throughput functional annotation tools, which attempt to systematically map a large number of genes and their gene ontology terms; further, statistical analyses identify enriched gene ontologies. These tools help in identifying the biological processes relevant to a specific phenomenon being explored. This is a knowledge-based approach (vide infra), which employs workflows to effectively link many tools present in the platform.

Another important example worth considering in this chapter is an attempt made using quantitative and systems pharmacology. Fang et al. [43] reported an in silico approach to predict the new targeted cancer therapy. The authors assembled a drug-target network containing 7314 interactions which are associated with 751 targets and 2388 natural products; further, they designed predictive network models. The algorithm used is based on substructure-drug-target network. A few new targets were predicted using the natural products—Genistein, Resveratrol, and Kaemferol, which were successfully validated. Using this information, the authors further identified new anticancer indications of a few known drugs—Metformin, Naringenin, and Disulfiram, suggesting mechanisms of drug action. The ab initio gene prediction software, chemgenome [44], is also an advanced bioinformatics tool, which along with Gene Translator tools, helps in computationally predicting genes and targets. These tools utilize information from prokaryotic and viral genomes effectively in terms of six reading frames and subsequently provide the associated protein sequences.

6.9.3 Pharmacoinformatics in CADD

In the field of pharmaceutical sciences, lots of data is being generated and hence many databases are being created. In addition to drug discovery, many other sub-fields of pharmaceutical sciences—namely, drug delivery, drug disposition, and clinical trials—also are employing the IT tools. Especially metabolism and toxicity are very complicated sciences, where data is more than the concepts derived from the data. Associated topics like drug regulatory issues, drug packaging, and drug transportation are also using IT tools. Pharmacoinformatics topic proved to be especially useful in carrying out drug discovery research during social distancing and lockdown periods, in the year 2020. Studies in pharmaceutical management are being carried out mostly employing IT tools only. Thus, the applications of IT in pharma field are progressing prosperously. In this section, we deal with only those topics which are leading to CADD.

As mentioned earlier, huge databases of chemicals and biological molecules are available. Selecting a few species from large chemical databases for therapeutic action is a challenge. Similarly, selecting a few druggable macromolecular targets from large biochemical databases is also a challenge. In addition, data is pouring-in from many fronts—genomics, proteomics, immunology, neurology, toxicology, and all the fields which are linked to pharmaceutical sciences. Further, data from specific therapeutic areas such as cancer, HIV, TB, diabetes, inflammation, viruses, anesthesiology, is also rising. Hence, specific informatics topics associated with all these subtopics of pharmaceutical sciences are emerging independently, however, are being used assimilatively. Figure 6.11 represents one way of integrating these topics

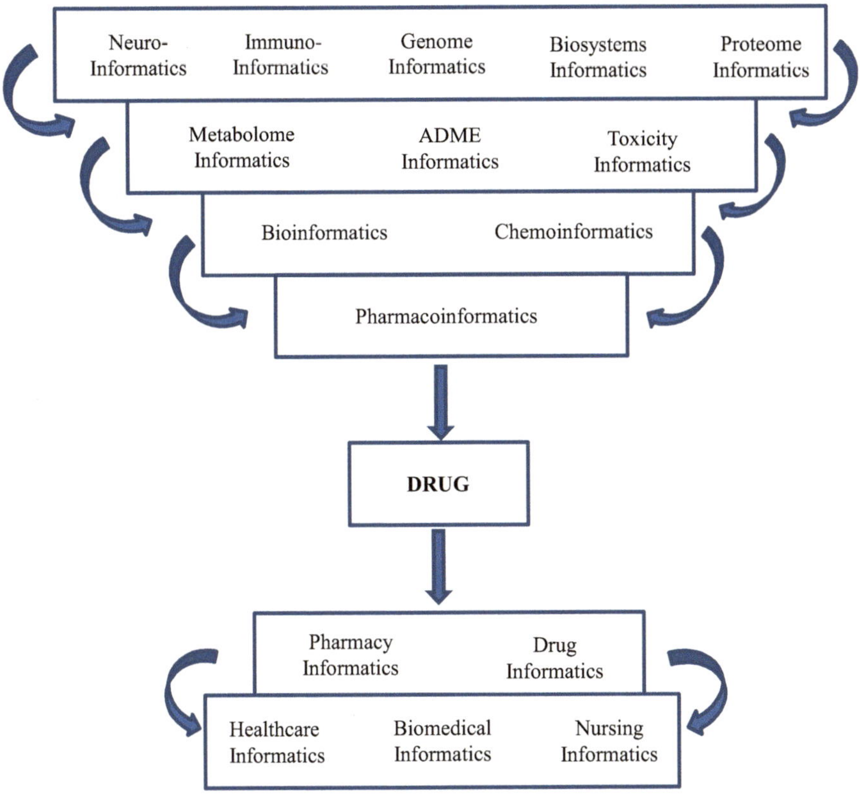

Fig. 6.11 Flow chart defining the flow of activities in the field of pharmacoinformatics

towards drug discovery and beyond—describing the flow of activities in pharmacoinformatics.

As a specific case of CADD research using pharmacoinformatics, the following example may be considered—the design and synthesis of humanTopoisomerase IIα (*h*TopoIIα) inhibitors as anticancer agents (Fig. 6.12). Guchhait and coworkers initially designed a few analogs (*N*-fused imidazoles) of the known lead compounds for the *h*TopoIIα inhibition [45]. The compounds were subjected to molecular modelling studies as a part of rational design. The possible preferred tautomeric states and the protonation states of the nitrogen-rich heterocyclic chemotype were evaluated using quantum chemical methods (B3LYP/6-31+G(d,p)). In the next stage, molecular docking studies were carried out to understand their interactions with the amino acids in the ATP-binding domain of *h*TopoIIα (using the crystal structure of *h*TopoIIα with AMPPNP (PDB id: 1ZXM). The synthesized compounds were subjected to biological evaluation, including ATP competitive binding kinetics study. The compounds which showed promising biological signature were further subjected to molecular dynamics analysis to establish the stability of the chemotype during the dynamical state of the macromolecule–ligand assembly. The binding free

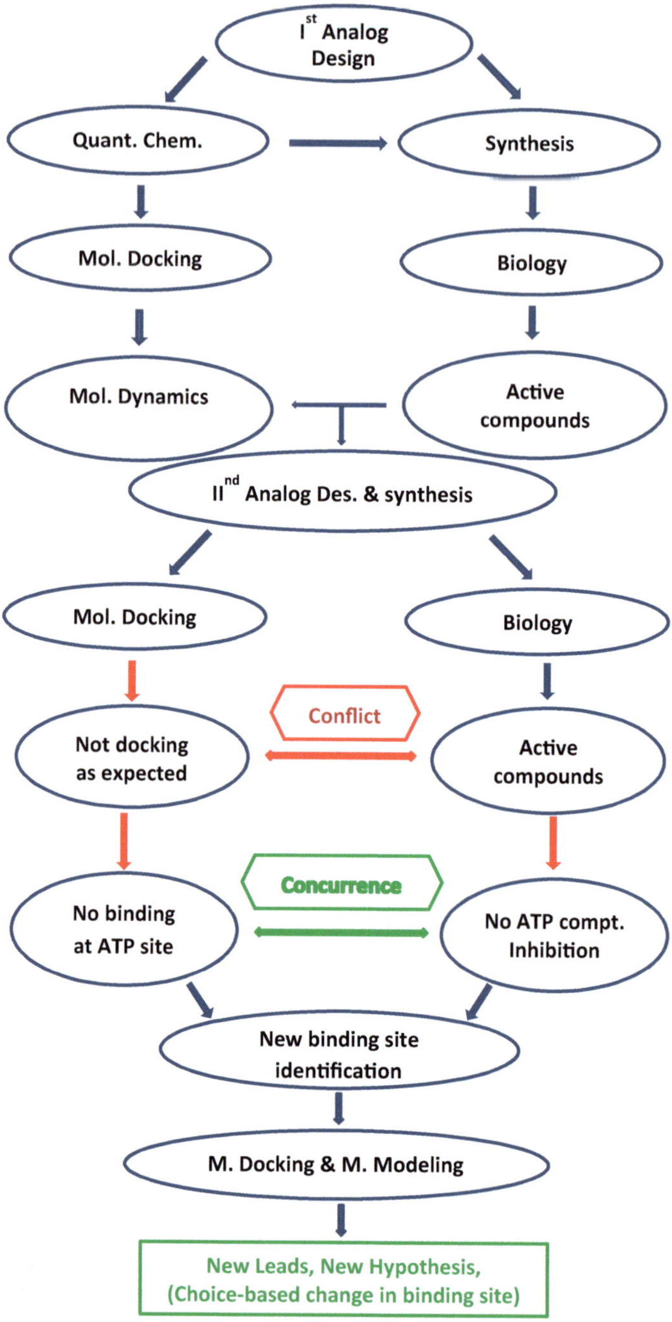

Fig. 6.12 A flow chart showing a thorough pharmacoinformatics effort integrated with experimental studies, which provided new lead molecules, new hypothesis, identification of unknown site of inhibitory action. During this work, all CADD techniques were employed, including bioinformatics, chemoinformatics, quantum chemistry, molecular modelling, and molecular dynamics. Database development, pharmacoinformatics tool development were also part of this work [46–49]

energies of the interaction between the ligand and the macromolecule were estimated using MMPBSA and MMGBSA analyses. Thus, first part of the work based on synthesis, rational design including CADD study, and biological evaluation was successfully carried out, in a highly interdependent manner [45].

Encouraged by the results from the first part of the work, Guchhait and coworkers designed new derivatives of the lead compound (analog design) and synthesized [46]. Simultaneously, this series of new compounds were subjected to molecular modelling analysis as well as the biological evaluation. The computational team and the biological team provided conflicting (non-correlated) results. The molecular docking analysis suggested that the compounds cannot bind successfully in the cavity of ATP-binding domain, because the size of each compound in this series is larger than the size of the ATP-binding domain. The biological experiments suggested that the set of new analogs showed good inhibitory potential. While these conflicting signals were perplexing, they provided new challenges and opportunities to the four teams (one pharmacoinformatics, one synthesis, two biology) working in the project. The challenge to the pharmacoinformatics team was to consider flexibility of ATP-binding site. The challenge to the biology team was to confirm whether ATP competitive binding is observable. The informatics team reconfirmed their earlier observation and the biology team proved that the new compounds are not providing the desired ATP competitive binding effect; thus, both teams obtained mutually agreeable results.

These results lead to fresh challenges—(i) If ATP competition is not the cause, what is the origin of inhibitory activity? (ii) If ATP-binding site is not the site of inhibitory action, what is the binding site of the new class of analogs? To answer such questions, extensive pharmacoinformatics analysis was carried out. The available information and data were analyzed thoroughly using the SBDD techniques. It was found that, there are at least six different sites of ligand binding possible. At least five different reaction steps can be blocked in the catalytic cycle of the *h*TopoIIα. From analogy, it was concluded that the designed ligands were probably bound in the Merbarone (a lead compound which was withdrawn from clinical studies) binding site. However, it was realized that Merbarone-binding site was not clearly established, the only available clue was the Merbarone probably binds at a site which is in close proximity with that of the Etoposide-binding site (a TopoII poison). By performing extensive molecular docking studies, the Merbarone-binding site and bind pose were identified. Further using molecular dynamics analysis, the conformational changes of *h*TopoIIα which can lead to the inhibitory activity of Merbarone were established. Subsequently, it was found, using molecular docking and molecular dynamics studies, that the new series of compounds indeed bind effectively in the Merbarone-binding site. Thus, "a switch in the site of inhibition of human topoisomerase IIα inhibitors" was realized using an integrative approach involving pharmacoinformatics, organic synthesis, and biological evaluation [45–49], and subsequently new leads were identified. This study lead to the introduction of a strategy of "choice-based change in the site of inhibition of macromolecules" for rapid drug discovery.

6.10 QSAR

The 3D structure of a drug molecule is one of the important deterministic features of its activity (as emphasized in earlier sections). This has been a dogma (a pretty reliable one) of drug discovery research, since the time Fischer proposed the "lock and key" hypothesis (in 1894) for enzymatic action. Based on this concept, several structure–activity relationships (SAR) were established. A small change in the chemical structure is expected to introduce specific and significant changes in the activity, which can be quantitatively measured, i.e., the SAR of drug action can be quantitative.

The relationship between a chemical functional group and the reactivity of the chemical species were well known. Hammett, in 1937, proposed a quantitative relationship between the reactivity of a chemical species to its structural features. Hammett equation, Taft equation, etc. became very reliable sources of estimating the SAR in chemistry. Hammett equation proposes a relation between substituent constant (σ) and the reactivity of a species, in which, the chemical substituent is remotely placed with reference to the center of chemical reaction. The σ constant includes the inductive effect and the mesomeric effect due to a given substituent. A reaction constant (ρ) was also considered; it is related to the charge on the chemical species during the rate-determining step of a chemical reaction. This parameter is useful in determining a reaction mechanism. Taft introduced a steric parameter (E_s) to account for the influence of the steric effects on a chemical reaction. Such quantitatively definable relations were traditionally known as linear free energy relationships.

The concept of linear free energy relationship was extended to the biological activity of chemical species by Hansch, which paved the way for the very well established subtopic of CADD, i.e., Quantitative Structure Activity Relationship (QSAR) [50–53]. Hansch equation deals with the electronic, hydrophobic, and steric influences of chemical substituents on the biological activity of a given chemical species. Slight variants of QSAR relevant for drug discovery community are QSPR (Quantitative Structure Property Relationship) and QSTR (Quantitative Structure Toxicity Relationship).

If the structural features of a series of chemical species can be quantitatively (QS) expressed and their biological activities (QA) can also be quantitatively expressed, it is possible to express their relation in quantitative terms (QR). In QSAR, it is not possible to *create* a relationship between the structural features and activity data; it is only possible to *identify* the relationship, if it exists. To identify such relationship, it is necessary to make many attempts to realize which structural features are relatable to the experimentally observed activity data. The relationship

(QR) between QS and QA is not a mathematical relationship (which is definitive), but it is a statistical relationship, and hence it is a correlation.

QS: How to express structural features of a chemical species in quantitative terms? This can be done by oversimplification, by losing a lot of information. For example, molecular weight is a quantitatively definable factor which can be obtained from the structure of a molecule. From the given structure, it is possible to obtain molecular weight, but from molecular weight, it is not possible to reconstruct the structure of a molecule. Hence, the QS adopts an oversimplification approach. In fact, molecular weight is a numerically expressible parameter of a molecule. Similarly, there are several numerically expressible parameters (of molecules) which can be estimated once the structure of the molecule becomes available. All such parameters describe the molecule in quantitative terms; hence they are termed as descriptors. It can be considered that once the structural information of any drug/chemical species becomes available, it is possible to estimate many descriptor values. No single descriptor value fully defines the structure, but collectively, many of the descriptors provide sufficiently clear information about the structure. Thus, descriptors provide information about the structure in quantitative terms and they are the most important part of QS.

More than 5000 different descriptors have been defined for QSAR analysis. Such descriptors can be classified into groups based on the method adopted in estimating these values. Important classes are listed below.

(i) *Physicochemical descriptors:* These numerical values of chemicals can be estimated by performing physical chemistry experiments. For example, molar refractivity (MR) is a descriptor which can be obtained from the experimentally measured refractive index (n). It can be calculated using Lorentz-Lorenz equation (Eq. (6.4)). Other descriptors are log P, dipole moment, vapor pressure, etc. These descriptors can be estimated using experiments. More importantly, they can also be calculated using structural information. For example, log P value can be experimentally determined by distributing a given molecule between organic and aqueous phases (octanol/water). Alternatively "calculated log P" value can be estimated using additivity principle (without carrying out any experiment) because log P of a molecule is the summation of the log P values of its constituent fragments.

$$MR = (M/\rho)(n^2 - 1)/(n^2 + 2) \tag{6.4}$$

where M is molecular weight, ρ is density of the system, and n is refractive index.

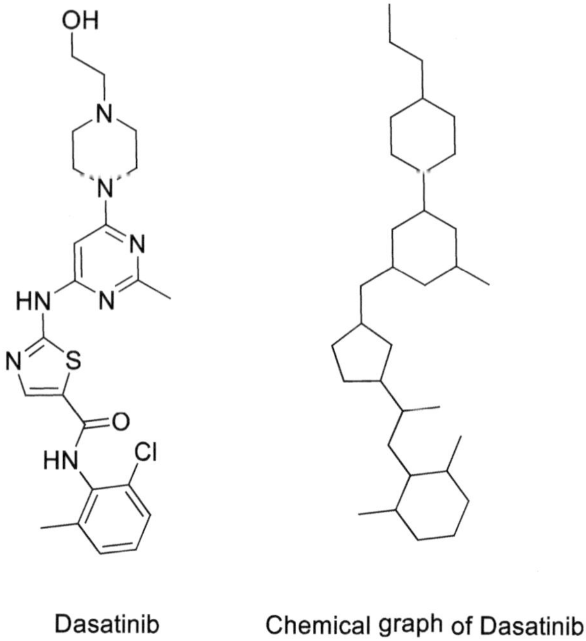

Fig. 6.13 2D structure and molecular graph of the anticancer drug Dasatinib

(ii) *Graph theoretical descriptors:* These are also known as topological descriptors, which can be estimated using the molecular graph of a molecule. For example, Wiener index is a descriptor which is a summation of the topological distances of the molecular graph of a drug (Fig. 6.13). A molecular graph can be evaluated using many principles of graph theory. Topochemical descriptors are values considering the chemical as well as topological characteristics and employing graph theory. There are many graph theoretically definable descriptors useful in QSAR.

(iii) *Indicator descriptors:* These descriptors indicate whether a specific feature is present or not in a given molecule. For example, for carboxylic acid indicator descriptor (CAID) the value is either 1 or 0 in a given molecule. CAID value for Ibuprofen is 1 and the same for Isoniazid is 0.

(iv) *Constituent descriptors:* These define the components present in the molecules. Examples are atom count, bond count, amino acid count, etc.

(v) *Quantum chemical descriptors:* The quantum chemical descriptors can be estimated only after performing quantum chemical analysis (ab initio or semi-empirical)—HOMO, LUMO energies, their differences, chemical potential, electrophilicity index, etc.

(vi) *Electronic descriptors*—which are associated with the electrostatic characteristics. For example, the charge of a molecule, partial atomic charge on specific atoms in a molecule, polarizability, calculated inductive effect values, Electron localization function, etc.

Not all descriptors are relevant all the time for QSAR analysis. In a typical classical QSAR equation identification process, about 2–10 descriptors are used. Hence, descriptor selection becomes a challenge. A few guidelines have been suggested for the same—this includes a set of rules of thumb as well as automated procedures of descriptor selection [52].

QA: Quantitative definition of biological activity in terms of numerical values can come from pharmacology and toxicology experiments (in vitro or in vivo). The biochemical activity produced by a specific drug on a chosen biomolecule may not be easy to estimate. Generally, the observed change is very small and often the necessary biochemical assay to estimate the value using biophysical measurements may not be available. In many cases, the manifestation of the biochemical change in a different but easily measurable reaction (among the cascade of possibilities) can be recorded. For example, the inhibition of GSK3 (Glycogen Synthase Kinase) enzyme can be estimated using FRET (Fluorescence Resonance Energy Transfer) experiments. The K_i, IC_{50}, and ED_{50} values, etc. are suitable as QA values. The values related to "the percentage reduction in inhibition" are not suitable as QA in QSAR. A lot of quality control should be adopted before choosing QA values. These values should be obtained using a specific biological assay on a series of molecules. If all the values are obtained from a single laboratory (using the same assay), it is preferable (even though published in more than one article). Values obtained from different biological assays (for a given enzyme inhibition study) should not be clubbed for QSAR purpose. QA data selection from in vivo experiments is quite difficult in comparison to that of in vitro experiments.

QR: This is relationship information, expressed in quantitative terms (obtained using statistical methods). Whether or not a correlation exists between a set of QS values and the corresponding set of QA values defines the QR. The variance, correlation coefficient, the cross-validated correlation coefficient, standard deviation, etc. statistical parameters can be used to define QR. To obtain these values, statistical methods like PLS (Partial Least Squares), MLR (Multiple Linear Regression), PCR (Principle Component Regression), GFA (Genetic Function Approximation), etc. are useful. A given quantitative relationship may be reliable, if not it may be improved. For the identification of a QR, the QA value of a molecule (on the left hand side of the equation) needs to be correlated with a linear combination of selected set of descriptors in QS (on the right hand side of the equation). A typical, classical QSAR equation appears as given in Eq. (6.5).

$$\begin{gathered} pIC_{50} = c1 * MR + c2 * ZI + c3 * \omega + c4 \\ n = 30, r^2 = 0.987, r^2_{\mathrm{LOO}} = 0.876, SEE = 0.123, r^2_{predict} = 0.765 \end{gathered} \quad (6.5)$$

It can be interpreted as follows. A set of 30 novel chemical species (n) belonging to a particular chemical series have been correlated with their experimentally observed IC_{50} values (Inhibitory constant values at 50% concentration) after converting the values using the equation pIC_{50} is $-\log$ (IC_{50}). Such a conversion is required because correlation is possible with the experimental data, only after the conversion. The molecular descriptors which are exhibiting correlation with the pIC_{50} values are molar refractivity (MR), Zagreb index (ZI, a graph theory-based descriptor), electrophilicity index value (ω, a quantum chemical descriptor). The $c1$, $c2$, $c3$, and $c4$ values are coefficient values, which represent the contribution from the molecular descriptors towards the observed correlation. The r^2 value is correlation coefficient, whose value should be close to 1, if the correlation is strong. The r^2_{LOO} is the cross-validated correlation coefficient obtained using Leave One Out method, this value also should be sufficiently high (close to 1) to consider the correlation as a reliable one. SEE is the Standard Estimated Error, whose value should be very small (close to 0), in any reliable QSAR equation.

As illustrated in the above example, the most important requirement before starting any QSAR identification exercise is the availability of a set of molecules, belonging to a specific chemical class for which the biological activity has been experimentally evaluated. The number of molecules in the series should be generally >30. In practice, the available set of molecules are divided into two sets—training set (about 25–60 molecules from a congeneric series) which are employed in the identification of the relationship and the test set (approximately 20% of the number of molecules in the training set) which are employed to verify whether the QSAR is reliable. The predicted biological activity values of the test set of molecules can be estimated and their deviation from the experimental values can be noted (typically the deviations should be less than 0.5 units). This data can be used to calculate the $r^2_{predict}$; if this value is sufficiently high, i.e., >6, the identified QSAR can be considered as a validated QSAR. To gain further confidence, the QSAR can be subjected to further validation using statistics or with the help of more test set molecules. The biological activities predicted using a validated QSAR equation can be safely believed; thus, QSAR method provides an opportunity to computationally estimate the biological activity values of a set of molecules, which are yet to be generated in a chemical laboratory. The synthesis of the molecules can be taken up only when their predicted biological activity is sufficiently satisfactory. Thus, QSAR is a very powerful tool in the hands of drug discovery scientist; it helps in taking informed decisions. Further advantages are saving time, money, and the effort of a drug discovery scientist. The related QSTR studies are even more strongly useful because they help in saving the animal life. The QSAR and QSTR methods are being utilized by the drug regulatory authorities also.

Classical QSAR: When 2D structural information of molecules is compulsory in the identification of a reliable QSAR equation (for example, to calculated graph theoretical descriptors), such a QSAR is known as 2D QSAR or classical QSAR. The following example illustrates the use of classical QSAR.

Norfloxacin is an antibacterial agent; it inhibits topoisomerase II (DNA gyrase) and topoisomerase IV [54–56]. Designing this drug involved the application of classical QSAR methods. QSAR analysis of 6-, 7-, or 8-monosubstituted l-ethyl-1,4-dihydroquinoline-3-carboxylic acids was carried out against biological data obtained from *Escherichia coli* (taking Nalidixic acid as lead molecule). The classical QSAR analysis showed that the antibacterial activities of the derivatives of 4-oxoquinoline-3-carboxylic acid were correlated well with steric factors of

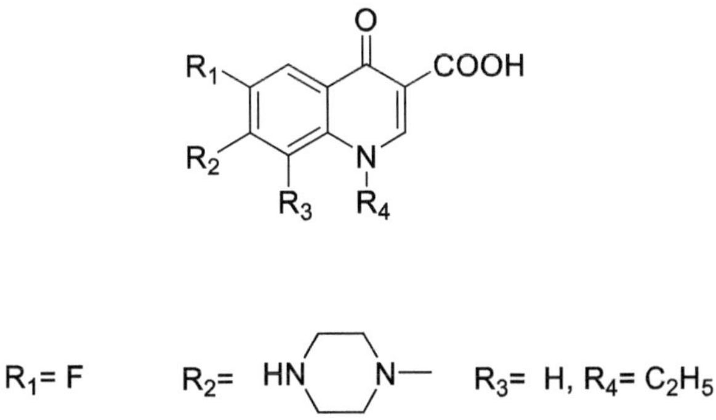

Fig. 6.14 4-oxoquinoline-3-carboxylic acid

substituents R1 and R3 (Fig. 6.14). The Hansch analysis revealed that the 6-fluoro- and 6-chloro-7-(l-piperazinyl) derivatives can be very potent. This combined SAR and QSAR study lead to the identification of the drug Norfloxacin (FDA approval in 1986).

In 1987, Klopman and coworkers performed computer automated structure evaluation of quinoline antibacterials. In this study, the molecular fragments of biological relevance were automatically generated and significance of these major fragments were evaluated on the basis of multivariate linear regression analysis based on the stepwise selection of a subset of descriptors. The observed correlation was expressed as in Eq. (6.6). Using this technique, the experimental and theoretical DNA gyrase inhibitory values for Norfloxacin were 2.50 and 1.78, respectively, very comparable to each other [56].

$$-\,log\ BA = 1.05 + 0.80\, n_1 F_l + 1.34\, n_2 F_2 - 0.48\, n_3 F_3 - 0.06\,(\,log\ P)^2 \quad (6.6)$$

where $R = 0.88$, $F_{(4,48)} = 42.27$, $S = 0.35$, $N = 53$, nF is the number of occurrences of the fragment within a molecular structure. BA represents binding affinity and P represents probability.

3D QSAR: This is of two types. One is classical 3D QSAR and the other is field-based 3D QSAR. In classical 3D QSAR, molecular descriptors are obtained from 3D structures of the molecules. For example, all the quantum chemical descriptors can be obtained only when the 3D structure of molecule becomes available. Also, parameters like molecular volume, Connolly surface area, etc. can also be estimated only when the 3D structure of the molecule becomes available. A QSAR equation obtained by employing such 3D descriptors is known as classical 3D QSAR.

Every molecule is associated with molecular fields. Such molecular fields of each molecule can be recognized and aligned using virtual grids. The numerical values associated with the steric and electrostatic effects experienced by a probe (e.g., methyl cation) at each grid point can be estimated. Collective influence of all these numerical values provides the field effect. The difference density maps obtained using these numerical values can be plotted as contour maps. This hypothesis was employed by Cramer in suggesting a nonclassical 3D QSAR technique CoMFA. Many other techniques like CoMSIA, COMMA, and AFMoC, which are field-based

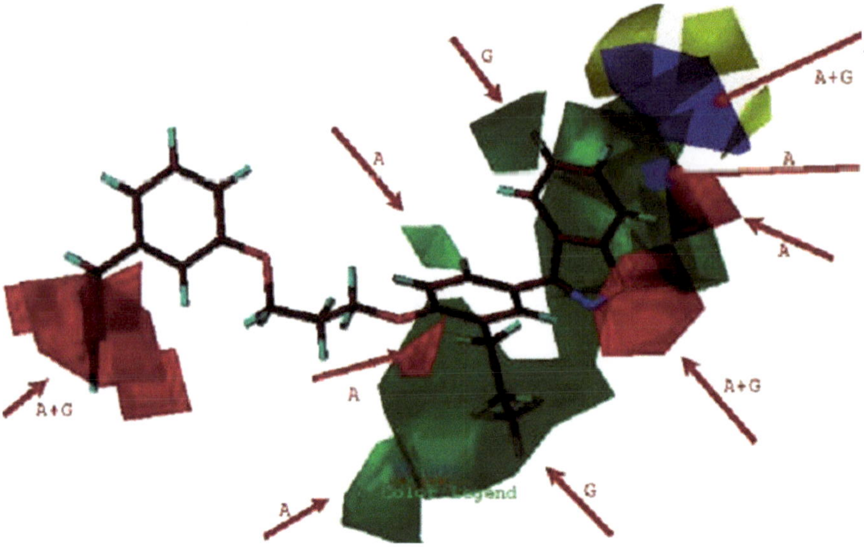

Fig. 6.15 Steric and electrostatic contour map for the dual model showing the contributions from each model. "A" depicts the contributions made by the α-model and "G" depicts the contributions made by the γ-model. Reproduced with permission from American Chemical Society, "Ref. [62]"

3D QSAR methods, were proven to be useful [57–61]. An example of the use of CoMFA method has been described below.

3D-QSAR methodology was employed to introduce the concept of "additivity" of fields by Khanna et al. to design dual PPAR activators [62]. Biological activities (pIC_{50}) noticed for the individual receptors were added to define additive activity on which CoMFA was performed. The fields resulting from the CoMFA model represent "additivity fields." A series of 5-aryl thiazolidinedione and oxazolidinedione derivatives acting as PPARα and PPARγ dual activators were employed to develop three CoMFA models: (i) α-model, (ii) γ-model, and (iii) dual-model. The basic skeleton and conformation of Rosiglitazone was modelled after its structure extracted from PDB structure (PDB code: 2PRG). U-shape of the ligand was maintained while performing energy minimization, and the 3D structures of all other ligands were generated by considering Rosiglitazone as the template. Alignment of all the molecules was carried out while maintaining the bioactive conformation noticed for Rosiglitazone. To numerically estimate the CoMFA field parameters, the steric and electrostatic parameters were calculated (125 each). Tripos force fields were employed to estimate the van der Waals and the Coulomb potential values (using an sp^3 carbon atom as a probe). Partial Least Squares (PLS) analysis was used for statistical evaluation of the CoMFA fields and the IC_{50} values. Leave-one-out (LOO)-based cross-validation of the QSAR results was carried out. The steric contour (green and yellow) and the electrostatic contour maps (red and blue) are plotted as in Fig. 6.15.

Based on the above studies, the chemical features essential for PPARγ activity includes an acidic head group, central aromatic region, and a lipophilic side chain. The TZD ring in the chosen molecules makes three important H-bonds with His323, Tyr473, and His449. Taking clues from these observations, the design of non-TZD PPARγ agonists was taken up. Computer-aided design of a novel series of PPAR ligands based on barbituric acid were designed [63–65]. The synthesis of about 14 novel molecules was carried out and their in vitro activities were measured using radioligand-binding assay. About six compounds were found to bind to the PPARγ with IC_{50} ranging from 0.1 to 2.5μM. Similarly, many compounds containing 2-hydroxy-1,4-naphthoquinone as "acidic head group" were designed and synthesized. Out of eight molecules, two molecules were found to be better than Pioglitazone (IC_{50} 0.7μM). Continuing this work, many new leads were identified [64, 65].

6.11 Pharmacophore Perception

Pharmacophore is a set of features of the drugs, which collectively are responsible for the observed therapeutic action. These features are not the chemical functional groups but are due to the chemical functional groups. For example, –OH group is not a pharmacophoric feature, but hydrogen bond donor character is a pharmacophoric feature (this feature may be due to the –OH group or any other group). The important pharmacophoric features are hydrogen bond donor center, hydrogen bond acceptor center, aliphatic hydrophobic center, aromatic hydrophobic center, acidic center, basic center, electrostatically positive center, and electrostatically negative center on the ligands. For each of the pharmacophoric centers on the ligands, there should be corresponding complementary pharmacophoric center on the macromolecule for effective binding.

Any expert drug discovery scientist can easily perceive (or comprehend) the features of a pharmacophore, due to his/her experience. Attempts were made to develop software to mimic this process of "pharmacophore perception". One technology, which was found to be useful in this context, was pharmacophore mapping. It should be distinguished that a pharmacophore is different from a pharmacophore map—i.e., pharmacophore is a set of features and pharmacophore map is a result of a computational effort to perceive a pharmacophore [66–69]. Mapping is a topic of mathematics whose concepts were employed in developing the CADD technology known as "pharmacophore mapping". Mapping mathematics is useful to identify if two groups of objects have anything in common by establishing one-to-one correspondence between the objects of these two groups. Figure 6.16 shows a description of the pharmacophoric features and pharmacophore map associated with glitazone class of molecules towards PPARγ agonistic activity.

To carry out pharmacophore map generation, we need several chemical classes of molecules, which are acting on a single target. This is the clear distinguishing point in comparison to QSAR. In QSAR, several compounds from a given chemical class

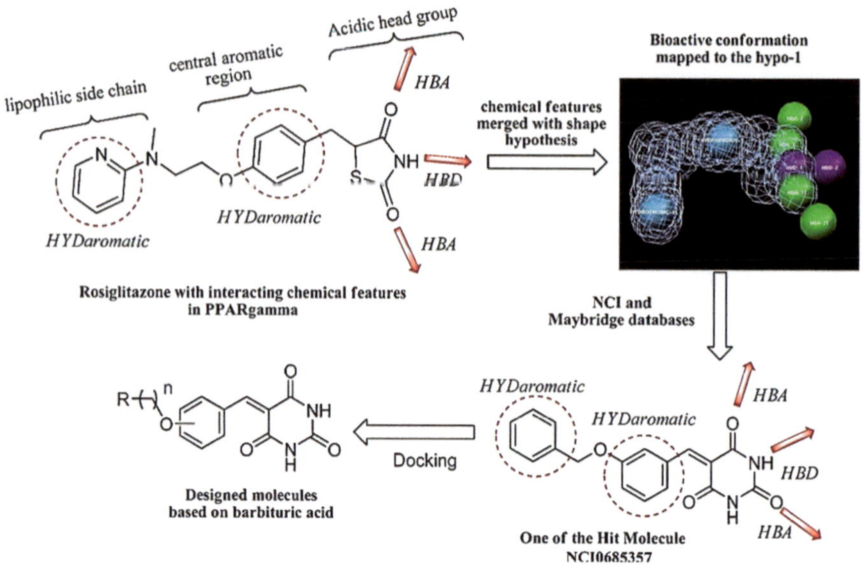

Fig. 6.16 Design of barbituric acid derivatives as PPARγ ligands employing pharmacophore generation and subsequent virtual screening. Reproduced with permission from Elsevier Science publishers, "Ref. [65]"

are required, whereas in pharmacophore mapping, the molecules should be from different chemical classes (there may be many entries from each class). Most active molecules from each chemical class are chosen to identify the pharmacophoric features. Subsequently, the common pharmacophoric features among these most active compounds need to be chosen. Thus, sufficient diversity as well as sufficient similarity is required among the molecules chosen for pharmacophore mapping. The diversity is in the form of chemical functional groups among the various series of molecules, and the similarity is in terms of commonality of the pharmacophoric features present in the most active compounds. Hence, the chosen set of molecules should exhibit "unity in diversity" type character. Also the chosen common features (among many features) of each of the most active compounds fall into the category of "common-minimum" set of features. This can be illustrated in the following example.

For the inhibition of the enzyme GSK3β (Glycogen Synthase Kinase 3β), many chemical classes of molecules were chosen. A CADD scientist may chose about six classes and group them as G1–G6. In group G1, all the molecules belong to a specific chemical class, from this group the most active compound M1 may be chosen and its pharmacophoric features can be noted, for example, 22 features. Similarly, from group G2, molecule M2 may be chosen for which the number of pharmacophoric features may be 20. The number of common pharmacophoric features present between the molecules M1 and M2 may be only about 15, after considering all possible conformations of M1 and M2. When the same exercise is

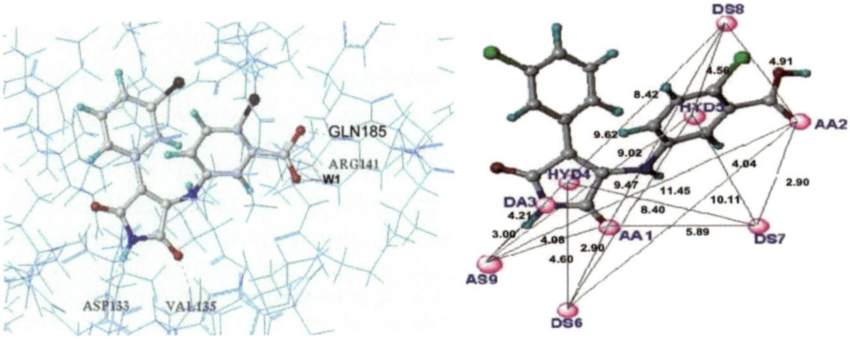

(a) (b)

Fig. 6.17 (**a**) Interactions shown by the 3-anilino-4-arylmaleimide in the active site of GSK-3β. (**b**) Pharmacophoric features and their distance (Å) relation (ligand point-site point) noticeable in the model along with a derivative of [3-anilino-4-arylmaleimide. Reproduced with permission from Springer Nature "Ref. [70]"

extended to M3 from G3, the number of common features among M1, M2, and M3 get further reduced to about 12. Subsequently, when this process is extended to M4, M5, and M6, respectively, from G4, G5, and G6, the number of common features among the chosen best molecules from diverse classes may remain only 6/8. Thus, the "unity in diversity" as well "common-minimum" feature selection becomes a part of the finally selected pharmacophore map. To perform the conformational analysis of the chosen molecules for the mapping purpose after superimposing the pharmacophoric features, the help of computers is required. Once a pharmacophore with about eight features has been identified, the distances among all these features, in 3D, can be noted along with tolerance limits. The above procedure was followed in the pharmacophore mapping technologies—DISCO (DIStance COmparision) and DISCOTECH (DIStance COmparision TECHnology). These platforms provide a 3D map of the pharmacophore; a typical example is given in Fig. 6.17. The distance between any two features in the map cannot be rigid, it should be partially variable; the permitted variance of the distance is known as the tolerance. For example, the distance between a particular hydrophobic feature and a specific hydrogen bond donor feature may be 7.8 Å and the tolerance limit is ±0.5 Å. This implies in a given molecule if these two pharmacophoric features are present within 7.8 ± 0.5 Å in 3D, in any part of the molecule, such set of features can be further considered [70].

There are many CADD technologies available to perform pharmacophore perception, each with a unique algorithm. Various variations in mapping procedures were adopted and some of the algorithms cannot be labelled as mapping technologies. Hence, "computational pharmacophore perception" is a more appropriate term instead of "pharmacophore mapping" [66–68].

Glycogen Synthase (GS) is a key enzyme in glycogenesis, GSK-3, it is a serine/threonine kinase which inactivates GS. Inhibition of GSK-3 is required to facilitate activation of glycogen synthase. A number of potent GSK-3 inhibitors such as Hymenialdisine, Paullones, and Indirubins were known to inhibit the GSK-3 enzyme in ATP competitive manner. Selective GSK-3β inhibition is a requirement as these substrates also inhibit Cyclin-Dependent Protein Kinases (CDKs), leading to side effects. DISCO software finds common feature pharmacophore model among a set of highly active compounds. This pharmacophore map represents the 3D arrangement of pharmacophore features common to the reported set of GSK3 inhibitors. Three important chemical databases (Maybridge, NCI, Leadquest) were searched to identify new lead compounds. The hits obtained from each database were subjected to molecular docking analysis using FlexX method and the crystal structure (PDB: 1Q4L) [70].

Generally, pharmacophore mapping is considered as an LBDD approach, in which the information regarding small molecules is sufficient to carry out mapping. The receptor-based (or structure based) pharmacophore mapping method is an SBDD method, because the structures of macromolecules are considered in performing the mapping of features. Generally, such a method can be employed when the information from the receptor site is available in clear terms and the information from the ligands is available in less quantity (or quality). Such an approach is less frequently used but it is also effective.

6.12 Molecular Docking

Docking word in English is associated with parking a ship in dockyard, which is an involved process. Molecular docking can be similarly considered as a parking process in which a drug/lead molecule needs to be parked inside the active site of a macromolecule; this is also a very complicated process. The most popular methods of molecular docking consider that the given macromolecular environment is rigid and the ligand being docked is flexible. This implies that the molecular docking technologies currently are not incorporating the induced fit effects (i.e., mutual change in the conformational preferences of the drug and the active site as they approach each other, so as to maximize the interaction between them).

Molecular docking is one of the most widely used CADD methods [71, 72]. The science and technology associated with this method is growing in strength; however, several issues associated with molecular docking are yet to be addressed. Incorporating induced fit effects routinely is one of the unsolved issues in molecular docking (though a few specific attempts were made to resolve this). Hence, molecular docking technology has an extremely high growth potential, in terms of both science and technology. Further, this technology is being used for the prediction of adverse effects of drugs, drug repurposing, polypharmacology as well as target fishing and profiling [73] (Fig. 6.18).

Docking algorithms: Molecular docking is a $3n$-dimensional problem because it has to take care of $3n - 6$ degrees of freedom due to vibrational motion within the molecule, as well as the three rotational motions and three translational motions into

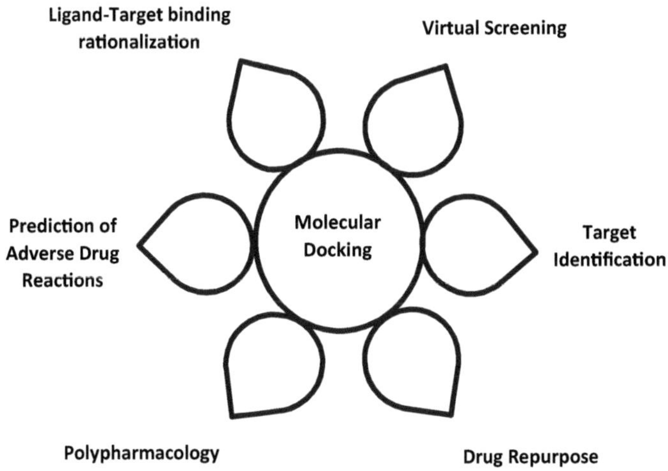

Fig. 6.18 A diagram showing the various applications of molecular docking

account in the active site cavity of the macromolecule. Several methods have been considered to carry out this exercise. The purpose is to identify the bioactive conformation of the drug inside the cavity of the macromolecule and to identify the effective interactions between the macromolecule and the small molecule. The final aim is to design many molecules which can exhibit similar interactions and similar conformational preferences.

Molecular docking site identification: The active site of a macromolecule may be formed due to many factors. In the case of receptors, it is the site where small molecules are recognized, received, and hosted. In the case of enzymes, the active site may be the substrate-binding site or allosteric-binding site or cofactor-binding site or DNA-binding site or catalytic triade or metal-binding site or a covalent bond forming site, etc. In the case of nucleic acids, the active site may be between the base pairs (for intercalator binding) or in the minor groove or in the major groove or a specific atom where a covalent bond may be formed with small organic species (e.g., methylation centers). These active sites may be reorganized to host the small molecules as in the case of receptors. Alternatively, the active site may be dynamically formed as the host approaches the site as in the case of intercalators.

In addition to the active sites where the small molecules can bind, there are several channels within the macromolecules through which the small molecules can traverse. For example, in GPCRs (G-protein Coupled Receptors), the receptor site is actually any place along a channel formed due to the parallel arrangement of α-helices. In case of enzymes, there are substrate-binding domains and catalytic domains; for example, in CYP3A4, the drug entry channel is characterized by five phenylalanine residues. Ion channels for metallic ions (potassium channel, sodium channel), anion transporters, cation transporters, etc. are all characterized by pathways, where the inhibitors may bind. These channels are not preorganized in a rigid manner to allow the travel; they are dynamically opened up as the small molecules approach to selectively allow a few small molecules.

Before performing molecular docking, the macromolecular structure should be prepared as defined in Sect. 6.7. The site at which the small molecule needs to be docked should be identified. This space may be deep inside the macromolecule (cave shaped, with a possible entry into the cave)—e.g., PPARγ (Peroxisome Proliferator-Activated Receptor). The active site may appear like a pocket, in which the small molecules may enter/exit with ease. The active site may be on the surface of the macromolecules (a shallow bowl shaped). This space may be between two monomeric units of a dimeric system. It may be a groove, in which a peptide/a protein/a nucleic acid may bind. Allosteric domain is present in many enzymes; this may be the site for docking. Similarly, many more ways of defining docking site are available.

To identify the docking site, experimental structures of the macromolecules obtained from PDB are useful, when a bound small molecule is available. It is possible to make an intelligent guess of the active site from the experimental mutagenic studies also. Fluorescence measurements can also be used to identify the sites. Quite often, such experimental details may not be available, especially when the macromolecular structure was obtained from molecular modelling methods like homology modelling. In such cases, molecular modelling methods can be used to identify the active site. One approach is to identify all the spaces by filling the entire macromolecular environment with virtual spheres of 1 cubic angstrom. Many spaces with various volumes become apparent. Among them, a few can be dropped because they are too small in comparison to the volume of a known ligand. A few more may be dropped because they are too disconnected to the known biochemistry of the macromolecules. Remaining spaces (handful of them) can be validated by trial and error method. Such a concept is being adopted by many software modules—e.g., SiteID and SiteMap. Similarly, a few software tools were developed to identify the pathways through which the small molecules may traverse—e.g., MolAxis.

Docking site preparation: The purpose of molecular docking technology is to computationally mimic the molecular recognition process between drug and macromolecule. Under natural conditions, the drug and macromolecule recognize each other and adopt a pose which maximizes the complementarity between them. In other words, both the molecules adopt a "comfortable pose" which complement each other's surface properties. The molecular docking algorithms should mimic this natural process of molecular recognition. Experimentally the only source of knowing such complementarity is the crystal structures of the macromolecules when crystalized with drugs. NMR experiments and fluorescence experiments may also provide some indirect information, but mostly at bulk level, but not at the atomic level. Molecular docking needs to consider translational motion (3 directions), rotational motion (3 directions) of the drug in addition to its vibrational motions ($3n - 6$), in the active site. The "work space" at which molecular docking should be carried out needs to be defined in advance. This is known as the receptor site preparation. The site may be a grid box or a spherical region, within which all the atoms of the drug should be positioned during docking. For example, the molecular docking software "Glide" defines the receptor site as a grid of two concentric rectangular boxes. While finding the best docking pose of the small molecule, the centroid of the drug should not go beyond the limits of the inner box and none of the atoms of the drug should go beyond the outer box. The centroid of these two boxes can be defined using Cartesian coordinates; it should be close to the centroid of the active site. The molecular docking software GOLD uses a spherical workspace for the docking purpose, which can be defined by supplying the Cartesian coordinates of the center of the sphere and the radius of the sphere.

Molecular docking algorithms: A molecular docking method can be considered as a conformational analysis exercise, because it is required to find the best conformation of the molecule which effectively occupies the space available within the active site. This conformational analysis is with space limitations and interaction constraints. The active site consists of many amino acids or functional groups; the molecule which is entering the active site should not "bump" into the atoms contributing to the active site—i.e., the atoms of drug molecule should maintain sufficient distance with reference to the constituent atoms of the active site. This is a sterical requirement. The atoms of the drug should be complementarily positioned with reference to that of the constituent atoms of the active site. For example, a negatively charged center of the drug should be positioned close to the positively charged center of the active site. Similarly, a hydrogen bond donor center of the drug should be positioned near a hydrogen bond acceptor center of the active site. The aliphatic hydrophobic site, the aromatic hydrophobic site, the sites with possible charge transfer, the sites suitable for salt bridge formation, etc. of the drug should, ideally, find complementary atoms in the active site of the macromolecule. The job of a molecular docking algorithm is to maximize the favorable interactions between drug and the macromolecule, within the conformational constraints of the drug and that of the space provided. For this purpose, several innovative algorithms were developed; these include—Incremental Construction (IC), Shape Matching (SM), Monte Carlo Simulations (MCS), Genetic Algorithms (GA), Evolutionary Programming (EP), and Tabu Search (TS).

The details of the algorithm IC are given below. This is an innovative procedure, in which a drug molecule is considered as a union of several fragments stitched in a desired order. Initially, the molecular docking of the small fragments in various sub-pockets of the active site is taken up. Based on this analysis, a best molecular fragment (lead fragment) and its complementary sub-pocket are chosen. The rest of the molecule is reconstructed in the active site in an orderly manner, while maintaining best complementarity between various fragments with the amino acids in the active site. Such a process is repeated with all important lead fragments. Thus, many poses of the drug inside the cavity can be obtained, which can be compared and ranked on the basis of scores (vide infra).

Scoring functions: Once a molecular docking pose is identified, the energy associated with the interaction between the macromolecule and the drug in that pose can be estimated. Ideally this value should be same as the free energy change (ΔG) associated with the interaction between the drug and the receptor and it should be negative as energy release is expected. However, in practice, the estimated value cannot be ΔG, because (i) induced fit effect is not taken into account in many cases, (ii) to estimate the interaction energies of each contact between drug and the active site, the time required is high, (iii) several poses need to be considered and several ΔG values need to be estimated, the whole process becomes expensive in terms of computer time, (iv) the best way of estimating ΔG is quantum chemical method, which is prohibitively expensive, even the semiempirical methods are not practical. Hence, in a molecular docking experiment, ΔG values are not estimated. Instead, using some empirical functions, each pose is "scored"—i.e., relative importance of the various poses of the drug inside the active site are estimated. The scoring functions can be based on (i) molecular mechanics methods or (ii) knowledge-based methods or (iii) empirical functions or (iv) machine-learning methods. Thus, the scores obtained cannot be inferred on absolute terms, but they should be considered only on relative terms. If one is comparing 30 different poses (between a pair of drug and macromolecule), the best among them may be chosen based on the estimated scores.

It is important to remember that the best score values of a series of molecules against a particular receptor may not follow the same order as that of the biochemically observed activity values. There is no one-to-one correspondence between the scores and activities because (i) the scores are obtained using some type of a fitting function, (ii) full dynamical state of the macromolecule is not considered during molecular docking, and (iii) while calculating the scores the pharmacodynamics factors are considered but the factors associated with pharmacokinetics are not incorporated.

The score value of a molecular docking pose using an empirical scoring function may be estimated as follows. First, the number and types of various interactions need to be counted. In a particular pose, a few interactions may be favorable (hydrogen bonding), a few may be unfavorable (hydrophobic–hydrophilic contacts). Each type of interaction carries a specific, previously assigned, empirical value. The total score can be obtained by summing up all the interaction energies.

In fact, scoring functions and molecular docking algorithms are interlinked. To identify the best complementarity, it is required to explore the poses with maximum number of favorable interactions. "The reliability of molecular docking depends on the accuracy of the adopted scoring function, which can guide and determine the ligand poses when thousands of possible poses of ligand are generated" [74–76]. To estimate the number and type of favorable interactions, scoring functions are useful. Hence, molecular docking algorithms are seamlessly integrated with the scoring functions. The rank ordering of various poses of the drug inside the active site is based on the best pose which produces highest score. The users, however, need to compare the result from molecular docking with the known biology, and chemistry of drug–macromolecule interactions before proceeding to the next step of the drug discovery process. Comparative Assessment of Scoring Functions (CASF) is an open access benchmark platform, which follows four metrics—"scoring power," "ranking power," "docking power," and "screening power" [76]; in this process, the scoring process is decoupled with docking process.

In practice, any molecular docking effort is associated with the following steps—(i) ligand 3D structure preparation (Sect.6.6), (ii) protein 3D structure preparation for molecular docking (Sect. 6.7), (iii) docking site identification (in 3D), (iv) redocking the ligand which is already known to occupy the same active site (if available), so as to validate the chosen docking algorithm, (v) dock a set of known ligands to understand the important interactions and possibly to know the mechanism of pharmacodynamics of the drug action, (vi) dock many molecules from a compound library to perform virtual screening, and (vii) finally suggest a few compounds from the chemical library as hits towards lead identification and lead optimization. While suggesting new hits, the comparison with the docking score and docking pose of the already known lead molecule can be taken as a guide. Only those molecules which show interactions similar to that of the lead compounds need to be considered. Also, it should be ensured that the docking scores are comparable to (or better than) that of the lead compound.

Case Study 2

Discovery of Anti-influenza drug Zanamivir. Influenza is a viral infection; the antiviral drugs become useful in combating this infection. Sialidase inhibitors are anti-influenza drugs. The enzyme sialidase cleaves *o*-sialic acid, at the end of the viral replication cycle, which further releases viral progeny leading to the infection of other cells. Inhibiting sialidase reduces the replication of virus and thus it was recognized as a drug target [77]. Colman and coworkers reported the active site of sialidase (PDB id: 7NN9) [78]. They identified the active site to contain many charged residues [79]. von Itzstein et al. [80] performed molecular docking studies to design inhibitors of sialidase. Grid-based calculations using a probe (protonated primary amine) were performed to identify a hotspot around the accessible region. A salt bridge formation with Glu119 was found to be an important requirement. The replacement of the hydroxyl group of sialic acid with basic guanidinyl group was found to be a useful structural modulation during this exercise [81]. This structure-based drug discovery effort finally lead to the identification of Zanamivir as a drug. Figure 6.19 shows the docking poses of Zanamivir in the active site of sialidase. In vivo analysis validated the results of these molecular docking studies. In 1999, Zanamivir was approved as an anti-influenza agent [82]

Similarly, many anti-HIV drugs, introduced in the past 20 years, were designed after thorough rational efforts employing molecular docking. During the rational design of HIV-1 integrase inhibitors, the metal-binding character of the beta-diketo acid (DKA) groups was exploited [83]. After replacing the DKA pharmacophore with naphthyridine carboxamide core also, the basic interactions were maintained [84]. Further, modification to *N*-alkyl hydroxypyridinone carboxylic acids (using molecular docking analysis) resulted in new analogs which exhibited good pharmacokinetic profile in rats [85]. Using molecular modelling tools, further modifications in this class of compounds led to the design of Raltegravir, a promising pyrimidinone carboxamide derivative [86].

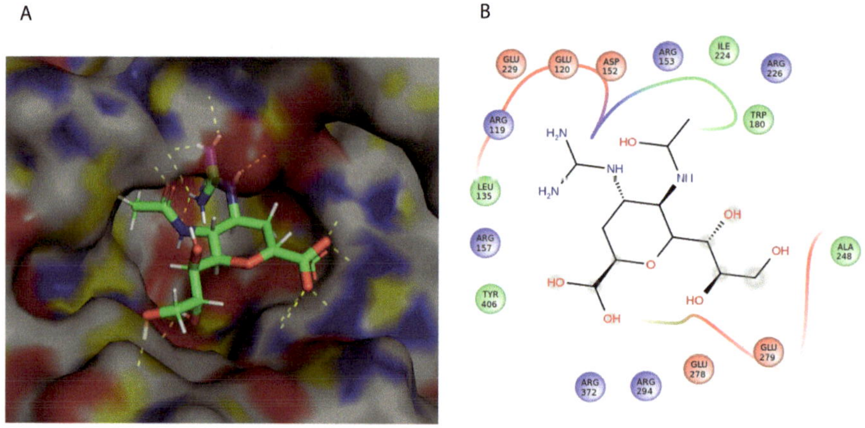

Fig. 6.19 (**a**) Ligand–protein complex of Zanamivir bound to influenza-A Sialidase enzyme (PDB id 5L17). Yellow dotted lines represent hydrogen bonds. (**b**) 2D diagram of the ligand–protein interaction showing a closer look at binding cavity. Zanamivir shows interactions with Arg119, Glu120, Leu135, Asp152, Arg153, Trp180, Ile224, Arg226, Glu229, Glu278, Glu279, Arg294, Arg372, and Tyr406 residues of sialidase

6.13 De Novo Drug Design

De novo drug design is an SBDD method [87]. When the structure of a therapeutic target (macromolecule) is available and when there is no idea about the structure of the small molecules which best fits in the active site of this macromolecule, De novo design is quite useful. Once the structure of the macromolecule is available, it is important to know the bioactive site, as described in Sect. 6.12. Next step is to consider a library of chemical fragments from a database of biologically relevant chemical functional groups. Molecular docking of all these fragments helps in shortlisting a few fragments which fit comfortably in various sub-pockets of the active site. A small set of these successful fragments can be stitched via linker fragments to generate a library of virtual molecules (LigBuilder is a known software to carry out this work). These designed ligands can be docked in the active site cavity to further select molecules on the basis of their docking score and docking pose. Thus, de novo design helps in identifying molecules, which may exhibit effective pharmacodynamic properties; a flow chart showing the activities relevant in de novo design are given in Fig. 6.20. This technique is also being used to design new scaffolds (of organic molecules) when a few classes of compounds are known, which do not show sufficient progress in drug discovery research.

Tacrolimus is an immunosuppressive drug (given to lower the risk of organ rejection after organ transplant), which inhibits the production of interleukin-2. It is a macrolide, which binds to a family of enzymes FKBPs and thus exhibits potential to treat neurodegenerative disorders including Parkinson's disease. A few lead molecules are known, which weakly bind to these enzymes. FKBP12 is one of the important targets; using its 3D structure (PDB id: IF40), a few lead molecules

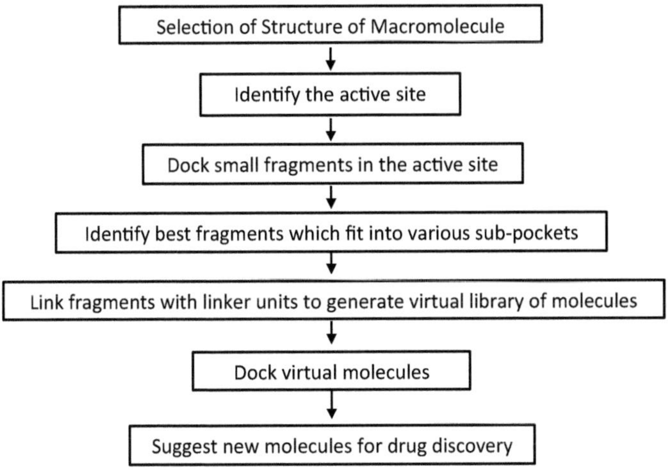

Fig. 6.20 A flow chart explaining the sequential set of activities involved in de novo design

were designed, employing the computational de novo drug discovery software LUDI. Initially, a few fragments, which form hydrogen bonds in various sub-pockets of FKBP12, were identified from a fragment library. This helped in the identification of a lead molecule with novel scaffold. These fragments were linked using spacers and about 57,120 virtual molecules were generated, which were subjected to energy minimization. In this set of virtual molecules, to improve hydrophobic contacts, two methyl groups were introduced in the designed molecules [88, 89]. Also, to further improve electrostatic interactions, a specific carbon atom was replaced with O/S atoms in the pyrrolidine ring (of the lead molecule). FKBP52 is very similar to FKBP12 in terms of sequence and structural similarity and it has better mechanistic significance in neuronal survival. Molecular docking of the 57,120 virtual molecules was carried out using the modelled 3D structure of FKBP52. The molecular docking software DOCK 4.0 was employed to perform docking analysis. The docking site was identified based on the similarity with the Tacrolimus-binding site of FKBP12. Three different scoring and ranking functions were used to screen molecules. In the first step, a scoring function was used which performs shape complementarity analysis, 10,000 compounds were selected. In the second step, an energy scoring function (based on van der Waals and electrostatic energies) was used and 1000 compounds were selected. In the third step, compounds with enhanced recognition of chemical complementarity were designed using Insight-II software. Finally, 43 compounds were selected for synthesis based on the available reagents, drug likeness, cost of generation, and synthetic viability. Biological assays identified one of the 43 as the potential lead molecule, thus providing proof of the concept [90, 91].

6.14 Virtual Screening

Virtual screening (VS) is one of the most important technologies associated with CADD [92–94]. Here, virtual implies that the chemical compounds are virtually defined using computational methods and the filters used in the screening process are based on computational techniques. The set of compounds employed from databases are known as virtual library of compounds, a few of these species may be experimentally known. The virtual filters are software-based analytical tools or a set of empirical rules, which can be employed as virtual screens. The result of the virtual screening is a set of hit molecules, which may or may not exist experimentally. Virtual Screening is the art of selecting a few compounds from large number of virtual chemical species. Each time a filter is applied during VS, a specific desirable property gets added to the selected set of molecules. The desirable property may be based on pharmacophore mapping or molecular docking or some empirical rules, or physicochemical parameters or toxicological indicators. There is no prescribed way of carrying out virtual screening; depending on the problem at hand, the work flow in virtual screening varies. The major goal is to identify a set of molecules, which carry a set of desirable properties, which can be taken to the next level of drug discovery research. The selected set of molecules may be purchased (if available) or synthesized in chemical laboratories. If the virtual screening strategy is successful, there is every chance of finding a novel lead molecule from a set of many compounds.

Generally, the initial starting point in VS includes millions of compounds (from a databank), the end point of the VS exercise contains handful of compounds. A data bank of virtually created molecules can also be used, in which case, the strategy employed to create the virtual library of molecules also adds a desirable therapeutic value to the set of molecules. In a few cases, several series of data expansion and data reduction steps are involved in the VS workflow. Each step adds a desirable property to the molecules being created or selected. Figures 6.21, 6.22, and 6.23 describe three different workflows, with specific examples.

Figure 6.21 describes the selection of compounds for GKS3 inhibition purpose. Initially, the 3D structures of ~350,000 compounds were chosen from three different databases—NCI, Maybridge, and Leadquest. The first screen was a pharmacophore map (developed as described in Sect. 6.11). This was followed by the filtration using Lipinski's rule of five. After these two filters, 5878 hits were identified. Further, restricting the number of rotatable bonds to less than eight decreased the number of compounds to 967. Only 61 compounds from this list were found to be comfortably docked in the active site of GSK3β. Finally, based on the docking scores and docking poses, nine compounds were selected for further modulation. These hit compounds carry the necessary pharmacophoric features, they show complementarity with the active site of the enzyme, they exhibit binding scores better than the lead compound, they follow Lipinski's rule of oral bioavailability, and they possess less than eight rotatable bonds. Taking clues from the above work and further carrying out de novo design as well as QSAR studies and electronic structure analysis, the authors designed many new species computationally, synthesized them and showed

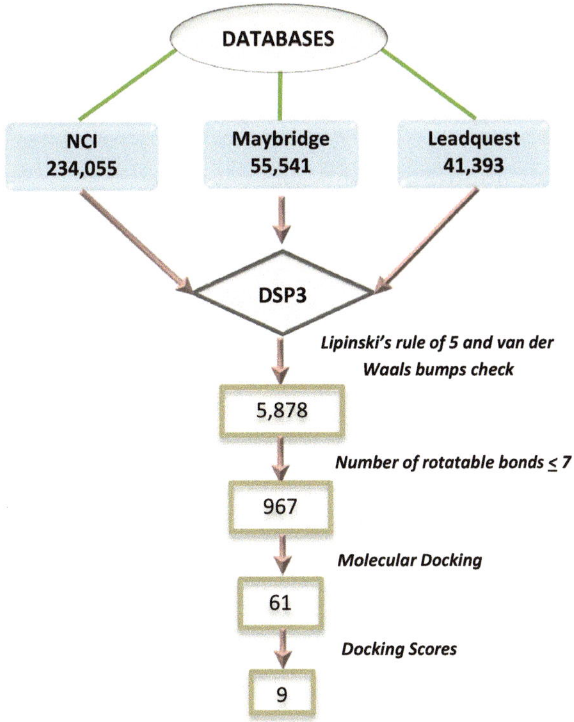

Fig. 6.21 The results of the virtual screening using DSP3 model

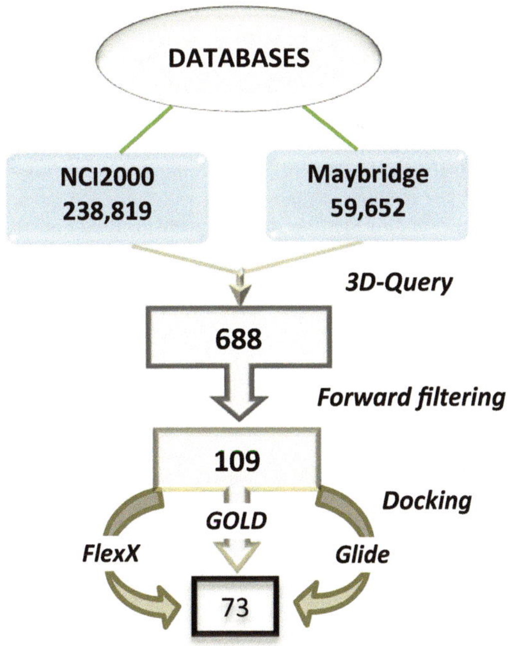

Fig. 6.22 Virtual screening for the identification of potential *Pf*DHFR inhibitors

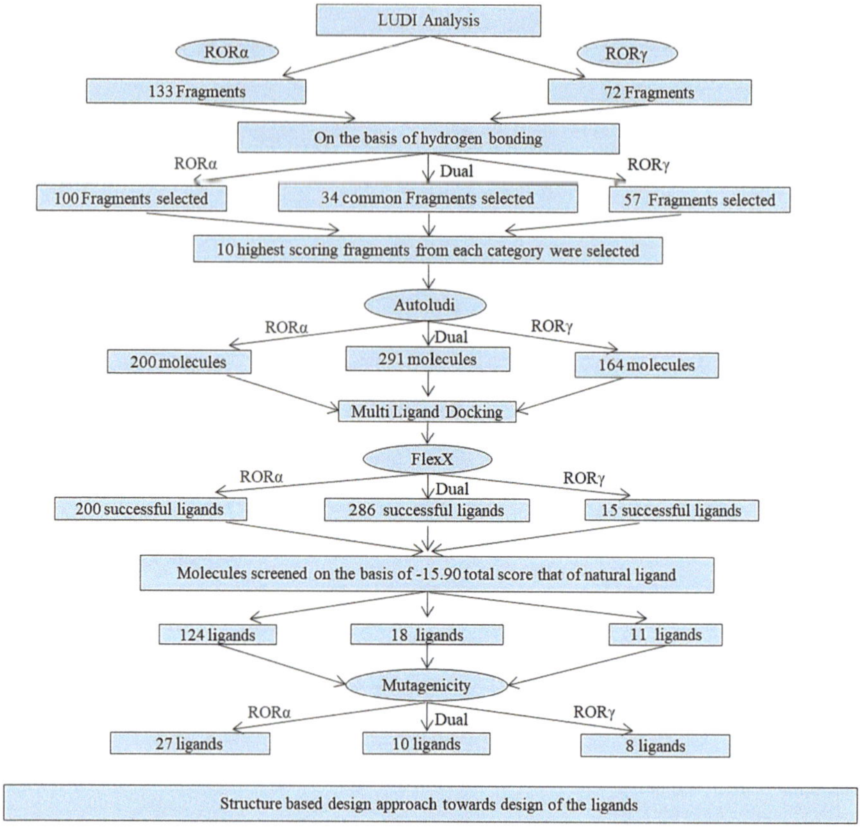

Fig. 6.23 Designing dual activators of RORα and RORγ. Initially, based on LUDI software, many fragments were found to fit in various sub-pockets of RORα and RORγ. After considering interactions, three classes of fragments were identified, namely, RORα selective, RORγ selective, and dual binders. Ten highest scoring fragments were selected to design molecules de novo. Molecular docking of these virtual molecules and mutagenicity-based screening lead to the identification of 10 ligands as dual activators of RORα and RORγ

them to be active towards GSK3β (with IC_{50} values in 2–85 nm range) and selective against CDK2 [70, 95–99]. Figures 6.22 and 6.23 provide two more examples of virtual screening.

6.15 Molecular Dynamics in Drug Design

Drugs, macromolecules and their complexes are dynamical species. The dynamism originates because of the forces acting on the atoms due to the $3N - 6$ degrees of freedom, which is manifested in the form of bond stretching, angle bending, and torsional angle motions. The description of the structures and interaction energies considered in Sects. 6.2–6.14 are based on the static state of these species; though this description is not the best, it provides sufficient information to the CADD scientist. However, if the dynamism can be considered, there is added advantage; thus, molecular dynamics methods were developed. These methods are dependent on the Newton's laws of motion and the time-dependent changes in the interactions between atoms.

Induced fit theory [1, 100–102] of drug receptor interactions define that—when a small molecule approaches a macromolecule, conformational change occurs in both. The structures of the macromolecule and the drug molecule are not readily prearranged to facilitate maximum interaction between them (as in the lock and key model); they undergo conformational changes to maximize the interactions. The drug and macromolecule are probably not in the energetically best conformational state (individually) in the complex, but the overall complex is in the energetically best arrangement due to induced fit between the drug and macromolecule. Further, these conformational changes trigger a few responses which may be the origin of the observed therapeutic action due to drug–receptor interactions. The same concept was further extended to consider the specific vs. nonspecific conformational changes during drug–receptor interactions (macromolecular perturbation theory) [4]. Further, it was postulated that the drug–receptor interactions may exist in equilibrium between activated and nonactivated states, whereas only agonists drive the conformation towards the activated state (activation-aggregation theory) [4]. In another extension of the concept, it was postulated that the macromolecules exist in two states—relaxed and tense (R & T), the relaxed state being the active state (two state theory). The characteristics of agonists, antagonists, partial agonists, and inverse agonists were defined with reference to these hypotheses [4]. One thing common is that drug–macromolecule complexes are always flexible. In the molecular docking efforts, the structures of the macromolecules are generally considered to be static (for convenience), which is not acceptable according to molecular dynamics methods.

Every atom in a molecule experiences force due to all other atoms, nearby atoms produce maximum effect. As a result of this force, the position of the atoms gets modulated (Δx) after a small change in time Δt. Simultaneously velocity also gets modulated ($\Delta x/\Delta t$). This change in position and change in velocity leads to change in force being exerted on the atoms, which in turn changes energy (ΔE) of the molecule. In a microcanonical ensemble, N, V, and E (N is number of particles, V is volume, and E is energy) are kept constant; hence the exchange of potential and kinetic energies takes place. Thus, iteratively all these parameters get modified, which can be measured. Changes happening after every femtosecond (fs) are important. Over a period of one picosecond, (1 ps = 1000 fs) many changes happen, though these changes are small. Over a period of a nanosecond, (1 ns = 1000 ps), again many more changes happen at the atomic level, though the system remains the same. These changes, which continuously happen in any molecule, are the characteristic features of the molecules.

When these changes are observed carefully in any system, it is possible to get information regarding the dynamical character of the system. If the observations are made after every 2 ps, total 5000 observations can be recorded over a period of 10 ns.

Under micro canonical conditions, since T (temperature) is not constant, with a change in temperature, further modulations take place, which provide further information. When a molecule is in equilibrium state, the Root Mean Squares Deviation (RMSD) of these changes turn out to be very small. It is generally considered that every possible change of a given drug molecule can happen within a period of 1 ns. For a macromolecular system, it requires about 10 100 ns to note all possible changes. When these changes are measured, it is possible to estimate the change in entropy (ΔS) also, which helps in estimating the changes in free energy (ΔG). Whereas the static models provide only the information regarding enthalpy (H), the dynamical models provide information regarding the free energy change ($\Delta G = \Delta H - T\Delta S$); this is a clear advantage.

The 3D structures of the drugs, macromolecules, and drug–macromolecule complexes can be defined in terms of Cartesian coordinates, and thus small changes in these coordinates (dx) with reference to small change in time (dt) can be measured; this further helps in observing the dynamical 3D state of a molecule. Recording observations after every few fs, is a very time-consuming process and it cannot be done without computers. In practice, to record these observations, after every few fs the dynamical process needs to be halted, and after recording the observations, it needs to be restarted. At the same time, new values of the velocity and acceleration need to be assigned to each atom. This period of few fs is known as "step size"; ideal step size to make observations about a drug–macromolecule complex is 2 ps. These observations need to be recorded many times over a period of long time, which is known as the "length of the run". There is no specific recommended value for the length of the run, because it is best to observe the system for a long time—as long as possible. In practical terms, the length of the run may be anywhere between 10 and 1000 ns. Analysis of the recorded changes, within the periodic boundary conditions, over the chosen length of the run helps in estimating many dynamical factors, including the estimation of ΔG. It is possible to estimate the binding free energy change due to the drug–macromolecule interaction, once the ΔG values of the drug, macromolecule, and the drug–macromolecule complex become available. Such an information is essential to estimate the relative binding affinity of a series of molecules and to take informed decisions required in CADD [103, 104].

$$\Delta G_{binding} = \Delta G_{drug\text{-}macromole} - \Delta G_{drug} - \Delta G_{macromolecule} \tag{6.7}$$

MD simulations cannot be routinely used for CADD purpose, because the computer time required to perform the simulations is very high. Also, the MD simulation methods are not technology-oriented methods—they provide a lot of scientific information. These scientific inputs can be used to explain the chemical/biochemical/biological phenomena observed but they cannot be used as quick predictive models. The following two case studies provide examples of the utilization of MD simulations to understand intricate scientific aspects, which eventually helped in designing inhibitors.

Case Study 3

Molecular dynamics evaluation of P218: an antimalarial drug candidate and the design of novel selective inhibitors of PfDHFR. P218 has been reported to be an antimalarial lead molecule, targeting PfDHFR; it is currently undergoing volunteer infection studies for prophylaxis. Yuthavong and coworkers developed P218 as an antifolate under the "Medicines for Malaria Venture (MMV)." It was reported that P218 selectively binds to PfDHFR, without influencing human DHFR (hDHFR). It was found worth to identify the reasons for the desirable selectivity being exhibited by P218. To evaluate this, molecular dynamics simulations (35 ns, using Amber software) were carried out on the complexes hDHFR-P218 and PfDHFR-P218 (four varieties—wild-type, quadruple mutant, double mutant-1, and double mutant-2). The MM-PBSA calculations were performed to estimate the binding free energy contribution which revealed that P218 has a higher binding affinity (~−30 to −35 kcal/mol) towards PfDHFR (all forms), whereas it exhibits a relatively weak binding affinity towards hDHFR (~−22 kcal/mol) (Table 6.2). Further, residue-wise interaction energies were estimated, which indicated that the origin of selectivity can be traced to Arg122 residue (Fig. 6.24). The MD simulation analysis verified that mutations do not influence the P218 binding affinity. Taking clues from this work, guanylthiourea (GTU) derivatives were found to exhibit effective PfDHFR inhibitory potential [105–107]

Case Study 4

Evaluation of GSK-3 active site using molecular dynamics simulations to design novel selective iminothiazolidin-4-one derivatives as selective GSK-3β inhibitors. GSK-3β is a serine threonine kinase, which was implicated in several physiological pathways and it is a druggable target for Alzheimer's and Type-II diabetes. To explore the origin of selectivity and with reference to the ATP competitive inhibition (GSK3β vs. CDK2), MD simulations and MMPBSA free energy evaluation were performed. The crystal structures of GSK-3β and CDK-2 in complex with 12 different reported ATP competitive inhibitors were chosen for this study. Based on the RMSD trajectories, it can be inferred that the complexes with CDK-2 showed higher level of fluctuations in comparison to GSK-3-inhibitor complexes. It was found that the electrostatic interaction energy plays crucial role in determining the selectivity against CDK-2. Per-residue energy decomposition analysis was carried out, which suggested that 13 amino acids are important for GSK3. Contrarily, only 8 amino acids are important for CDK2 inhibitory action. Lys85 and Thr138 (in GSK-3) make favorable interactions, while Lys33 and Asp86 (in CDK2) interact unfavorably with the chosen 12 chemicals. Arg141 was found to be very crucial for selective binding. Lys89 in CDK-2 was not found to interact with selective GSK-3 inhibitors. This highlighted the importance of Lys85, Thr138, Arg141 and the electrostatic energy in selective inhibition of GSK-3. Figure 6.25 shows the interaction energies of the six selected ligands with GSK-3β. It was found that the ligands L1 and L2 show strong stabilizing effects due to electrostatic as well as van der Waals forces. However, ligand L3 exhibits stabilizing van der Waals force, but destabilizing electrostatic force. Alternatively, ligand L4 exhibits stabilizing electrostatic force but

Table 6.2 The estimated binding free energies between the lead compound P218 and DHFR (five types). The energy values are in kcal/mol. Reproduced with permission from Ref. [107] The data given in bold are the experimental values or the final binding free energy values

Method	Contribution	*h*DHFR	wt*Pf*DHFR	qm*Pf*DHFR	dm1*Pf*DHFR	dm2*Pf*DHFR
	K_i (nM)	**2841 ± 319**	**0.51 ± 0.03**	**0.53 ± 0.13**		
MM	VDW	−45.36	−49.06	−40.29	−47.79	−46.78
	ELE	−131.46	−185.44	−242.81	−208.07	−178.86
	GAS	−176.82	−234.50	−283.10	−255.86	−225.65
PBSA	PBSUR	−3.59	−3.75	−3.59	−3.68	−3.77
	EPB	132.68	178.08	226.96	204.38	169.17
	PBELE	1.22	−7.36	−15.85	−3.69	−9.69
	PBSOL	129.07	174.33	223.37	200.71	165.40
	PBTOT	−47.75 ± 4.14	−60.17 ± 4.69	−59.73 ± 5.94	−55.16 ± 5.94	−60.24 ± 4.19
	−TΔS	25.61 ± 3.35	25.27 ± 2.37	25.58 ± 2.79	25.63 ± 2.44	25.31 ± 2.59
	ΔG_{bind}	**−22.14**	**−34.90**	**−34.15**	**−29.53**	**−34.93**

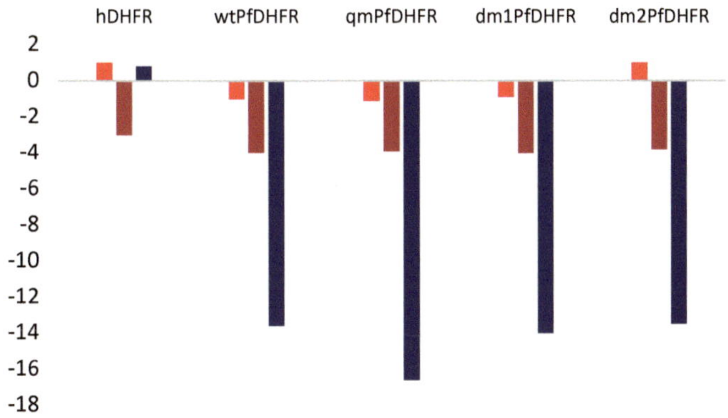

Fig. 6.24 Residue-wise interaction energies of the five types of DHFR. Phe (red), Gln (brown) and Arg (blue). The amino acids Phe and Arg are providing destabilizing effect in *h*DHFR. Arg is stabilizing in all *Pf*DHFR complexes

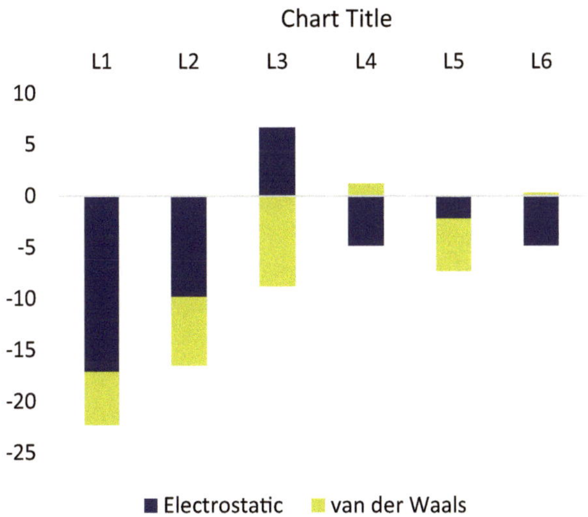

Fig. 6.25 The binding affinities of six GSK3 inhibitors were estimated using molecular dynamics simulations. Negative values indicate stabilizing effect and positive values indicate destabilizing effect. Accordingly, L3 exhibits destabilizing electrostatic interactions, whereas L4, L6 experience mild destabilizing van der Waals forces. All other interactions are stabilizing. The values along y axis are ΔG values in kcal/mol

destabilizing van der Waals force. The results provided in-depth knowledge of the structure–affinity relationship. Based on the above study, new iminothiazolidin-4-one derivatives were designed as selective GSK-3β inhibitors [105, 106]

6.16 Artificial Intelligence in Drug Design

The term Artificial Intelligence (AI) became very popular early in the field of computer science (~60 years ago), mostly due to the promise it was holding, however, not delivered. Though the "dream part" of the objectives of AI were never realized, continuous efforts in this field lead to the development of new concepts and technologies. A few of these concepts found application in chemistry, biology, and pharmaceutical sciences; in this section, we delve on these possibilities. It should be remembered that AI technologies are not yet routinely practical in CADD; they provide several clues, which may be useful. In comparison to the topics like LBDD and SBDD, the topic AIDD (Artificial Intelligence in Drug Design) is in its nascent state [108–110]. Even today many scientists working in AIDD are skeptical while making brave and pioneering efforts. Coaching computers the cognitive abilities is the major driving force behind AI.

AI emerged from the mathematics of logic, Boolean algebra, set theory, etc., in an attempt to make computers intelligent. On one hand it is associated with the technology of robots and on the other hand it provided many software platforms. Initially logical programming platforms were developed (PROLOG—programming logic and LISP—List processing) on par with general programing languages like FORTRAN and C++. Later several innovative technologies for machine learning (ML) were developed—Artificial Neural Networks (ANN), Support Vector Machines (SVM), Genetic Algorithms (GA), Fuzzy systems, pattern recognition tools, and classifiers. Currently the most progressive subtopic of AIDD is "deep learning." In the field of CADD, the AI methods are emerging complementary to the molecular modelling-based methods. As of today, it appears that the data intense operations (being offered by AIDD) appear to be different from those from the atomic-level operations (being offered by molecular modelling); however, the overlaps are increasing with an exponential growth in this field.

Machine learning (ML) became synonymous to AI. At one time, ML was a component of AI. However, the large progress made in ML and the numerous applications of ML available today are giving the impression that AI means ML. Especially in the field of CADD, other topics of AI (e.g., expert systems) became less prominent in comparison to ML, as MLDD (Machine Learning in Drug Design) has emerged as a successful endeavor. Machine learning and data mining are useful in providing many solutions such as classification, regression, clustering, dimensionality reduction, reinforcement learning, deep learning, and anamoly detection. Further, the several subtopics of machine learning include ANN, DNN (Deep Neural Networks), GA, SVM, Bayesian Networks, Decision trees (DT), Logistic Regression (LR), *k*-NN (*k* nearest neighbors), and Navie Bayesian (NB). Even the knowledge-based methods are integrated with machine-learning methods. Hence, in practice, there is very limited difference between artificial intelligence and machine-learning methods today. This is all possible because of the advancements in computational statistics. In this section, the applications of knowledge-based methods, genetic algorithms, ANN methods, and the SVM methods in drug design are introduced.

In simple terms, using ML methods, software models can be developed, which learn a few facts from the data provided. In fact, the hardware does not learn anything in ML—i.e., ML is not robotics process, in which it appears that robots have indeed learned some things very similar to humans. ML is a computer-aided technique through which some software component can be trained to grasp knowledge about a set of data. ML models infer patterns from the supplied data and make reasonable predictions on the new data. It is not possible to ascertain what the model learned in terms of derivitization or in terms of heuristic reasoning. Similar to the fact that human brain captures a few facts by observing, the ML models learn a few factors associated with the data provided to it. We cannot know what an individual synapse (a junction between two nerve cells) in human brain captures. Similarly, we cannot ascertain which component of an ML model learned which factors of the supplied data. The ML models learn using a statistical/probabilistic/optimization processes (but not a deterministic process). Hence, whether the learning process is complete—can only be established by testing the ML model. The available data needs to be initially divided into training set and test set (similar to QSAR methods). During training process the model learns, during testing also the model learns, while using the model also the model can be made to learn. With an increase in the data supplied to it, the ML model's quality increases. One of the advantages of ML methods is they can find patterns from noisy data also.

6.16.1 *Knowledge-Based Systems (KBS) in Drug Design*

Knowledge Base (KB) is a database of knowledge—either in the form of heuristic terms or numerical terms. Information related to any particular topic is stored in the form of Fact Bases (FB) in KB. The relations between various factual data points in Fact Bases are stored in the form of Rule Bases (RB). The rules need to be interpreted effectively connecting the elements in the Fact bases. Such an effort leads to the development of Expert systems (ES). First successful examples of knowledge bases are expert systems. To develop expert systems, facts from the experience of the experts need to be collected along with the simple rules of interpreting the same. Several intricate details are available with experts, who gained experience of interpreting the same to achieve the necessary results. Quite often, the experts may not consider a few facts as a part of their knowledge. It is the duty of the "expert system developer" to seek such details from the experts and prepare the appropriate fact bases and rule bases. The complexity of expert systems increase with an increase in the relations (expressed in the form of rules) among the facts. The expert systems served only limited sets of users.

With the advent of internet and explosion of knowledge, Knowledge Bases got transformed into Knowledge-Based Systems (KBS). Searching through internet is an application of KBS. Logical reason is the underlying connecting principle behind

KBS. In KBS, the knowledge does not emerge from one or two experts. Facts, rules and relations can be gathered from many sources, all should be factual and interpretable. Meta databases are one type of KBS [111–113].

In the case of CADD, the calculation of several descriptors for QSAR purpose can be considered as a knowledge-based system. For example, the molar refractivity of a given molecule can be calculated from the data associated with the molar refractivity values of its fragments (because it follows an additivity principle). The knowledge of the various fragments and their MR values are stored in a knowledge-based location and the MR values of any molecule can be calculated (on the fly).

Knowledge-based scoring functions (for molecular docking purposes) rely on the statistical observations of intermolecular contacts collected in large 3D structural databases. Potential mean forces (distance dependent) among various subunits of molecules are derived from the chemical and macromolecular databases (CCSD and PDB), because many such interactions occur frequently among the small and macromolecules. DrugScore, SmoG, Smog2016, ASP, DSX, IT-Score, DLIGAND, etc. are some of the known scoring functions which are derived from the KBS methods. For example, DrugScore scoring function has been derived from the crystal data, employing the distance-dependent pair potentials from nonbonded interactions. The scoring functions DrugScore^{PDB} and DrugScore^{CSD} are derived from the crystal data from CSD and PDB, respectively. Klebe and coworkers established that DrugScore^{CSD} provides relatively more satisfactory results [112]. To predict protein–protein interactions, a distance-dependent knowledge-based scoring function was developed (ITScore-PP). The crystal structures of 851 dimeric protein complexes were employed to derive this scoring function parameter [113]. Several ADME–Tox property-based quantitative values are also being estimated using knowledge-based methods [111].

6.16.2 Genetic Algorithms in Drug Design

Genetic Algorithms (GA) technique follows an algorithm which adopts the natural selection procedure over a period of generations. The optimization procedure adopted in GA includes the consideration of several generations of solutions, involving many candidate solutions (population) in each generation. All the alternate solutions provided in the first generation are not optimal. As the data moves from one generation to the next generation, (i) a few solutions get eliminated as per a fitness function (selection), (ii) a few get added (by crossover process), and (iii) a few get mutated (in a systematic manner). In this way, the solutions in the next generation (iteration) get better and the number of candidate solutions gets reduced marginally. When this procedure is continued iteratively for a few generations, only the fittest solutions remain, from which making appropriate choice becomes practical (Fig. 6.26a). The population size in the first generation, fitness function, crossover process, and the mutation procedure need to be decided based on the nature of the problem being handled [114, 115].

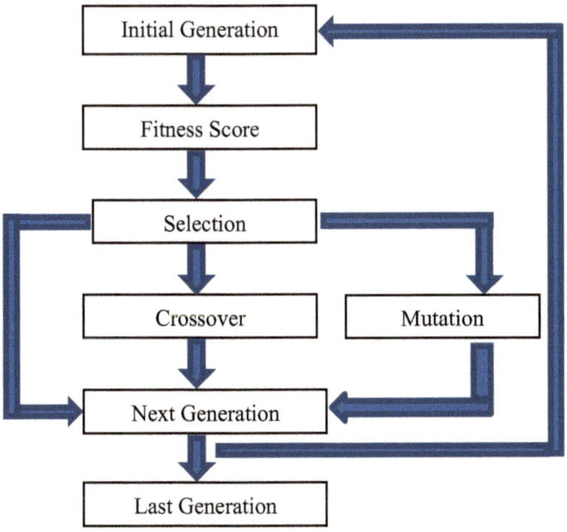

A)

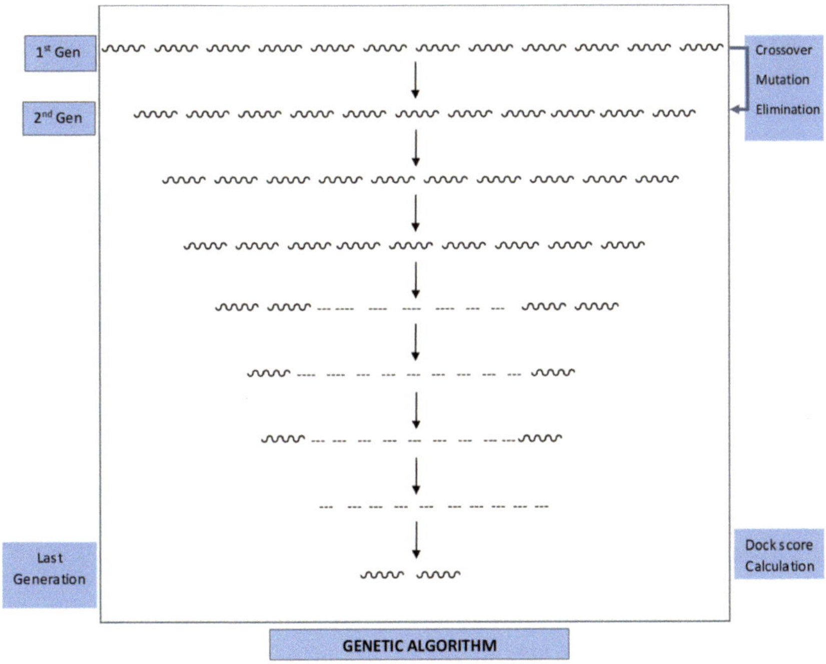

B)

Fig. 6.26 Workflow in genetic algorithms in two different representations [**a**, **b**]. [**b**] In molecular docking using GA approach, several conformers are considered initially, which gradually get reduced after several cycles

Genetic function approximation in QSAR is an alternate method to regression analysis [116]. Following this method, many models are built from which software provides a choice after considering generations of models. It can make automatic choice of descriptors for QSAR. It helps in discovering combinations of descriptors which leads to an observed correlation. It has inherent features to avoid overfitting. In the case of QSAR, descriptor selection can be done with the help of GA. In the initial generations several descriptors can be considered. Descriptor selection towards an effective correlation can be done after considering several generations of QSAR analysis. Statistical methods like MLR/PLR provide the statistical significance. In such exercise, the important caution is to ensure that the chosen descriptors are not completely random and the descriptors are not dependent. Ensemble methods were also integrated with GA-QSAR methods.

Sun et al. [117] effectively combined QSAR and GA methods to generate in silico prediction models to design inhibitors of methyl transferase. 134 known analogs were used to generate QSAR models, in which the descriptors obtained from quantum chemical methods were employed. A combination of MLR and GA was used to obtain the statistical significance. GA was specifically applied to select the descriptors among many. The descriptors—polarizability, TCI (topological charge indices), IP (Ionization potential), and number of primary aromatic amines in a molecule—were identified as the major contributors to the observed correlation. This analysis further identified nine substructures to be crucial in defining the SAR, which helped in designing novel species as inhibitors of methyl transferase. Similarly, antimalarial agent design using GA-QSAR was also reported in designing inhibitors of *N*-Myristoyltransferase [118]. GA-MLR methods were also employed to predict tyrosinase inhibitor design [119] as well as in cancer drug design [120].

GA in molecular docking: Molecular docking is a conformational search process inside a constrained environment of the active site of macromolecules. Several molecular docking algorithms based on GA have been developed. In general, these procedures follow a strategy, involving several generations of conformers. In the very first generation, many conformers of a given drug are considered paying attention to the torsional angles of the rotatable bonds. The conformers in the second generation will be created based on the three genetics processes—selection, mutation, and crossover, which are implemented, respectively, as follows—(i) Based on some knowledge-based energy function, a few conformers are eliminated from the first generation, (ii) the torsional angles of some selected conformers from the first generation are modified randomly to generate new conformers, and (iii) the torsional angles found in a few conformers will be exchanged with that of others. It will be ensured that the number of conformers in the second generation will be less than that of the first generation using some specific criteria (minimum 3% less). It is quite possible that on-an-average, the conformers present in the second generation are more suitable than that of the first generation—mainly because some rationale was employed in choosing the conformers of the second generation. The same procedure can be adopted to create a new generation of conformers using the three criteria (mutation, selection and crossover). This procedure can be continued for many generations until a set of 20/30 conformers are found in the latest generation. This

GA procedure ideally provides more reliable conformers in the latest generation, in comparison to those of all the previous generations. The conformers in the latest generation can be subjected to energy minimization to further choose the most reliable ones. Thus, several poses of the drug inside the cavity of the macromolecule can be identified using the automated procedure. The most suitable one can be subsequently chosen on the basis of known chemistry and biology. The software AutoDock uses the GA method for molecular docking purpose. Figure 6.26b shows a schematic representation of the GA-based molecular docking exercise.

6.16.3 Machine-Learning Methods in Drug Design

Computer software can be trained to grasp the important characteristics associated with drugs, nondrugs and allow the software to select drugs from millions of chemical species. RF, SVM, logic regression, NB, ANN, etc. methods are successful ML methods. This can be applied to specific classes of drugs like anticancer drugs. This was one of the earliest applications of machine-learning methods in drug design [121]. An example was provided by the work of Garg and Verma in the prediction of blood brain barrier permeability of drugs using ANN methods [122]. Such efforts are known as machine learning in drug design (MLDD). Major drawback in MLDD is to get the information regarding the logic applied by the software in making distinction between drugs and nondrugs and to correlate the same with the medicinal chemist's knowledge [123–126]. As a result, the drug discovery scientists were not very much convinced about the applications of AIDD, though the prediction by software was found to be reliable. Over the past few years, the applications of the ML in drug design are increasing exponentially and thus the skepticism is decreasing. In this section, a few examples of MLDD in drug design are included.

Estimating the scores during molecular docking using machine-learning methods is one of the successful ventures in MLDD. Classical scoring functions in molecular docking are useful but suffer from a few drawbacks. Especially, the nonlinear dependency of non-covalent interactions is ignored. Advantage of these improved methods is that the functional relation between drug and receptor is inferred in a statistical manner from the data. Wójcikowski et al. reported the development of RF-Score-VS, which was obtained from random forest training procedure [127]. The trained random forest network of 102 targets grasp the ability to perform virtual screening. The authors reported that this scoring function substantially improved the virtual screening analysis. A scoring function to improve the performance of molecular docking software AutoDock Vina was proposed using the RF-based machine-learning approach [128]. The reported limitations of these scoring functions are due to overfitting the data and absence of applications in new targets.

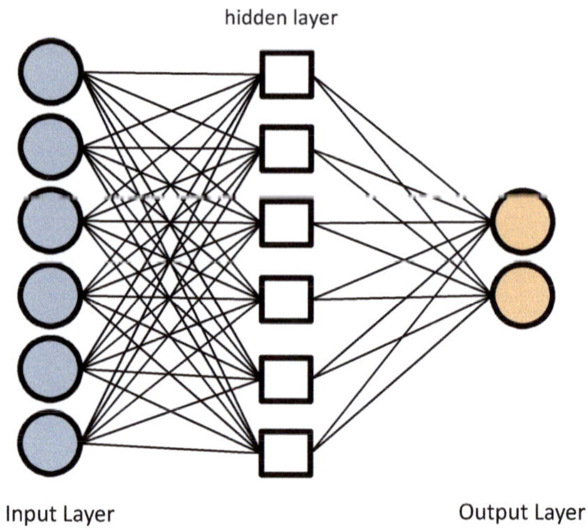

Fig. 6.27 Diagrammatic representation of an Artificial Neural Network (ANN)

6.16.3.1 Artificial Neural Networks (ANN)

Taking clues from the neural networks in human brain, the AI technology ANN was introduced [121]. ANNs found applications in classification initially; however, they are being innovatively used to teach intelligence to computers. A typical ANN consists of several layers of neurons. The first layer is known as an input layer and the last layer is known as output layer. Including one or two hidden layers, a typical neural network may appear as shown in Fig. 6.27. Each neuron is more like a synapse in natural neural network; it carries the ability to decide whether to send a signal or not to the next layer of neurons. Signals from neurons get multiplied with weights before getting transmitted to neurons in the next layer. The set of weights and the neurons in a given ANN grasp information regarding the system being studied. Each ANN needs to be trained using specific training procedure. Once an ANN is trained it carries the trained weights, which can very effectively produce the expected result as per the training provided to the network.

In case of an ANN trained for chemical purposes, the input layer may contain the physicochemical parameters or descriptors. Initially, an ANN may be trained with the help of a specific expected output using about 100–200 molecules. Training a neural network can be done using supervised methods or unsupervised methods. The popular algorithms for training an ANN are feedforward and backpropagation. Once trained, the network of weights and neurons carry the ability to reproduce the results as per the training provided. A well-trained ANN can be tested using a few more test set molecules (20–50 molecules) to verify whether the network is providing the expected results or not. A validated ANN can be used confidently many times, to predict properties of many molecules, which are not known. Once trained, the network carries the necessary information and predictive power. It is not possible to pinpoint where the information is present, the network as a wholesome unit carries the information. A trained network will give the reliable answer; however, it cannot give the reason behind the answer provided—this is the biggest limitation of this approach.

Improvements in ANN lead to Deep Neural Networks (DNN), Recurrent neural networks (RNN), Convolutional Neural Networks (CNN), Deep Convolutional Neural Networks (DCNN), Message Passing Neural Networks (MPNN), Feed-forward Neural Networks (FNN), Molecular Graph-Based Neural Networks (MGNN), etc. These improved methods are now finding place in CADD in a big way, at least the promise is very high. Especially the availability of GPU (Graphics Processing Unit)-based cluster computers facilitated the growth of DNN-based approaches. Considering that DNN found many applications in drug discovery [129–142], the next section describes the recently reported applications, as examples of ANN and all its improvements.

6.16.3.2 Deep Neural Networks

DNN is quite like ANN, but it contains multiple hidden layers. Nonlinear relations between the input parameters and the output can be effectively grasped by DNN during training. Mostly they follow the feedforward approach. The neurons in DNN multiply the inputs and weights and take a decision to give an output signal in binary

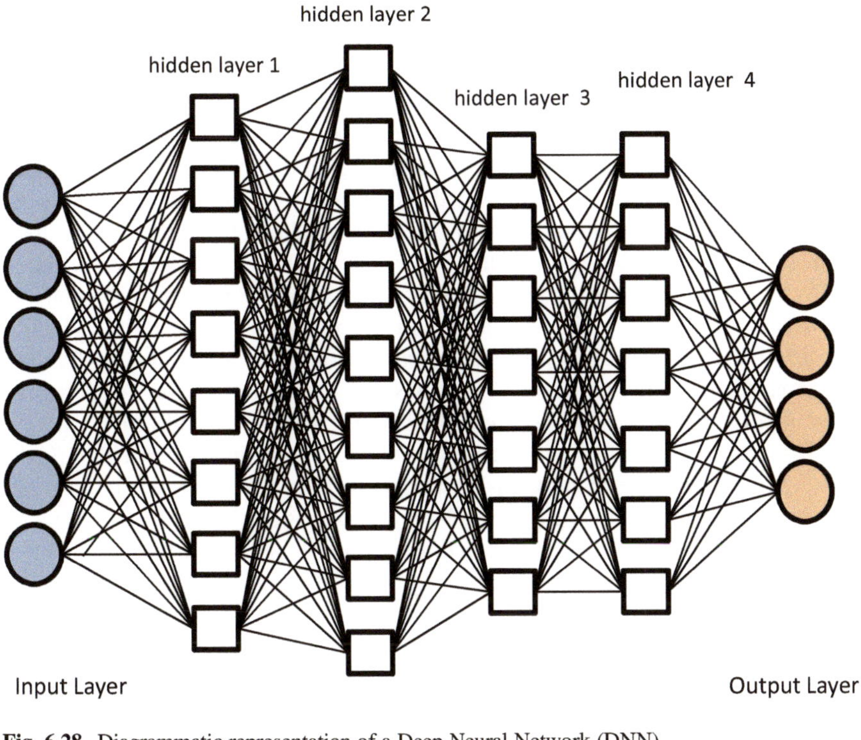

Fig. 6.28 Diagrammatic representation of a Deep Neural Network (DNN)

form (0 or 1). The weights get improved (in fact get adjusted) till the time the desired output is reached (during training). Computation time required to train a DNN is much larger than that of an ANN. Figure 6.28 shows a general representation of a DNN. Overfitting is a problem in DNN, which can be overcome with increased number of samples used in training. CNN and MGNN are neural networks which can be effectively used to interpret the 2D structural information of the drugs; they are tuned to perform image recognition.

Several reviews were published [129–136] which report the application of DNN in de novo molecular design, suggesting synthetic routes, prediction of binding affinity, estimation of activity properties, and evaluation of ADMET properties. Some of the advantages of DNN over other NN models is that DNN can handle thousands of parameters in the input layer—preselection of descriptors is not a requirement in this approach. The feature deduction happens during the DNN training process. Input layer can accept descriptors obtained from 2D/3D structures as well as from the molecular fingerprints. It is possible to fine-tune the performance of a DNN by modulating the number of layers, number of nodes in each layer, the activation function, and other characteristics. Multitasking ML is practical using DNN.

For example, Priyakumar and coworkers [141] recently introduced a Deep Learning model for "Energy estimation and Geometry Optimization of small Molecules (BAND-NN)". This framework follows a classical force field approach to predict atomization energy; it predicts potential energy surfaces of small molecules. This transferable and molecule-size ML model was designed to predict atomization energies of equilibrium as well as nonequilibrium structures over the reaction space in addition to configurational and conformational spaces. Geometry optimization of drugs using this model is an added advantage. Several benchmark studies were reported, which established that DNN performs better than RF, SVM, etc. methods. Similarly, a few studies were performed establishing that the multitask DNN performs better than single-task DNN [131]. DNN methods perform better with an increase in the sample size—i.e., the larger the number of chemotypes to train a DNN, the better is its applicability (this factor may also be considered as a limitation of DNN as the models with small datasets may not perform satisfactorily). Identifying new chemical structures carrying desirable molecular properties (clogP, drug-likeness, etc.) is being carried out using DNN. Big data analysis using AI techniques is an important aspect, which is being effectively carried out using DNN.

Mayr et al. reported the introduction of a DeepTox pipeline (Fig. 6.29) to predict toxicity of chemicals and drugs using Deep Learning approach [139]. This deep learning was facilitated by big data available on toxicological behavior of chemicals. Since, big data requires large computational resources, GPU clusters were used to develop the pipeline. Many types of features were estimated and classified into static (topological features, surface area, van der Waals volume, atom counts, existence of substructures) and dynamic data (fingerprints) categories. Standardization of data was done before taking up model building. The data set of 12,707 compounds (from Tox21 Challenge set) were reduced to 8694 fragments after normalization and merging. Model validation was carried out using cluster cross-validation approach.

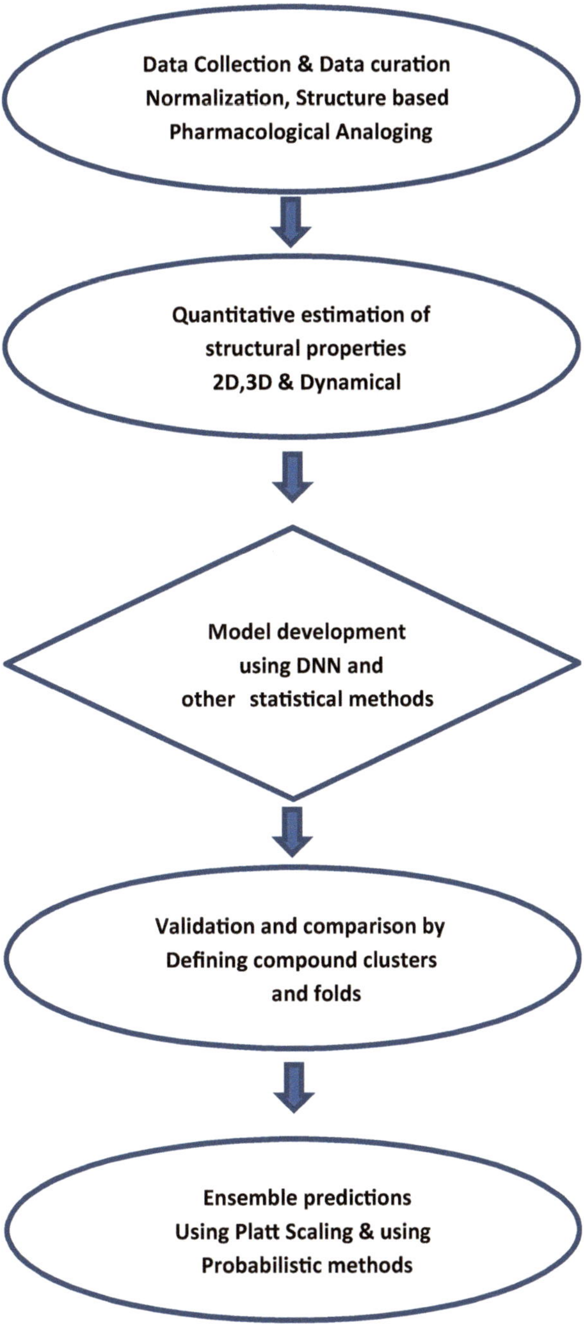

Fig. 6.29 Various steps involved in toxicity prediction protocol adopted using Deep Learning approach by Hochreither and coworkers

Ensemble predictions were made using Platt scaling approach. The DNN model outperformed the other models based on SVM, RF, etc.

Rodríguez-Perez and Bajorath employed multitask DNN methods for classifying kinase inhibitors as highly potent and weakly potent species [143]. As many as 19,030 potent inhibitors were reported to be active against 103 different human kinases. This work combined information of many compounds and targets and hence it was suitable for multitask DNN model development. It was found that the chemical features of kinase inhibitors can be effectively used for the classification purpose using DNN approach. The compounds exhibiting $pIC_{50} < 10$ nM were considered as highly potent and compounds exhibiting $pIC_{50} > 1000$ nM were considered as weakly potent. Kinase targets for which five positive inhibitors and five negative inhibitors reported were chosen for the model development. The input data was based on fingerprints (e.g., ECFP4-based model contained 1127 bits). Out of the many DNN generated, one model contained three hidden layers consisting of 2000, 1000, and 100 neurons. Back propagation algorithm was used to train the DNN. Overfitting of the DNN was handled using an algorithm (dropout—25%) which was specifically known for this purpose. This model performed better than many other alternatives considered for the classification of kinase inhibitors [144].

6.16.3.3 Support Vector Machines in Drug Design

Support Vector Machines (SVMs) are ML techniques; they are useful for building classifiers. The data being generated by drug discovery scientists is very complex and varied. SVMs can be used to recognize subtle patterns and classify the data appropriately. SVMs can be trained using supervised/unsupervised learning methods; when the data is labelled, supervised learning is practical; otherwise unsupervised learning is the only option. When the data can be classified into two classes, linear methods can be adopted. When the data needs to be classified into more than two classes, nonlinear kernel algorithms are useful. Typically, an SVM identifies a hyperplane which can classify the given data. A hyperplane is a virtual plane which can separate two or more classes of compounds with the help of support vectors. A support vector is a datapoint in a class which is closest to the hyperplane. Figure 6.30 shows a diagram explaining the hyperplane, SV, and an SVM for linear classifier of two classes. Prioritization of chemicals for virtual screening is being actively carried out using SVMs. The ability to address nonlinear problems is one of the main advantages of SVMs.

Kinnings et al. [145] showed that the SVMs can be used to establish the correlation between experimentally known binding affinities with that of the values predicted from molecular docking. The authors constructed one regression model and one classification model. A multiple-planar SVM procedure was adopted to overcome the influence of negative data. A new scoring function was introduced by the authors for the inhibitors of InhA (a target of isoniazid in treating MTB). This work lead to the conclusion that a few PDE (Phophodiesterase) inhibitors can be repurposed to target InhA.

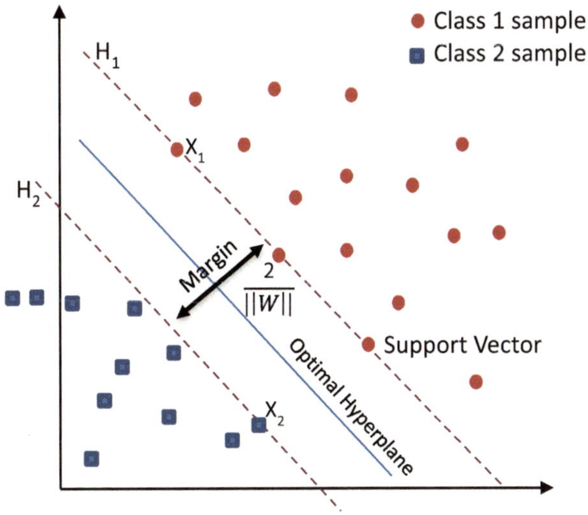

Fig. 6.30 Support Vector Machine (SVM)

6.16.4 Applications of AIDD

Many varieties of methods are being adopted in AIDD, quite often, integrated efforts using more than one of the AIDD methods are being adopted in specific research topics [146, 147]. It is not yet possible to establish that a particular method is more useful over others in AIDD. In fact, integration of AIDD with CADD techniques (namely molecular docking, QSAR, virtual screening, pharmacophore mapping, de novo design) is being practiced. For example, in drug repurposing, biomarker development, toxicity prediction, metabolism prediction, classification of drugs vs. nondrugs, many AIDD efforts are being employed. In addition to the above discussion, a few research efforts involving RF, NB analysis are also being considered [124, 148–154]. Recently, Luo et al. [150] performed studies on drug adverse effects by evaluating >1200 drugs against 600 proteins. The binding modes were identified using AutoDock Vina and ML methods were used to link docking scores to predict side effects. 1533 putative adverse reactions were recorded from this exercise. DNN, RF, KNN, and SVM techniques are being used for toxicity prediction also. For example, several novel descriptors are being calculated (based on AUC) and are being employed in QSTR and QSAR. ML methods are also being used to perform research in exploring biological networks. However, only a few examples are available in which the results from AIDD effort was taken to experimental laboratory of drug discovery scientist till now. The state of the art was recently presented by Schneider and Clark in an elegant review [154]. Given below are two examples, which involve the application of AIDD, presented as case studies.

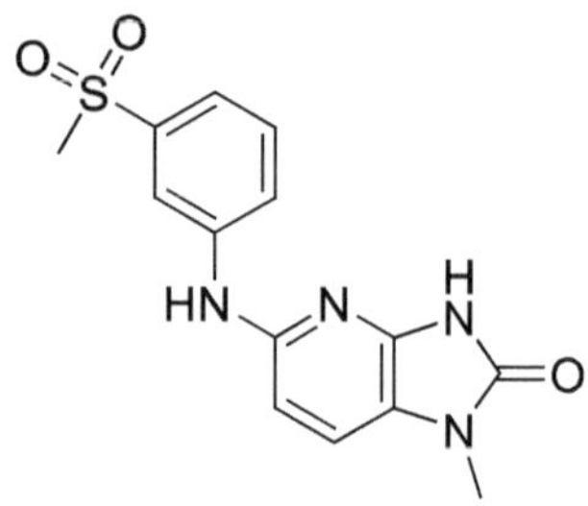

Fig. 6.31 2D structure of molecule designed using AIDD and CADD methods

Case Study 5

Polykovskiy et al. proposed an AIDD architecture "entangled conditional adversarial Autoencoder" and established its application in identifying JAK3 (Janus Kinase 3) inhibitor, in vitro analysis of which showed sufficient activity [155]. The inhibition of JAK3 was found to be important for drug discovery in the areas of vitiligo, psoriasis, alopecia, and rheumatoid arthritis. Selective inhibitor is required for JAK3 as against JAK2 inhibition. This team initially developed several disentanglement techniques using ML. For the model development, a set of 1.8 million molecules were preselected from ZINC database. Canonical SMILES notation was used to create input data of tokens (atoms) in these molecules. Two ANN models—one consisting of two hidden layers with 128 neurons each and the other consisting of three hidden layers with 128 neurons each—were generated. The generation of new structural analogs was carried out using MACCS binary fingerprints. Chemoinformatics analysis was carried out estimating Tanimoto similarity and Hamming distance analyses on the newly designed molecules. A total of 300,000 molecules were generated, which were subjected to several in silico filters. The binding affinity of the molecules was estimated using molecular docking. 5000 molecules were further subjected to molecular dynamics analysis and 100 molecules were selected. Many of these compounds were synthesized after considering synthetic accessibility (computational) and finally subjected to in vitro analysis. This pharmacoinformatics effort lead to the identification of a compound (Fig. 6.31) as a marginally selective JAK3 inhibitor with 7μM IC_{50} value.

Case Study 6

Schneider and coworkers [156] adopted generative ML approach to design anticancer peptides. An RNN (Recurrent Neural Network) was developed on anticancer peptides. The computational model captured the two important features of the amphipathic helices—the ampipathic character and the overall positive charge. A total of 1000 peptides (6–28 amino acid length) were computationally designed. Twelve of them were selected for synthesis after performing pharmacophore similarity analysis; ten of them were found to be active and six of them were found to exhibit selective anticancer activity.

Overall Message from This Chapter

CADD was an emerging topic 30 years ago with limited techniques; however, it has grown into a mature topic with wide variety of techniques available now. A lot of science is being employed in rationally designing compounds involving innumerable varieties of virtual flow charts. A few of the CADD methods have been standardized and are being routinely used, but many emerging CADD technologies are also being adopted at various stages of drug discovery. This chapter provided a glimpse of well-accepted as well as evolving CADD techniques. Though a lot of research is being carried out using CADD methods, it has not been propagated thoroughly among all the drug discovery scientists. The reasons can be traced to the fact that many of these methodologies require thorough understanding of the science associated with them. The number of new approaches being introduced to carry out pharmacoinformatics research are many. This fact can be considered as an advantage as well as disadvantage—it is advantage because all these and many more are required; disadvantage is that too many are being introduced, keeping track is becoming difficult. Another aspect to be considered is that a few of the techniques address the needs of medicinal chemists and a few others converse well with pharmacology experts; however, too few are addressing the needs of experts in the fields of pharmaceutics/pharmaceutical analysis/pharmaceutical technology/toxicology/metabolism/primacy practice/clinical trials, etc. There is no denial of the fact that the efforts in pharmacoinformatics shall penetrate all walks-of-life of professionals in pharmaceutical sciences, but at the same time the existing gaps need to be bridged. This chapter made one more attempt in this direction.

Acknowledgements The author thanks all his students (MS as well as Ph.D. students over the past 20 years) who helped in shaping the ideas covered in this article. Most of the text is based on the lectures delivered by the author to his students and based on the interactions he had with them. This article was developed to address the pharmacy students, who are willing to learn the topic as freshers.

References

1. Murcko MA. What makes a great medicinal chemist? A personal perspective. J Med Chem. 2018;61:7419–24.
2. Cohen NC. Guidebook on molecular modeling in drug design. India: Elsevier; 2014.
3. Cavasotto CN, editor. In silico drug discovery and design: theory, methods, challenges, and applications: CRC Press; 2017.
4. Silverman RB, Holladay MW. The organic chemistry of drug design and drug action: Academic Press; 2014.
5. Merz KM Jr, Ringe D, Reynolds CH, editors. Drug design: structure- and ligand-based approaches: Cambridge University Press; 2010.
6. Reddy MR, Erion MD. Free energy calculations in rational drug design: Springer; 2001.
7. Doble M. Drug design: basics and applications: McGraw Hill; 2011.
8. Roche VF, Zito SW, Lemke TL, Williams DA. Foye's principles of medicinal chemistry: Wolters Kulwer; 2020.
9. Klebe G. Drug design: methodology, concepts, and mode-of-action: Springer; 2013.
10. Roy K, editor. In silico drug design: Academic Press; 2019.
11. Luciano G. Essential computer graphics techniques for modeling, animating, and rendering biomolecules and cells: a guide for the scientist and artist. CRC Press; 2019.

12. Bharatam PV, Patel DS, Iqbal P. Pharmacophoric features of biguanide derivatives: an electronic and structural analysis. J Med Chem. 2005;48:7615–22.
13. Bharatam PV, Amita. Negative hyperconjugative interactions in S-nitrosothiols: a theoretical study. Tet Lett. 2002;43:8289–91.
14. Leach A. Molecular modelling: principles and applications. 2nd ed: Pearson Education; 2009.
15. Goodman JM. Chemical applications of molecular modelling: Royal Society of Chemistry; 2004.
16. Jaladanki C, Shaik A, Bharatam PV. Biotransformation of isoniazid by cytochromes P450: analyzing the molecular mechanism using density functional theory. Chem Res Tox. 2017;30:2060–73.
17. Chourasiya SS, Kathuria D, Nikam SS, Ramakrishnan A, Khullar S, Mandal SK, Chakraborti AK, Bharatam PV. Azine-hydrazone tautomerism of guanylhydrazones: evidence for the preference toward the azine tautomer. J Org Chem. 2016;81:7574–83.
18. Kathuria D, Chourasiya SS, Wani AA, Singh M, Sahoo SC, Bharatam PV. Geometrical isomerism in Guanabenz free base: synthesis, characterization, crystal structure and theoretical studies. Cryst Growth Des. 2019;19:3183–91.
19. Hubbard RE, Neidle S, editors. Structure-based drug discovery: an overview: Royal Society of Chemistry; 2006.
20. Ghosh AK, Gemma S. Structure-based design of drugs and other bioactive molecules: tools and strategies: Wiley-VCH; 2014.
21. Mohan CG, editor. Structural bioinformatics: applications in preclinical drug discovery process: Springer; 2019.
22. Jayaram B, Dhingra P, Mishra A, Kaushik R, Mukherjee G, Singh A, Shekhar S. Bhageerath H: A homology ab initio hybrid server for predicting tertiary structures of monomeric soluble proteins. BMC Bioinf. 2014;16:S7.
23. Kathuria D, Bankar AA, Bharatam PV. What's in a structure? The story of biguanides. J Mol Struct. 2018;1152:61–78.
24. Patel DS, Bharatam PV. Novel⊕ N (← L)$_2$ species with two lone pairs on nitrogen: systems isoelectronic to carbodicarbenes. Chem Comm. 2009;9:1064–6.
25. Gasteiger J, Engel T, editors. Chemoinformatics: a textbook: Wiley-VCH; 2003.
26. Engel T, Gasteiger J, editors. Chemoinformatics: basic concepts and methods: Wiley-VCH; 2018.
27. Leach AR, Gillet VJ. An introduction to chemoinformatics: Springer; 2007.
28. Bajorath J. Chemoinformatics for drug discovery. New York: John Wiley & Sons; 2013.
29. Gaur AS, et al. Assessing therapeutic potential of molecules: molecular property diagnostic suite for tuberculosis (MPDSCTB). J Chem Sci. 2017;129:515–31.
30. Mount DW. Bioinformatics: sequence and genome analysis: CBS Publishers, 2005.
31. Ramsden J. Bioinformatics: an introduction: Springer; 2015.
32. Ewens WJ, Grant GR. Statistical methods in bioinformatics: an introduction: Springer; 2005.
33. Böckenhauer HJ, Bongartz D. Algorithmic aspects of bioinformatics: Springer; 2007.
34. Polanski A, Kimmel M. Bioinformatics. Springer, 2007.
35. Sperschneider V. Bioinformatics: problem solving paradigms: Springer; 2008.
36. Loging WT. Bioinformatics and computational biology in drug discovery and development: Cambridge Univ Press; 2016.
37. Roy K, editor. Multi-target drug design using chem-bioinformatic approaches: Springer; 2018.
38. Larson RS, Oprea T, editors. 3rd ed. Bioinformatics and drug discovery: Humana Press, 2019.
39. Shamanskiy VA, Timonina VN, Popadin KY, Gunbin KV. ImtRDB: a database and software for mitochondrial imperfect interspersed repeats annotation. BMC Genomics. 2019;20:295.
40. Babbi G, Martelli PL, Casadio R. PhenPath: a tool for characterizing biological functions underlying different phenotypes. BMC Genomics. 2019;20:548.
41. Sundararajan Z, Knoll R, Hombach P, Becker M, Schultze JL, Ulas T. Shiny-Seq: advanced guided transcriptome analysis. BMC Res Notes. 2019;12:432.

42. Huang DW, Sherman BT, Tan Q, Kir J, Liu D, Bryant D, Guo Y, Stephens R, Baseler MW, Lane HC, Lempicki RA. DAVID bioinformatics resources: expanded annotation database and novel algorithms to better extract biology from large gene lists. Nucleic Acids Res. 2007;35: W169–75.
43. Fang J, Wu Z, Cai C, Wang Q, Tang Y, Cheng F. Quantitative and systems pharmacology. 1. In silico prediction of drug-target interaction of natural products enables new targeted cancer therapy. J Chem Inf Model. 2017;57:2657–71.
44. Singhal P, Jayaram B, Dixit SB, Beveridge DL. Prokaryotic gene finding based on physicochemical characteristics of codons calculated from molecular dynamics simulations. J Biophys. 2008;94:4173–83.
45. Baviskar AT, Madaan C, Preet R, Mohapatra P, Jain V, Agarwal A, Guchhait SK, Kundu CN, Banerjee UC, Bharatam PV. N-fused imidazoles as novel anticancer agents that inhibit catalytic activity of topoisomerase IIα and induce apoptosis in G1/S phase. J Med Chem. 2011;54:5013–30.
46. Baviskar AT, Amrutkar SM, Trivedi N, Chaudhary V, Nayak V, Guchhait SK, Banerjee UC, Bharatam PV, Kundu CN. Switch in site of inhibition: a strategy for structure-based discovery of human topoisomerase IIα catalytic inhibitors. ACS Med Chem Lett. 2015;6:481–5.
47. Tripathi N, Guchhait SK, Bharatam PV. Pharmacoinformatics analysis of merbarone binding site in human topoisomerase IIα. J Mol Graph Model. 2019;86:1–18.
48. Tripathi N, Shaikh N, Bharatam PV, Garg P. HToPred: a tool for human topoisomerase II inhibitor prediction. Mol Inform. 2019;38:e1800046.
49. Tripathi N, Deb R, Guchhait SK, Bharatam PV. Chapter 2. Pharmacoinformatics studies on human topoisomerase II: exploring the mechanism of enzyme inhibition. In: Kumar R, Sing S, editors. Topoisomerase inhibitors. New York: Nova Science; 2017.
50. Hansch C, Leo A, Hoekman DH. Exploring QSAR: hydrophobic, electronic, and steric constants: American Chemical Society; 1995.
51. Puzyn T, Leszczynski J, Cronin MT, editors. Recent advances in QSAR studies: methods and applications: Springer; 2012.
52. Kubinyi H. QSAR: Hansch analysis and related approaches: John Wiley; 2019.
53. Kubinyi H, Folkers G, Martin YC, editors. 3D QSAR in drug design: recent advances: Springer; 1998.
54. Albrecht R. Development of antibacterial agents of the nalidixic acid type. Pro Drug Res. 1977;21:9–104.
55. Koga H, Itoh A, Murayama S, Suzue S, Irikura T. Structure-activity relationships of antibacterial 6,7- and 7,8-disubstituted 1-alkyl-1,4-dihydro-4-oxoquinoline-3-carboxylic acids. J Med Chem. 1980;23:1358–63.
56. Klopman G, Macina OT, Levinson ME, Rosenkranz HS. Computer automated structure evaluation of quinolone antibacterial agents. Antimicrob Agents Chemother. 1987;31:1831–40.
57. Cramer RD, Patterson DE, Bunce JD. Comparative molecular field analysis (CoMFA). Effect of shape on binding of steroids to carrier proteins. J Am Chem Soc. 1988;110:5959–67.
58. Klebe G, Abraham U, Meitzner T. Molecular similarity indices in a comparative analysis (CoMSIA) of drug molecules to correlate and predict their biological activity. J Med Chem. 1994;37:4130–46.
59. Robinson DD, Winn PJ, Lyne PD, Richards WG. Selforganizing molecular field analysis: a tool for structure-activity studies. J Med Chem. 1999;42:573–83.
60. Silverman BD, Platt DE. Comparative molecular moment analysis (CoMMA): 3D-QSAR without molecular superposition. J Med Chem. 1996;39:2129–40.
61. Gohlke H, Klebe G. DrugScore meets CoMFA: adaptation of fields for molecular comparison (AFMoC) or how to tailor knowledge-based pair-potentials to a particular protein. J Med Chem. 2002;45:4153–70.
62. Khanna S, Sobhia ME, Bharatam PV. Additivity of molecular fields: CoMFA study on dual activators of PPARα and PPARγ. J Med Chem. 2005;48:3015–25.

63. Sundriyal S, Bharatam PV. Important pharmacophoric features of pan PPAR agonists: common chemical feature analysis and virtual screening. Eur J Med Chem. 2009;44:3488–95.
64. Sundriyal S, Viswanad B, Bharathy E, Ramarao P, Chakraborti AK, Bharatam PV. New PPARγ ligands based on 2-hydroxy-1,4-naphthoquinone:Computer-aided design, synthesis and receptor-binding studies. Bioorg Med Chem Lett. 2008;18:3192–5.
65. Sundriyal S, Viswanad B, Ramarao P, Chakraborti AK, Bharatam PV. New PPARγ ligands based on barbituric acid: virtual screening, synthesis and receptor binding studies. Bioorg Med Chem Lett. 2008;18:4959–62.
66. Guner OF, editor. Pharmacophore perception, development, and use in drug design: The Book Service; 2000.
67. Mahato S. A feature matching approach for pharmacophore modelling: concepts and applications: Scholars Press; 2016.
68. Langer T, Hoffmann RD, editors. Pharmacophores and pharmacophore searches: Wiley-VCH; 2006.
69. Guner O, Clement O, Kurogi Y. Pharmacophore modeling and three dimensional database searching for drug design using catalyst: recent advances. Curr Med Chem. 2004;11:2991–3005.
70. Patel DS, Bharatam PV. New leads for selective GSK-3 inhibition: pharmacophore mapping and virtual screening studies. J Comput Aided Mol Des. 2006;20:55–66.
71. Vlachakis DP, editor. Molecular docking: Intech Open; 2018.
72. Azevedo WF Jr, editor. Docking screens for drug discovery: Humana; 2019.
73. Pinzi L, Rastelli G. Molecular docking: shifting paradigms in drug discovery. Int J Mol Sci. 2019;20:E4331.
74. Li J, Fu A, Zhang L. An overview of scoring functions used for protein–ligand interactions in molecular docking. Interdiscip Sci Comput Life Sci. 2019;11:320–8.
75. Jain AN. Scoring functions for protein-ligand docking. Curr Protein Pept Sci. 2006;7:407–20.
76. Su M, Yang Q, Du Y, Feng G, Liu Z, Li Y, Wang R. Comparative assessment of scoring functions: the CASF-2016 update. J Chem Inf Model. 2019;59:895–913.
77. Von IM. The war against influenza: discovery and development of sialidase inhibitors. Nat Rev Drug Discov. 2007;6:967–74.
78. Varghese JN, Laver WG, Colman PM. Structure of the influenza virus glycoprotein antigen neuraminidase at 2.9 a resolution. Nature. 1983;303:35–40.
79. Colman PM, Varghese JN, Laver WG. Structure of the catalytic and antigenic sites in influenza virus neuraminidase. Nature. 1983;303:41–4.
80. Itzstein VM, Wu WY, Kok GB, Pegg MS, Dyason JC, Jin B, Phan TV, Smythe ML, White HF, Oliver SW. Rational design of potent sialidase-based inhibitors of influenza virus replication. Nature. 1993;363:418–23.
81. Itzstein VM. A computational procedure for determining energetically favorable binding sites on biologically important macromolecules. J Med Chem. 1985;28:849–57.
82. Hayden FG, Osterhaus AD, Treanor JJ, Fleming DM, Aoki FY, Nicholson KG, Bohnen AM, Hirst HM, Keene O, Wightman K. Efficacy and safety of the neuraminidase inhibitor zanamivir in the treatment of influenza virus infections. GG167 influenza study group. N Engl J Med. 1997;337:874–80.
83. Winquist AG, Fukuda K, Bridges CB, Cox NJ. Neuraminidase inhibitors for treatment of influenza A and B infections. CDC. U.S. Department of health and human services. 1999;48:1–9.
84. Fan X, Zhang FH, Al-Safi RI, Zeng LF, Shabaik Y, Debnath B, Sanchez TW, Odde S, Neamati N, Long YQ. Design of HIV-1 integrase inhibitors targeting the catalytic domain as well as its interaction with LEDGF/p75: a scaffold hopping approach using salicylate and catechol groups. Bioorg Med Chem. 2011;19:4935–52.
85. Hazuda DJ, Felock P, Witmer M, Wolfe A, Stillmock K, Grobler JA, Espeseth A, Gabryelski L, Schleif W, Blau C, Miller MD. Inhibitors of strand transfer that prevent integration and inhibit HIV-1 replication in cells. Science. 2000;287:646–50.

86. Zhuang L, Wai JS, Embrey MW, Fisher TE, Egbertson MS, Payne LS, Guare JP, Vacca JP, Hazuda DJ, Felock PJ, Wolfe AL, Stillmock KA, Witmer MV, Moyer G, Schleif WA, Gabryelski LJ, Leonard YM, Lynch JJ, Michelson SR, Young SD. Design and synthesis of 8-hydroxy-[1,6]naphthyridines as novel inhibitors of HIV-1 integrase in vitro and in infected cells. J Med Chem. 2003;46:453–6.
87. Schneider G. De novo molecular design. Weinheim: Wiley-VCH Verlag Gmb H & Co. KGaA; 2013.
88. Christner C, Wyrwa R, Marsch S, Küllertz G, Thiericke R, Grabley S, Schumann D, Fischer G. Synthesis and cytotoxic evaluation of cycloheximide derivatives as potential inhibitors of FKBP12 with neuroregenerative properties. J Med Chem. 1999;42:3615–22.
89. Dragovich PS, Barker JE, French J, Imbacuan M, Kalish VJ, Kissinger CR, Knighton DR, Lewis CT, Moomaw EW, Parge HE, Pelletier LA, Prins TJ, Showalter RE, Tatlock JH, Tucker KD, Villafranca JE. Structure-based design of novel, urea-containing FKBP12 inhibitors. J Med Chem. 1996;39:1872–84.
90. Gopalakrishnan R, Kozany C, Gaali S, Kress C, Hoogeland B, Bracher A, Hausch F. Evaluation of synthetic FK506 analogues as ligands for the FK506-binding proteins 51 and 52. J Med Chem. 2012;55:4114–22.
91. Wu YQ, Wilkinson DE, Limburg D, Li JH, Sauer H, Ross D, Liang S, Spicer D, Valentine H, Fuller M, Guo H, Howorth P, Soni R, Chen Y, Steiner JP, Hamilton GS. Synthesis of ketone analogues of prolyl and pipecolyl ester FKBP12 ligands. J Med Chem. 2002;45:3558–68.
92. Alvarez J, Shoichet B. Virtual screening in drug discovery: CRC Press; 2005.
93. Kontoyianni M. Docking and virtual screening in drug discovery: Springer; 2017.
94. Sotriffer, editor. Virtual screening: principles, challenges, and practical guidelines: Wiley-VCH; 2011.
95. Adane L, Patel DS, Bharatam PV. Shape and chemical feature-based 3D-pharmacophore model generation and virtual screening: identification of potential leads for P. falciparum DHFR enzyme inhibition. Chem Biol Drug Des. 2010;75:115–26.
96. Arfeen M, Bhagat S, Patel R, Prasad S, Roy I, Chakraborti AK, Bharatam PV. Design, synthesis and biological evaluation of 5-benzylidene-2-iminothiazolidin-4-ones as selective GSK-3β inhibitors. Eur J Med Chem. 2016;121:727–36.
97. Arfeen M, Bharatam PV. Design of glycogen synthase kinase-3 inhibitors: an overview on recent advancements. Curr Pharm Des. 2013;19:4755–75.
98. Patel DS, Bharatam PV. Selectivity criterion for pyrazolo[3,4-b]pyrid[az]ine derivatives as GSK-3 inhibitors: CoMFA and molecular docking studies. Eur J Med Chem. 2008;43:949–57.
99. Dessalew N, Bharatam PV. Structure based de novo design of novel glycogen synthase kinase 3 inhibitors. Bioorg Med Chem. 2007;15:3728–36.
100. Koshland DE Jr. Key-lock theory and the induced fit theory. Angew Chem Int Ed Engl. 1994;33:2375–8.
101. Xu M, Lill MA. Induced fit docking and the use of QM/MM methods in docking. Drug Discov Today Technol. 2013;10:411–8.
102. Weikl TR, Paul F. Conformational selection in protein binding and function. Protein Sci. 2014;23:1508–18.
103. Gervasio FL, Spiwok V. Biomolecular simulations in structure-based drug discovery: Wiley-VCH Verlag GmbH & Co. KGaA; 2018.
104. Sotriffer CA. Molecular dynamics simulations in drug design: Springer; 2006.
105. Bhagat S, Arfeen M, Adane L, Singh S, Singh PP, Chakraborti AK, Bharatam PV. Guanylthiouea derivatives as potential antimalarial agents: synthesis, in vivo and molecular modeling studies. Eur J Med Chem. 2017;135:339–48.
106. Arfeen M, Patel R, Khan T, Bharatam PV. Molecular dynamics simulation studies of GSK-3β ATP competitive inhibitors: understanding the factors contributing to selectivity. J Biomol Struct Dyn. 2015;33:2578–93.
107. Abbat S, Jain V, Bharatam PV. Origins of the specificity of inhibitor P218 toward wild-type and mutant *Pf*DHFR: a molecular dynamics analysis. J Biomol Struct Dyn. 2015;33:1913–28.

108. Zhang Y, Rajapakse JC. Machine learning in bioinformatics: Wiley; 2008.
109. Zupan J, Gasteiger J. Neural networks in chemistry and drug design: an introduction: Wiley-VCH; 1999.
110. Speck-Planche A, editor. Multi-scale approaches in drug discovery: from empirical knowledge to in silico experiments and back: Elsevier; 2017.
111. Ghose AK, Herbertz T, Salvino JM, Mallamo JP. Knowledge-based chemoinformatic approaches to drug discovery. Drug Discov Today. 2006;11:1107–14.
112. Velec HFG, Gohlke H, Klebe G. DrugScoreCSD knowledge-based scoring function derived from small molecule crystal data with superior recognition rate of near-native ligand poses and better affnity prediction. J Med Chem. 2005;48:6296–303.
113. Huang SY, Zou X. An iterative knowledge-based scoring function for protein-protein recognition. Proteins. 2008;72:557–79.
114. Devillers J. Genetic algorithms in molecular modeling: Academic Press; 1996.
115. Douguet D, Thoreau E, Grassy G. A genetic algorithm for the automated generation of small organic molecules: drug design using an evolutionary algorithm. J Comput Aided Mol Des. 2000;14:449–66.
116. Pérez-Castillo Y, Lazar C, Taminau J, Froeyen M, Cabrera-Perez MA, Nowé A. GA(M)E-QSAR: a novel, fully automatic genetic-algorithm-(Meta)-ensembles approach for binary classification in ligand-based drug design. J Chem Inf Model. 2012;529:2366–86.
117. Sun G, Fan T, Sun X, Hao Y, Cui X, Zhao L, Ren T, Zhou Y, Zhong R, Peng Y. In silico prediction of O^6-methylguanine-DNA methyltransferase inhibitory potency of base analogs with QSAR and machine learning methods. Molecules. 2018;23:2892.
118. Santos-Garcia L, de Mecenas FM, Musilek K, Kuca K, Ramalho T, da Cunha E. QSAR study of N-myristoyltransferase inhibitors of antimalarial agents. Molecules. 2018;23:2348.
119. Bazl R, Ganjali MR, Derakhshankhah H, Saboury AA, Amanlou M, Norouzi P. Prediction of tyrosinase inhibition for drug design using the genetic algorithm–multiple linear regressions. Med Chem Res. 2013;22:5453–65.
120. Ambrose GO, Afees OJ, Kalu UJ, Oshireku AW, Todimu AD, Oluwasegun OE, Olatoye T, Rebecca FR, Precious A. QSAR model generation of phthalazinones as poly (ADP-ribose) polymerase inhibitors by the genetic algorithm and multiple linear regression (GA-MLR) method: a ligand-based approach for Cancer drug design. J Proteomics Bioinform. 2018;11:10.
121. Terfloth L, Gasteiger J. Neural networks and genetic algorithms in drug design. Drug Discov Today. 2001;6:102–8.
122. Garg P, Verma J. In silico prediction of blood brain barrier permeability: an artificial neural network model. J Chem Inf Model. 2006;46:289–97.
123. Varnek A, Baskin I. Machine learning methods for property prediction in cheminformatics: Quo Vadis? J Chem Inf Model. 2012;52:1413–37.
124. Lo YC, Rensi SE, Torng W, Altman RB. Machine learning in chemoinformatics and drug discovery. Drug Discov Today. 2018;23:1538–46.
125. Lima AN, Philot EA, Trossini GHG, Scott LPB, Maltarollo VG, Honorio KM. Use of machine learning approaches for novel drug discovery. Expert Opin Drug Discov. 2016;11:225–39.
126. Khamis MA, Gomaa W, Ahmed WF. Machine learning in computational docking. Artif Intell Med. 2015;63:135–52.
127. Wójcikowski M, Ballester PJ, Siedlecki P. Performance of machine-learning scoring functions in structure-based virtual screening. Sci Rep. 2017;7:46710.
128. Li H, Leung KS, Wong MH, Ballester PJ. Improving AutoDock Vina using random forest: the growing accuracy of binding affinity prediction by the effective exploitation of larger data sets. Mol Inform. 2015;34:115–26.
129. Ekins S. The next era: deep learning in pharmaceutical research. Pharm Res. 2016;33:2594–603.
130. Panteleev J, Gao H, Jia L. Recent applications of machine learning in medicinal chemistry. Bioorg Med Chem Lett. 2018;28:2807–15.

131. Chen H, Engkvist O, Wang Y, Olivecrona M, Blaschke T. The rise of deep learning in drug discovery. Drug Discov Today. 2018;23:1241–50.
132. Bajorath J. Data analytics and deep learning in medicinal chemistry. Future Med Chem. 2018;10:1541–3.
133. Jing Y, Bian Y, Hu Z, Wang L, Xie XQS. Deep learning for drug design: an artificial intelligence paradigm for drug discovery in the big data era. AAPS J. 2018;20:58.
134. Gawehn E, Hiss JA, Schneider G. Deep learning in drug discovery. Mol Inf. 2016;35:3–14.
135. Gawehn E, Hiss JA, Brown JB, Schneider G. Advancing drug discovery via GPU-based deep learning. Expert Opin Drug Discov. 2018;13:579–82.
136. Zhang L, Tan J, Han D, Zhu H. From machine learning to deep learning: progress in machine intelligence for rational drug discovery. Drug Discov Today. 2017;22:1680–5.
137. Ma J, Sheridan RP, Liaw A, Dahl GE, Svetnik V. Deep neural nets as a method for quantitative structure–activity relationships. J Chem Inf Model. 2015;55:263–74.
138. Ghasemi F, Mehridehnavi A, Pérez-Garrido A, Pérez-Sánchez H. Neural network and deep-learning algorithms used in QSAR studies: merits and drawbacks. Drug Discov Today. 2018;23:1784–90.
139. Mayr A, Klambauer G, Unterthiner T, Hochreither S. Deep Tox: toxicity prediction using deep learning. Front Environ Sci. 2016;3:80.
140. Nam Y, Kim M, Chang HS, Shin H. Drug repurposing with network reinforcement. BMC Bioinf. 2019;20:383.
141. Laghuvarapu S, Pathak Y, Priyakumar UD. BAND NN: a deep learning framework for energy prediction and geometry optimization of organic small molecules. J Comput Chem. 2020;41:790–799.
142. Smith JS, Nebgen BT, Zubatyuk R, Lubbers N, Devereux C, Barros K, Tretiak S, Isayev O, Roitberg AE. Approaching coupled cluster accuracy with a general-purpose neural network potential through transfer learning. Nat Commun. 2019;10:2903.
143. Rodríguez-Perez R, Bajorath J. Multitask machine learning for classifying highly and weakly potent kinase inhibitors. ACS Omega. 2019;4:4367–75.
144. Maltarollo VG, Kronenberger T, Espinoza GZ, Oliveira PR, Honorio KM. Advances with support vector machines for novel drug discovery. Expert Opin Drug Discov. 2019;14:23–33.
145. Kinnings SL, Liu N, Tonge PJ, Jackson RM, Xie L, Bourne PE. A machine learning-based method to improve docking scoring functions and its application to drug repurposing. J Chem Inf Model. 2011;51:408–19.
146. Hessler G, Baringhaus KH. Artificial intelligence in drug design. Molecules. 2018;23:2520.
147. Mahé P, Vert JP. Virtual screening with support vector machines and structure kernels. Comb Chem High Throughput Screening. 2009;12:409–23.
148. Svetnik V, Liaw A, Tong C, Culberson JC, Sheridan RP, Feuston BP. Random forest: a classification and regression tool for compound classification and QSAR modeling. J Chem Inf Comput Sci. 2003;43:1947–58.
149. Rogers D, Brown RD, Hahn M. Using extended-connectivity fingerprints with Laplacian-modified Bayesian analysis in high-throughput screening follow-up. J Biomol Screen. 2005;10:682–6.
150. Luo H, Fokoue-Nkoutche A, Singh N, Yang L, Hu J, Zhang P. Molecular docking for prediction and interpretation of adverse drug reactions. Comb Chem High Throughput Screen. 2018;21:314–22.
151. Wu Y, Wang G. Machine learning based toxicity prediction: from chemical structural description to transcriptome analysis. Int J Mol Sci. 2018;19:2358.
152. Camacho DM, Collins KM, Powers RK, Costello JC, Collins JJ. Next-generation machine learning for biological networks. Cell. 2018;173:1581–92.
153. Chen B, Harrison RF, Papadatos G, Willett P, Wood DJ, Lewell XQ, Greenidge P, Stiefl N. Evaluation of machine-learning methods for ligand-based virtual screening. J Comput Aided Mol Des. 2007;21:53–62.

154. Schneider G, Clark DE. Automated de novo drug design: are we nearly there yet? Angew Chem Int Ed. 2019;58:2–14.
155. Polykovskiy D, Zhebrak A, Vetrov D, Ivanenkov Y, Aladinskiy V, Mamoshina P, Bozdaganyan M, Aliper A, Zhavoronkov A, Kadurin A. Entangled conditional adversarial autoencoder for de novo drug discovery. Mol Pharm. 2018;15:4398–405.
156. Grisoni F, Neuhaus CS, Gabernet G, Müller AT, Hiss JA, Schneider G. Designing anticancer peptides by constructive machine learning. Chem Med Chem. 2018;13:1300–2.

Chapter 7
Pharmacological Screening: Drug Discovery

Kumar V. S. Nemmani

7.1 Pharmacological Screening

7.1.1 Background

Traditionally, new treatments have been discovered through random testing of natural products, primarily parts/extracts of plants and herbs, on humans. Lately, new drugs were discovered by screening extracts from natural sources on animals, subsequently on humans. The testing of agents on living beings obviously is highly expensive and often involves ethical issues. This approach of screening compounds on living systems in modern terms is called phenotypic or physiology-based screening. In this screening approach, the target is unknown and involves physiologic/phenotypic readout end points. Although the approach is fraught with difficult deconvolution strategies to identify molecular target, it is time consuming and proven to yield first-in-class compounds. In contrast to the traditional drug discovery, modern drug discovery (Mechanism-Based Approach) is facilitated with the advent of new screening technologies of miniaturization and automation of biological assays against a molecular target. As a result, high-throughput screening (HTS) platforms have been evolved to screen a large number of compounds typically in miniatured and automated biological assays. With the current technology platforms, several thousands of compounds can be screened in weeks if not months [1, 2]. The basic premise of screening is that biological assays are reproducible, reliable, robust, and biologically relevant. The activities of HITs (initial actives or leads) in HTS platforms are typically confirmed in semi-manual/manual primary and secondary assays. Further characterization is done in preclinical and clinical studies. This target-based strategy is highly successful in developing best-in-class compounds (Table 7.1).

K. V. S. Nemmani (✉)
Shri Vishnu College of Pharmacy, Bhimavaram, Andhra Pradesh, India

R. Poduri (ed.), *Drug Discovery and Development*,
https://doi.org/10.1007/978-981-15-5534-3_7

Table 7.1 Traditional vs. modern drug discovery: advantages and limitations

	Traditional Discovery (Function based)	Modern Discovery (Target based)
Mode of action	Unknown	Known
HTS amenable	Difficult	Easy
Read out	Physiological parameter	Target modulation
Rational design	No	Yes
Copy-ability	Low	High
Rational improvement (over fist-in-class)	Low	High
Key advantage	Integrated response	Rational design
Key limitation	Low throughput	Selection of target

7.2 In Vitro Pharmacological Screening

7.2.1 High-Throughput Screening

High-throughput screening (HTS) is the process with which large numbers of compounds can be tested, in an automated fashion, for activity as inhibitors (antagonists) or activators (agonists) of a particular biological target.

The main goal of HTS is to screen large number of compounds with diverse chemical structures and identify potent HITs (compounds that affect the target in desired manner). The more the potency of the HIT, the more likely that it will exhibit specificity and selectivity and thereby exhibit desired activity and less likely to have adverse effects (undesired effects). High-throughput screening efforts that led to good number of HITs with decent potency and diverse structures give flexibility for medicinal chemists to choose HITs that have decent patent space and amenability to synthesize analogs around the structure. Thus, HTS has become an important exercise across all major pharmaceutical research organizations [3]. Pharmaceutical organizations basically follow one of the following screening approaches based on the need or availability of resources or time [4].

Compound libraries screening approaches [4]

1. Random Screening:
   Screening of the entire compound library against the drug target
2. Focused or Knowledge-Based Screening:
   Selecting from the library subsets of molecules with potential activity at the target protein
3. Fragment Screening:
   Making very small molecular weight compound libraries (screened at high concentrations)
4. Combinatorial Chemistry Libraries:
   Compounds libraries synthesized in microtiter plates (multiplexes)

7.2.1.1 Sample Preparation

Typically, samples for HTS screening are prepared in microtiter plates (MTPs). The reason behind preparing samples in MTP are (1) samples are plated in array format, (2) low sample volume depending on number of wells/plate, (3) ease of handling and storage, (4) ease of identifying a compound in the plate with the address, and (5) ease of making duplicate plates from *Stock Plates* or *Mother Plates*. The selection of format (size) of the plate depends on the assay sensitivity and volume requirements for robust, reliable, and reproducible results during the entire campaign of screening for a particular target. Typical formats include 96 (100μL), 384 (10–50μL), 1536 (5–10μL), or 3456 (2–5μL) well/plate. Campaigns are usually done at a single concentration between 1 and 20μL. Samples from the combinatorial chemistry often are mixtures of samples in a well. In some cases, screening plates are prepared as 5–20 compounds/well. The logic for the pooling samples is to increase throughput and reduce costs, reagent requirements, and time. However, this strategy is fraught with higher rates of false positives and negatives and necessitates deconvolution to find active agent/s in a given mixture of a well. Thus, this deconvolution strategy is time taking and losing its flavor nowadays as data of pooled samples often of poor quality.

7.2.1.2 Establishment of a Biological Assay Suitable for Miniaturization and Automation

The most critical points that should be considered in the assay development are robustness and quality of the assay, as sample volume is very low and the detection signal. Assays that are generally run on HTS can be classified into solution-based biochemical assays and cell-based assays. Homogenous "*mix and measure*" assays are preferred over cell-based assays for HTS as they avoid filtration, separation, and wash steps, which can be time-consuming and difficult to automate. All cell-based assays may not be feasible to set up on HTS platform as they involve multiple complex steps [5, 6]. Solution-based assays are dependent on radioactive (radioligand binding assays, scintillation proximity assay—SPA) or fluorescence (fluorescence resonance energy transfer—FRET; fluorescence polarization—FP, homogenous time resolved fluorescence—HTRF) detection principles to quantify the interaction of test compounds with the biological target (Table 7.2).

Some of the commonly used second messenger assays on HTS platform are:

1. cAMP, PKA
2. Calcium levels (fluorescence imaging plate reader FLIPR)
3. Reporter gene assays that measure responses at transcriptional/translational level
4. Cell proliferation/inhibition assays measuring cell number growth or inhibition

Table 7.2 Biochemical vs. cellular assays: advantages and disadvantages

Parameter	Biochemical Assays (Primary Assays)	Cellular Assays (Secondary Assays)
Test Systems	Isolated protein, recombinant proteins	Tissues, isolated cells, target protein overexpressed cell lines, etc.
Measurement	Measures ability of compound to interact with the target (e.g., Binding assays—GPCRs, kinases, ion channels) Measures function of target protein (Examples—Kinase/ATPase assays, protease assays, protein interaction assays)	Measures function of protein in the context of cell (e.g., changes in second messengers, transcriptional levels, protein levels and localization) Measures functional consequence of target engagement in the living system (e.g., Cell proliferation/apoptosis, secretion of cytokines)
Assay development (3Rs) – Relevance – Robustness – Reproducibility	Physiologically less relevant Highly robust Highly reproducible	Physiologically more relevant Robust Reproducible (variability high)
Cost	Less	High
Practicality (Feasibility)	Easy to setup assays	Needs expertise in setting up assays
Automation	Easy (as often homogenous assays)	Difficulty as often involve multiple steps (heterogeneous assays)
Advantages	Simple, less complex, less noise	Physiologically relevant, can simultaneously assay for compound characteristics such as permeability, cytotoxicity
Disadvantage	Less physiologically relevant	Off-target effects of compound

Factors Considered for Assay Development

1. Pharmacological importance of the assay:
   Ability to identify compounds with the desired mechanism of action
2. Reproducibility:
   Reproducible across assay plates, screen days, and the length of the drug discovery program
3. Quality:
   Pharmacology of the standard compound(s) falls within predefined limits
4. Effect of compound in the assay:
   Not sensitive to the concentrations of solvents used in the assay

7.2.1.3 Configuration of a Robotic Workstation

Screening large number of compounds per day is feasible only with automated robotic systems that autonomously manage multiple steps on multiple plates simultaneously. Robotic platforms for high-throughput screenings range from simple automated liquid handling machines to multidimensional workstations performing multiple functions. This is usually achieved with the support of one or more mechanical arms. Typically, a robotic system manages microplates from station to station for several steps such as reagent addition, mixing, incubation, and detection steps [7].

7.2.1.4 Acquisition and Handling of Data

An HTS-dedicated plate reader can measure hundreds of plates in a single day, generating a considerable amount of data points. Consequently, data management is a critical point in automated high-throughput screening, given the large number of compounds tested, from a wide variety of chemical libraries and the necessity to correlate and compare results are from different screening campaigns. Dedicated data analysis and management platforms intervene data originated from different screening campaigns with compound structures, as well as performed assays to each other, facilitating the extraction of detailed information from different perspectives [8].

7.2.2 Primary Screening Assays (Biochemical/Target-Based Screening Assays)

Biochemical/ target-based screening generally refers to the use of a purified or partially purified target protein to screen small molecule compound libraries to find candidates for drug development. In principle, it is the simplest experimental procedure for this purpose allowing the uncomplicated evaluation of the interaction between a target and the test molecule. Biochemical assay procedures, as they are inherently simple, are more robust and easier to automate (refer Table 7.2). Membrane permeability, serum binding and compound metabolism are common variables in cell-based assays and in in vivo studies but are not assay variables in biochemical studies. Often, such complexities can prevent the identification of weak inhibitors that might allow a drug discovery or medicinal chemistry program. However, the superior performance of a biochemical approach requires isolation of the target and reconstitution of its activity in a way that can be easily monitored and quantified.

7.2.3 Secondary Screening Assays (Cell/Functional-Based Screening Assays)

With primary biochemical data in hand and HIT compounds identified, screening strategy may include cell-based and other more complex secondary assays to predict compound behavior better on living system. Functional bioassays employing isolated organs to measure and quantify the response to compound activation have made great contributions to pharmacological science and early drug discovery. These classical pharmacological tests provide important and useful data linking in vitro results to in vivo effects. However, they have been in limited use as the test assays for compound screening due to very low throughput, high costs, and requirement of organs. In modern era, functional cell-based assays employing recombinant cell lines mimicking the pharmacological bioassays are preferred as they circumvent the mentioned shortcomings above. They provide cost-efficient protocols to screen a large number of compounds, which are amenable to miniaturization in microplate formats/test tubes and contribute to minimizing the use of organs. In addition, cells expressing human protein offer the advantage of mimicking the signaling pathways that may exist in normal/ abnormal human cells. Often cell signaling pathways are rich in source of targets for intervention using small molecules (Table 7.2).

The Refinement Process

- Generating dose–response curves in the primary assay for each hit and quantifying their potency
- Screening the suitable hits in a secondary assay
- Analyzing the positive HITs for structure–activity relationship (SAR) and identifying the essential elements in the structure that may contribute the activity (Pharmacophore)
- Further evaluating the compounds using in vitro tests with regard to absorption, metabolism, stability, physicochemical properties, specificity screening assays, etc.

7.2.4 *Specificity Screening Assays*

Specificity screening assays consist of in vitro assays on a broad range of potential drug targets to determine whether the compound is appropriately specific for desired target or not. Compounds screening for specificity (across receptors) and selectivity (across receptor subtypes) should be done as early as possible to minimize failure at later stages and reduce costs. Compounds which are not highly specific and selective are likely to exhibit many unwanted side effects. The range of targets included in such counter screenings strategy depends much on compound chemical characteristics and type of molecular target. Contract research organizations (such as Eurofins, CEREP SA, DiscoveRx) offer battery of assays—mainly binding assays—but also range of functional assays—designed to detect affinity for wide range of GPCRs, enzymes, transporters, and ion channels. Ideal leads are the ones which are as high as 100-fold more potent in cell-based efficacy assay over that of specificity and selectivity assays.

Pharmacological screening—flow diagram and criteria

7.3 Drug Metabolism and Pharmacokinetics (DMPK) Screening

7.3.1 In Vitro ADME (Absorption, Distribution, Metabolism, Elimination) Studies

Compounds that exhibit desired in vitro efficacy parameters (potency and selectivity) are selected for the following in vitro ADME studies. These studies are relatively rapid and cost effective. The data serve as surrogates and indicators of pharmacokinetic profile of compounds in in vivo studies. Compounds that are optimized for desired profile (Fig. 7.3) are selected for in vivo pharmacokinetic studies.

In Vitro Studies [9]

- Solubility
  - An important parameter for compound formulation, absorption, bioavailability
- Lipophilicity (Log P)
  - Plays an important role in absorption, distribution (Central nervous system), fat tissues, organs, metabolic fate, clearance, etc.
- Stability (Biological matrices—Plasma, Gastric fluids, Intestinal fluids)
  - Important parameter in design and selection of assays, predicting compound levels in in vivo studies
- Absorption/Permeability (Caco-2; PAMPA models)
  - Colon carcinoma (Caco-2) cell permeability assay—Useful for determining oral absorption potential and mechanism(s) of transport
  - Parallel Artificial Membrane Permeability Assay (PAMPA), a high-throughput assay for predicting passive, transcellular intestinal absorption
  - Assays aid in predicting gastrointestinal absorption of compounds
- Protein Binding (Dialysis method)
  - Plays critical role in distribution, metabolism, and elimination, thereby efficacy of compound
- CYP Inhibition Profiling
  - hCYP isoenzymes—1A2, 2B6, 2C9, 2D6, 3A4 are used in the studies to determine whether compound has the potential to inhibit the enzymes
  - Useful information to determine the potential drug–drug interactions
- CYP Reaction Phenotyping (with purified recombinant hCYPs and UDP-glucuronosyltransferase 1 and 2—UGTs)
  - Identifies enzymes responsible for metabolism of compounds
  - Aid in determining potential drug–drug interactions

- Microsomal Stability—Liver microsomes (mouse, rat, dog, monkey, human)
  - Microsomes (mainly consist of endoplasmic reticulum) stability studies determines metabolic fate of compounds
- Metabolite Profiling and Identification
  - Helps in predicting metabolic fate of a compound
  - Useful in selecting species/strain in efficacy/toxicity studies
- PXR Induction
  - Indicates the compound induction potential of CYP
- hERG Inhibition
  - Determines cardiac liability potential
- Cytotoxicity Using Cultured Cell Lines (Hep G2 cells)
  - Determine the cytotoxic potential of a compound

7.3.2 *In Vivo* Pharmacokinetic Studies

Mammals are used in screening of compounds and are shown in Table 7.3. Mice, rats, and dogs are among the most frequently used species for screening small molecules. Whereas primates (e.g., monkeys, marmosets) are primarily used in screening of large molecules (e.g., Biologics). Mammals used are inbred or outbred in origin. The outbred mammals (e.g., Swiss mice, Sprague Dawley rats, and Wistar rats) are regularly used for screening due to availability, low cost, ease of in-house breeding, etc. Similar to studies in humans, testing on animals is regulated and permission from Institutional Animal Ethics Committee (IAEC) is mandatory to

Table 7.3 Commonly used species in pharmacological screening

Species	Scientific Use
Rodents	
Mice	Cancer, transgenic studies (KO/Ki), diabetes and obesity, CNS disorders, etc.
Rat	Toxicological , pharmacokinetic, analgesic, anti-inflammatory, gastric ulcer, CNS disorders studies, etc.
Guinea pig	Allergies, respiratory (bronchodilators, antihistamines), vaccines, TB, etc.
Hamsters	Antiviral, COPD (smoke inhalation), etc.
Gerbils	Stroke, epilepsy, heart diseases, auditory studies, etc.
Non rodents	
Rabbits	Atherosclerosis, eye disorders, skin, pyrogen testing, etc.
Dog	Diabetes, cardiovascular, GI (ulcerative colitis), organ transplantation, etc.
Cat	Neurological (sensory, vision, epilepsy), aging, spinal cord injury studies, etc.
Monkey	CNS (psychopharmacology), GIT, menstrual cycle, fertility, PK studies, etc.
Pig	Cardiovascular studies

ensure the Principles of 3R—Reduction, Refinement, and Replacement—in planning experiments. The choice of appropriate species models depends on question to be addressed in the study. It requires understanding of species-specific physiology and similarity with regard to the target organ, metabolic pathways, as well as financial, regulatory, and ethical considerations.

Footnote: Replacement: Methods which avoid or replace the use of animals
Reduction: Methods which minimize the number of animals used per experiment
Refinement: Methods which minimize animal suffering and improve welfare.

In Vivo Pharmacokinetic Studies [9]

1. Preliminary Intravenous and Oral PK study on rat or mice (relevant species)—To estimate PK parameters, absolute bioavailability, comparative PK, urine or faces recovery
2. Dose-dependent study in rat or mice—To assess linear or nonlinear PK characteristics of compound
3. Mechanistic studies (Bile duct cannulated rats, portal vein cannulated rats or mice)—To estimate percentage dose excretion in urine, bile, or feces
4. Multiple dose studies—To determine accumulation or autoinduction potential of compound
5. Tissue distribution studies—To determine tissue distribution of target engagement by the compound

7.4 In Vivo Pharmacological Screening

7.4.1 Primary In Vivo Screening Model

Preliminary screening studies are usually executed in the disease model at a single dose for selected compound/s. Commonly used species and strains in screening for various indications are given in Tables 7.3 and 7.4. The disease animal model should be validated prior to the routine screening of good number of molecules. The active molecules may be further studied at least with three doses in the selected model to confirm the efficacy and determine potency (MED-Minimum effective dose, ED_{50}—Dose at which 50% of response achieved). Various compound and species/strain related factors to be considered in designing screening protocols are detailed below.

Footnote:

Inbred strain: Defined as a product of over 20 generations of brother-sister mating, which results in individuals that are 98% identical to each other.

Outbred strain: Genetically heterogeneous and are often produced by breeding systems that minimize inbreeding

The term “stock” is used to denote an outbred population of mice while the term “strain” is used to denote an inbred population

Table 7.4 Commonly used mouse and rat strains in pharmacological screening

Strain	Therapeutic Indication
Mouse strains	
Non obese diabetes (NOD) mice	Type 1 diabetes
DBA/1J mice	Arthritis
db/db mice	Type 2 diabetes
ob/ob mice	Type 2 diabetes, obesity, hyperlipidemia
Nude mice	Solid tumors
SCID mice	Hematological cancers
Rat strains	
Long-Evans rat	Behavioral studies
Goto-Kaizaki rat	Type 2 diabetes complications
Zucker diabetic fatty rats	Type 2 diabetes, hyperlipidemia
Lewis rat	Arthritis, autoimmune disorders
Brown Norway rat	Asthma, transplantation studies
Dark agouti rat	Autoimmune disorders, transplantation

7.4.1.1 Compound-Related Parameters for In Vivo Screening

Selection of Compound/s

Compound/s which fulfill the criteria of HIT as per in vitro screening strategy and initial lead as per preliminary DMPK studies.

Dose Selection

Based on the initial PK parameters wherein the expected concentrations in plasma are at least three times more than ED_{50} in in vitro studies. In case of screening compounds for anticancer activity, the expected concentrations in plasma (preferably unbound concentration) should be at least three times more than IC_{90} in cytotoxic assays. It is important to note that free drug concentration (unbound concentration) in blood or tissue is the one that can bind with the target and elicit response [10]. So free drug concentration should be considered in dose selection. If target tissue is known, sufficient free drug concentration in tissue should be aimed for achieving proper screening of molecules for efficacy. The unbound concentrations in tissues can be quantified from ex vivo measurements in tissue homogenates, using traditional techniques such as Rapid Equilibrium Dialysis (RED) in a similar manner to measure free drug concentration in plasma [11]. It should also be kept in mind that whether in vitro assay test has serum or albumin in the assays. If serum or albumin is present in the assay, necessary corrections to be done for unbound concentrations.

The Time of Dose of Compound

The time of administration of compound should be selected in such a way that peak concentrations of the compound in circulation should reach prior to time of efficacy parameter estimation.

Duration of Treatment

Dose regimen (dose, number of times of administration/day, number of days of treatment) selection should be based on the PK parameters of the compound and disease progression/development of animal model.

Treatment Paradigm

Prophylactic or therapeutic treatment strategy depends on the questions to be addressed in the study (prior treatment is to verify whether the compound minimizes the development/induction of the disease; therapeutic treatment strategy is to verify whether the compound delays the progression/reverses the disease).

7.4.1.2 Animal Model-Related Parameters In In Vivo Screening

Species Selection

The species used in the screening should mimic human in terms of anatomy and physiology.

Disease Induction

The pathogenic factors for disease development/induction should be similar to that of human (e.g., chemical/diet/loss of specific region)

Ethical Issues

To minimize the ethical concerns, if any, during animal disease induction should be envisaged and due optimizations are necessary for successful screening of compounds (e.g., body weight loss, tumor size development in allograft/xenograft studies/concerns/precautions, sterilization conditions, need of air handling units, etc.).

Number of Animals

Sufficient number of animals in each group is to be allotted to ensure power of the study.

Validation of Animal Model

The animal model should be validated using an appropriate positive control or standard compound which has the ability to bind to the target of interest and elicit reproducible effect.

7.4.2 Secondary In Vivo Studies

The selected active molecules in the primary screening studies are further evaluated in diverse additional animal models to expand the profile of the compound for comparing the activity and safety profile w.r.t. standard/reference treatments. The studies should be planned to determine the onset of activity of the molecule, duration of activity, as well as sustenance of the activity, post withdrawal of the treatment. Also, the studies should aim at finding the site of action, potential mechanism of action, plausible biomarker of efficacy, etc. In addition, compound levels in plasma and tissues may be quantified to model the pharmacokinetic and pharmacodynamic relationship.

7.4.3 Target Engagement and Biomarker

7.4.3.1 Target Engagement

Target engagement in drug discovery refers to interaction of compound with the target protein in living system. It is very critical from efficacy and safety points of view that the compound should engage the target in desired tissue and at desired time [12]. In an effort to improve R&D productivity, AstraZeneca has come up with a revised strategy "5R Framework" [13]. The revised strategy is to focus decision-making on five technical determinants (the right target, right tissue, right safety, right patient, and right commercial potential). Undesirable target tissue engagement may lead to safety issues. Thus, right target and right tissue engagement by compound are very much essential in any drug discovery and development program.

7.4.3.2 Biomarkers

Biomarkers are factors that are objectively measured and evaluated as indicators of normal biological processes or pathological processes, and/or as indicators of pharmacological responses to therapeutic intervention [14]. Coupling of target engagement studies with biomarker evaluation increases the success rate of drug discovery and development programs, and thereby accelerate the availability of new drugs [15]. Characteristics of an ideal biomarker are given below. In a BRAF inhibitor (MAPK) program, the investigators at Novartis measured p-MEK and p-ERK as biomarkers to indicate modulation of $BRAF^{V600E}$. The critical analysis of the effect-concentration relationship indicated that p-MEK: plasma analysis data better correlate with the target engagement as biomarker, whereas p-ERK: plasma analysis as efficacy biomarker (i.e., tumor shrinkage). These findings clearly underscore the closeness (proximity) of the drug target to the biomarker in a signaling cascade as the most appropriate marker for the study [12].

Characteristics of an Ideal Biomarker

1. The selected biomarker is ideally a direct measure of the target modulation by the compound.
2. Time-course of onset, duration, and offset of biomarker mimics the target modulation by the compound.
3. There should be a direct link between biomarker and disease and the dynamic range of biomarker response should relate to efficacy readout in the animal model.
4. Suitable assay system should be available that accurately and precisely measure the subtle changes in biomarker levels.
5. Ideally in vitro (cell-based assay) and in vivo mechanisms are similar, and the biomarker is the same.
6. Ideally the selected biomarker in preclinical studies should be the same as in potential clinical studies for better translation.

7.4.4 *Use of Genetically Engineered Mice in Screening and Lead Characterization*

Species differences in the molecular target protein characteristics (size, sequence, shape, etc.) influence the interaction of therapeutic compound which may limit the utility of wild-type mice as a preclinical species/strain for efficacy and safety testing. To overcome this limitation, mice expressing the human drug target protein can be generated. Genetically, humanized mice have potential for use in a wide variety of research applications, including efficacy and safety testing. Transgenic technology has been widely used as a method to overexpress human genes. The term "transgenic mice" refers to mice carrying exogenous DNA that has integrated within the genome and is expressed "in trans" (i.e., not within its native genetic locus). mPGES-1 knock-in mice are an excellent example of the use of humanized mice for drug efficacy and safety application [16] and relieves pyrosis and pain in preclinical models of inflammation. Microsomal prostaglandin E synthase-1 (mPGES-1) is a terminal prostaglandin E_2 (PGE_2) synthase in the cyclooxygenase pathway. Inhibitors of mPGES-1 may block PGE_2 production and relieve inflammatory symptoms. To test whether inhibition of mPGES-1 can block inflammation, a selective PGES-1 inhibitor, MF63 *(human mPGES-1 enzyme,* $IC_{50} = 1.3$ *nM; do not inhibit mouse or rat enzyme)*, was tested in in vitro and in animal models of inflammation in mice. MF63 strongly inhibited human mPGES-1 in human cells in vitro and exhibited minimal activity against mouse mPGES-1. As expected, treatment of mPGES-1 transgenic (knock-in) mice with MF63 blocked PGE_2 synthesis and reduced inflammatory responses.

Genetically engineered mouse models (GEM models) have been increasingly used in drug discovery especially in clinical nominated candidate characterization in addition to new target identification and validation. The knock-in/knock-out mice contain an inactive gene inserted/ deleted using embryonic stem cells or more recently CRISPR. These mice typically used to confirm whether the nominated molecules activity in vivo is contributed mainly from the desired target. The presence of inactive or loss of target protein may result in no activity, following the treatment with the compound [17].

7.5 Lead Optimization Studies

Lead optimization is basically an iterative process following identification of HITs from high-throughput screening wherein medicinal chemists, biologists, and toxicologists work together in order to optimize pharmacodynamic properties such as efficacy, potency, and selectivity in in vitro and in vivo studies, physicochemical properties, pharmacokinetic properties (absorption, distribution, metabolism, and elimination), and preliminary toxicological aspects of HIT molecule (Fig. 7.1). This step is most exciting, time-consuming, and labor-intensive exercise. In silico or virtual screening is a different approach used for lead generation. The virtual chemical libraries are docked into binding sites of target proteins with known structure. Steric and electrostatic complementarity (the fit) between the compounds and target protein are quantified and given a docking score. The compounds are ranked according to docking score and prioritized for subsequent in vitro and in vivo studies for optimization of parameters. The assays used at this stage in in vitro are basically profiling compound with respect to selectivity and specificity to the target of interest. In addition, in vitro hERG assay (assay to test potential to interact with potassium channels) is performed to minimize potential cardiovascular hazard in humans. Parallel to in vitro profiling of pharmacodynamic parameters with respect to potency, specificity, selectivity, and safety, in vitro assays for the prediction of pharmacokinetic parameters should be performed (In Vitro Studies in Sect. 7.3.1).

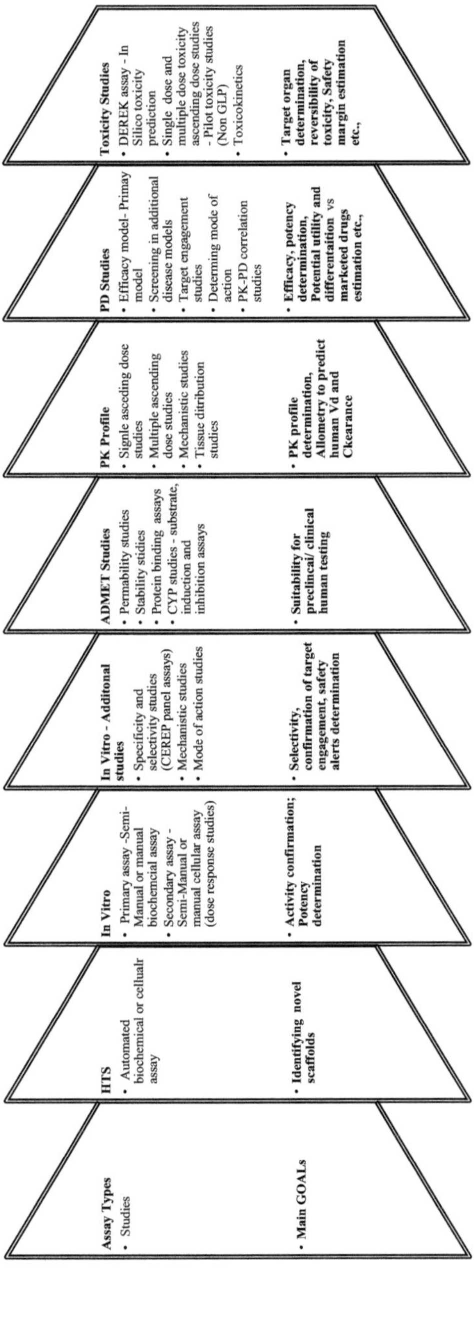

Fig. 7.1 Screening steps and their goals

Compounds that met the desired in vitro profile are subjected to in vivo models profile for pharmacokinetic parameters (In Vivo Studies in Sect. 7.3.2) and pharmacodynamic profile [18]. These studies are often conducted on few selected molecules as they require large quantities of compound. Optimized compounds with desired in vitro and in vivo profile are subjected to preliminary toxicity studies which include Ame's mutagenicity test, cytotoxicity test in HepG2 cells, and toxicokinetic studies (single ascending dose and multiple ascending dose studies). It is important to be stringent on the criteria of pharmacokinctic studies, pharmacodynamic studies and safety at early stage to be line with the sage, *Fail cheap and Fail early*. Although these studies are independent to each other, they are commonly performed simultaneously to save time and resources [19, 20, 21] (Fig. 7.2). Selected candidate(s) for development is subjected to large-scale synthesis in GLP compliant labs for nonclinical regulatory toxicology studies and clinical development (Fig. 7.3).

7.6 Case Study: Discovery of OSI-906 (Linsitinib), A Dual Inhibitor of Insulin-Like Growth Factor-1 and Insulin Receptors

Insulin-like growth factor-1 receptor/insulin receptor (IGF-1R/IR) pathway is critical for promotion and survival of several types of cancer. Both IGF-1R and IR are transmembrane receptor kinases and consist of two alpha units (extracellular) linked with two transmembrane beta units (intracellular) containing cytoplasmic tyrosine

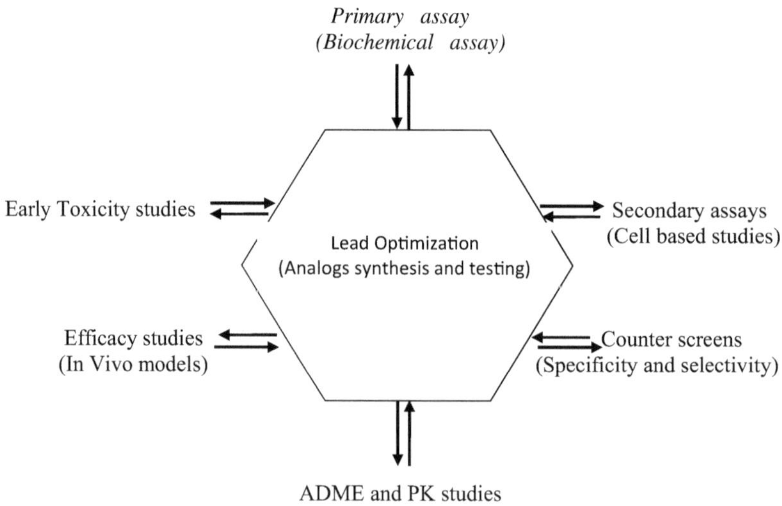

Fig. 7.2 Lead optimization strategy

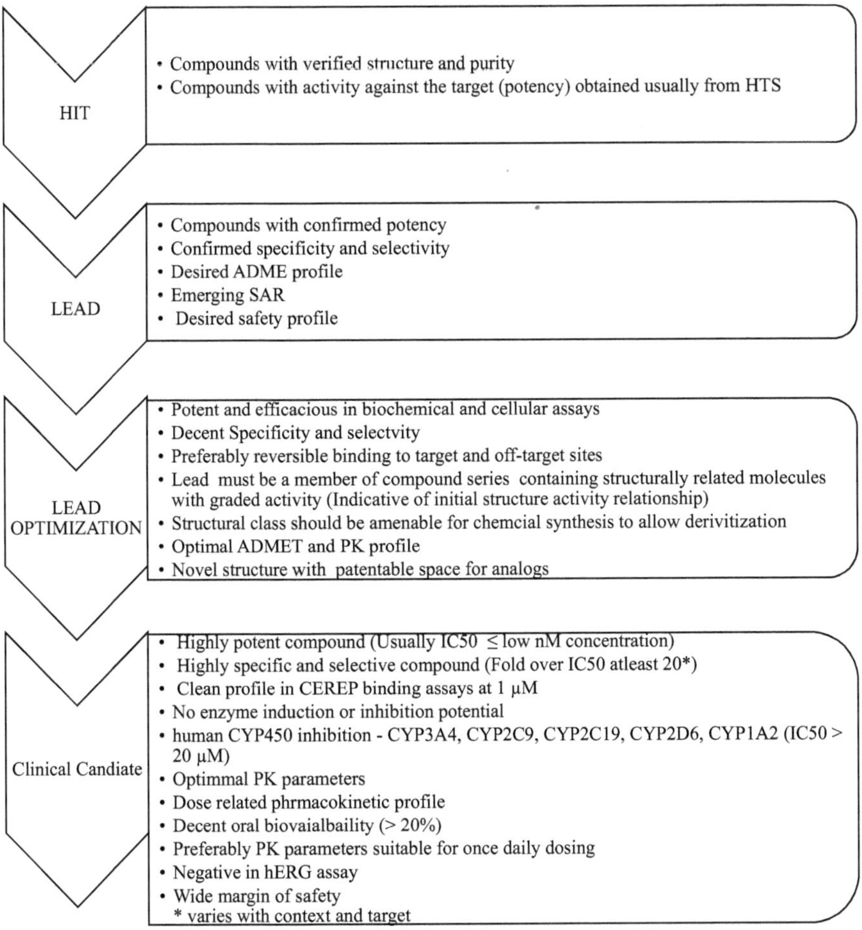

Fig. 7.3 Characteristics of HIT, lead and clinical candidate

kinase activity. IGF-1R and IR can exist as homodimers and heterodimers. IGF-1 and IGF-2 are ligands for IGF-1R; insulin is ligand for IR whereas IR can be activated by insulin or IGF-2 (Fig. 7.4). The pathway has been exploited for discovering novel agents against cancers using mAbs (monoclonal antibodies) and small molecule kinase inhibitors [22]. Various important screening activities in the discovery of OSI-906 (Linsitinib) as a dual kinase inhibitor are shown in Table 7.5.

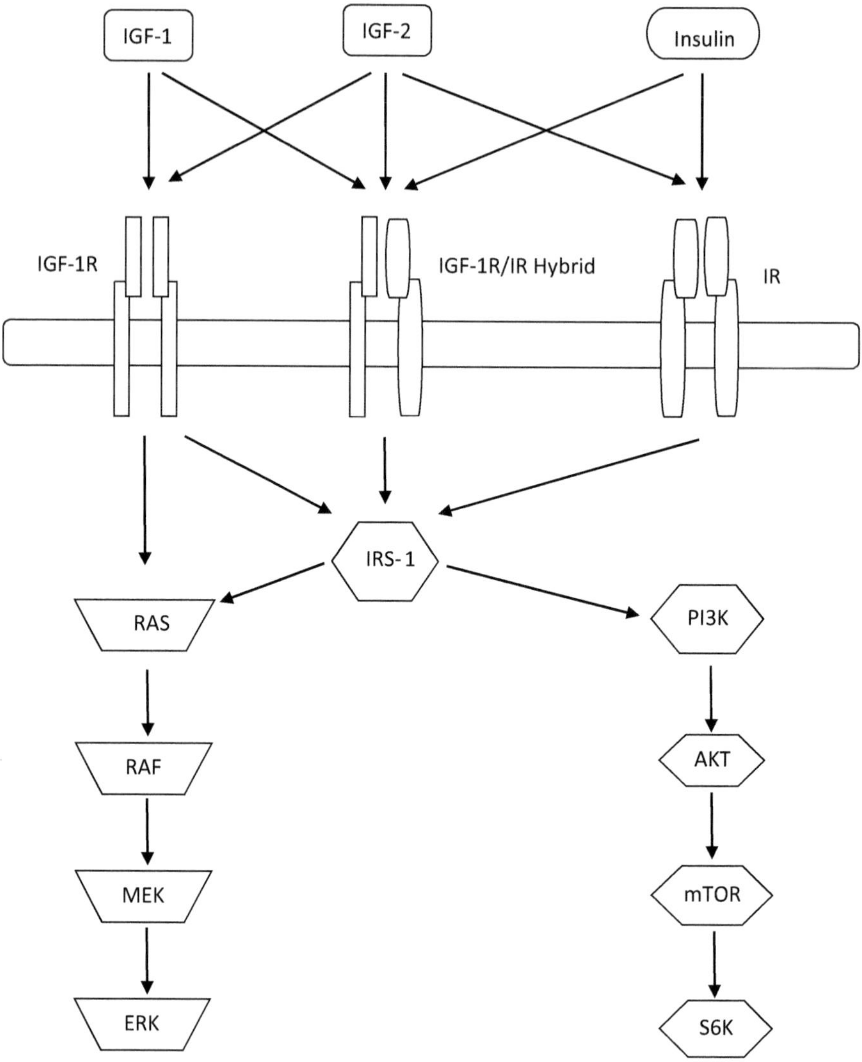

Fig. 7.4 IGF-1R/IR signaling pathway

Table 7.5 Case study: discovery of OSI-906, a highly potent, selective, and orally bioavailable dual IGF-1R and IR inhibitor

Screening Strategy [23]	Discovery Team Approach	Progress/Outcome
In vitro assays		
Primary Assay (Biochemical assay) GST-tagged recombinant kinase domain of hIGF-RSubstrate: Poly (Glu:Tyr—4:1)ATP: 100 μmol/L Assay type: ELISA	Structure-based drug design and empirical medicinal chemistry	Compounds with activity below 1μM discovered
Secondary Assay (Cellular assay) Full length hIGF-1R over-expressing NIH-3T3 cells (LISN cell line) Inhibition of IGF-1 stimulated receptor autophosphorylation Assay type: ELISA	Additional compounds synthesized based on structure activity relationships and screened in the secondary assay	Compound 1a identified with IC_{50} of 606 nM in the biochemical assay [24]
	Additional SAR studies to improve potency	Compound 15c identified with IC_{50} of 166 nM in biochemical assay and 191 nM in cellular assay (Initial lead compound)
Preliminary PK studies in mice	To understand PK properties of the series	Favorable oral bioavailability (47%) but with high clearance (93 mL/min/kg)
In vitro specificity profiling assays(against 15 purified protein kinases from tyrosine and serine/threonine kinase families)	To gain insight to the specificity and selectivity profile of the series	Less than 50% inhibition observed at 10μM Compound 15 and 16 series considered as early leads with increased potency and desired selectivity
	Improvement in drug metabolism and pharmacokinetic (DMPK) properties needed to select compound for clinical development—Additional analoging studies initiated based to binding determinants from the X-ray cocrystal structure of IR and compound 15a	Compound 17 with 20-fold increase in potency in IGF-1R identified [25]
	Different salt forms of compound 17 derived and used different formulations for studying DMPK parameters	Favorable oral bioavailability (77%) and clearance (41 mL/min/kg) observed in mouse
	Additional analoging efforts to identify advanced leads using binding determinants from the X-ray cocrystal structure of IGF-1R with compound 17	Efforts lead to PQIP and AQIP as advanced leads with improved potency in biochemical and cellular assays (PQIP—IC_{50}—24 and 35 nM, respectively; AQIP—

(continued)

Table 7.5 (continued)

Screening Strategy [23]	Discovery Team Approach	Progress/Outcome
		IC_{50}—19 and 20 nM, respectively) [26, 27]
hCYP3A4 inhibition assay	To determine potential drug interaction with CPY3A4 substrate/inhibition drugs	IC_{50} μM—8.3 and >20 for PQIP and AQIP, respectively indicating decent potential safety
Microsomal stability (ER)	Human, moue, rat and dog	Low stability translated in to lower in vivo clearance (PQIP—19 mL/min/kg; AQIP—24 mL/min/kg) and good oral bioavailability (PQIP—100%; AQIP 100%)
DMPK Profile	Intravenous and oral studies	Despite the positive drug-like properties both compounds displayed a low Cmax and high Vss (PQIP—2.2μM and 8 L/kg, respectively; AQIP—3.4μM and 10 L/kg, respectively)
	Focused efforts for analoging using PQIP and AQIP	Best in the series identified as OSI-906 with desirable potency, efficacy and overall ADME/DMPK properties [28]
Hepa-1 cell line (Hepatoma cell line) Inhibition of pIR	To determine potency against hIR	IC_{50} in Hepa-1 cell line—39 nM
GEO cell line (Colorectal cell line) Naturally harboring IGF-1R and IGF-2 autocrine loop cell line	LISN and GEO cell line develops tumor in mice—Hence can be used for proof of concept study	IC_{50} in GEO cell line—24 nM
Downstream Signaling protein phosphorylation inhibition n LISN/GEO cell line	Serves as proof of concept and PD biomarker in in vivo studies	p-ERK1/2 (IC_{50}—28 nM) and p-70S6K (IC_{50}—60 nM)
Functional Cellular assays (CellTiterGlo™ luciferase-based cellular ATP assay kit (Promega) Anti-proliferative activity against various cancer cell line—Colorectal cancer (HT-29, Colo205, GEO, SW620, SW480), Pancreatic cancer (BxPC3), NSCLC (H358, H292, H322, Calu1, H441 and DU4475) and Breast cancer (MCF-7, SKBR3)	To determine the sensitivity against variety of cancer cell lines	IC_{50} s range—Colorectal cancers (21–191 nM), Pancreatic cancer (150 nM), NSCLC (86–830 nM) and Breast cancer (177–635 nM)—Suggesting that dual inhibitor active against various cancers

(continued)

Table 7.5 (continued)

Screening Strategy [23]	Discovery Team Approach	Progress/Outcome
Recceptor Tyrosine Kinase Assay Signal realized by capture array—Inhibition of phosphorylation of IR and IGF-1R	To determine the activity against IR and IGF-1R	OSI-906 at 1μM fully inhibited both IR and IGF-1R phosphorylation
Protein Binding Study (Ultracentrifugation technique)	To estimate free drug levels	Mouse 98.4%, human 96.7%
In Vitro -Selectivity Profile (against 88 additional purified protein kinases representing the tyrosine and serine/threonine kinase families) A follow study conducted to determine IC_{50} against key subset of kinases	To determine specificity and selectivity of compound	Compound did not inhibit more than 50% at 1μM indicating high specificity and selectivity IC_{50} found to be more than 10μM
PK profile in mice, rat and dog	Allometric extrapolation to human to predict PK profile	Desirable PK profile observed in each species Mice (5 mg/kg, p.o.: BA—98%; 5 mg/kg, i.v.: Cl—12 (mL/min/kg; Vss – 2.05 L/kg) Rat (12.5 mg/kg, p.o.: BA—74%; 5 mg/kg, i.v.: Cl—4 (mL/min/kg; Vss—0.79 L/kg) Dog (5 mg/kg, p.o.: BA—64%; 2.5 mg/kg, i.v.: Cl—39 (mL/min/kg; Vss—4.3 L/kg)
In Vivo studies Antitumor activity—LISN xenograft model in mice (Tumor growth inhibition—TGI)	To determine efficacy in mice	OSI-906 (25 and 75 mg/kg, o.d., p.o., × 12 days) exhibited significant antitumor activity (25 mg/kg—TGI –60%; 75 mg/kg—Tumor regression 55%).
Antitumor activity—GEO and Colo-205 colorectal xenograft models in mice	To profile the efficacy in additional models	OSI-906 (60 mg/kg, o.d., p.o., × 14 days) exhibited significant antitumor activity (60 mg/kg—Tumor growth inhibition ≥ 90%).
PK-PD correlation study—LISN xenograft study Drug exposure (PK)-Inhibition of IGF-1R phosphorylation (PD)	To determine time-course of activity and its correlation with plasma levels To predict exposures required for sustained inhibition in humans	Maximal inhibition of IGF-1R phosphorylation (80%) achieved between 4 and 24 h with plasma levels of 26.6 and 4.77μM, respectively. This is inconsistent with robust antitumor activity

(continued)

Table 7.5 (continued)

Screening Strategy [23]	Discovery Team Approach	Progress/Outcome
		observed in the LISN xenograft model. Such as correlation was observed in the GEO and Colo-205 xenograft studies. IGF-1R and IR cell + human plasma IC50 values along with PK-PD study can be utilized in predicting OSI-906 exposures required for sustained inhibition in humans.

References

1. Hertzberg RP, Pope AJ. High-throughput screening: new technology for the 21st century. Curr Opin Chem Biol. 2000;4(4):445–51. https://doi.org/10.1016/s1367-5931(00)00110-1.
2. Wölcke J, Ullmann D. Miniaturized HTS technologies—uHTS. Drug Discov Today. 2001;6 (12):637–46. https://doi.org/10.1016/s1359-6446(01)01807-4.
3. Macarron R, Banks MN, Bojanic D, Burns DJ, Cirovic DA, et al. Impact of high-throughput screening in biomedical research. Nat Rev Drug Discov. 2011;10(3):188–95. https://doi.org/10.1038/nrd3368.
4. Bleicher KH, Böhm HJ, Müller K, Alanine AI. Hit and lead generation: beyond high-throughput screening. Nat Rev Drug Discov. 2003;2(5):369–78. https://doi.org/10.1038/nrd1086.
5. Sundberg SA. High-throughput and ultra-high-throughput screening: solution- and cell-based approaches. Curr Opin Biotechnol. 2000;11(1):47–53. Review. https://doi.org/10.1016/s0958-1669(99)00051-8.
6. Silverman L, Campbell R, Broach JR. New assay technologies for high-throughput screening. Curr Opin Biotechnol. 1998;2(3):397–403. https://doi.org/10.1016/s1367-5931(98)80015-x.
7. Michael S, Auld D, Klumpp C, Jadhav A, Zheng W, et al. A robotic platform for quantitative high-throughput screening. Assay Drug Dev Technol. 2008;6(5):637–57. https://doi.org/10.1089/adt.2008.150.
8. Brideau C, Gunter B, Pikounis B, Liaw A. Improved statistical methods for hit selection in doiigh-throughput screening. J Biomol Screen. 2003;8(6):634. https://doi.org/10.1177/1087057103258285.
9. Chung TDY, Terry DB, Smith LH, Sittampalam GS, Grossman A, et al. In vitro and in vivo assessment of ADME and PK properties during lead selection and lead optimization—guidelines, benchmarks and rules of thumb. Assay guidance manual [Internet]. Bethesda, MD: Eli Lilly & Company and the National Center for Advancing Translational Sciecompounds; 2004.
10. Smith DA, Di L, Kerns EH. The effect of plasma protein binding on in-vivo efficacy: misconceptions in drug discovery. Nat Rev Drug Discov. 2010;9:1229–939.
11. Banker MJ, Clark TH. Plasma/serum protein binding determinations. Curr Drug Metab. 2008;9:854–9. https://doi.org/10.2174/138920008786485065.
12. Mulvihill J, Elizabeth B. Accounts in drug discovery: Case studies in medicinal chemistry. In: Barrish JC, Carter PH, PTW C, Zahler R, editors. RSC drug discovery series No. 4: Royal Society of Chemistry; 2011.

13. Mulvihill MJ, Cooke A, Rosenfeld-Franklin M, Buck E, Foreman. Discovery of OSI-906: a selective and orally efficacious dual inhibitor of the IGF-1 receptor and insulin receptor. Future Med Chem. 2009;1(6):1153–71. https://doi.org/10.4155/fmc.09.89.
14. Wu J, Li W, Craddock BP, Foreman KW, Mulvihill MJ, Ji QS, Miller WT, Hubbard SR. Small-molecule inhibition and activation-loop trans-phosphorylation of the IGF1 receptor. EMBO J. 2008;27(14):1985–9. https://doi.org/10.1038/emboj.2008.116.
15. Ji QS, Mulvihill MJ, Rosenfeld-Franklin M, Cooke A, Feng L, et al. A novel, potent, and selective insulin-like growth factor-I receptor kinase inhibitor blocks insulin-like growth factor-I receptor signaling in vitro and inhibits insulin-like growth factor-I receptor dependent tumor growth in vivo. Mol Cancer Ther. 2007;6(8):2158–67. https://doi.org/10.1158/1535-7163.MCT-07-0070.
16. Mulvihill MJ, Ji QS, Coate HR, Cooke A, Dong H, et al. Novel 2-phenylquinolin-7-yl-derived imidazo[1,5-a]pyrazines as potent insulin-like growth factor-I receptor (IGF-IR) inhibitors. Recent Results Cancer Res. 2007;172:59–76. https://doi.org/10.1016/j.bmc.2007.10.061.
17. Armold L, Mulvihill MJ. US Pat 7 534 797 B2; 2009.
18. Tuntland T, Ethell B, Kosaka T. Implementation of pharmacokinetic and pharmacodynamic strategies on early research phases of drug discovery and development at Novartis institute of biomedical research. Front Pharmacol. 2014;5:1–16. https://doi.org/10.3389/fphar.2014.00174.
19. Morgan P, Brown DG, Lennard S, Arrowsmith J. Feltner et al. Impact of a five-dimensional framework on R&D productivity at AstraZeneca. Nat Rev Drug Discov. 2018;17(3):167–81. https://doi.org/10.1038/nrd.2017.244.
20. Colburn WA. Biomarkers in drug discovery and development: from target identification through drug marketing. J Clin Pharmacol. 2003;43:429–41. https://doi.org/10.1177/0091270003252480.
21. Kraus VB. Biomarkers as drug development tools: discovery, validation, qualification and use. Nat Rev Rheumatol. 2018;14(6):354–62. https://doi.org/10.1038/s41584-018-0005-9.
22. Xu D, Rowland SE, Clark P, Giroux A, Côté B, et al. MF63 [2-(6-chloro-1H-phenanthro[9,10-d]imidazol-2-yl)-isophthalonitrile], a selective microsomal prostaglandin E synthase-1 inhibitor, relieves pyresis and pain in preclinical models of inflammation. J Pharmacol Exp Ther. 2008;326(3):754–6. https://doi.org/10.1124/jpet.108.138776.
23. Zuberi A, Lutz C. Mouse models for drug discovery, can new tools and technology improve translational power? ILAR J. 2016;57:178–85. https://doi.org/10.1093/ilar/ilw021.
24. Vogel HG. Drug discovery and evaluation. Pharmacological assays. Berlin, Heidelberg: Springer; 2002. p. 1–18.
25. Lombardino JG, Lowe JA 3rd. The role of the medicinal chemist in drug discovery—then and now. Nat Rev Drug Discov. 2004;(10):853–62. https://doi.org/10.1038/nrd1523.
26. Kerns EH, Li D. Pharmaceutical profiling in drug discovery. Drug Discov Today. 2003;8(7):316–23.
27. Giersiefen H, Hilgenfeld R, Hillisch A. Modern methods in drug discovery: an introduction. Switzerland: Springer; 2009. p. 1–18.
28. Wang Y, Ji QS, Mulvihill M, Pachter JA. Inhibition of the IGF-I receptor for treatment of cancer. Kinase inhibitors and monoclonal antibodies as alternative approaches. Bioorg Med Chem. 2008;16(3):1359–75. https://doi.org/10.1007/978-3-540-31209-3_5.

Chapter 8
Drug Target Identification and Validation

Srinivas Gullapalli

8.1 Introduction

In the process of drug discovery and development, the pharmacólogical 'drug target' plays a pivotal role along with the therapeutic drug and the purported disease itself. In the words of the Nobel laureate 'Paul Elhrich', who was known for his countless contributions to the field of pharmacology, '*corpora non agunt nisi fixate*—drugs will not act unless they are bound' [1]. In other words, the entity to which a drug binds to elicit therapeutic effect can be called as 'drug target'.

Thus, a drug to show its pharmacological effects first needs to bind to its biological target at a molecular protein level. One important insight to keep in mind is that both the molecular switch aka 'drug target' and the physiological process/pathological disease state that gets regulated/modulated by the drug are endogenous to the living system or body. However, the drug which elicits physiological effects or provides symptomatic relief, treats, cures the pathological disease is an external agent (a small molecule chemical substance or a biologic entity).

8.2 What Is a 'Drug Target'?

As per Pharmacology/Biochemistry/Molecular Biology, a drug target is defined as '*a biological entity (usually a protein or gene) associated with a physiological process or a pathological disease state, that interacts with and whose activity is modulated by a potential drug substance/compound*'.

S. Gullapalli (✉)
Drug Discovery Research Consultant, Kalyan, Mumbai, Maharashtra, India

R. Poduri (ed.), *Drug Discovery and Development*,
https://doi.org/10.1007/978-981-15-5534-3_8

8.2.1 What Makes a 'Drug Target' Count?

It is very important to know the properties that make a 'drug target' an attractive one. ***A 'drug target' needs to be relevant to the disease phenotype and should be amenable to therapeutic modulation in order to be called as 'good and attractive'.*** *At the same time, the good 'drug target' needs to have a good* ***'therapeutic index/ window' or a positive bias for beneficial therapeutic effect(s) over side effects and preferably be selective / specific for its desired pharmacological effects.***

The basic premise for a biological 'drug target' is that it is to be amenable for modulation by a therapeutic agent, defined by a term called as 'druggable'—be it a small molecule new chemical entity (NCE) or a large molecule new biologic entity (NBE). A 'druggable' target is a protein, peptide or nucleic acid with activity that is amenable for modulation by a drug, which can be a small molecular weight new chemical entity (NCE) or a large molecular weight new biologic entity (NBE) such as an antibody or a recombinant protein.

8.2.2 Druggable May Not Necessarily Mean a Good 'Drug Target'

The ability of a protein to bind a drug substance at the required binding site might make it druggable, but does not necessarily make it a potential drug target, for that honour belongs only to the proteins that are also linked to disease pathology.

The potential drug targets that the pharmaceutical industry want to exploit have to be part of both druggable human genome/proteome and are also found to be related to the specific disease pathology with proper validated proofs.

So, what makes a 'druggable target' a good/promising/ideal one are the set of properties it is ordained with [2, 3].

Properties of a Promising/Good/Ideal Biological 'Drug Target'

1. First of all, the drug target needs to have a safety profile with no target-based severe adverse events and side effects.
2. The drug target needs to have a confirmed beneficial role in the pathophysiology of a disease and/or is disease-modifying (e.g. verified by knock-in/knock-out phenotypic data).
3. Target expression is desired to be tissue selective (if not specific) and not distributed evenly and abundantly throughout the body.
4. The target protein 3D-structure needs to be available to assess druggability with a clear link to disease pathology.
5. The target needs to have favourable 'assayability' and amenable for high-throughput screening techniques.
6. The availability of target/disease specific biomarkers to facilitate monitoring of therapeutic and other pharmacological effects.

7. The target needs to have a favourable safety/adverse effects profile with predictable phenotypic data (e.g. in k.o. mice or genetic mutation databases).
8. The proposed target also needs to have a favourable intellectual property (IP) status (free to operate on target with no blocking patents).

8.3 Drug Target Classes

The biological drug target being a protein or gene product belong to different classes broadly known as receptors (G-protein coupled receptors (GPCR) also called as 7-Trans Membrane (7-TM) receptors), enzymes, ion channels, nuclear receptors, kinases and various other miscellaneous types. These main drug target types alone account for 44% of all human protein targets (GPCRs 12%, ion channels 19%, kinases 10% and nuclear receptors 3%) and various kinds of enzyme targets constitute around 27% of total targets of human druggable genome.

Moreover, among the small molecule approved drugs per target, these privileged families constitute almost 70% of marketed drugs (GPCRs 33%. ion channels 18%, nuclear receptors 16%, kinases 3–6%). Enzymes are the second largest groups of target proteins in the human genome that also constitute second most class of marketed approved treatments [2, 4–6] (Table 8.1).

Even among the total human protein drug targets, the small molecule drug targets comprise a majority of 82% approximately with the remaining 18% human proteins exclusively providing biologic drug targets. And accordingly, of the total approved drugs a majority of 84% drugs are of small-molecule-based drugs with remaining 16% drugs mainly comprising various types of biological drugs [5].

The advent of technological advancement in the drug discovery and development has given plethora of opportunities in modulation of 'drug targets' by various modes of action and technologies (Table 8.2).

Among the approved drugs, across different target types, most of them are highly specific/selective modulators of one particular drug target protein following the paradigm of 'one drug and one target'. However, all the approved small molecule kinase inhibitor drugs are either multi-target drugs or combination products

Table 8.1 Major human protein drug target families—with proportion of the entire drug target proteome and fraction of the approved drugs targeting those families [5]

Drug target type	% of human protein drug targets	% Approved small molecule drugs
GPCRs (7TM)	12%	33%
Enzymes	27%	20%
Ion channels	19%	18%
Kinases	10%	3%
Nuclear receptors	3%	16%
Others	29%	10%

Table 8.2 Drug target classes, technologies and modes of action

Drug target classes and their modes of action for NCEs, NBEs and nucleic acids		
Drug	Drug target class	Mode of action
Small molecule new chemical entities (NCEs)	Receptors	Agonists, antagonists, inverse agonists, modulators, allosteric modulators, sensitizers
	Enzymes	Inhibitors, activators
	Transcription factors	Inhibitors, activators
	Ion channels	Inhibitors, openers
	Transport proteins	Inhibitors
	Protein–protein complexes	Inhibitors
	Nucleic acids (DNA) binding	Alkylation, complexation, intercalation
	Miscellaneous proteins	Inhibitors, activators
New biologic entities (NBEs)		
	Extracellular proteins	Antibodies
	Transmembrane receptors	Recombinant proteins
	Cell surface receptors	Antibody-drug conjugates (ADCs)
	Substrates and metabolites	Enzymatic cleavage agents
Nucleic acids	RNA	RNA interference (RNAi), siRNA, miRNA, shRNA agents

suggesting that 'polypharmacology' is an emerging scenario of treating some of the complex diseases. The highly conserved kinase catalytic binding domain being one of the reasons for the multi-kinase targeted drugs [7].

Having defined a biological drug target and the characteristics that make a druggable protein into a good drug target—it is important to have an understanding of drug discovery and development process before going into various drug target identification and validation techniques and methodologies.

8.4 A Perspective on Drug Discovery

In the past one century, there was a dramatic change in the process of drug discovery and in the practice of medicine.

For many years the drug discovery was based on ethno-botanical knowledge and depended on various traditional systems of medicine. Wherein, the pursuit used to start with testing herbal extracts or their pure constituents in various animal models for their biological effects based on scientific literature out of basic research. And such active moieties were made using semi-synthetic/synthetic chemical methodologies. This whole process used to be traditionally called as *'forward pharmacology'* or *'phenotypic screening'* or *'phenotypic discovery'*.

In this process, traditionally the drug-like substance(s) or chemical compound (s) was/were made to undergo extensive phenotypic assay or disease model screening studies to find out a prospective disease therapy area and specific pharmacological outcome effect. Mostly, in the phenotypic discovery the purported biologic target was identified at the culmination of comprehensive screening of the drug-like product. And once the candidate biological drug target was identified, more comprehensive studies would be taken up to find out mechanism of drug action, to characterize the safety profile in various pharmacological and toxicological studies to qualify it for regulatory submission [2, 8–11].

The process of identifying or deciphering the biologic target retrospectively for the efficacious compound or drug substance is known as *'Target Deconvolution'* [8, 12, 13]. However, in principle *'there is no regulatory requirement to know the molecular target of a drug or clinical candidate since all that matters in the end is that a drug is safe and efficacious'*. In fact, there are a number of approved drugs for which the mechanism of action is unknown.

However, the drug development process will be greatly facilitated if the target is known since this enables rational design of new molecules with improved potency and safety profiles based on the initial lead. The advent of cell, molecular and biochemical techniques, advancements in human genomics and proteomics in the last four decades has completely revolutionized the very drug discovery and development approach. In the present scenario, drug discovery has now become a hypothesis-driven target-based discovery powered with computerization and automation processes. The *'target-based discovery process'* is also known as *'Reverse Pharmacology*', wherein the process starts with the identified drug target for a potentially unmet medical condition and the rationally designed drug moieties were then screened against a series of target-based pharmacological/biological screens.

Unlike the phenotypic discovery model, the target-based discovery process is greatly facilitated as the known drug target enables the rational drug design of new molecules with improved potency and safety profiles [2–5, 10, 14].

Furthermore, among these two approaches, the target-based drug discovery approach has been highly successful. In the past two decades (since 1999), of the total 326 new drugs approved by US FDA, more than 80% of drugs got discovered through target-based drug discovery process as per the report of Center for Drug Evaluation and Research (CDER) on annual new drug approvals till 2018. The below figure gives a comparative scheme of these two approaches (Fig. 8.1).

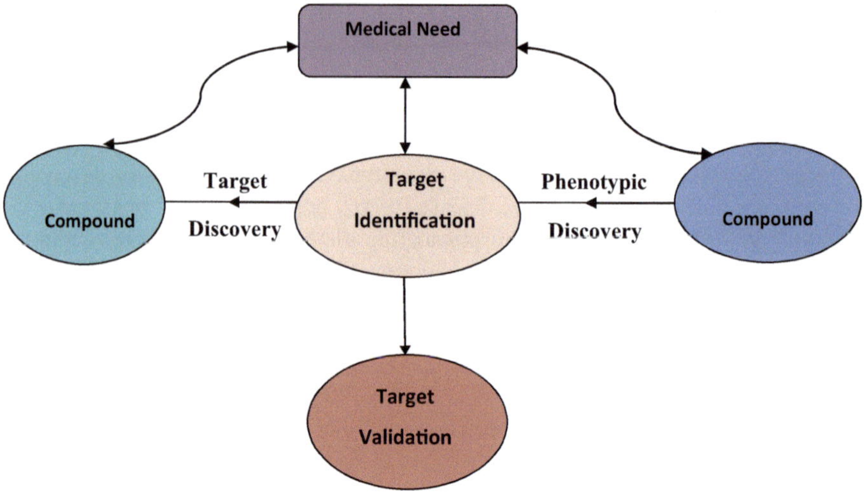

Fig. 8.1 Comparative schematic of phenotypic discovery and target-based discovery

8.5 Phenotypic Discovery Vs. Target-Based Discovery

The phenotypic discovery approach usually starts with the characterization of compound(s)/drug substance(s) in various array of phenotypic biochemical, cell-based, tissue-based and animal-model-based screens to find out on which specific biological/disease assay type the drug molecule exerts desired effects.

The phenotypic approach consists of characterizing the compound / substance across various biological target proteins including entire pathway of known signalling molecular proteins. In this process, the drug substance's biological/pharmacological effect is first determined, followed by characterizing the specific 'drug target protein' or the mechanism of action responsible for the observed phenotypic responses. This retrospective approach of identifying the drug target protein is often called as 'Target Deconvolution'.

One of the greatest advantages of phenotypic approach is the ability of discovering a 'first-in-class' drug moiety. Another advantage is the ability of demonstrating the drug efficacy in native cell or tissue biological environment rather than a purified target in a biochemical or heterologously expressed cell-based screens. However, the cost and availability of wide array of complex assay methodologies involving cell-based and tissue-based phenotypic screens in the throughput format may pose potential challenges in this approach. However, major technological advances, automation and miniaturization of three-dimensional cell models with physiological relevance have made a definite possibility for resurgence for phenotypic-based drug discovery screening [15].

The drug target-based approach normally gets started with an established biological target for the chosen unmet medical condition and rationally designed drug molecules would be subjected to target-based screening assays to identify hit and

Table 8.3 Comparative profile of target-based and phenotypic discovery approaches

	Target-based approach	Phenotypic approach
Molecular target	Known / starting point	To be identified / culminating point
Cost	Lower	Higher
Speed	Rapid	Moderate
Ease of SAR	Easier	Difficult
Screening methods	Target based	Activity based
Mode of action	Predefined	Needs to be assessed
Optimizing hit/lead	Easier	Difficult
Safety-Tox	Predictable	Needs to be assessed
Drug outcome	Best-in-class	First-in-class

lead candidates. The knowledge of the underlying molecular mechanism makes the whole approach a highly focused one and makes the target-based screening methods easier to carry and less expensive to develop with high throughput and thus makes it highly advantageous over the traditional phenotypic approach. Target-based drug discovery approach can adapt various technologies such as genomic, computational modelling, crystallography, biophysical, biochemical and binding kinetics to unravel exactly how a drug interacts with the target protein of interest. These technologies greatly facilitate the development of structure-activity relationship (SAR) and early biomarkers to make the whole approach faster and thus enabling the discovery of best-in-class molecule [14]. The below depicted table gives a prospective comparison of these two processes (Table 8.3).

8.6 Drug Target Identification

Irrespective of the approach, the drug target identification plays a pivotal role in the drug discovery and development process. Accordingly, different methodologies are being adapted in discovery or identification of the drug target. The successful drug target identification basically depends on three key factors, namely profound understanding of disease pathology, knowledge of various underlying molecular mechanisms and the availability of supportive technologies and predictive models. Accordingly, it requires confluence of expertise in biological disease pathology, molecular biology and chemical biology.

The *drug target identification methodologies* can be broadly classified into two main approaches of

'Chemical Proteomic-Based Approaches/Direct Biochemical Methods'

and

'Functional Genomic Approaches'.

On the other hand, based on the type of discovery approach they can be classified as 'phenotypic target deconvolution techniques and direct 'target discovery' techniques. Apart from these, computational methodologies such as web-based database searching and systems biology-bio-informatics-based data integration and network analysis can also be adapted for searching the drug target information [2, 8, 12, 13, 16, 17].

8.6.1 Chemical Proteomic-Based Approaches/Direct Biochemical Methods

The retrospective drug target identification underlying the 'phenotypic discovery' which is also known as 'target deconvolution' consists of mainly five *'chemical proteomic-based approaches'*, namely affinity chromatography, expression cloning techniques, protein microarray, reverse transfected cell microarray and biochemical suppression.

8.6.1.1 Affinity Chromatography

Affinity purification by chromatography is the most widely used technique to isolate specific target proteins from a complex proteome. Typically, the small molecules identified in a phenotypic screen are immobilized onto a solid support linker that can be used to isolate bound protein targets by incubating with protein extracts. Unbound proteins are removed by washing, leaving only bound proteins—with a strong enough affinity to the ligand. The bound proteins are then eluted using buffer conditions that interfere with the target protein–ligand interaction. The target protein is then identified by using 2D-gel electrophoresis/liquid chromatography—tandem mass spectrometry or immunodetection methods.

8.6.1.2 Expression Cloning Techniques

The expression cloning techniques utilize a library of cDNAs inserted into cloning vectors to express a library of proteins. A small molecule–protein interaction can be detected by adding a tagged small molecule onto a *phage display, mRNA display* or *three-hybrid system* followed by affinity purification. In a sense, expression cloning techniques are similar to typical affinity purification because they also require chemical modification to attach a tag. However, when the target of interest is of low abundance or is unstable, expression cloning can be an excellent alternative. The bound phage/mRNA/three-hybrid system which display the target protein are eluted and amplified using bacterial host cells. DNA sequencing is then used to identify the protein target.

8.6.1.3 Protein Microarray

The protein microarray is a high-throughput analysis of the molecular interactions between the target and drug candidate. The protein targets to be analysed are first purified and subsequently immobilized onto a glass slide. The resulting array features numerous potential targets (fixed at specific positions). It is incubated with a labelled version of the small molecule drug candidate. The array is then washed thoroughly to remove any unbound molecules. After washing, any remaining labelling signal 'spots' indicate successful drug–protein binding. The location of the labelled drug–protein complex can then be mapped to the specific protein target fixed at that position.

8.6.1.4 Reverse Transfected Cell Microarray

The 'reverse transfected' cell microarray is also referred to as 'living' microarrays, as live cells transfected with cDNAs are used rather than proteins. The transfected cells express specific cDNAs at different locations on the array, and the array is consequently covered in cell clusters that overexpress specific proteins at specific positions. Here also, the cell microarray is incubated with a labelled version of the small molecule drug candidate, which enables detection of the target protein.

8.6.1.5 Biochemical Suppression

Biochemical suppression is an alternative strategy for identifying drug targets and signalling pathway components. This approach does not depend on a molecule's affinity for its biological target but rather depend on its ability to inhibit an activity of a specific protein measured by the use of an assay system. However, because most activity-based probes (ABPs) are designed to target a specific enzyme class, ABPs are particularly useful for phenotypic screening and lead optimization where a specific enzyme or enzyme family is suspected to be involved in a certain disease state or pathway. Firstly, a protein extract is mixed with a molecule that is known to inhibit the activity of interest. If a fraction is identified for the small molecule's inhibitory activity; further rounds of fractionation are performed to purify the suppressor activity.

8.6.2 Functional Genomic Approaches

Here, the target identification and characterization begin with identifying the function of a possible therapeutic target (gene/protein) and its role in the disease. Next-generation genome sequencing and sophisticated genome-wide functional methods in last few decades have led to a significant increase in the identification of novel drug target candidates and understanding of the relevance of these genomic and molecular changes for the diseases.

The drug target identification can be achieved by using several molecular tools such as DNA (nucleic acid) microarray, RNA interference, clustered regularly interspaced short palindromic repeat-associated systems (CRISPAR)-Cas9 endonucleases, antisense oligonucleotides, zinc finger endonucleases among others for probing the disease-relevant pathology related to biological targets [2, 11, 18–22].

The advent of DNA microarray technology has greatly facilitated the study of variation of genetic expression profiles of normal versus disease tissue samples screened in one go against the whole of the human transcriptome. In a typical test protocol, the cDNA samples (from the extracted mRNA samples) from the tissues/cells labelled with different fluorochromes are incubated on to the microarray chip for hybridization over several hours. The unbound material is washed and the microarray chip is laser-scanned by a robot, and data is processed, visualized and analysed for the interpretation of the variations in target gene expressions in disease cell/tissue samples [23–25].

The advances in molecular biology have greatly facilitated the genetic perturbations to modulate the gene expression combined with phenotypic observations as a powerful method to study in both 'in vitro' cellular systems (both yeast-based and mammalian-based cells) and 'in vivo' (genetically modified mouse models) settings [26–33].

Such gene expression regulations can be achieved in many ways for the purpose of drug target identification and some of them also for the purpose of target validation:

- Deletion of the gene (knock down), as in the haploinsufficiency profiling in yeast method
- Mutating the gene (randomly chemical induced mutations)
- Binding to the DNA with the so-called zinc finger proteins
- Binding to the mRNA with antisense RNA
- Silencing the gene with siRNA, shRNA, RNAi, miRNA
- Gene-editing CRISPR-Cas9 endonucleases, or
- Overexpression or up-regulation by transfection of cells by cDNA or plasmids or viral vectors

An outlook on the above methodologies is being discussed in the cited references in much more detail.

8.6.3 Computational Approaches of Drug Target Search/ Identification

Drug target searching or identification can also be done through various web-accessible databases as mentioned below [34].

Utility	Uniform Resource Locator (URL)
Human metabolome data	http://www.hmdb.ca
In silico target identification	http://www.dddc.ac.cn/pdtd/
Pathway analysis	http://www.genome.jp/kegg/
	http://www.geneontology.org
	http://www.reactome.org
	http://www.pantherdb.org
	http://www.biocarta.com
	http://www.ingenuity.com/
Chemogenomic data	http://www.ebi.ac.uk/chembldb
	http://pubchem.ncbi.nlm.nih.gov
Drug target database	http://www.drugbank.ca
Protein data bank	http://www.pdb.org
Disease-specific target database	http://thomsonreuters.com/metacore
Pharmacogenomic data	http://www.pharmgkb.org
Multi-level drug data	http://r2d2drug.org/DMC.aspx
Comparative toxicogenomic database	http://ctdbase.org
Target-toxin database	http://www.t3db.org
Protein expression information	http://www.proteinatlas.org
Therapeutics target database	http://bidd.nus.edu.sg/group/cjttd/

8.7 Drug Target Validation

Target validation is the process of demonstrating the functional role of the identified target in the disease phenotype. The aim of the preclinical drug target validation is to increase the confidence on a particular drug target. The purpose of the target validation is to prove that the particular molecular target is of key or causative for pathogenic or symptomatic mechanisms in a disease.

Target validation shows that a molecular target is directly involved in a disease process, and the modulation of that target is likely to have a therapeutic effect.

The drug target validation has got pivotal importance in drug discovery and development process, as around a half of the total drug failures in Ph 2 and Ph 3 clinical trials were because of the insufficient efficacy [35, 36].

The validation of a drug's efficacy and safety in numerous pre-clinical disease-relevant cell models and animal models is extremely valuable and challenging, as the real validation of a drug target safety and efficacy is only possible in a patient population in a clinical trial setting. The target validation has three important parameters of reproducibility of intended efficacy, further characterizing the target binding sites and reconfirmation of ligand (drug)–target correlation by the introduction of variations. The drug target variations could be modulation of efficacy by changing the drug affinity towards the target, non-variability of drug efficacy across different cell or tissue types and modulation/loss of ligand efficacy by way of mutations of binding domain of the target protein [2, 37, 38].

The target validation and characterization can be achieved by several approaches:

- Genetic manipulation of target genes (in vitro) cellular systems like yeast, non-mammalian species (zebra fish, *C. elegans*)
  - Knocking down the gene (shRNA, siRNA, miRNA), knocking out the gene (CRISPR, ZFNs), knocking-in the gene (viral transfection of mutant genes)
- Genetic manipulation of target genes (in vivo) mammalian embryonic cell systems (genetically engineered mouse models (GEMMs)
  - Transgenic, knock-out, knock-in mouse models; RNAi knock-outs, inducible RNAi mice, conditional RNAi mice, gene editing by CRISPR-Cas
- Antibodies—interacting the target with high affinity for blocking further interactions
- Radio-ligand binding studies for receptor binding site characterization
- Chemical genomics—approaches against genome encoding protein

Even though several 'in vitro' methods of drug target validation are available, the clinically relevant 'in vivo' models of pharmacodynamic efficacy, mechanism of action, biomarkers, safety and toxicity models are of more relevance. Such animal models (be it non-mammalian, wild-type or genetically engineered animal models) are invaluable for prediction of drug target efficacy/safety in human disease setting.

Advent of technological advancements in the fields of genomics and cell and molecular biology has revolutionized to produce wide array of genetically engineered mouse models (GEMMs) of human disease—be it from transgenic mouse models by random transgenesis to targeted transgenesis, recombinase-mediated cassette exchange (RMCE) method to combined targeted transgenesis with tetraploid blastocysts method, engineered nuclease-assisted GEMMs to various RNAi-based models of targeted and conditional GEMMs [39].

Finally, for successful drug target identification, validation and for the overall success, a right framework approach throughout the drug discovery and development process is very essential (as recently proposed in a comprehensive internal review by pharma major Astra Zeneca for a decade long period of their internal programme) [40].

A six 'Right' framework approach of: the right target, the right patient, the right tissue, the right safety, the right commercial potential and another crucial sixth factor of 'right culture' of encouraging effective decision-making based on these technical determinants (Table 8.4).

Table 8.4 The six right framework—summary of the six-dimensional framework for a potential successful drug discovery and development

Six right parameter	Salient features
Right target	• Strong link between target and disease • Differentiated efficacy • Available and predictive biomarkers • Predictable safety pharmacology
Right tissue	• Adequate bioavailability and tissue exposure • Definition of PD biomarkers • Clear understanding of preclinical and clinical PK/PD • Understanding of drug-drug interactions • Predictability of best indication and right pharmacology
Right safety	• Differentiated and clear safety margins • Understanding of secondary pharmacological risks • Understanding of reactive metabolites, genotoxicity • Understanding of target liability and drug interactions
Right patients	• Identification of the most responsive patient population • Maximizing the potential success of clinical efficacy • Definition of risk-benefit for a given population
Right indication	• Differentiated value proposition vs future standard care • Personalized health-care strategy, biomarkers and diagnostics • Focus on market accessibility
Right culture	• Inter-disciplinary team cohesiveness • Effective and impassionate decision-making-based on technical aspects

So, adapting such a right framework of especially the right 'drug target' identification and validation approaches along with right safety, right patient population and right disease indication will definitely decrease the attrition and improve the overall success rate of drug discovery and development.

References

1. Bosch F, Rosich L. The contributions of Paul Ehrlich to pharmacology: a tribute on the occasion of the centenary of his Nobel prize. Pharmacology. 2008;82(3):171–9.
2. Gashaw I, Ellinghaus P, Sommer A, Asadullah K. What makes a good drug target? Drug Discov Today. 2011;16(23–24):1037–43.
3. Hopkins AL, Groom CR. The druggable genome. Nat Rev Drug Discov. 2002;1(9):727.
4. Rask-Andersen M, Almén MS, Schiöth HB. Trends in the exploitation of novel drug targets. Nat Rev Drug Discov. 2011;10(8):579.
5. Santos R, Ursu O, Gaulton A, Bento AP, Donadi RS, Bologa CG, Karlsson A, Al-Lazikani B, Hersey A, Oprea TI. A comprehensive map of molecular drug targets. Nat Rev Drug Discov. 2017;16(1):19.
6. Sriram K, Insel PA. G protein-coupled receptors as targets for approved drugs: how many targets and how many drugs? Mol Pharmacol. 2018;93(4):251–8.
7. Li YH, Wang PP, Li XX, Yu CY, Yang H, Zhou J, Xue WW, Tan J, Zhu F. The human kinome targeted by FDA approved multi-target drugs and combination products: a comparative study from the drug-target interaction network perspective. PLoS One. 2016;11(11):e0165737.

8. Cho YS, Kwon HJ. Identification and validation of bioactive small molecule target through phenotypic screening. Bioorg Med Chem. 2012;20(6):1922–8.
9. Eder J, Herrling PL. Trends in modern drug discovery. In: Nielsch U, editor. New approaches to drug discovery. Cham: Springer; 2015. p. 3–22.
10. Hughes JP, Rees S, Kalindjian SB, Philpott KL. Principles of early drug discovery. Br J Pharmacol. 2011;162(6):1239–49.
11. Schenone M, Dančík V, Wagner BK, Clemons PA. Target identification and mechanism of action in chemical biology and drug discovery. Nat Chem Biol. 2013;9(4):232.
12. Lee J, Bogyo M. Target deconvolution techniques in modern phenotypic profiling. Curr Opin Chem Biol. 2013;17(1):118–26.
13. Lomenick B, Olsen RW, Huang J. Identification of direct protein targets of small molecules. ACS Chem Biol. 2010;6(1):34–46.
14. Croston GE. The utility of target-based discovery. San Diego, CA: Taylor & Francis; 2017.
15. Moffat JG, Vincent F, Lee JA, Eder J, Prunotto M. Opportunities and challenges in phenotypic drug discovery: an industry perspective. Nat Rev Drug Discov. 2017;16(8):531.
16. Terstappen GC, Schlüpen C, Raggiaschi R, Gaviraghi G. Target deconvolution strategies in drug discovery. Nat Rev Drug Discov. 2007;6(11):891.
17. Xia X. Bioinformatics and drug discovery. Curr Top Med Chem. 2017;17(15):1709–26.
18. Andrade E, Bento A, Cavalli J, Oliveira S, Freitas C, Marcon R, Schwanke R, Siqueira J, Calixto J. Non-clinical studies required for new drug development-part I: early in silico and in vitro studies, new target discovery and validation, proof of principles and robustness of animal studies. Braz J Med Biol Res. 2016;49(11).
19. Fellmann C, Gowen BG, Lin P-C, Doudna JA, Corn JE. Cornerstones of CRISPR–Cas in drug discovery and therapy. Nat Rev Drug Discov. 2017;16(2):89.
20. Mohr SE, Smith JA, Shamu CE, Neumüller RA, Perrimon N. RNAi screening comes of age: improved techniques and complementary approaches. Nat Rev Mol Cell Biol. 2014;15(9):591.
21. Moore JD. The impact of CRISPR–Cas9 on target identification and validation. Drug Discov Today. 2015;20(4):450–7.
22. Roti G, Stegmaier K. Genetic and proteomic approaches to identify cancer drug targets. Br J Cancer. 2012;106(2):254.
23. Dufva M. DNA microarrays for biomedical research. Humana Press; 2009.
24. Jayapal M, Melendez AJ. DNA microarray technology for target identification and validation. Clin Exp Pharmacol Physiol. 2006;33(5–6):496–503.
25. Kurimoto K, Saitou M. Single-cell cDNA microarray profiling of complex biological processes of differentiation. Curr Opin Genet Dev. 2010;20(5):470–7.
26. Chakraborty C, Sharma AR, Sharma G, Doss CGP, Lee S-S. Therapeutic miRNA and siRNA: moving from bench to clinic as next generation medicine. Mol Ther Nucleic Acids. 2017;8:132–43.
27. Check E. RNA interference: hitting the on switch. Nature. 2007;448:855–8.
28. Giaever G, Flaherty P, Kumm J, Proctor M, Nislow C, Jaramillo DF, Chu AM, Jordan MI, Arkin AP, Davis RW. Chemogenomic profiling: identifying the functional interactions of small molecules in yeast. Proc Natl Acad Sci. 2004;101(3):793–8.
29. Ho CH, Piotrowski J, Dixon SJ, Baryshnikova A, Costanzo M, Boone C. Combining functional genomics and chemical biology to identify targets of bioactive compounds. Curr Opin Chem Biol. 2011;15(1):66–78.
30. Jost M, Weissman JS. CRISPR approaches to small molecule target identification. ACS Chem Biol. 2018;13(2):366–75.
31. Li L-C, Okino ST, Zhao H, Pookot D, Place RF, Urakami S, Enokida H, Dahiya R. Small dsRNAs induce transcriptional activation in human cells. Proc Natl Acad Sci. 2006;103 (46):17,337–42.
32. Walke DW, Han C, Shaw J, Wann E, Zambrowicz B, Sands A. In vivo drug target discovery: identifying the best targets from the genome. Curr Opin Biotechnol. 2001;12(6):626–31.

33. Wang S, Sim TB, Kim Y-S, Chang Y-T. Tools for target identification and validation. Curr Opin Chem Biol. 2004;8(4):371–7.
34. Katsila T, Spyroulias GA, Patrinos GP, Matsoukas M-T. Computational approaches in target identification and drug discovery. Comput Struct Biotechnol J. 2016;14:177–84.
35. Arrowsmith J, Miller P. Trial watch: phase II and phase III attrition rates 2011–2012. Nat Rev Drug Discov. 2013;12:569.
36. Arrowsmith J. Trial watch: phase III and submission failures: 2007–2010. Nat Rev Drug Discov. 2011;10:87.
37. Gullapalli S, Ramarao P. L-type Ca2+ channel modulation by dihydropyridines potentiates κ-opioid receptor agonist induced acute analgesia and inhibits development of tolerance in rats. Neuropharmacology. 2002;42(4):467–75.
38. Lindsay MA. Target discovery. Nat Rev Drug Discov. 2003;2(10):831.
39. Lee H. Genetically engineered mouse models for drug development and preclinical trials. Biomol Ther. 2014;22(4):267.
40. Cook D, Brown D, Alexander R, March R, Morgan P, Satterthwaite G, Pangalos MN. Lessons learned from the fate of AstraZeneca's drug pipeline: a five-dimensional framework. Nat Rev Drug Discov. 2014;13(6):419.

Chapter 9
Genetics and Drug Discovery

Aruna Poduri and Amit Khanna

9.1 Introduction

Every living being is inherited with a set of features that is not matched to anyone else in this world. This makes each and every residing individual different from one another. This uniqueness is due to the genetic makeup of an individual. It is encoded by the molecules present in the cells of our own body known as the DNA [deoxyribonucleic acid]. DNA is the genetic material of all living organisms that is inherited from the previous generations. Species of the same phylum look like each other, as their genetic material overlaps to an extent but not completely and share a similar growing phases of life. For example, genetic makeup in the subphylum: vertebrates and in the class species of reptiles have similar DNA among them. But the genetic makeup of the reptiles is different from amphibians, another class of species in the same subphylum, vertebrates. Also, there are differences between males and females. All these differences make each species distinct from one and another. Here is another example, despite 99% genetic overlap between humans and chimpanzees [1], it is easy to identify these two species. We humans are bipedal, walking with our two legs, less body hair, big size of the brain, more sharp features like nose, distinct lips, and many more features that distinguish us from chimpanzees. All these features are inherited from our ancestors.

Among humans itself, there are enormous variations, starting with the color and texture of the hair and eye shape and color, color of the skin, height, weight, and other features, which sets every human different from one and another. All these observable characteristics are transmitted biologically, whereas others come from our hereditary. Like for example, the color of the hair is inherited from our parents

A. Poduri (✉)
Department of Biology, Stanford University, Palo Alto, CA, USA

A. Khanna
Dovetail Genomics, Scotts Valley, CA, USA

R. Poduri (ed.), *Drug Discovery and Development*,
https://doi.org/10.1007/978-981-15-5534-3_9

but the language we learn comes from the traditional hereditary. Therefore, many characteristics of humans are a combination of biological inheritance and influences from the environment. Like for instance, our weight is governed partially by genetics and also by environmental factors, starting with the amount of food intake, workout routines, and several others factors. Overall, the traits are inherited from our ancestors and also acquired from environmental factors, both these factors makeup the genetics of every living being. The study of biology that deals with the concerns of biological inheritance and factors influenced by environment is known as genetics.

Genetics is based on the principle where the units of hereditary govern the traits that are passed on during reproduction from parents to their children. These units of inheritance are called as genes. In 1866, Gregor Mendel was the first scientist who noticed and identified the transmission of genes. He is also known as the father of genetics. He studied the seven features of the pea plants including height, color, and shape of pea pod and seed, and position and color of the flower. He found that these hereditary units, genes, which he named them as factors, are transmitted from parents to progeny. His work *"Experiment on Plant Hybridization"* was published in the year 1866. Recently, scientists have highlighted his work again [2–4]. This work by him was referred as classical genetics, which focused on the results of the reproducible actions and also later was recognized as Mendelian inheritance. The advancement of techniques and tools has provided more insightful information about genes and shifted the classical genetics towards modern genetics. Modern genetics talked more about the function of genes and how dysfunction of the genes can lead to a variety of diseases. Modern genetics shed light on these genes and how they play as important factors for drug discovery.

Recognizing the molecules and understanding their mechanisms involved in human diseases are the first critical steps for the development of new therapeutic drugs. These novel therapeutic drugs are developed based on the scientific research performed and data generated by in vitro, in vivo, and genome-wide association studies [5–7]. This work is known as translational research where scientific discoveries are moving from laboratory's bench side and reaching to bedside of the clinics to diagnose, prevent, and treat the human diseases. Translational research helps in bringing a new therapy, drug, or technique into the clinic for use in general population. However, there are several roadblocks in generating an accurate drug target. For instance, studies have shown that targets identified by in vivo studies fail to reproduce in clinical trails and researchers have to refocus and bring in new drugs [7–9]. One of the major reasons for this failure is due to interindividual response to drug therapy that relies on the differences in the genetic makeup of the individual. For example, statins drugs are prescribed to lower lipid levels; however, there is a degree of variability to this therapy [10–13]. These studies have shown variability in the results to a drug response ranging from a benefit to a complete lack of response. Likewise, goes with the drug side effects that can be an array ranging from none to high number of adverse response events [14]. Today, scientific community is aiming to understand and minimize this variability and deliver optimized health care to individuals. Therefore, there is an immense need to generate work to provide more information about target identification, validation, and their prospective in clinical efficacy. Future studies should focus on identifying underlying mechanisms involved in disease and develop more preclinical assays to investigate and validate various disease related multi-parameters. This chapter gives a brief overview information about DNA as the genetic material, genetic mutations, and genetic approaches used for drug discovery, and some of the challenges in the field of drug discovery.

9.2 Genetic Material: DNA, Replication, Transcription, and Translation

Every living being has a defined structural organization and it is beneficial to study the fundamental architecture of the organism. The cell is the smallest self-sufficient active unit of any living entity. Even a bacterium, which is structurally a very small single cell, has all capabilities of performing the independent functions required to survive life. A human cell typically consists of cell membrane that encloses cytoplasm, holding variety of functioning units called as the organelles. There are number of organelles in a human cell, starting with endoplasmic reticulum, Golgi apparatus, lysosomes, vesicles, vacuole, centrosome, ribosomes, mitochondria, and nucleus.

In the nucleus of every cell, genetic information is stored in the DNA molecules that are arranged in thread-like assemblies known as the chromosomes. These chromosomes are visible under microscope when stained with specific dyes during cell division. Each chromosome has a center point called the centromere, which divides the chromosome into two sections called as the arms. The arms are labeled as p [short arm] and q arm [long arm]. The number of chromosomes varies from species to species. Humans have 23 pairs of chromosomes making a total of 46 in number. Twenty-two pairs of chromosomes are similar in both males and females called as autosomes, whereas the last pair known as sex chromosomes is different in male and females. Joe Hin Tjio and Albert Levan were first scientists to identify the number of chromosomes as 46, in the year 1956 [15–17]. The number of chromosomes remains constant in the cells for every specific organism. Each chromosome is made up of DNA, which is tightly coiled around the proteins called histones. Histones are alkaline in nature and positively charged, which makes them get easily associated with negatively charged DNA. These histones help to package the DNA into a compact form that fits in the cell's nucleus. Both histones and DNA combine together to form the chromatin present within the chromosome and control gene expression and transcriptional regulation of the cell [18]. Deviations in the chromatin structure are associated with changes in histone and DNA's function causing histone modifications and DNA methylation, leading to cancer and cardiovascular complications [19–23].

In the early 1950s, scientists focused to understand the structure of DNA in detail. In 1953, James Watson and Francis Crick are the two scientists who demonstrated that DNA is made up of two chains of molecules, each twisted around the other to form the double-stranded helix. It is a right-handed helix structure. Thereafter, the structure of DNA is also known as Watson-Crick DNA structure [24]. The structure of DNA demonstrated by them has transformed the field of the biology. They also suggested that DNA undergoes replication and governs the hereditary unit of life.

The double helix DNA identified by Watson-Crick is right-handed where the individual chain follows a clockwise pathway as it moves forward. They are also called as A- and B-form DNA. A left handed DNA has also been observed called as Z-DNA, also runs in clockwise manner. Nucleotides are subunits of each strand of the DNA that consists of a base attached to phosphorylated molecule of the deoxyribose sugar. Four bases are present in DNA—adenine [A], guanine [G], thymine [T], and cytosine [C]. Adenine and guanine are purine bases whereas cytosine and thymine are pyrimidines bases. Purines are four nitrogen containing double ring structures and pyrimidines are single ring nitrogen molecules. It is demonstrated that at any position, if one DNA strand has adenine, it always pairs with thymine [T] base by two hydrogen bonds. The other pair is between guanine and cytosine with three hydrogen bonds. These hydrogen bonds hold the helix together. If one strand has a G base, it will always pair with C base. This pairing of T with A and G with C is known as complementary. Complementary base pairing means that each base along one strand of the DNA is matched with a base in opposite position on the other strand. In every case, complement of adenine [A] is thymine [T] and complement of guanine [G] is cytosine [C]. This pairing is also known as Watson-Crick pairing. Another scientist, Erwin Chargaff, developed a tool, which measures the amounts of each base present in the DNA. He described that the amount of adenine equals that of thymine, and similarly the amount of guanine equals that of cytosine. Finally, he mentioned that the amount of purine bases equals that of pyrimidines bases. These are also known as Chargaff's rules [25]. He used this method to identify the DNA content from different sources. He showed that the percent of guanine and cytosine varies from species to species but it is constant in all cells of an organism and among the species. A single individual chain is called as single-stranded DNA and double-stranded DNA is also called as duplex DNA. DNA structure is also called as B-form of DNA, where each strand takes a complete turn in 34Å. The form of DNA the bases are planar and perpendicular to the long axis of the double helix. The spiraled structure of DNA is not symmetrical dividing DNA into two grooves. One groove is known as the major grove and other one is minor groove. In nature, protein molecules interact and contact with the DNA present in either minor or major groove or it can be in both the grooves.

In addition, to the bases, the other units present in the DNA molecule are deoxyribose sugar and phosphoric acid. Each base linked to the deoxyribose sugar is called as a nucleoside and when a phosphate group is also attached to this base and sugar unit, this complete entity is known as a nucleotide. These nucleotides together join to form the two polynucleotide strands. The bonds between sugar molecules attached to an adjacent nucleotide through phosphate groups are known as phosphodiester bonds. In double-stranded DNA, individual DNA strand has directionality and polarity. The two polynucleotide strands are having opposite polarity and run in anti-parallel direction, required for DNA to replicate and transcribe. In general, polarity and direction is determined by the position of the nucleotides on the single DNA strand. One strand is known as the 3′ [pronounced as three prime] strand and other end is the 5′ [pronounced as five prime] strand.

DNA makes copies of itself through cell division; Watson and Crick discovered this process. Copying of single DNA strand into two complementary strands is called as replication. The first step in this process of DNA replication is to disengage the double helix structure of DNA molecule. This process requires helicases, an enzyme that breaks the hydrogen bonds that hold the bases together. This separation of the two strands generates a Y-shaped structure called a replication fork. These two separated DNA strands act as a template for making the new progeny strands of DNA. The orientation of one strand is in 3′ to 5′ direction; in other words, this strand is towards the replication fork, also known as the leading strand. The other strand is oriented in 5′ to 3′ direction, away from the fork structure, and is called as the lagging strand. Due to this orientation of two strands, they are replicated differently. In the end of this process, individual double-stranded DNA is synthesized with two copies of similar sequences. An example is described below where Step 1 is the parental sequence that undergoes replication and in Step 2 [2a and 2b] shows synthesis of two double-stranded DNA sequences. During replication process, the nucleotides are joined to the 3′ end of the synthesizing template. The newly produced sequences [shown in italics below] are generated from the parental template below:

Step 1:	5′-CACTGGAC-3′	
	3′-GTGACCTG-5′	
Step 2:	(a) 5′-CACTGGAC-3′	(b) *5′-CACTGGAC-3′*
	3′-GTGACCTG-5′	3′-GTGACCTG-5′

The genetic information is passed on from DNA to RNA [riboxynucleic acid] and generation of RNA from DNA is recognized as transcription. The first step in transcription procedure is double helix DNA strand opens up with the help of enzymes, where one strand acts as template for the synthesis of a complementary strand of RNA. This synthesized RNA molecule is called as a transcript. For RNA, Watson-Crick base sequences remain the same, with an exception that instead of thymine [T], another base, uracil pairs with adenine [A]. Like DNA, RNA also has polarity which starts with 5′ to 3′ end and nucleotides are joined towards the 3′ end of the synthesizing RNA template. Thus, 5′ end of RNA is produced initially and the process of transcription runs along the DNA template in the 3′ to 5′ direction. A certain number of nucleotides encode a particular gene. For every gene to be transcribed there is an initiation and termination sites and certain nucleotide sequences are associated with these sites. The RNA structure is comparable to DNA but not completely matching. The major difference between RNA and DNA is ribose sugar instead of deoxyribose. Till date, three major types of RNA are identified. The first one is messenger RNA [mRNA], which transports the genetic information from DNA to polypeptide sequence generation. Second one is ribosomal RNA [rRNA], located in the ribosomes organelle, where synthesis of polypeptide chain takes place. Third, transfer RNA [tRNA] deciphers the mRNA sequence and works at specific sites located in the ribosomes. The mRNA is scanned in units of three adjacent nucleotides at a time, called as codons or genetic codon. Each codon represents a specific amino acid, and tRNA, a third form of RNA, identifies this codon. The function of this RNA form is to decipher the mRNA sequence and works at a specific site located in the ribosomes. For example, if an mRNA sequence starts with AUG codon, it represents the amino acid, methionine. Another example, given is an mRNA sequence: UCCACUCAUAAUGCU. After decoding this mRNA sequence, it reads like this: UCC: serine, ACU: threonine, CAU: histidine, AAU: asparagine, GCU: alanine. In general, for nearly all organisms, the codon-AUG represents methionine, also known as start codon for polypeptide chain generation. This is the common start codon for many organisms. The genetic codes: UAA, UAG, and UGA are recognized as the termination codons. These termination codons release the polypeptide sequences from the ribosomal site and stop the translation process. In the end, RNA sequence ultimately leads to the synthesis of a polypeptide chain, also known as protein molecule. The transfer of information from DNA to RNA and finally to the protein is known as the central dogma of biology.

Proteins are macromolecules involved in a variety of biological and metabolic actions in the cells. These proteins are critical for the production and breakdown of these polypeptide chains, which generate the energy required for the different activities of the cells. Like for example, the structural proteins support the cell structure and shape. Another set of proteins are the cell membrane proteins; these are mainly receptors that regulate the movement of molecules moving in and out of the cell. This is essential for maintaining many cellular activities like generation of cAMP and other signaling molecules. Generally, a protein comprises amino acid sequences. There are 20 amino acids known till date. Multiple combination of these 20 amino acids leads to a polypeptide chain. For example, hemoglobin in humans comprises 4 polypeptide chains, 2 chains of alpha subunits with 141 amino acids each and 2 chains of beta subunits with each having 146 amino acids. Hemoglobin protein is found in the red blood cells of humans and its main function is to transport oxygen through the body [26–28].

9.3 Genetic Mutations and Deficient Proteins

Any change or an error in the reading frame of the DNA sequence can lead to a genetic mutation. An abnormality, which occurs during the synthesis of DNA, is referred as a genetic mutation. Many types of mutations have been observed that lead to changes in DNA nucleotide sequence. These mutations can be a substitution, deletion, or insertion of nucleotide bases. The substitutions are also called as point mutations, and are the most common type of mutations. This occurs when a purine base is substituted with another purine, known as a transition change. Another is transversion, when a purine is substituted for a pyrimidine or vice versa. These point mutations occur in the DNA sequences that code proteins making them silent or missense or nonsense mutations. As name suggests, silent mutations are recognized where a base substitution will not change the amino acid production and function. For example, the codon AAA is modified and instead AAG codon is incorporated into the polypeptide chain, although there is change in the nucleotide, but both lead to same amino acid synthesis, lysine. This change in polypeptide chain generation where there is no effect on the structure and function of protein is known as silent mutation. Second, when a base change results in a codon that differs in amino acid sequence synthesis, leading into a different polypeptide chain, the structure or function of the synthesized polypeptide sequence is altered. This results in deleterious effects on the structure and function of the protein. Changes in the amino acid sequences can be very important for the protein function. This is known as missense mutation. For example, subjects with sickle cell anemia are results of missense mutation where glutamic acid, a hydrophilic amino acid, is replaced with valine, a hydrophobic amino acid, at the sixth position in beta-globin chain of hemoglobin protein. This single replacement of an amino acid results into deformed red blood cell formation that are unable to carry enough oxygen as a normal red blood cell [29–31]. Sickle cell anemia is an autosomal recessive disease, and according to CDC [Center for Disease Control and Prevention] reports (https://www.cdc.gov/ncbddd/sicklecell/data.html), it affects 1 in 13 for African Americans and one of the most common inherited blood diseases in the United States, where it affects nearly 100,000 individuals. When a nucleotide base substitution results in truncating a polypeptide chain production, leading to the generation of non-functional protein, it leads to nonsense mutations. These mutations end up in generating stop codons. Examples of stop codons are UAA, UAG, and UGA [32]. These codons lead to the termination of the protein synthesis [32, 33]. For example, people with cystic fibrosis are unable to synthesize cystic fibrosis transmembrane conductance regulator [CFTR] protein; synthesis of this polypeptide is stopped prematurely, causing a truncated and a non-functional protein. This due to an introduction of stop codon and cell identifies this as faulty and destroys itself [34–36]. Research is aimed to introduce compounds that can override the premature stop signals; a complete CFTR protein is synthesized [36]. Deletion or insertions result in the frame shift in the reading of DNA sequence. Deletion or insertion mutations can occur either with an absence or addition of more than one nucleotide bases. Either of these mutation forms can have a variety of potential disease outcomes. In Fragile X syndrome, fragile X mental retardation 1 [FMR1] gene translates into fragile X mental retardation [FMRP] protein [37–39]. This protein is commonly found in brain and important for cognitive and female reproductive system. This syndrome is caused due to CGG triplet repeats in the gene where it is repeated with more than 200 times [39–41]. This abnormally expands CGG nucleotide segments and turns off the FMRP production. Deficiency of this protein shows signs of disruptive nervous system. Similarly, patients with Huntington's disorder have repeated trinucleotides of CAG that add glutamines to the translated protein and these CAG repeats range from 36 to 120 times [42, 43]. Mutation is caused in the huntingtin gene and plays important role in the nerve cells of the brain [44]. Other forms of the mutations are exposure to environmental factors like chemicals and radiations. Studies have shown exposure to UV radiation can induce cancer [45–48]. In another study, exposure to excessive use of tobacco smoke can lead to DNA mutations causing lung or neck or throat cancer [49–51]. In addition, to environmental cues, DNA replication and recombination are prone to errors. DNA polymerase enzyme inserts new nucleotides during the replication process; the rate of insertion can lead to addition of inappropriate nucleotides. This damage to DNA modifies the structural configuration. In general, a proof reading enzyme identifies this fault and corrects the error but some of faults survive the process, causing a potential genetic disease. Scientists have demonstrated several mechanisms to repair DNA. These mechanisms are present in prokaryotes and eukaryotes and conserved during the evolution. Cells have an incredible feature to sense and repair different types of damage that can occur during replication or caused by environmental factors. DNA is involved in cell division and DNA repair is strictly controlled by different checkpoint mechanisms of the cell cycle [52]. These mechanisms ensure that the DNA is undamaged in the process of DNA replication and cell division. Abnormality in checkpoints mechanisms can produce mutations. For example, xeroderma pigmentosum [XP] is a genetic disorder described by sensitivity to ultraviolet rays from sunlight and associated with a reduced ability to repair DNA damage caused by this exposure. Other diseases that have photosensitivity issues are Cockayne syndrome [CS] and trichothiodystrophy [TTD]. Individuals with these XP, CS, and TTD have defects in the nucleotide excision repair pathway [53–55]. There are many other genetic disorders: ataxia telangiectasia, Bloom syndrome, Werner syndrome, and Fanconi anemia, which are caused by failure to repair in the DNA damage.

Enzymes are the molecules known to activate any biological reactions in the cell. Later, these enzymes were described as proteins. Dr. Archibald Garrod in 1908 described "one gene-one enzyme" statement based on his work on alkaptonuria disease [56]. An inborn error of metabolism happens due to inheritance of a dysfunctional enzyme. In this genetic disorder, subjects excrete alkaptonuria molecules in the urine. Previous name of homogentisic acid was alkapton; hence this disease is named as alkaptonuria. The rate of incidence is 1 in 100,000 subjects [57]. The major symptom of this inherited disorder is urine of the subject turns black due to oxidation of homogentisic acid, and thus also known as black urine disease. Another disorder of inborn errors of metabolism is phenylketonuria [58]. In this disease, the enzyme phenylalanine hydroxylase is absent, required for the breakdown of an essential amino acid, phenylalanine. Deficiency of this enzyme leads to building up of phenylalanine in the body, which can harm central nervous system causing mental retardation in children. The rate of incidence is reported to be 1 in 10,000 in United States, fairly a common disorder. Other inborn errors of metabolism are identified, which can be a result of defects in the biochemical or metabolic pathways affecting fats, carbohydrates, or decreased organelle function causing severe medical conditions.

9.4 Genetic Testing

With the progression in the field of molecular genetics, now genetic testing is making its way into the clinics. The genetic disorders can be identified as early in life with a possibility of early diagnosis and treatment that can change the lifestyle of an individual. Physicians in the clinics are communicating and providing an option to the patients to go for genetic testing [59]. Genetic testing is defined as an evaluation of human DNA or RNA or chromosomes or proteins using various scientific techniques. This is to identify any kind of abnormalities that are inherited from parents. This testing can detect disorders in DNA or RNA which are co-inherited and associated with the disease causing gene or assess the products of the genes: proteins [testing of biochemical molecules] or investigate the entire chromosome [cytogenetic testing]. Most of the genetic testing requires a blood specimen from the individual. In other words, they are invasive testing, however with recent advances in the scientific technologies, non-invasive procedures are coming up. These include, cell-free fetal DNA techniques. These testings hold lot of potential but have their limitations. Genetic testing serves a range of objectives, which includes diagnostic, predictive and carrier testing, prenatal and newborn screening [59, 60]. Predictive evaluation is used to identify the status and risk of the predisposition of the genetic disorder in the members of the families. The individual who received two gene copies from parents is known as the affected member in the family, having the symptom/s for the genetic disorder. An individual can be asymptomatic but he or she is the carrier in the family for an autosomal recessive or recessive X-linked disease. In general, with this genetic testing, the predictive or carrier seeks the information and are advised for reproductive choices. Prenatal testing is done to check the genetic status of pregnancy for the risk of any inherited disease. For prenatal testing besides blood specimens, mother's serum, amniotic fluid, chronic villus samples, and periumbilical blood are collected for testing. Newborn screening is usually performed to check for any inborn errors of metabolism or hearing loss or any other deficiencies with new babies. The results of the testing yield useful information but at times multiple rounds of testing is suggested, also family members can be included in it. Cost and coverage of genetic testing is also important to be considered. Several points are kept into consideration before going into genetic testing [59–61]. This testing plays a huge role in the clinical organization and individual decision-making. Results of genetic testing have implications not only on the patient but also on the family members. This chapter will not go into details of specific considerations that need to be taken during, before, and after genetic testing.

9.5 Genetic Approaches in Drug Discovery

To treat disorders, scientists have developed many drugs over the decades. However, there are still many diseases that have no cure or treatment till date. Therefore, the need of the hour is to do more research and discover drug targets that translate into new therapies to improve human health. The first step in drug discovery is to identify the molecules and elucidate their mechanisms involved in the development of the disease. The process of drug discovery has evolved over the years. Conventionally, it starts with identifying the target involved in a specific disease, and then search begins for the chemical molecule that can inhibit or modify the function of the target. After multiple rounds of successful testing in cell culture, ex vivo, and animal models, the therapy is introduced into human clinical trails [62, 63]. At these trails, many drug targets fail to reproduce the results as found in cell culture and animal studies. Further, aiming to discover drug targets, scientist initiated The Human Genome Project to have a deeper view about the human genome, where they sequenced the complete genome [70, 71]. Now, with the completion of project and published in early 2000, scientists are focusing to examine the information printed in the DNA sequence, and how it interacts and functions in the human body [70]. Over the years, many techniques have revolutionized and improved our understanding about genetics and helped in the process of drug discovery. These methods have the capacity to perform high-throughput screening, including polymerase chain reaction [PCR] and next-generation sequencing. Briefly, some of these current methods are discussed below.

One of the main tools to study about DNA was the discovery of PCR [64, 65]. This was discovered by Kary Mullis in 1984 and won Nobel Prize in Chemistry in the year 1993. This tool has revolutionized many scientific laboratories worldwide and today it is an indispensable approach of molecular biology to investigate about DNA and RNA. The basic principle of PCR is exponential amplification of a particular section of the DNA that results in several million copies of the DNA region. The PCR technique depends upon thermal cycling, with the steps of denaturing, annealing, and extension of the template strand, and final step involves elongation at low temperatures for an unlimited time. After the introduction of PCR, it has been widely used for many applications, including mutation detection, cloning of a gene, sequencing, forensic applications, genotyping, study methylation, paternity testing, mutagenesis, identify the pathogens, fingerprints for DNA profiling, and in agriculture [66–68]. Another method based on PCR is real-time PCR, also known as quantitative PCR [qPCR]. This tool has advantage over PCR, where we can oversee the amplification of the target DNA segment in real time. The qPCR is now a standard method to examine mRNA quantitation that is quick, sensitive, and produced reproducible results. The principle of this approach is to develop complementary DNA from RNA template using reverse transcription reaction and fluorescent probes. The aim is to measure fluorescence intensity, which is directly proportional to the quantity of cDNA present in the sample. These approaches enabled us to detect many pathogenic organisms in a specimen, including bacteria and virus and different parasites causing infectious diseases, assisting in efficient diagnosis and treatment. These genetic approaches are also useful in diagnostic and biochemical analysis and used for drug development to treat cancer and other diseases.

Sequencing of DNA has provided enormous information about the order of the nucleotides present in a cell as well as in the organism. For example, the accomplishment of the Human Genome Project has given complete sequences of many genes and the sequencing approaches have greatly enhanced our understanding about DNA and RNA. This project has also led to an era of rapid, reasonable, accurate genome sequencing method to decipher molecular mechanisms and identified potential drug candidates. In 1977, Frederick Sanger and coworkers developed the method of DNA sequencing that relies on chain termination of dideoxynucleotides, a type of deoxynucleotide triphosphates [dNTPs] that have hydrogen molecule instead of 3′ hydroxyl group, also known as Sanger sequencing method and being used by the scientific community for more than 40 years now. Sanger sequencing is considered as the gold standard to study about DNA. DNA sequencing became a key approach in the field of molecular biology, evolution, medicine, and forensics [69]. Recently, a more rapid sequencing method has evolved for large-scale genome sequencing—known as next-generation sequencing [NGS], with automated analysis and only requires a very low amount of DNA/RNA as input. However, the challenging part of this new approach is the data analysis, which can be complex due to huge generation of DNA sequence reads. However, with the progress in science, this method is improving and becoming the face in identifying potential candidates for biomarker and drug discovery [72, 73].

Another breakthrough to study about DNA or genes is the invention of microscopes. Microscopy-based techniques have provided enormous amount of information about the cell behavior and characteristics. Microscope is an instrument where we can view objects with the help of single or multiple lenses otherwise cannot be noticed with the naked eye. Along with the lenses, an imaging and lighting system and a specimen support stage make up a microscope. With the current development in the technology, there is a camera attached and the image is seen on the computer in addition to the eyepiece lens. Previously, microscopes were used to study about the characteristics of microorganism, pathogens, animal cell, and plant cells. Currently, there is a wide range of microscopes used to view and examine about a variety of specimens, starting from chromosomes to nucleus of the cell, from bacterial cell to a cancerous cell. More than an instrument, microscope is now used as an approach to identify many genes and study their expression or co-expression of multiple genes in the cells or organs of the human body. This approach has provided strong information about the exact location of the gene in the tissue or the organ. Various time course studies have demonstrated change of the gene expression in the resident tissue and also able to track the expression of gene if migrating from one area to another in the human organ/s. A wide range of microscopes are available starting with a simple stereomicroscope to a complex electron microscope and a confocal microscope. These complex microscopes have helped to study the specimen and to highlight the structure and shape of it using different contrasting effects. In addition, these advanced tools have also provided information not only in 2D level but also in 3D state. Some of the examples of microcopy are bright-field, phase-contrast, dark-field, and fluorescence. One of the applications of fluorescence microscopy is fluorescence in situ hybridization [FISH], which uses fluorescent probes that attach to a specific nucleic acid sequence in the cell with a high level of complementarity. These probes are either single-strand DNA or RNA templates that have complementarity to the region of interest in a tissue. FISH can identify the presence or absence of DNA or RNA sequences on the chromosomes and helps us to detect any chromosomal abnormalities [74, 75]. FISH has helped to detect genetic diseases such as Down syndrome, acute lymphoblastic leukemia, Angelman's syndrome, and Prader Willi syndrome [75, 76]. In recent years, scope of live cell imaging using time-lapse microscopy has increased to monitor the cell growth and morphology, and to study various biological processes such as migration and development of cells, cell trafficking, rare biological events, and time course of gene expression. Live cell imaging has helped in drug discovery, optimization and validation at a single cell level and to study different mechanisms involved in the development of disease.

Flow cytometry plays a critical role in drug discovery and a great approach to study about DNA or protein and also about the characteristic of a particular cell. It is a multiparameter tool, records various measurements of the cell. Cells are labeled with fluorescent dyes and goes through different lasers, and then the resulting fluorescent signals are captured and studied. It can identify if the target molecule is present on the surface of the cell or present in the intracellular system [77, 78]. Conjugated dyes interact with the DNA in a relative and a linear fashion, which aids in DNA quantitation providing information about diploid cell, is at rest or synthesizing DNA phase or premitotic or mitotic phase. Also number of chromosomes can be identified if it is abnormal or not. It can also test if the cell is undergoing apoptosis [79, 80].

These genetic tools have provided us a chance to examine a comprehensive functional analysis about human genome and filling the holes in the scientific knowledge about the activities happening at the cellular and the molecular level. Still, each approach has both set of benefits and restrictions that should be considered into account before choosing the correct approach to answer a specific scientific research question and in the path of drug discovery.

9.6 Future of Genetics and Drug Discovery

Over the past few years the progression of scientific approaches has shed more light about genes function and their biological pathways that are relevant to normal physiology and in disease. Currently, scientists are focused to identify and validate key targets and develop new medicines for incurable diseases. Now scientific research is reorienting and integrating multidisciplinary approaches such as genetics, bioinformatics, cell and molecular biology, pharmacology, biochemistry, and medicinal chemistry to identify and validate new drug targets through different tools. The potential of genetics and its approaches for new therapies will be realized if all these disciplines are integrated into the process of drug discovery at all stages—starting from generation of hypothesis to assessment of clinical studies. The aim is to expedite the testing of new drug candidates in humans and optimize therapy for each and every patient. The translation of genetic information into disease knowledge, confirm detailed molecular mechanisms of the targets, and new drug discoveries will make the upcoming years as an era of biomedical revolution.

References

1. Wong K. The 1 percent difference. Genome comparisons reveal the DNA that distinguishes Homo sapiens from its kin. Sci Am. 2014;311(3):100.
2. Abbott S, Fairbanks DJ. Experiments on plant hybrids by Gregor Mendel. Genetics. 2016;204 (2):407–22.
3. van Dijk PJ, Weissing FJ, Ellis THN. How Mendel's interest in inheritance grew out of plant improvement. Genetics. 2018;210(2):347–55.
4. Zhang H, Chen W, Mendelism SK. New insights from Gregor Mendel's lectures in Brno. Genetics. 2017;207(1):1–8.
5. Plenge RM, Scolnick EM, Altshuler D. Validating therapeutic targets through human genetics. Nat Rev Drug Discov. 2013;12(8):581–94.
6. Folkersen L, Biswas S, Frederiksen KS, Keller P, Fox B, Fleckner J. Applying genetics in inflammatory disease drug discovery. Drug Discov Today. 2015;20(10):1176–81.
7. King EA, Davis JW, Degner JF. Are drug targets with genetic support twice as likely to be approved? Revised estimates of the impact of genetic support for drug mechanisms on the probability of drug approval. PLoS Genet. 2019;15(12):e1008489.
8. Nelson MR, Tipney H, Painter JL, Shen J, Nicoletti P, Shen Y, et al. The support of human genetic evidence for approved drug indications. Nat Genet. 2015;47(8):856–60.
9. DiMasi JA, Feldman L, Seckler A, Wilson A. Trends in risks associated with new drug development: success rates for investigational drugs. Clin Pharmacol Ther. 2010;87(3):272–7.
10. Poduri A, Khullar M, Bahl A, Sehrawat BS, Sharma Y, Talwar KK. Common variants of HMGCR, CETP, APOAI, ABCB1, CYP3A4, and CYP7A1 genes as predictors of lipid-lowering response to atorvastatin therapy. DNA Cell Biol. 2010;29(10):629–37.
11. LaRosa JC. Statins and risk of coronary heart disease. JAMA. 2000;283(22):2935–6.
12. Puccetti L, Acampa M, Auteri A. Pharmacogenetics of statins therapy. Recent Pat Cardiovasc Drug Discov. 2007;2(3):228–36.
13. Mangravite LM, Thorn CF, Krauss RM. Clinical implications of pharmacogenomics of statin treatment. Pharmacogenomics J. 2006;6(6):360–74.
14. Kitzmiller JP, Mikulik EB, Dauki AM, Murkherjee C, Luzum JA. Pharmacogenomics of statins: understanding susceptibility to adverse effects. Pharmacogenomics Pers Med. 2016;9:97–106.
15. Tjio JH. The chromosome number of man. Am J Obstet Gynecol. 1978;130(6):723–4.
16. Trask BJ. Human cytogenetics: 46 chromosomes, 46 years and counting. Nat Rev Genet. 2002;3(10):769–78.
17. Ford CE, Hamerton JL. The chromosomes of man. Acta Genet Stat Med. 1956;6(2):264–6.
18. Hammond CM, Stromme CB, Huang H, Patel DJ, Groth A. Histone chaperone networks shaping chromatin function. Nat Rev Mol Cell Biol. 2017;18(3):141–58.
19. Valencia AM, Kadoch C. Chromatin regulatory mechanisms and therapeutic opportunities in cancer. Nat Cell Biol. 2019;21(2):152–61.

20. Park S, Kim GW, Kwon SH, Lee JS. Broad domains of histone H3 lysine 4 trimethylation in transcriptional regulation and disease. FEBS J. 2020;287(14):2891–902.
21. Papait R, Serio S, Pagiatakis C, Rusconi F, Carullo P, Mazzola M, et al. Histone methyltransferase G9a is required for cardiomyocyte homeostasis and hypertrophy. Circulation. 2017;136(13):1233–46.
22. Mongelli A, Atlante S, Bachetti T, Martelli F, Farsetti A, Gaetano C. Epigenetic signaling and RNA regulation in cardiovascular diseases. Int J Mol Sci. 2020;21(2):509.
23. Han P, Li W, Yang J, Shang C, Lin CH, Cheng W, et al. Epigenetic response to environmental stress: assembly of BRG1-G9a/GLP-DNMT3 repressive chromatin complex on Myh6 promoter in pathologically stressed hearts. Biochim Biophys Acta. 2016;1863(7 Pt B):1772–81.
24. Watson JD, Crick FH. Molecular structure of nucleic acids; a structure for deoxyribose nucleic acid. Nature. 1953;171(4356):737–8.
25. Vischer E, Chargaff E. The separation and quantitative estimation of purines and pyrimidines in minute amounts. J Biol Chem. 1948;176(2):703–14.
26. Steensma DP, Shampo MA, Kyle RA. Max Perutz and the structure of hemoglobin. Mayo Clin Proc. 2015;90(8):e89.
27. Gell DA. Structure and function of haemoglobins. Blood Cells Mol Dis. 2018;70:13–42.
28. Huisman TH. The structure and function of normal and abnormal haemoglobins. Baillieres Clin Haematol. 1993;6(1):1–30.
29. Williams TN, Thein SL. Sickle cell anemia and its phenotypes. Annu Rev Genomics Hum Genet. 2018;19:113–47.
30. Desforges JF, Wang MY. Sickle cell anemia. Med Clin North Am. 1966;50(6):1519–32.
31. Strouse J. Sickle cell disease. Handb Clin Neurol. 2016;138:311–24.
32. Tuite MF, Stansfield I. Termination of protein synthesis. Mol Biol Rep. 1994;19(3):171–81.
33. Beutler E, Gelbart T, Han JH, Koziol JA, Beutler B. Evolution of the genome and the genetic code: selection at the dinucleotide level by methylation and polyribonucleotide cleavage. Proc Natl Acad Sci U S A. 1989;86(1):192–6.
34. Hull J. Cystic fibrosis transmembrane conductance regulator dysfunction and its treatment. J R Soc Med. 2012;105(Suppl 2):S2–8.
35. Derichs N. Targeting a genetic defect: cystic fibrosis transmembrane conductance regulator modulators in cystic fibrosis. Eur Respir Rev. 2013;22(127):58–65.
36. Pettit RS, Fellner C. CFTR modulators for the treatment of cystic fibrosis. P & T. 2014;39(7):500–11.
37. Bardoni B, Schenck A, Mandel JL. The fragile X mental retardation protein. Brain Res Bull. 2001;56(3-4):375–82.
38. Aryal S, Klann E. Turning up translation in fragile X syndrome. Science. 2018;361(6403):648–9.
39. Jin P, Warren ST. Understanding the molecular basis of fragile X syndrome. Hum Mol Genet. 2000;9(6):901–8.
40. Peprah E. Fragile X syndrome: the FMR1 CGG repeat distribution among world populations. Ann Hum Genet. 2012;76(2):178–91.
41. Loesch D, Hagerman R. Unstable mutations in the FMR1 gene and the phenotypes. Adv Exp Med Biol. 2012;769:78–114.
42. Moncke-Buchner E, Reich S, Mucke M, Reuter M, Messer W, Wanker EE, et al. Counting CAG repeats in the Huntington's disease gene by restriction endonuclease EcoP15I cleavage. Nucleic Acids Res. 2002;30(16):e83.
43. Genetic Modifiers of Huntington's Disease Consortium. Electronic address: ghmhe, Genetic Modifiers of Huntington's Disease Consortium. CAG repeat not polyglutamine length determines timing of Huntington's disease onset. Cell. 2019;178(4):887–900.e14.
44. Guo Q, Bin H, Cheng J, Seefelder M, Engler T, Pfeifer G, et al. The cryo-electron microscopy structure of huntingtin. Nature. 2018;555(7694):117–20.
45. Narayanan DL, Saladi RN, Fox JL. Ultraviolet radiation and skin cancer. Int J Dermatol. 2010;49(9):978–86.

46. Saladi RN, Persaud AN. The causes of skin cancer: a comprehensive review. Drugs Today. 2005;41(1):37–53.
47. Jhappan C, Noonan FP, Merlino G. Ultraviolet radiation and cutaneous malignant melanoma. Oncogene. 2003;22(20):3099–112.
48. Soehnge H, Ouhtit A, Ananthaswamy ON. Mechanisms of induction of skin cancer by UV radiation. Frontiers in bioscience: a journal and virtual. Library. 1997;2:d538–51.
49. Jethwa AR, Khariwala SS. Tobacco-related carcinogenesis in head and neck cancer. Cancer Metastasis Rev. 2017;36(3):411–23.
50. Zhang ZF, Morgenstern H, Spitz MR, Tashkin DP, Yu GP, Hsu TC, et al. Environmental tobacco smoking, mutagen sensitivity, and head and neck squamous cell carcinoma. Cancer Epidemiol Biomarkers Prev. 2000;9(10):1043–9.
51. Koyanagi YN, Matsuo K, Ito H, Wakai K, Nagata C, Nakayama T, et al. Cigarette smoking and the risk of head and neck cancer in the Japanese population: a systematic review and meta-analysis. Jpn J Clin Oncol. 2016;46(6):580–95.
52. Branzei D, Foiani M. Regulation of DNA repair throughout the cell cycle. Nat Rev Mol Cell Biol. 2008;9(4):297–308.
53. Kraemer KH, Patronas NJ, Schiffmann R, Brooks BP, Tamura D, DiGiovanna JJ. Xeroderma pigmentosum, trichothiodystrophy and Cockayne syndrome: a complex genotype-phenotype relationship. Neuroscience. 2007;145(4):1388–96.
54. Lehmann AR. DNA repair-deficient diseases, xeroderma pigmentosum, Cockayne syndrome and trichothiodystrophy. Biochimie. 2003;85(11):1101–11.
55. Berneburg M, Lehmann AR. Xeroderma pigmentosum and related disorders: defects in DNA repair and transcription. Adv Genet. 2001;43:71–102.
56. Piro A, Tagarelli A, Tagarelli G, Lagonia P, Quattrone A. Archibald Edward Garrod: the physician father of biochemistry. Metabolism. 2009;58(4):427–37.
57. Mistry JB, Bukhari M, Taylor AM. Alkaptonuria. Rare Dis. 2014;1(1):e27475.
58. Williams RA, Mamotte CDS, Burnett JR. Phenylketonuria: an inborn error of phenylalanine metabolism. Clin Biochem Rev. 2008;29(1):31–41.
59. Pagon RA, Hanson NB, Neufeld-Kaiser W, Covington ML. Genetic testing. West J Med. 2001;174(5):344–7.
60. Botkin JR. Ethical issues in pediatric genetic testing and screening for current opinion in pediatrics. Curr Opin Pediatr. 2016;28(6):700–4.
61. Burke W, Tarini B, Press NA, Evans JP. Genetic screening. Epidemiol Rev. 2011;33 (1):148–64.
62. Plenge RM, Scolnick EM, Altshuler D. Validating therapeutic targets through human genetics. Nat Rev Drug Discov. 2013;12(8):581–94.
63. Stitziel NO, Kathiresan S. Leveraging human genetics to guide drug target discovery. Trends Cardiovasc Med. 2017;27(5):352–9.
64. Saiki RK, Scharf S, Faloona F, Mullis KB, Horn GT, Erlich HA, Arnheim N. Enzymatic amplification of beta-globin genomic sequences and restriction site analysis for diagnosis of sickle cell anemia. Science. 1985;230(4732):1350–4.
65. Mullis KB, Faloona FA. Specific synthesis of DNA in vitro via a polymerase-catalyzed chain reaction. Methods Enzymol. 1987;155:335–50.
66. Valones MA, Guimarães RL, Brandão LA, de Souza PR, de Albuquerque Tavares Carvalho A, Crovela S. Principles and applications of polymerase chain reaction in medical diagnostic fields: a review. Braz J Microbiol. 2009;40(1):1–11.
67. Bej AK, Mahbubani MH, Atlas RM. Amplification of nucleic acids by polymerase chain reaction (PCR) and other methods and their applications. Crit Rev Biochem Mol Biol. 2008;26(3–4):301–34.
68. Zhu H, Zhang H, Xu Y, Laššáková S, Korabečná M, Neužil P. PCR past, present and future. Biotechniques. 2020;69(4):317–25.
69. Heather JM, Chain B. The sequence of sequencers: the history of sequencing DNA. Genomics. 2016;107(1):1–8.

70. International Human Genome Sequencing Consortium. Finishing the euchromatic sequence of the human genome. Nature. 2004;431(7011):931–45.
71. Venter JC, Adams MD, Myers EW, Li PW, Mural RJ, Sutton GG, et al. The sequence of the human genome. Science. 2001;291(5507):1304–51.
72. Xuan J, Yu Y, Qing T, Guo L, Shi L. Next-generation sequencing in the clinic: promises and challenges. Cancer Lett. 2013;340(2):284–95.
73. Richard McCombie W, McPherson JD, Mardis ER. Next-generation sequencing technologies. Cold Spring Harb Perspect Med. 2019;9(11):a036798.
74. Cui C, Shu W, Li P. Fluorescence in situ hybridization: cell-based genetic diagnostic and research applications. Front Cell Dev Biol. 2016;4:89.
75. Ratan ZA, Zaman SB, Mehta V, Haidere MF, Runa NJ, Akter N. Application of fluorescence in situ hybridization (FISH) technique for the detection of genetic aberration in medical science. Cureus. 2017;9(6):e1325.
76. Teshima I, Chadwick D, Chitayat D, Kobayashi J, Ray P, Shuman C, Siegel-Bartelt J, Strasberg P, Weksberg R. FISH detection of chromosome 15 deletions in Prader-Willi and Angelman syndromes. Am J Med Genet. 1996;62(3):217–23.
77. Sklar LA, Carter MB, Edwards BS. Flow cytometry for drug discovery, receptor pharmacology and high-throughput screening. Curr Opin Pharmacol. 2007;7(5):527–34.
78. Chang RL, Yeh CH, Albitar M. Quantification of intracellular proteins and monitoring therapy using flow cytometry. Curr Drug Targets. 2010;11(8):994–9.
79. Wlodkowic D, Skommer J, Darzynkiewicz Z. Flow cytometry-based apoptosis detection. Methods Mol Biol. 2009;559:19–32.
80. Riccardi C, Nicoletti I. Analysis of apoptosis by propidium iodide staining and flow cytometry. Nat Protoc. 2006;1:1458–61.

Chapter 10
Discovery and Development of Stem Cells for Therapeutic Applications

Arun H. S. Kumar

10.1 Brief Background on Cells

A cell is the smallest structural and functional unit of any multicellular organism. Considerable diversity is observed in the microanatomy, biochemistry, and physiology of cells both within and between living organisms. It is estimated there are ~220 different cell types identified so far. Understanding the diversity in the structure and function of the cells has been the subject of extensive ongoing research, which has contributed to numerous advancements in therapeutics. Every organism in the animal kingdom develops from a single cell stage, wherein a single cell multiplies to form several cell types, which eventually undergo differentiation to acquire specialized structure and function. This fundamental concept of a single cell giving rise to several other cell types has been the foundation stone in our efforts to understand and develop stem cell for therapeutics. However not all cells which can divide and multiply can be classified as stem cells. Instead, stem cells are defined as undifferentiated cell/s of a multicellular organism, which have the potential to indefinitely give rise to cells of either same and/or other type/s. The ability of the cells to multiply indefinitely, stay unspecialized, and potentially differentiate into specialized cells is the key criteria in defining any cell as a stem cell. When a stem cell divides, it has the potential to remain as a stem cell or it may become another cell type with a more specialized function. Based on the source of the stem cells, they can be classified into the following three major categories (Fig. 10.1):

1. Embryonic stem cells (ESC).
2. Adult stem cells (ASC).
3. Induced pluripotent stem cells (IPS).

A. H. S. Kumar (✉)
Stemcology, School of veterinary medicine, University College Dublin, National University of Ireland-Dublin, Dublin 4, Ireland
e-mail: arun.kumar@ucd.ie

R. Poduri (ed.), *Drug Discovery and Development*,
https://doi.org/10.1007/978-981-15-5534-3_10

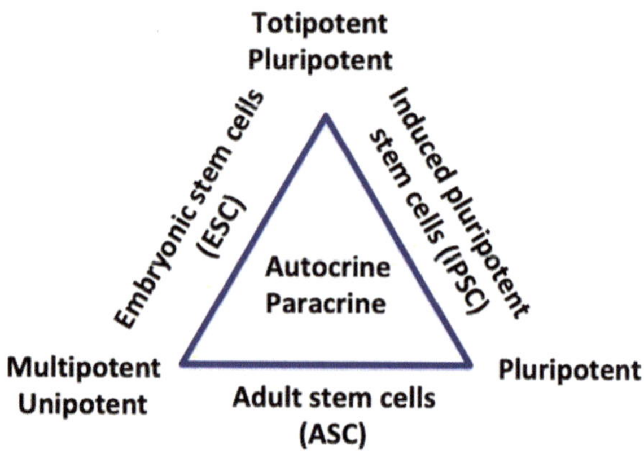

Fig. 10.1 The categories of stem cells, their nature and effects

The embryonic stem cells are isolated from the inner mass of blastocysts in a developing embryo. The adult stem cells are isolated from blood circulation, bone marrow, or tissue-specific niche areas in an individual. While the induced pluripotent stem cells are genetically engineered by expressing 3–4 sets of genes (Sox-2, OCT-4, c-Myc and Klf-4) in any cell type. The stem cells may be either:

1. Totipotent (which can develop into all cell types),
2. Pluripotent (which can develop into all cell types except trophoblastic cells),
3. Multipotent (which can develop into many cell types), or
4. Unipotent (which can develop into only a specific cell type) (Fig. 10.1).

Progenitor cells is another terminology, which is extensively used in the literature; essentially these are adult stem cells. The response of the stem cells could be either direct (differentiating into the desired cell type) or indirect (by secretion of factors, i.e., paracrine effect). Cells are known to secrete many factors, which may influence the same (autocrine effect) or other (paracrine effect) cell types. The knowledge of the factors secreted from cells is proving to be helpful in development of stem cell therapy.

10.2 Rationale for Therapeutic Use of Stem Cells

Understanding the rationale for therapeutic use of stem cells requires a gross know-how of anatomy, physiology, and pathology. A multicellular organism is made up of several cell types, which form various types of tissues, which in turn form various organs and organ systems. In a simplistic explanation there are structures (anatomy) that are associated with specific functions (physiology). Any abnormality with either the structure and/or function is what is defined as the pathology (disease), which requires appropriate measures to correct the structural and functional defects (Fig. 10.2). The currently available therapeutics predominantly target correction of

Fig. 10.2 The rationale for stem cell therapy and evolution of stemcology

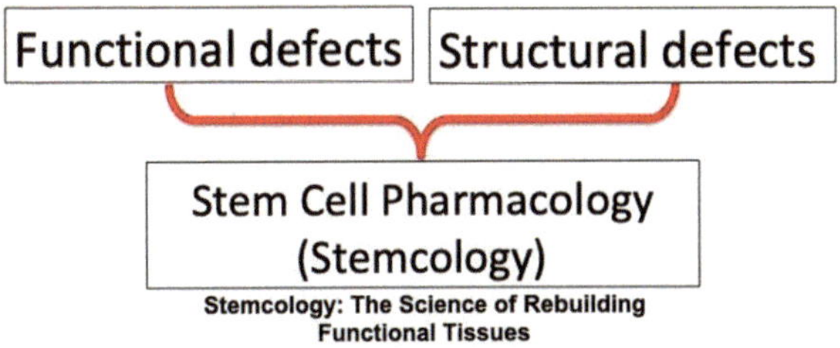

functional defects while the repair of structural defects can be achieved to a certain extent by surgical procedures. Correcting the functional defects also seems to enhance endogenous potential to restore the structural defects. The discovery of stem cells offers an additional evidence-based approach and probably an optimal approach to correct the structural defects. The best approach to correct structural defects is by using similar structures and stem cells offer the feasibility to make available structural units for repairing structural defects. The potential of stem cell is twofolds, one by direct contribution to rebuilding the structure and second by mobilizing the endogenous repair mechanisms by paracrine effects. It is also feasible to combine stem cell treatment with established therapeutics to optimally and effectively correct structural and functional defects, an approach I prefer to name as stem cell pharmacology or stemcology in short. Cell therapy in the broadest definition is not new in clinical medicine, for instance blood transfusion can be viewed as a simplest form of cell therapy. Blood transfusion based on compatibility matching is a time-tested approach, which has established its therapeutic merit in terms of both safety and efficacy. The discovery of the stem cells can be seen as a new dimension to the blood transfusion albeit with refinement to the nature of cells transferred. Several studies have demonstrated the safety of stem cell therapy although with limited efficacy, nevertheless this has not limited the quick transition of the stem cell therapy into clinical applications. The major work in the future should hence focus on improving the therapeutic efficacy while retaining the established safety features of stem cell therapy.

10.3 Techniques Helpful in Stem Cell Research

Cell culture techniques and cell storage facility are very fundamental requirement in stem cell research. Besides cell culture and storage, techniques to identify cell surface markers and purify single cells from a pool of mixed cells are essential. For this, flow cytometry and fluorescence-activated cell sorting (FACS) techniques specifically at single cell resolution have proved to be very useful (Fig. 10.3). Also recent research has indicated the benefits of 3D cell culture approach compared to the classical 2D cell culture techniques. Live cell-imaging techniques together with labeling the cells for intracellular and surface marker are also necessary for stem cell research. Good cell culture practices (GCCP) are an absolute must for reliable valid

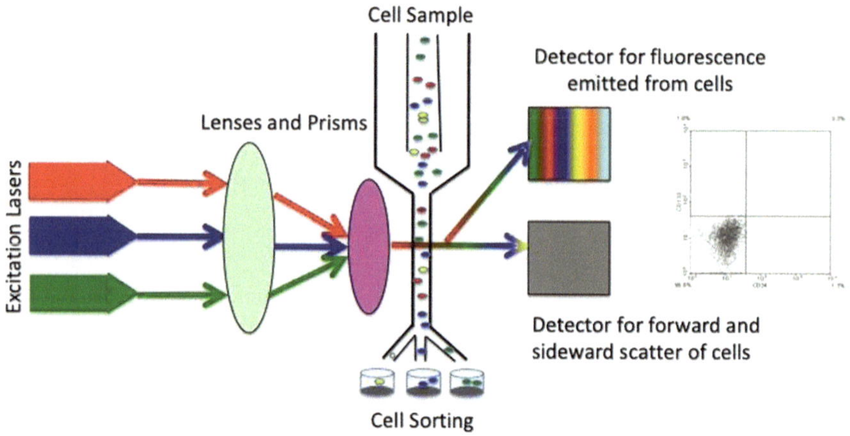

Fig. 10.3 An illustration of flow cytometry for detecting and separating single cells

and reproducible stem cell research. Besides adopting strict aseptic working conditions, stem cells should be regularly assessed for any potential contaminants (bacterial, viral, mycoplasma) and must always be labeled appropriately. It is essential to note that cells in culture are likely to change their characteristics; hence it is a good practice to regularly validate the purity profile of the cells with established markers. Similar quality control with cell culture consumables is also necessary. It is absolutely must to establish a validated criteria to define the stability and quality of the stem cells and as part of the GCCP the validated criteria should be strictly implemented. Specific emphasis should be on cell culture media used, the sourcing of serum, essential nutrients supplemented, antibiotics used, pH/temperature conditions, oxygen/carbon dioxide levels, cell subculture techniques, and the cell culture surface/matrix used. All these listed factors either individually or in combination can significantly influence the stem cell phenotype/genotype and hence a strict quality control of these factors is necessary.

Good cell culture practices (GCCP) should be based on the following major principles:

1. A clearly defined in vitro system for cell culture practice and complete know-how of relevant factors, which could affect the in vitro system.
2. Adopting strict quality assurance of all materials and methods adopted using standard operating protocols.
3. Effective documentation and tracking of all activities related to the materials and methods used.
4. Periodic review and evaluation of all cell culture activities by an independent review body for quality compliance and adherence.
5. Applying very high standards of health and safety issues.
6. Implementing all ethical principles in compliance with relevant legislations.

7. Provision in place for continued education and training of personnel on points 1–6 above.

10.4 Source of Stem Cells

Classically the adult stem cells are sourced from blood samples, bone marrow, umbilical cord, or tissue biopsies (Fig. 10.4). In many tissues the local presence of stem cells serves as endogenous repair system to repair or replace the cells which are lost due to routine wear and tear throughout life. The embryonic stem cells are sourced from spare embryos produced in the *in vitro* fertilization clinics. While the induced pluripotent stem (IPS) cells can be generated by expressing three to four sets of genes, i.e., Sox-2, OCT-4, c-Myc, and/or Klf-4, in any cell type (usually fibroblasts). Delivery of stemness related gene (Oct-3/4, Sox2, Klf4, and c-Myc quartet) has been reported to reprogram adult somatic cells into becoming embryonic stem cells with high degree of plasticity. Due to the limited numbers of adult vascular progenitors, interest is focused on IPS technology to generate vascular progenitors from the adult cells. Recently, beneficial effects of therapy with fibroblasts transduced with human stemness factors OCT3/4, SOX2, KLF4, and c-Myc were reported in a mouse acute MI model. Improvement in ejection fractions, fractional shortening, and systolic wall thickness was shown suggesting the potential of iPS in repair of acute myocardial infarction. Additionally these cells also had differentiation potential into endothelial ($CD31^+$), smooth muscle, and cardiomyocytes, indicting their multilineage potential. More recently, it is reported that only 2 transcription factors (*OCT4*/*SOX2* or *OCT4*/*KLF4*) are enough to generate iPS cell lines and iPS cells expressing OCT4/SOX2/NANOG/LIN28 markers had cardiomyocyte differentiation potential.

Bone marrow is rich source of several types of progenitors which could be broadly classified into three categories, i.e., hematopoietic, mesenchymal, and tissue/cell type specific progenitors. It is still not clear if a common progenitor for all these progenitors exists in adults. Hematopoietic stem cells give rise to white

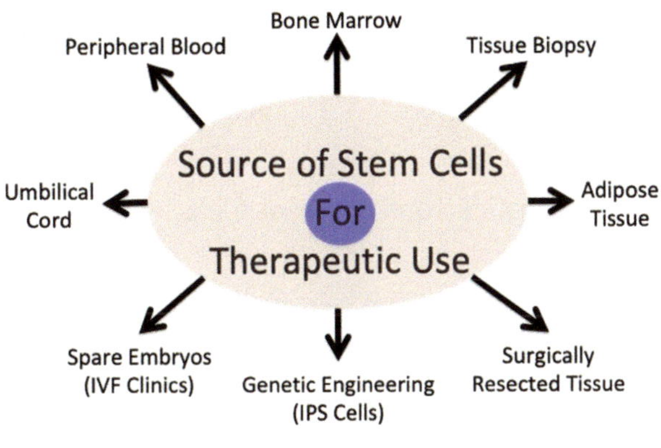

Fig. 10.4 Sources of stem cells for therapeutic use

blood cells (leukocytes), red blood cells (erythrocytes), and platelets (thrombocytes). While the mesenchymal stem cells have the capability to differentiate into osteoblasts, chondrocytes, adipocytes, myocytes, and many other types of cells. BM-MNC has been evaluated in several clinical trial for their therapeutic potential in acute myocardial infarction and as well as peripheral artery disease. However the responses with the BM-MNC administration have been mixed which could be attributed to the heterogeneous population of cells observed in the marrow niche which may have countervailing effects. Till date $CD34^+$, $CD133^+$, and BM-MNC cells with or without G-CSF-mediated mobilization have been evaluated for benefits in patients with acute myocardial infarction. While a few studies have failed to observe any improvement with BM-MNC therapy, a significant improvement in left ventricular function is reported following bone marrow-derived $CD34^+$, $CD133^+$, or BM-MNC cells.

Irrespective of the source of stem cell, it is essential to retain the purity of stem cells for achieving reliable therapeutic outcomes. The purity of stem cells can be achieved by separating the cells to single cell resolution using techniques such as FACS and then deriving single cell clones from them. As mentioned before, cells are likely to change their phenotype while in culture; hence regular monitoring of the cell purity becomes mandatory. The sourcing and use of stem cells is highly regulated, with some nations having strict restrictions on any research involving use of stem cells. The ethics concerning the use of stem cells is a very complex issue, which varies between different regions. Hence there is huge diversity in the extent to which scientific and clinical advancement can progress depending on the geographical location. While it is beyond the scope of this chapter to look into the differences in various regulations on the use of stem cells globally, in the table below I have selected a few guidelines applicable to responsible use of stem cell for research advancement and/or clinical development:

Organization	Webpage link
International Society for Stem Cell Research	http://www.isscr.org/
ICMR and DBT: National Guidelines for Stem Cell Research 2017	http://www.icmr.nic.in/guidelines/Guidelines_for_stem_cell_research_2017.pdf
US National Academy of Sciences	https://nas-sites.org/stemcells/

Most regulations used worldwide are within the remit of these guidelines.

10.5 Applications of Stem Cells

The application of stem cell in biomedicine is as diverse as the nature of these cells and it continues to evolve as we progress in understanding the stem cells (Fig. 10.5). The simplest application is to replace damaged cells in the system by transferring the stem cell by either intravenous route or seeding the stem cells locally within the region of interest in the tissue. For example, in case of myocardial infarction (heart

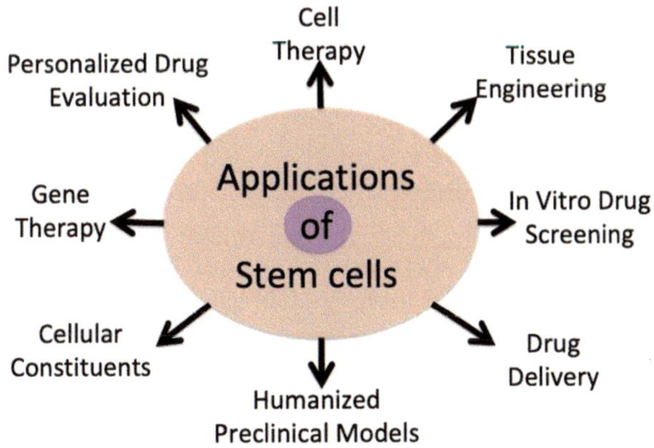

Fig. 10.5 The current and potential applications of stem cells

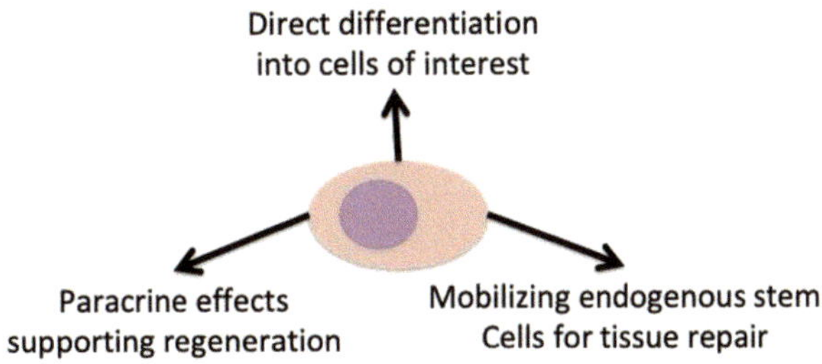

Fig. 10.6 The major mechanisms of action of stem cells

attack), the heart will lose substantial number of cardiomyocytes post myocardial infarction. Here the objective of developing stem cell therapy for myocardial infarction will be to replace the lost cardiomyocytes potential cardiomyocyte-specific stem cells. The approach to replace the lost cardiomyocytes may be achieved by either administering the cardiomyocyte-specific stem cells locally (via the coronary artery) or systemically (any intra-arterial administration) and/or by directly injecting the stem cells into cardiac muscle near the region of the damage. In either of the approach, the potential benefits of the stem cells are achieved by

1. Direct differentiation of injected stem cells into cardiomyocytes and hence replacing the missing structural component of the heart.
2. Indirectly by release of various secreted factors from the injected stem cells, which potentially favor endogenous stem cells to mobilize into the site of tissue injury and repair the damaged tissue.
3. The secreted factors from the injected stem cells may create a favorable biochemical environment for tissue repair at the site of tissue injury (Fig. 10.6).

In the recent times the cell base approach has gained increasing interest due to their potential to treat the underlying injuries associated with cardiac and vascular disease both early and late in the disease process. There is strong evidence for the beneficial role of adult progenitor cells in several preclinical studies, which has extended progenitor cell-based therapy from bench to early controlled clinical studies. To date, bone marrow mononuclear cells, endothelial progenitor cells, and skeletal muscle myoblasts have been clinically evaluated for their benefits in acute myocardial ischemia, peripheral vascular disease, and ischemic injuries. Collaterally several cell types have been and continued to be identified based on cell surface marker profiles, initial source of isolation, cell culture conditions, and their differentiation potential without or with cytokine preconditioning. However like for any new technology, there exists technical limitations and considerable refinement is needed to overcome these limitations to realize its immense therapeutic potential in future.

10.6 Applications of Stem Cells in Developing Novel Therapeutics

As stem cells are not defined as drugs, I would refrain from calling the process of identifying and developing stem cell for therapeutic application as drug discovery/ development process, I would rather prefer to name it as therapeutic discovery and development. The therapeutic discovery and development of stem cell has gained considerable momentum over the past two decades and have quickly transitioned from preclinical evaluation to clinical applications for the therapy of variety of chronic diseases. The considerable advancement in stem cell isolation and culture techniques has supported the quick clinical transition. Broadly the therapeutic discovery and development involves:

1. Identifying the stem cells which have potential to differentiate into the cell type of therapeutic interest (e.g., cardiomyocytes, neurons, blood vessels, smooth muscle cells, endothelial cell, insulin producing cell).
2. The identified stem cells are validated and clearly defined based on surface marker or biochemical profiles.
3. Testing the safety and efficacy of the stem cells in a suitable preclinical model.
4. Seeking regulatory approval for testing the safety and efficacy of the stem cells in humans.

After successfully passing through all these phases, unlike for classical drugs, the stem cells cannot become readily available for clinical use in off-the self format because the cells will have to be cultured and the cell culturing process needs strict quality control to generate valid and reliable source of stem cell for therapeutic application. This is the most important process for effective delivery of stem cell therapeutics. Stem cells by their very nature are very dynamic and are likely to change during the cell culture process and hence maintaining a strict quality control remains to be challenging. The stem cells used for therapy could be either autologous (i.e., derived from the same patient and cultured to increase their number) or allogeneic (i.e., sourced from other healthy individuals and cultured to increase their number). The autologous stem cells have the merit of optimal integration over to allogeneic stem cells; however allogeneic stem cells are likely to have better efficacy than autologous stem cells under disease conditions. There are also a few stem cell products that are approved and available for off-the-self use; however I am skeptical on their merit in achieving optimal therapeutic efficacy. To date, only one stem cell-based product has been approved by the US FDA, which consists of blood-forming stem cells derived from cord blood for limited use in patients with defects in the production of blood. Factors such as quality of stem cells, the number of cell passages, cell culture conditions, and the genetic profile of the donor can all influence achieving optimal therapeutic efficacy from stem cell therapy.

The administration of stem cells has a very wide therapeutic application which ranges from diseases associated with structural defects (cardiovascular, renal, hepatic, gastrointestinal, nervous system defects) to autoimmune disorders (systemic sclerosis, type I diabetes, systemic lupus erythematous, rheumatoid arthritis, myasthenia gravis), inflammatory conditions, and more recently certain forms of cancer (Fig. 10.2). Several ongoing clinical trials are evaluating the potential of stem cells for various therapeutic applications (Table 10.1). Irrespective of the therapeutic area for which the administration of stem cells is intended, the pharmacodynamics and pharmacokinetics of the administered stem cells are highly variable and often poorly predictable. Unlike the classical drugs which have fixed target/s and predicted distribution patterns, achieving such fixed target/s and predicted distribution with stem cells is a challenge due to the vary dynamic nature of stem cells. Many of the studies, which have evaluated the distribution patterns of stem cells, have found the cells predominantly lodged in tissue niches such as liver and spleen within few hours following their systemic administration. Despite this limitation of not localizing at the target site, the stem cells are known to evince additional protective mechanisms by paracrine effects. Understanding true pharmacokinetic features of stem cell is also technically challenging. Most often the cells will have to be labeled with any detectable marker to facilitate their tracking in pharmacokinetic studies. It is not clear if the unlabeled stem cells would distribute differently from the labeled cells. Most pharmacokinetic evaluations are performed under physiological conditions or in controlled modified disease conditions, which may not be reflective of the patient-specific pathology. It is very likely that the patient-specific pathology may also have a diverse influence over the distribution of administered stem cells. Hence in my opinion deciphering the true pharmacokinetics of administered stem cells is an unmet technical challenge.

In contrast to the limitations with pharmacokinetics, the pharmacodynamic evaluations of stem cells are relatively reliable and technically less challenging. There are two major approaches to screening of investigational therapeutics, i.e., (1) target-based screening and (2) phenotype-based screening. Over the last few decades the chemical biology-based drug discovery process has heavily relied on the target-based screening process; however with shift in focus towards biologicals-based drug discovery process, the focus is shifting towards a phenotype-based screening process. Although in my view a blended approach involving both target- and phenotype-based screening will be appropriate for efficient approach to assessing efficacy and safety of investigational therapeutics. The discovery and development

Table 10.1 List of ongoing clinical trials registered in www.clinicaltrials.gov using stem cells

Type of stem cells	Clinical condition	Country
Adipose-derived mesenchymal stem cells	Sexual hormone deficiency	Vietnam
Adipose-derived mesenchymal stem cells	End stage renal disease (ESRD) Vascular access complication	USA
Adipose-derived mesenchymal stem cells	Pulmonary hypertension	China
Adipose-derived stem cell	Vestibulodynia	Belgium
Adipose-derived stem cells	Androgenetic alopecia	Egypt
Adipose tissue-derived mesenchymal stem cells	Knee osteoarthrosis	Jordan
Adipose-cord mesenchymal stromal cells	Ulcerative colitis (UC)	China
Adipose-derived mesenchymal stem cells	Knee osteoarthritis Hip osteoarthritis Glenohumeral osteoarthritis Osteoarthritis	Poland
Adipose-derived stem cells	Perianal fistula Crohn disease	USA
Adult allogeneic mesenchymal cells from adipose tissue	Acute graft-versus-host disease	Spain
Allogeneic adipose-derived stem/stromal cells	Xerostomia due to radiotherapy Hyposalivation Xerostomia	Denmark
Allogeneic hematopoietic stem cell transplantation Allogeneic bone marrow transplantation	Hematopoietic and lymphoid cell neoplasm	USA
Allogeneic hematopoietic stem cells	Primary myelofibrosis Secondary myelofibrosis	USA
Allogeneic hematopoietic stem cells	Multiple myeloma Plasma cell leukemia Extramedullary plasmacytoma	China
Allogeneic hematopoietic stem cells	Acute lymphoblastic leukemia Acute myeloid leukemia Aggressive non-Hodgkin lymphoma	USA
Allogeneic hematopoietic stem cells	Acute myeloid leukemia	USA
Allogeneic hematopoietic stem cells	B acute lymphoblastic leukemia BCR-ABL1 fusion protein expression Minimal residual disease	USA
Allogeneic hematopoietic stem cells	B acute lymphoblastic leukemia	USA
Allogeneic human cardiosphere-derived stem cells	Pulmonary arterial hypertension (PAH)	USA
Allogeneic human mesenchymal stem cells	Alzheimer dementia	USA

(continued)

Table 10.1 (continued)

Type of stem cells	Clinical condition	Country
Allogeneic mesenchymal stem cells	Tibial closed diaphyseal fractures	Iran
Allogeneic mesenchymal stem cells	Major depression Alcohol use disorder	USA
Allogeneic mesenchymal stem cells	Cystic fibrosis	USA
Allogeneic mesenchymal stem cells	Diabetes mellitus	USA
Allogeneic mesenchymal stem cells	Heart failure Ischemic cardiomyopathy	Greece
Allogeneic stem cell	Myelodysplastic syndrome Chronic myelomonocytic leukemia	Germany
Allogeneic stem cell	Leukemia, B cell lymphoma, Hodgkins lymphoma, non-Hodgkins lymphoma, B cell	USA
Allogeneic stem cells	Hodgkin lymphoma	USA
Allogeneic stem cells	Heart failure	Denmark
Allogeneic stem cells	Bone marrow fibrosis	Germany
Allogeneic stem cells	Leukemia, acute lymphoblastic Acute myeloid leukemia Mixed-lineage acute Leukemias	Russia
Allogeneic umbilical cord Mesenchymal stem cell	Ketoacidosis, diabetic	China
Allogeneic umbilical cord mesenchymal stem cells	Acute myocardial infarction	Taiwan
Autologous adipose-derived stem cells stem cells	Diabetic foot ulcer	Poland
Autologous adipose-derived stem cells stem cells	Skin Scar Cutis Laxa	Poland
Autologous adipose-derived mesenchymal stem cells	Knee osteoarthritis	China
Autologous adipose-derived mesenchymal stem cells	Knee osteoarthritis Cartilage degeneration	China
Autologous aldehyde dehydrogenase-bright stem cells	Coronary artery disease	USA
Autologous BMAC	Peripheral nerve injury upper limb	USA
Autologous bone marrow concentrate Adipose-derived stromal vascular fraction	Osteoarthritis	USA
Autologous bone marrow-derived mesenchymal stem cells	Erectile dysfunction	Korea
Autologous bone marrow-derived stem cells	Optic neuropathy	Jordan

(continued)

Table 10.1 (continued)

Type of stem cells	Clinical condition	Country
Autologous bone marrow-derived stem cells	Premature ovarian failure	Jordan
Autologous bone marrow-derived stem cells	Rheumatoid arthritis Osteoarthritis, knee Osteoarthritis, hip	Jordan
Autologous bone marrow-derived stem cells	Multiple sclerosis	Jordan
Autologous bone marrow mono-nuclear cells	Liver cirrhosis, biliary	Vietnam
Autologous bone marrow stem cells	Autism	Vietnam
Autologous bone marrow stem cells	Spinal cord injury	Brazil
Autologous bone marrow stem cells	Liver diseases	China
Autologous bone marrow stem cells	Lumbar disc herniation	China
Autologous bone marrow-derived CD34+, CD133+, and CD271+ stem cell	Retinitis pigmentosa	Jordan
Autologous bone marrow-derived cells	Heart failure	Germany
Autologous CD123CAR-CD28-CD3zeta-EGFRt-expressing T lymphocytes	Adult acute myeloid leukemia in remission Acute biphenotypic leukemia Early relapse of acute myeloid leukemia	USA
Autologous CD34-selected peripheral blood stem cells	Crohn's disease	USA
Autologous CD34+ cells	Wiskott-Aldrich syndrome	France UK
Autologous endothelial progenitor cells	Anterior wall myocardial infarction	Canada
Autologous endothelial progenitor cells	Hypertension, pulmonary	USA
Autologous enriched CD34+ cell fraction that contains CD34+ cells	Anemia, sickle cell	USA
Autologous hematopoietic cells	Leukemia, myeloid	South Korea
Autologous hematopoietic pro-genitor cells	Multiple myeloma Amyloidosis	USA
Autologous hematopoietic stem cell	Crohn's disease Inflammatory bowel diseases Gastroenteritis	Brazil
Autologous hematopoietic stem cell	Systemic lupus erythematosus	Germany
Autologous hematopoietic stem cell	HIV infection Mature T cell and NK cell non-Hodgkin	USA

(continued)

Table 10.1 (continued)

Type of stem cells	Clinical condition	Country
	lymphoma Plasmablastic lymphoma	
Autologous hematopoietic stem cell	Mantle cell lymphoma Recurrent diffuse large B cell lymphoma Recurrent follicular lymphoma	USA
Autologous hematopoietic stem cells	Autoimmune disease Neurologic autoimmune disease	USA
Autologous hematopoietic stem cells	Acute myeloid leukemia	USA
Autologous hematopoietic stem cells	Systemic scleroderma	USA
Autologous lung stem cells	Idiopathic pulmonary fibrosis	China
Autologous mesenchymal stem cells	Urticaria Autoimmune diseases Immune system diseases Skin diseases	Turkey
Autologous mesenchymal stem cells	Musculoskeletal pain Knee osteoarthritis Cartilage injury Cartilage degeneration	USA
Autologous mesenchymal stem cells	Epilepsy	Jordan
Autologous mesenchymal stem cells	Chronic myocardial ischemia	France
Autologous mesenchymal stem cells	Asthma	USA
Autologous mesenchymal stem cells	Critical limb ischemia	Ireland
Autologous mononuclear bone marrow cells	Idiopathic dilated cardiomyopathy	Spain
Autologous mononuclear cells	Hypoplastic left heart syndrome	USA
Autologous multilineage potential cells (AMPC)	Acute myeloid Leukemia recurrent	Thailand
Autologous muscle-derived cells for gastrointestinal repair (AMDC-GIR)	Oropharyngeal dysphagia	USA
Autologous regenerative cells of adipose tissue	Rectovaginal fistula	Russia
Autologous stem cells	Non-obstructive azoospermia	Jordan
Autologous stem cells	Lymphoma	China
Autologous stem cells	Waldenström macroglobulinemia	China
Autologous stem cells	T cell non-Hodgkin lymphoma	USA
Autologous stem cells	Very high risk neuroblastoma	France
Bone marrow-derived autologous stem cells	Duchenne muscular dystrophy	Jordan
Bone marrow-derived stem cells	Cerebral palsy	Jordan

(continued)

Table 10.1 (continued)

Type of stem cells	Clinical condition	Country
Bone marrow stem cell	Alzheimer disease Alzheimer dementia Vascular dementia	USA
Bone marrow stem cell	Neurologic disorders Nervous system diseases Neurodegenerative diseases	USA
Bone marrow stem cells	Ischemia	Brazil
CD34 positive stem cell	Blood and marrow transplantation	USA
Cellgram™ (bone marrow-derived mesenchymal stem cells)	Alcoholic liver cirrhosis	USA
Cord blood stem cells	Hematological malignancy Acute leukemia in remission Acute lymphoblastic leukemia in remission	USA
Cord blood-derived expanded allogeneic natural killer cells	Non-myeloablative TCR alpha/beta-depleted Haploidentical hematopoietic stem cell transplantation Recurrent lip and Oral cavity carcinoma Recurrent malignant endocrine neoplasm	USA
Corneal epithelial stem cells	Dry eye syndromes Dry eye Ocular inflammation	USA
DCLK1 positive stem cell	Barrett's esophagus Esophageal adenocarcinoma	USA
Embryonic stem cell-derived retinal pigment epithelium	Retinitis pigmentosa	France
Embryonic-like stem cell	Premature ovarian failure	China
Ex vivo cultivated limbal stem cell pool	Corneal injuries Corneal burns Corneal scars and opacities	India
Gene-modified CD34+ hematopoietic stem cells	X-linked severe combined immunodeficiency XSCID SCID-X1 Gamma C-deficient SCID	USA
Haploidentical hematopoietic stem cell	Sickle cell-thalassemia disease Thalassemia	USA
Haploidentical stem cells	Neuroblastoma Ewing sarcoma Rhabdomyosarcoma	South Korea
Hematopoietic stem cells (enriched)	Renal failure	USA
Hematopoietic allogeneic stem cells	Mitochondrial neurogastrointestinal encephalomyopathy (MNGIE)	USA
Hematopoietic stem cell	Type 1 diabetes mellitus	Mexico
Hematopoietic stem cell	Sickle cell disease	USA
Hematopoietic stem cell	Inflammatory bowel diseases	China

(continued)

Table 10.1 (continued)

Type of stem cells	Clinical condition	Country
Hematopoietic stem cellCD34+	Metachromatic leukodystrophy Adrenoleukodystrophy	China
Hematopoietic stem cells	Amyotrophic lateral sclerosis	Mexico
Hematopoietic stem cells	Alpha thalassemia major Hemoglobinopathy; with thalassemia Hemoglobinopathies	USA
Hematopoietic stem cells	Multiple sclerosis	USA
Hematopoietic stem cells	Chronic inflammatory demyelinating polyneuropathy	USA
Human neural stem cells	Secondary-progressive multiple sclerosis	Italy, Switzerland
Human spinal cord-derived neural stem cells	Spinal cord injury (SCI)	USA
IPSC	Diabetes complications Diabetic retinopathy	USA
Limbus-derived stem cells	Corneal scars and opacities	India
Longeveron mesenchymal stem cells (LMSCs)	Aging frailty	USA
MCRcI® stem cells	Coronary artery disease	Taiwan
Mesenchymal stem cell	Renal artery stenosis Ischemic nephropathy Renovascular disease Chronic kidney disease	USA
Mesenchymal stem cell	Aplastic anemia	China
Mesenchymal stem cell	Cardiomyopathy Heart failure	USA
Mesenchymal stem cells	Hematopoietic stem cell transplantation	Turkey
Mesenchymal stem cells	Bronchopulmonary dysplasia	China
Mesenchymal stem cells	Diabetes mellitus, type 1 Diabetes mellitus, insulin-dependent	Iran
Mesenchymal stem cells	Rotator cuff tear	Czech Republic
Mesenchymal stem cells	Primary biliary cirrhosis	China
Mesenchymal stem cells	Inflammatory bowel diseases	Jordan
Mesenchymal stem cells	Full thickness rotator cuff tear	USA
Mesenchymal stem cells	Acute-on-chronic liver failure	China
Mesenchymal stem cells	Lung transplant reject Bronchiolitis obliterans	USA
Mesenchymal stem cells	Systemic lupus erythematosus	USA
Mesenchymal stem cells	Chronic obstructive pulmonary disease	USA
Mesenchymal stem cells	Hemorrhagic stroke Intracerebral hemorrhage	USA
Mesenchymal stem cells	Malignant ovarian Brenner tumor Ovarian clear cell adenocarcinoma Ovarian endometrioid adenocarcinoma	USA

(continued)

Table 10.1 (continued)

Type of stem cells	Clinical condition	Country
Mesenchymal stem cells	Myocardial infarction	Spain
Mesenchymal stem cells	Graft-versus-host disease Poor graft function Low donor T cell chimerism	Belgium
Mesenchymal stem cells	Alzheimer's disease	USA
Mesenchymal stem cells	Multiple organ dysfunction syndrome	China
Mesenchymal stem cells	Diabetes mellitus, type 1	USA
Mesenchymal stem cells	Pediatric liver transplantation	Germany
Mesenchymal stem cells	Nonunion fracture	France
Mesenchymal stromal cells	Chronic graft-versus-host disease	China
Neural stem cells loaded with an oncolytic adenovirus	Glioma Anaplastic astrocytoma Anaplastic oligodendroglioma	USA
Stem cells from human exfoliated teeth	Type1/2diabetes	China
Umbilical cord blood	Acute myeloid leukemia Acute lymphoblastic leukemia in remission Myelodysplastic syndromes Other acute Leukemias	USA
Umbilical cord-derived CD362-enriched mesenchymal stem cells	Acute respiratory distress syndrome	UK
Umbilical cord-derived mesenchymal stem cells	Parkinson's disease	China
Umbilical cord-derived mesenchymal stem cells	Parkinson disease	Jordan
Umbilical cord-derived mesenchymal stem cells	Acute respiratory distress syndrome	China
Umbilical cord-derived mesenchymal stem cells	Infertility, female endometrium	China
Umbilical cord-derived mesenchymal stem cells	Bronchopulmonary dysplasia	Taiwan
Umbilical cord mesenchymal stem cell	Lumbar discogenic pain	China
Umbilical cord mesenchymal stem cell	Osteoarthritis, knee	Indonesia
Umbilical cord mesenchymal stem cells	Bronchopulmonary dysplasia	Vietnam
Umbilical cord mesenchymal stem cells	Peripheral vascular disease Ischemia Diabetic foot	China
Umbilical cord mesenchymal stem cells	Thin endometrium Intrauterine adhesion	China
Umbilical cord mesenchymal stem cells	Rheumatoid arthritis	Panama
Umbilical cord mesenchymal stem cells	Psoriasis	China

(continued)

Table 10.1 (continued)

Type of stem cells	Clinical condition	Country
Umbilical cord mesenchymal stem cells	Spinal cord injuries	China
Umbilical cord mesenchymal stem cells	Infarction, middle cerebral artery Infarction, anterior cerebral artery Cerebral infarction	China
Umbilical cord mesenchymal stem cells	Uterine scar	USA
Very small embryonic-like stem cells (VSEL)	Organic erectile dysfunction	China

of stem cells for therapeutic application will benefit from adopting a phenotype-based screening approach, as often the intended application of stem cells would be to treat diseases, which will have multiple etiologies. A phenotypic outcome may eventually be a consequence to the influence of stem cells on multiple targets, which may actually be the case with the pharmacodynamic behavior of stem cells. While identifying the specific targets of the stem cells will be helpful to refine the safety profile of the therapy, this should not be the primary focus as it can reduce the success rate of advancing the stem cells for therapeutic applications. Phenotype-based screening systems can be developed both in vitro (cell-based assays) and in vivo (humanized preclinical models) for the evaluation of stem cell therapeutics. Another advantage of phenotype-based evaluation of stem cell therapeutics is that it offers a system-based medicine rather than symptom-based medicine, which in my view is holistic and probably more effective.

10.7 Examples for Stem Cells Used in Cardiovascular Therapeutics

Endothelial progenitor cells (EPC) are cell with potential to differentiate into functional endothelial cells (EC). EPC were initially identified and isolated in 1997 by Asahara et al. EPCs are the most widely investigated cells and probably the most controversial in the cell therapy field. Since their first identification as $CD34^+$ mononuclear cells (expressing endothelial and hematopoietic markers) in the peripheral circulation, several version EPCs have been reported (based on the differences in culture conditions and cell isolation protocols) by a diversified group of investigators and consequently have different functions attributed to them including their role in neovascularization. Possibly, subsets of $CD34^+$ cells that express CD133 and Flk-1 are the phenotypical and functional markers of EPC that play a role in postnatal angiogenesis. However, expression of the surface markers described above does not include stem cells from other sources (mesenchymal or even in the vessel wall) or their progenitor cells. Several populations of bone marrow-derived, circulating, or tissue-resident cells also are reported to possess properties of EPC.

These subpopulations are characterized by the expression of CD14, together with CD34 or VEGFR-2 suggesting their myeloid lineage. A recent study suggests that while the early EPC (displayed expression of monocytic marker CD14, CD11b, CD11c) facilitate the EC formation by secreting several cytokines, the late EPC ($CD14^-$ and strongly expressed markers of mature endothelial cells) are the ones which can differentiate into functional endothelial cells. Contradicting this, in vitro-cultured bone marrow cells have reported to consist of a small population of low-density lipoprotein (LDL) uptake-positive cells expressing endothelial markers, which demonstrated a phagocytosis function similar to monocyte/macrophages and expression of the panleukocyte surface marker (CD45) and monocyte marker (CD14). These LDL uptake-positive cells have poor proliferative capacity and die out gradually in long-term culture suggesting that these cells are actually monocytes/macrophages that can express some endothelial markers but are not "true endothelial progenitor cells" (EPCs). Such controversies are not uncommon in stem cell research.

Several preclinical studies have provided evidence to the role of EPC in postnatal vasculogenesis and hence their potential to treat complications associated with acute myocardial infarction and tissue ischemia. Based on the promising outcomes in the preclinical studies EPC have been extensively investigated in the clinical setting both as therapeutics as well as diagnostic and prognostic marker. For instance, circulating EPC numbers and function were significantly reduced in patients with coronary artery disease and among diabetic patients with peripheral artery disease. EPC levels were also negatively correlated with the degree of carotid stenosis and the stage of tissue ischemia. Although benefits from EPC therapy post myocardial infarction is only 6–8% improvement in left ventricular ejection fraction, weather this effect can be potentiated by co-administration of pharmacological agents or other vascular progenitor cells would be interesting to investigate.

Circulating progenitor cells (CPC) are heterogeneous population of cells in the peripheral blood which can give rise to EPC, smooth muscle progenitor cells (SPC) as well as cardiac progenitors and have CD34 as a common marker defining them. Several subtypes of these have been identified in the human peripheral blood, i.e., $CD34^+$, $CD34^+/CD133^+/VEGFR\text{-}2^-$, $CD34^+/CD133^+/VEGFR\text{-}2^+$, and $CD34^+/CD133^-/VEGFR\text{-}2^+$ cells. Generally it is believed that CPC area derived from bone marrow but their peripheral source (such as spleen and local tissue niche) cannot be excluded. There are also other less well-defined subpopulations, such as $CD34^+/CD117^+$, $CD34^+/CXCR4^+$, $CD34^+/CD38^+$, and $CD34^+/CD45^+$ cells, which follow the same time course as EPC. It is interesting to note that all of these cells are reported to be mobilized into peripheral circulation postacute myocardial infarction. Bone marrow-derived mononuclear cells (BM-MNC) are another subcategory of CPC, which when delivered in chronic myocardial infarction patients (followed until 6 years post therapy) resulted in significant improvement in left ventricular ejection fraction, left ventricular systolic volume, and ventricular contractility, while treatment with peripheral blood CPC resulted only in reduced left ventricular ejection fraction, left ventricular systolic volume, and ventricular contractility. Most cell therapy in patients have been with a single does administration of the progenitor cells and the benefits observed are only modest in most of the clinical trials. It would be interesting to evaluate if repeated administration of progenitor cells (Pharmacokinetics needs to be worked out) can enhance the benefits observed.

Endothelial colony-forming cells (ECFC) are endothelial lineage (not hematopoietic lineage) late outgrowth cells derived from adherent peripheral blood mononuclear cells (MNC) cultured for 6–21 days in endothelial conditions; they display cobblestone morphology and can form blood vessels when administered in vivo. They are characterized by the expression of CD133/CD34/CD146 markers. These are essentially EPC with robust proliferative and vasculogenesis potential but are very rare in the peripheral blood (1:1 million of the peripheral blood mononuclear cells). The attractive feature of ECFC is their genomic stability and cryo preservability. Although they have been isolated from hypertensive patients on antihypertensive medication and from patients with acute myocardial infarction, their therapeutic potential is yet to be explored. These cells display the ability to ingest acetylated low-density lipoprotein (acLDL) and bind Ulex Europeus Agglutinin-1 (UEA-1), in addition to staining positive for vascular endothelial growth factor receptor-2 (VEGFR-2) and CD31. Furthermore, the circulating concentration of these cells has been inversely associated with the Framingham risk score for adverse cardiovascular health outcomes in affected patients.

Skeletal myoblasts are isolated from the skeletal muscle biopsies and cultured using myoblast basal growth medium for about 11–12 doubling time (average doubling time of 29 h). Skeletal myoblasts are characterized by the expression of CD56 and are mononuclear. Skeletal myoblasts are evaluated in several clinical trials for the treatment of myocardial infarction. Skeletal myoblasts therapy into the ischemic myocardium is reported to improve left ventricular end-diastolic volume, end-systolic volume, and ejection fraction and enhance viability and revascularization of the myoblast-implanted segments. Moreover long-term follow-up studies have not reported any tumor potential of these cells; however the increased risk of arrhythmias and postoperative episodes of sustained ventricular tachycardia is still a cause for concern. The incidences of arrhythmias are attributed to the lack of expression of gap junction proteins by the differentiated myotubes and are only observed in the early postoperative period. However in a multicentric study the incidence of arrhythmia was not different from the placebo group. While a few clinical trials have failed to observe any beneficial effects on the LV function post treatment with skeletal myoblasts.

Mesenchymal progenitor cells have the differentiation potential to osteogenic, adipogenic, chondrogenic, and myofibrogenic lineages. They are generally involved in fibrosis and are associated with osteogenic/calcification potential. Compared with the use of mesenchymal progenitor cells, cell transplantation with endothelial progenitor cells after myocardial infarction resulted in better neovascularization and contractility. This suggests that angiogenesis is an important mechanism in attenuating the progression of left ventricular dysfunction after myocardial infarction. However preclinical studies have reported cardiovascular benefits with use of mesenchymal progenitor cells. For instance, their ability to differentiate into endothelial phenotype was reported to enhance vascular density, and improve heart function in a rat cellular cardiomyoplasty model.

The discovery of progenitor cells of myeloid origin has offered the exciting prospect of merging classical concepts of myeloid cell biology in atherosclerosis with evolving concepts of myeloid cell plasticity and endothelial/smooth muscle cells differentiation within the injured vessel wall. Myeloid cells are all non-lymphoid cell types, including monocytes/macrophages, dendritic cells (DC) (although DC can also be derived from a lymphoid precursor), granulocytes, erythrocytes, and megakaryocytes. Monocytes/macrophages have certainly been implicated in neovascularization and are able to mimic endothelial cells (EC), forming branched columns and vascular-like structures in vitro. When murine long-term BM culture (highly enriched in macrophage population) was introduced to chick embryonic heart tissue, it was observed that some BM cells trans-differentiate into cardiomyocytes and vascular-like cells. Monocyte-derived dendritic cells, a cell type closely related to macrophages and a well-known specialized antigen presenting population, also differentiate into EC-like cells when stimulated with angiogenic factors. In vivo, a leukocyte population co-expressing dendritic cell (DC) and endothelial cell markers has been found to infiltrate tumor and assemble neovasculature. C-kit^{+}CD45^{-} progenitor cells from the murine liver show neovascularization capacity in vivo with strong colony-forming capacity and marked capacity to undergo further vascular differentiation. Subsequent systemic infusion of isolated c-kit^{+}CD45^{-} progenitor cells into a model of hind limb ischemia demonstrated that the neovascularization capacity of these cells was even greater than that of c-kit^{+}CD45^{+} cells. Significant contribution of bone marrow-derived vascular progenitor cells towards neovascularization and re-endothelialization is reported in early stage atherosclerosis associated with vascular injury. CD34^{+} and Flk1^{+} bone marrow-derived progenitor cells contribute to tissue repair by differentiating into endothelial cells, vascular smooth muscle cells, hematopoietic cells, and possibly other cell types. CD133^{+} cells, which are reported to differentiate into both endothelial and smooth muscle cells, play potential role in the remodeling process of pulmonary arteries in chronic obstructive pulmonary disease. These monocyte (CD34^{+}/CD14^{+}) derived EPC can also express a range of pro-angiogenic factors such as hepatocyte growth factor (HGF), VEGF, and GCSF, providing one mechanism whereby these cells may contribute to neoangiogenesis when implanted in vivo. However, their in vivo efficiency of integration into neovasculature remains unclear. While labeled CD14 monocytes do not integrate into the neovasculature of ischemic muscle, these cells successfully integrate in the presence of CD34^{+} cells. This suggests that pro-angiogenic cues are necessary for in vivo differentiation of these cells. The specific role of CD34^{+} cells in facilitating endothelial differentiation of CD14 monocytes and the importance of secreted pro-angiogenic factors or other differentiation effects are still unknown. However, some investigators have recently shown that EPC derived from CD14^{+} monocytes do possess in vivo vasculogenic capacity, but it is unclear whether such vasculogenesis is CD34 dependent, is a universal monocyte trait, or pertains to a small subset of the CD14^{+} monocytes. Indeed, there is increasing evidence that even hematopoietic stem cells may participate in tissue repair through myeloid intermediates and these findings generate considerable debate as to whether monocyte-derived cells represent an endothelial progenitor cell class per se or whether these differentiation events represent a common myelomonocytic response in vivo. For instance, there is a large amount of data to suggest that tissue-resident endothelial cells often share classic monocytic markers such as CD14, CD32, CD80, and CD86. Although no direct differentiation of myeloid lineage cells to SMC-like cells has been reported, a reverse transformation has also been observed. After cholesterol loading mouse aortic smooth muscle cells can acquire macrophage like functionality and phenotype. Interestingly, SMC in human atherosclerotic plaques express fractalkine receptor CX3CR1 and undergo chemotaxis to its ligand, CX3CL1. CX3CR1 is a marker expressed by monocytes, macrophages, and dendritic cells and their progenitors, and the expression of CX3CR1 and CX3CL1 appears to be important in the pathogenesis in atherosclerosis, diabetes mellitus, and transplant vascular disease.

10.8 Applications of Stem Cells in Developing Tissue-Engineered Products

Stem cells can also be used for tissue engineering process. Stem cells can be trans-differentiated into specific tissues or organs of interest for therapeutic applications as tissue grafts or organ transplants (Fig. 10.5). For example, blood vessels (arteries and veins) have been developed for use as vascular grafts. Likewise, skin grafts are developed for treating wounds or use in cosmetic surgery. IPS cells have particularly been useful in developing tissue-engineered products for therapeutic applications. More recently the tissues or organs developed using stem cells have also been evaluated for their utility in safety/efficacy assessment of drugs or chemicals (elaborated below). Stem cells from mesenchyme origin have the potential to be differentiated into osteogenic cells leading to generation of artificial bone tissue, which can be used to repair several types of orthopedic disorders. Stem cells may also be cultured for the production of extracellular matrix, which has many applications in cosmetic and regenerative surgery. Stem cells have also been used to develop bioengineered scaffolds, which can be used to design cardiac valves with superior performance. Stem cell-based production of retinal-pigmented epithelium has been used to restore the lost photoreceptors and improve vision in experimental model of macular degeneration. A few phase 1 clinical trials have also validated the role of stem cell-derived retinal-pigmented epithelium in restoring vision. Additionally fully functional cornea and lens are also developed by tissue engineering of both autologous and allogeneic stem cells. Tissue engineering of skeletal muscles is yet another area where stem cells have promising application in treating muscle wasting disorders such as Duchenne's muscular dystrophy. Multinucleated myofibers derived from adult muscle stem cells when juxtaposed to damaged-muscle have shown to trigger self-renewal and facilitate muscle repair. Stem cell-based tissue engineering can also lead to developing artificial organs for transplant. This approach is extensively investigated to develop functional heart, pancreas, and kidneys. Although relatively few clinical trials have attempted stem cell-based tissue replacement therapy, nevertheless it remains an aspiration specifically for therapeutic applications in cardiac, renal, and hepatic structural disorders for which currently there is very limited alternative.

Stem cell can also be engineered for drug/gene delivery applications (Fig. 10.5). Stem cells with proven safety profile are ideal candidates for therapeutic delivery of drugs/gene. Safety and target-specific delivery remain the major advantages of stem cell-based delivery over to other delivery systems. Another application of stem cells in relation to tissue engineering is to use the stem cells to coat the surgical implants to enhance their biological integration. For example, coating of the cardiac implants with endothelial cell layer is reported to potentially improve the efficacy of integration of these implants and hence enhance their safety profile.

Various vascular grafts can be developed from smooth muscle progenitor cells. Smooth muscle progenitor cell are cells with potential to differentiate into a functional smooth muscle cells (SMC). CD14/CD105 double positive populations from human peripheral blood mononuclear cells (PBMC) are reported to be precursor of SMC, and indeed atherosclerotic patients contain significantly higher circulating levels of these cells. These precursors, when maintained under specific growth factor-supplemented media, appear spindle-shaped and expressed α-smooth muscle actin in addition to a $CD34^{-}$, $CD45^{+}$, $CD14^{+}$, and $CD105^{+}$ profile. Human mononuclear cells isolated from buffy coat seeded on collagen type 1 matrix yield outgrowth cells in endothelial growth medium (EGM-2) or EGM-2 and platelet-derived growth factor BB. Selection in platelet-derived growth factor BB-enriched medium caused rapid outgrowth and expansion of smooth muscle outgrowth cells (SOC) to >40 population doublings in a 4-month period. These SOCs were positive for smooth muscle cell-specific alpha smooth muscle actin (alphaSMA), myosin heavy chain, and calponin on immunofluorescence and Western blotting and were also positive for CD34, Flt1, and Flk1 receptor but negative for Tie-2 receptor expression, suggesting a potential bone marrow angioblastic origin. In contrast, endothelial outgrowth cells (EOCs) grown in EGM-2 alone and the initial MNC population were negative for these smooth muscle-specific markers. Integrin alpha5beta1 expression by FACS and Western blotting was significantly increased in SOCs compared with EOCs, and this was confirmed by eightfold greater adhesion of SOC to fibronectin, an effect that could be decreased using an alpha5beta1 antibody. Finally, SOC showed a significantly greater in vitro proliferative potential compared with EOCs of similar passage. The SOC hierarchy was determined based on in vitro clonogenic and proliferative potential. Myeloid lineage of smooth muscle-like cells in vasculogenic regions in vivo was also demonstrated in diseased vessel of human subjects who had undergone gender-mismatched cardiac transplantation. Primary high proliferative potential smooth muscle outgrowth cells (HPP-SOC) with myeloid phenotype (CD68 and CD14 positivity) expanded in culture from human peripheral blood mononuclear cells (PBMC) and recipient-derived chimeric smooth muscle cells and participate in vasculogenesis. However, HPP-SOC in vitro are distinct in being negative for several myeloid markers such as CD11b, CD13, and CD33, and CD45 surface antigens and chimeric SMC in vivo show no evidence of cell fusion propensity suggesting a possible myeloid subpopulation origin for smooth muscle outgrowth cells in blood and vasculogenic smooth muscle-like cells in the intima and adventitial microvasculature of diseased arteries. Using human gender-mismatched bone marrow transplant subjects it was shown that donor-derived SMC were observed in plaque, hence validating the hypothesis of BM-derived progenitors as a source of atherosclerotic-associated SMC, although this data by no means excludes the possibility of blood or other organs as sources of SMC. Given contradictory observations among the investigators, it is more probable that SMC in atherosclerotic lesions can have diverse origins under different pathophysiological conditions. Nevertheless the association of SMC progenitors with atherosclerotic disease envisages exploring their role as either potential therapeutic target or diagnostic marker of atherosclerosis progression. A characterization-based study suggested that a CD14/CD105 double positive population from human peripheral blood mononuclear cells (PBMC) as a precursor of SMC, and indeed atherosclerotic patients contain significantly higher circulating levels of these cells. These precursors, when maintained under specific growth factor-supplemented media, appear spindle-shaped and expressed α-smooth muscle actin in addition to a $CD34^-$, $CD45^+$, $CD14^+$, and $CD105^+$ profile. It is likely that apparent differences in cell immunophenotype in this field are as much due to dissimilar culture conditions, as much as lineage variation between precursor cells and heterogeneity within intimal SMC populations.

10.9 Applications of Stem Cells in Developing Screening Tools for Investigational Therapeutics

The potential application of stem cells in drug discovery includes compound profiling, screening, target identification and validating, safety evaluation, toxicity testing, and disease modeling (Fig. 10.5). It is envisaged that stem cells will offer improved model of human disease and drug adverse reactions than that seen in currently used animal models. Specifically tissue/organ-specific drug safety evaluation will be much refined in stem cell-based models that can be developed either in vitro, ex vivo, or in vivo. This shoudl be specifically considered as less than 12% of the drugs moving the clinical phase of development ever get approved, with an estimated budget of Euro 2.5 billion per every new drug approved. One of the reasons for this low rate of efficiency is the unreliable nature of preclinical animal model in drug evaluation. Hence a move towards humanized preclinical evaluation is considered to be useful in improving the rate of success in the clinical evaluation phase. Stem cell-derived tissues, organoids, or humanized animal models have the potential to offer the appropriate screening tools for drug discovery process. Recent advances in induced pluripotent stem cell and microfluidics technologies have considerably advanced the organ-on-a-chip systems, which have enormous applications in drug discovery strategies in future. Three-dimensional (3D) bioprinted model of intramembranous ossification using bone-cartilage interaction and metaplastic transformation of cartilage into bone was recently developed to be used as functional tissue systems for studying and evaluating novel therapeutics for osteochondral defect repair. Likewise, using human embryonic-like endothelial cells (ECs) derived from human pluripotent stem cells differentiated under arterial flow conditions, a screening platform was developed for high-throughput screening of novel therapeutics. This platform can potentially be used as a replacement to animal models for high-throughput testing of thc effect of substances on vascular development. Recently retina on a chip, a novel microphysiological model of the human retina integrating more than seven different essential retinal cell types derived from hiPSCs, has been developed for drug testing. This model has the potential to promote drug development and provide new insights into the underlying pathology of retinal diseases. Our understanding of IPS stem cells has opened a new avenue for human relevant disease modeling and safety assessment. Often the preclinical data derived from animal models have failed in clinical transition stage, hence triggering the need for developing humanized preclinical models. For example, human pancreatic beta-like cells and neuronal progenitor cells derived from IPS cells are being used for evaluating novel therapeutics for type II diabetes and fragile X syndrome, respectively. The feasibility to adopt these humanized preclinical models into high content or high-throughput screening systems is particularly making them reliable and suitable for modern era of technology-driven drug discovery process fortified with artificial intelligence. These test platforms being patient specific adds an additional advantage to the validity of the test results and supports an element of personalized medicine. The above are only few selected examples of screening platforms which can be developed using advances in stem cell research.

Cardiotoxicity, hepatotoxicity, and neurotoxicity together account for over 80–90% of drug failures in the phase 1 clinical trial. Recent development in 3D printing technology has facilitated the use of stem cells in 3D printers, with which it is feasible to print various tissues or organs. Using the 3D printers, a system consisting of the human IPS-derived cardiomyocytes, hepatocytes, and neurons can be printed onto a microfluidics device and used for high content analysis of the novel compounds for cardiotoxicity, hepatotoxicity, and neurotoxicity screening. Although technically challenging it is feasible to create all the organ systems in a single microfluidics platform for efficacy and safety assessment. Such a platform will be valuable in reducing, refining, and potentially replacing the use of animals in preclinical drug discovery and development.

10.10 Preclinical Safety Evaluation of Stem Cells

A thorough preclinical assessment of products safety is essential for translation of any product for clinical applications. Use of stem cells is no exception to this rule and the regulatory guidelines relevant to cell and gene therapy (CGT) products will be applicable here. The objective of the preclinical safety evaluation of CGT products is to provide a proof-of-principle of pharmacological and safety effects predictive of human response not only prior to clinical trials but also throughout the clinical development. Safety assessment of CGT products should be performed under GLP conditions. Ideally preclinical models replicating the specific disease may be more appropriate over to healthy animal models to assess the efficacy and safety of CGT products. Moreover preclinical models replicating the specific disease are better at profiling the risk benefit analysis of investigational CGT products. Another advantage of using the disease-specific preclinical models is the potential identification of activity risk biomarkers, which will prove to be useful in monitoring progress in clinical trails. The nature of the safety studies will broadly depend on:

1. Nature of the cell product (autologous vs. allogeneic)
2. The proliferation/differentiation ability of the stem cells
3. Immunogenic nature of stem cells
4. Degree of cell manipulation (genetic or non-genetic)
5. Intended route of cell administration
6. Duration of exposure or pharmacokinetics of stem cells
7. Formulation of the stem cell product
8. Purity of the stem cells

The pharmacokinetics evaluation of the stem cells is specifically of interest as this helps to identify target organs where the stem cells are likely to lodge post systemic administration and hence rationalize the histopathological evaluation of the target organ/s. One of the most important considerations in preclinical safety evaluation of the stem cells is to account for their inherent physical, chemical, biological, and mechanical variability and its potential impact on human response. A simple approach in safety evaluation of stem cells may be to perform single dose intravenous response with increasing number of stem cells until an apparent toxicity is observed. The animals (usually SCID mice) used for the preclinical safety assessment should be monitored for 14 days post intravenous administration of stem cells and must be sacrificed on day 15 for a through hematological and histopathological examination. Based on this study, a No Observable Adverse Effect Level (NOAEL) for stem cells should be established together with assessing the tissue distribution of the stem cells. Following this, a repeat dose toxicity study lasting for 28 days or 13 weeks (depending on the pharmacokinetics of the stem cells) should be conducted with evaluation of all toxicity parameters (including tumorogenicity) and complete histopathological profiling. The potential safety concerns with the use of stem cells include:

1. Reactions at the site of administration
2. Growth of tumors
3. Differentiating into inappropriate cell types and its functional consequences
4. Anaphylactic reactions
5. Platelet activation and formation of thrombus (especially in microvasculature)
6. Embolization of microvasculature

Due to the extensive diversity and variations in the nature of the stem cells and the very many uncontrollable confounding factors, which can influence the phenotype and genotype of stem cells, it is always a challenge to reliably establish the acute and chronic safety profile of stem cells.

10.11 Pharmacological Modulation of the Stem Cells

Stem cells are themselves not immune to structural and functional defects. This has led to the evolution of stem cell-based diseases as briefly explained before. While the correction of the structural defects in stem cells will only be possible using other stem cells, in contrast the functional defects in stem cells can be corrected using established pharmacological agents. This has led to advent of a new field of pharmacology, which specifically aim at restoring, modulating, or enhancing the function of stem cells. In the recent years several studies have reported the positive influence of currently used pharmacological agents on the beneficial effects of marrow-derived progenitor cell therapy in animal models and clinical cases of vascular remodeling and hence achieving optimal vascular repair. For instance, G-CSF-based PB-MNC transplantation was reported as a feasible treatment for the ischemic hind limb, which was superior to using PB-MNC alone without G-CSF. Similarly PPAR gamma agonist such as pioglitazone was reported to increase the number and function of endothelial progenitor cells (EPC) in patients with coronary artery disease. Exploring the effects of currently used drugs on various types of stem cells is perhaps necessary to further augment the therapeutic utility of these drugs. 3-Hydroxy-3-methylglutaryl coenzyme A (HMG-CoA) reductase inhibitors (statins) are widely used to decrease cholesterol synthesis and are well established to reduce cardiovascular diseases. Recently, it has been reported that statins mobilize endothelial progenitor cells ($CD34^+/CD144^+$ phenotype) from bone marrow during the first 4 weeks, which could help to prevent cardiovascular diseases. Statins are also reported to influence the differentiation from mononuclear cells to EPC, with an inhibitory effect on differentiation of smooth muscle progenitor cells. Such selective effects of statin to promote the formation of EPC will be very valuable in the context of vascular pathologies associated with endothelial dysfunction (myocardial infarction, stroke, hypertension, diabetes vasculopathy, etc.), hence further meriting the use of statin for preventing cardiovascular diseases. Human apoA-I transfer is reported to increase the number of circulating EPC, enhance their incorporation into allografts, promote endothelial regeneration, and attenuate formation of neointima, and hence targeting apoA-I may prove to be valuable in treating arteriosclerosis. The senescence of EPC is reported to increase in spontaneously hypertensive rats, DOCA-salt hypertensive rats, and hypertensive patients. Further EPC had low telomerase activities and their number negatively correlated with the organ damage in the hypertensive patients. The correlation of EPC phenotype and genotype with disease progression has several implications, including the feasibility to use EPC as a prognostic marker of disease and secondly to use established drugs such as statins to rectify the defective EPC and thus prevent the potential organ damage induced by defective EPC. Vascular endogenous erythropoietin receptor system is also reported to play an important role in angiogenesis in response to hind limb ischemia through upregulation of VEGF/VEGF receptor system, both directly by enhancing neovascularization and indirectly by recruiting EPC and BM-derived pro-angiogenic cells. Exogenous erythropoietin (Epo) treatment was shown to inhibit neointimal hyperplasia after arterial injury in an NO-dependent (Akt/eNOS phosphorylation and NO synthesis) manner and by mobilizing EPC to the neo-endothelium. Estrogen was also reported to contribute to therapeutic re-endothelialization by promoting EPC migration and proliferation via estrogen receptors and PI3K pathway which provides a novel insight and treatment strategy for vascular repair. Physiologically postnatal vasculogenesis is reported to involve estrogen-regulated bioactivity of BM-derived EPC, predominantly through the ER alpha eNOS dependent effect. Although most of the studies so far have looked into the effects of the pharmacological agents on EPC mobilization and recruitment, it would be interesting to study the effects of various pharmacological agents on all categories of stem cells (i.e., 1. embryonic stem cells (ESC) 2. adult stem cells (ASC), and 3. induced pluripotent stem cells (IPS)) including evaluating the effects of pharmacological agents on phenotypic and genotypic differentiation of the stem cells.

10.12 Future Directions for Stem Cells in Drug Discovery and Development

Although the discovery and development of stem cells continues to see its rise and fall, the technology per se has lot of potential to offer to the therapeutics discovery and development process. Although the current focus on using stem cells is largely towards developing human-specific/personalized valid and reliable screening platforms for investigational therapeutics, the momentum in directly using the stem cells for therapeutic application is slowly gaining. Also as we refine our understanding of the biology of stem cells, new domains continue to emerge. For instance, discovery and development of investigational therapeutics, which can activate and mobilize endogenous stem cells to achieve therapeutic benefit, has resulted in establishing a new therapeutic class of drugs. The prevalence of endogenous stem cells is now a well-established fact, and as we continue to understand their physiological significance, it is not a surprise to know that some of the diseases may be a consequence to defects in these endogenous stem cell populations. This has given rise to a new category of diseases, which can be classified as stem cell disorders. The stem cell disorders is a paradigm shift from our initial understanding of using stem cells for therapeutics to rather directly targeting them using novel therapeutics. The understanding of stem cell disorders can also pave the way for developing diseased stem cells as a biomarker to assess disease severity and/or prognosis of the disease following therapeutic intervention. The possibility of dysfunctional stem cell under any disease pathology is also a major setback for autologous stem cell therapy, which is perhaps the reason for more research into advancing allogeneic stem cell therapy.

Despite the current advances in the cell culture techniques, maintaining reliable quality of stem cells in adequate numbers remains a major challenge in therapeutic development of stem cells. Hence future research to further advance cell culture techniques to effectively meet these challenges are required. Some of these challenges may also be in replicating the whole human body using stem cells as currently it is limited to single tissue/organ type. Although it is less likely that stem cell platforms will fully replace the utility of animal model in drug discovery process, nevertheless we continue to observe the refinement stem cell platforms have to offer. One area where stem cells have a better edge over animal models is their potential to offer patient-specific personalized screening tools (patient-specific stem cell-derived organoids/tissues) for drug efficacy and safety assessment process. Also the patient-specific stem cells can be used to grow intracellular constituents such as mitochondria for therapeutic application (Mitochondrial replacement therapy), which may not be feasible by animal models. Without doubt the research on stem cells is deemed to progress and its potential application in therapeutics should be continued to be explored.

Suggested Readings

1. Clover AJP, Kumar AHS, Caplice NM. Monocyte recruitment via the CX3CR1 receptor influences burn wound healing. Br J Surg. 2011;98(52):56.
2. Clover AJP, Lane O'Neill B, Kumar AHS. Analysis of public attitudes towards the use of autologous and allogenic cells in tissue engineering products for burns. Br J Surg. 2011;98 (52):56.
3. Clover APJ, Kumar AHS, Caplice NM. Deficiency of CX3CR1 delays burn wound healing and is associated with reduced myeloid cell recruitment and decreased sub-dermal angiogenesis. Burns. 2011;37(8):1386–93.
4. Achberger K, et al. Merging organoid and organ-on-a-chip technology to generate complex multi-layer tissue models in a human retina-on-a-Chip platform. elife. 2019;8:pii: e46188. https://doi.org/10.7554/eLife.46188.
5. Aijaz A, Vaninov N, Allen A, Barcia RN, Parekkadan B. Convergence of cell pharmacology and drug delivery. Stem Cells Transl Med. 2019;8(9):874–9.
6. Alonzo M, Anil Kumar S, Roman B, Tasnim N, Joddar B. 3D bioprinting of cardiac tissue and cardiac stem cell therapy. Transl Res. 2019;211:64–83.
7. Kumar AHS, et al. Stent-based vascular cell delivery for therapeutic angiogenesis. Biomaterials. 2014;35(32):9012–22.
8. Kumar AHS, et al. Role of CX3CR1 receptor in monocyte/macrophage driven neovascularization. PLoS One. 2013;8(2):e57230. https://doi.org/10.1371/journal.pone.0057230.
9. Kumar AHS. Precision editing the cells and engineering or re-engineering life, the era of geno-pharmacology/therapeutics has begun!! J Nat Sci Biol Med. 2016;7(1):1–3.
10. Borlongan CV. Concise review: stem cell therapy for stroke patients: are we there yet? Stem Cells Transl Med. 2019;8(9):983–8. https://doi.org/10.1002/sctm.19-0076.

11. Breathwaite E, et al. Scaffold-free bioprinted osteogenic and chondrogenic systems to model osteochondral physiology. Biomed Mater. 2019;14:065010. https://doi.org/10.1088/1748-605X/ab4243.
12. Doyle B, et al. Progenitor cell therapy in a porcine acute myocardial infraction model induces cardiac hypertrophy, mediated by paracrine secretion of cardiotrophic factors including TGFß1. Stem Cell Dev. 2008;17(5):941–51.
13. Hynes B, Kumar AHS, et al. Potent EPC conditioned media related anti-apoptotic, cardiotrophic and proangiogenic effects post myocardial infarction are mediated by IGF-1. Eur Heart J. 2013;34(10):782–9.
14. Chabrat A, et al. Pharmacological transdifferentiation of human nasal olfactory stem cells into dopaminergic neurons. Stem Cells Int. 2019;2019:2945435. https://doi.org/10.1155/2019/2945435.
15. Mummery C, et al. Stem cells, scientific facts and fiction. 2nd ed. London: Academic Press; 2014.
16. Clover AJ, Kumar AHS, et al. Allogeneic mesenchymal stem cells, but not culture modified monocytes, improve burn wound healing. Burns. 2015;41(3):548–57.
17. Eskes C, et al. Good cell culture practices & in vitro toxicology. Toxicol In Vitro. 2017;45 (3):272–7. https://doi.org/10.1016/j.tiv.2017.04.022.
18. Gleeson BM, et al. Bone marrow-derived mesenchymal stem cells have innate procoagulant activity and cause microvascular obstruction following intracoronary delivery: amelioration by anti-thrombin therapy. Stem Cells. 2015;33(9):2726–37. https://doi.org/10.1002/stem.2050.
19. Blau HM, Daley GQ. Stem cells in the treatment of disease. N Engl J Med. 2019;380:1748–60.
20. Huang CL, et al. Synthetic chemically modified mRNA-based delivery of cytoprotective factor promotes early cardiomyocyte survival post-acute myocardial infarction. Mol Pharm. 2015;12 (3):991–6.
21. Alhaider IA, et al. Date palm (Phoenix dactylifera) fruits as a potential cardioprotective agent: the role of circulating progenitor cells. Front Pharmacol. 2017;8:592. https://doi.org/10.3389/fphar.2017.00592.
22. Schmeckpeper J, et al. Lentiviral tracking of vascular differentiation in bone marrow progenitor cells. Differentiation. 2009;78(2–3):169–76.
23. O'Sullivan JF, et al. Potent long-term cardioprotective effects 1 of single low dose insulin-like growth factor-1 (LD-IGF-1) treatment post myocardial infarction. Circ Cardiovasc Interv. 2011;4(4):327–35.
24. O'Sullivan JF, et al. Multidetector computed tomography accurately defines infarct size, but not microvascular obstruction after myocardial infarction. J Am Coll Cardiol. 2012;61(2):208–10.
25. Slack JMW. The science of stem cells. Hoboken, NJ: Wiley; 2017.
26. Martin K, et al. Differential endothelial coverage, response to injury and neointimal integration of CX3CR1/smooth muscle-like cells after carotid or femoral arterial injury. J Vas Res. 2013;50 (3):200–9.
27. Khot A, et al. Measurement and quantitative characterization of whole-body pharmacokinetics of exogenously administered T cells in mice. J Pharmacol Exp Ther. 2019;368(3):503–13. https://doi.org/10.1124/jpet.118.252858.
28. Lewandowski J, Kurpisz M. Techniques of human embryonic stem cell and induced pluripotent stem cell derivation. Arch Immunol Ther Exp. 2016;64(5):349–70.
29. Műzes G, Sipos F. Issues and opportunities of stem cell therapy in autoimmune diseases. World J Stem Cells. 2019;11(4):212–21.
30. Rao KS. Basics of stem cells and preclinical testing. Biol Eng Med Sci Rep. 2017;3(1):17–20.
31. Robert L, Anthony A. Handbook of stem cells. 2nd ed. London: Academic Press; 2013.
32. Rosa S, et al. A high-throughput screening method to identify compounds displaying human vascular embryonic toxicity. Curr Protoc Stem Cell Biol. 2019;50(1):e93. https://doi.org/10.1002/cpsc.93.
33. Rowe RG, Daley GQ. Induced pluripotent stem cells in disease modelling and drug discovery. Nat Rev Genet. 2019;20(7):377–88. https://doi.org/10.1038/s41576-019-0100-z.

34. Turner EC, et al. A novel selectable ISL-1 positive progenitor cell reprogrammed to expandable and functional smooth muscle cells. Stem Cells. 2016;34(5):1354–68.
35. Wang J, et al. Engineered skeletal muscles for disease modeling and drug discovery. Biomaterials. 2019;221:119416. https://doi.org/10.1016/j.biomaterials.2019.119416.

Chapter 11
Pharmacokinetics: Theory and Application in Drug Discovery and Development

Pradeep Sharma, Nikunjkumar Patel, Bhagwat Prasad, and Manthena V. S. Varma

11.1 Introduction

Pharmacokinetics (PK) is the science that describes the time course of drug concentration in the body resulting from administration of a particular drug dose. PK in its simplest form is how the body processes the drug after administration, i.e. absorption, distribution, metabolism and excretion (ADME) of drug. Understanding of these processes play a very significant role in selection and development of new chemical entity (NCE) as prospective drug.

The two most common routes of drug administration are intravenous (i.v.) injection (placing drug directly into blood/systemic circulation by injecting into veins) and extravascular peroral route (oral ingestion of drug into gastrointestinal tract and subsequent entry into systemic circulation). As shown in Fig. 11.1, after i.v. injection or oral administration, PK of drug is composite of various sequential and parallel processes—namely absorption (intake of drug into systemic circulation from port of entry or route of drug administration), distribution (general transport

P. Sharma (✉)
Clinical Pharmacology and Quantitative Pharmacology (CPQP), Clinical Pharmacology & Safety Sciences, R&D, AstraZeneca, Cambridge, UK
e-mail: Pradeep.Sharma@astrazeneca.com

N. Patel
Modelling and Simulations Group, Simcyp Division, Certara UK Limited, Sheffield, UK
e-mail: nikunjkumar.patel@certara.com

B. Prasad
Department of Pharmaceutical Sciences, Washington State University, Spokane, WA, USA
e-mail: bhagwat.prasad@wsu.edu

M. V. S. Varma
Medicine Design, Pfizer Worldwide Research, Development and Medical, Groton, CT, USA
e-mail: Manthena.V.Varma@pfizer.com

R. Poduri (ed.), *Drug Discovery and Development*,
https://doi.org/10.1007/978-981-15-5534-3_11

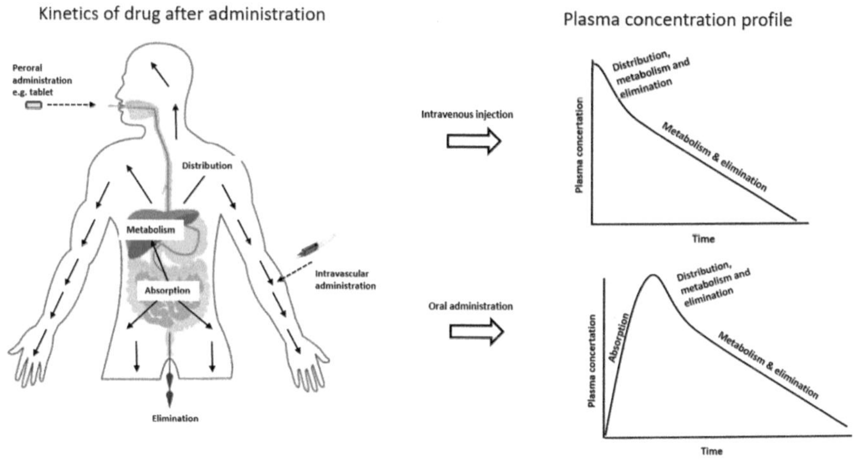

Fig. 11.1 Typical stages/process of drug pharmacokinetics after oral or intravenous administration of drug: absorption, distribution, metabolism and elimination and how they contribute to plasma concentration profile. For purpose simplicity, plasma profile is shown for drugs that show good distribution in peripheral body compartments (biphasic kinetics) and hence distribution phase is more legible in plasma concentration profile

and spreading over of drug into various organ systems of body), metabolism (subjected to breakdown by biocatalysts or enzymes) and finally elimination (exit or excretion from body).

This chapter is focussed on describing key PK concepts, processes (ADME), and how these are quantified/visualised by various PK parameters. Subsequently, it provides in-depth understanding of application of PK principles during different stages of drug discovery and development, with special emphasis on types of studies conducted, interpretation of data and contextualisation to drug projects.

11.2 Pharmacokinetics: Quantitative Aspects

There are two major approaches for quantitative study of various kinetic process of drug PK, model-based approach (pharmacokinetic modelling) and model-independent approach (also called as non-compartment analysis, NCA). In model-based approach, mathematical models are used to describe the changes in drug concentration in the body with time. This is traditional and most commonly used approach for pharmacokinetic characterisation of drugs. There are two most common model-based approaches, namely, compartment modelling and physiologically based PK (PBPK) modelling. All model-based approaches use differential and algebraic mathematical equations to describe spatial and temporal distribution of drug in a body.

In compartment modelling approach, the body is represented as a series of interconnected compartments arranged either in series or parallel to each other. These compartments are not real physiological or anatomical region but hypothetical mathematical compartments. Tissues of similar drug distribution characteristics are lumped together as one compartment. Drug transfer between these compartments is generally described by first-order kinetics. The central compartment includes blood and the highly perfused organs and tissues such as heart, brain, lungs, liver and kidney. In these organs, the administered drug usually equilibrates rapidly. Peripheral compartment(s) include(s) those organs that are less well-perfused such as adipose and skeletal muscle, and therefore the administered drug will equilibrate more slowly in these organs. The number of compartments needed to describe a PK of drug is decided by fitting observed plasma concentration data to different compartment models utilising non-linear regression methods. In contrast to compartment modelling, the PBPK approach considers body as a composite of real physiological organs and kinetics of drug transfer between them is modelled using mathematical expressions. Thus, modelling takes into account various physiological parameters of organs like blood flow, organ size, expression of proteins, etc. This approach is described in greater detail in later part of chapter in Sect. 11.7.3.3.

In non-compartment analysis, drug kinetics or plasma concentration data is analysed by model-independent approaches. This type of analysis relies upon algebraic equations to estimate PK parameters, making the analysis less complex than compartmental methods.

Following sections describe simple one-compartment models after i.v. and oral administration to introduce key PK parameters and terminology. Two compartment models or multi-compartment models are out of scope of this chapter but reviewed elsewhere [1].

11.2.1 Compartment Modelling

11.2.1.1 Intravenous Bolus Administration

The simplest PK data consists of a typical plasma concentration versus time graph of drug after single i.v. dose (bolus) and is shown in Fig. 11.2. This can be described by 'One-Compartment Open Model' (Fig. 11.2), i.e. the body acts like a single, homogenous compartment in which the drug can enter or leave the body easily (i.e. the model is 'open' for the drug movement). In this one-compartment model, all drug administration occurs directly into the central compartment (the site of measurement of drug concentration, usually plasma), and distribution of drug is considered to be instantaneous throughout the volume. This model is described by two parameters, namely, apparent volume of distribution (V_d) and elimination rate constant (k_e).

The volume of distribution (V_d) is defined as a hypothetical volume in which the total amount of drug in the body is required to be dissolved in order to reflect the

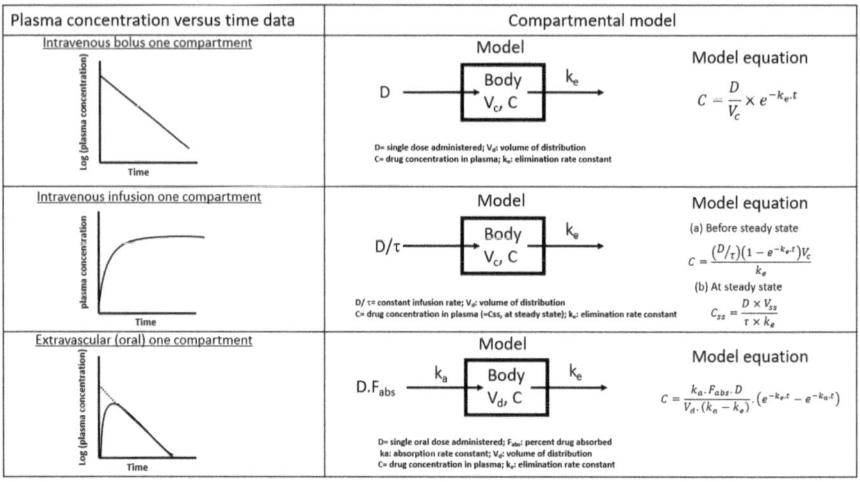

Fig. 11.2 Characteristics of one-compartment model after i.v. or oral administration

drug concentration in plasma. Volume of distribution (V_d) relates the amount of drug in the body to the concentration of drug in the blood or plasma.

$$V_{\mathrm{d}} = \frac{\text{Amount of drug in the body at time } t\ (A_t)}{C_{\text{plasma}} \text{at time } t} \tag{11.1}$$

Soon after an i.v. drug administration, plasma concentration is maximal (C_0), and assuming no drug elimination or distribution, the amount of drug in the body would be equal to the administered dose. Applying the definition of a V_d (Eq. 11.1), the initial volume of distribution (denoted as V_c) is

$$V_{\mathrm{c}} = \frac{\text{Dose}}{C_0} \tag{11.2}$$

The volume of distribution defined in Eq. (11.2) above considers the body as a single homogeneous pool (or compartment) of body fluids. For practical reasons, C_0 is estimated by extrapolation of plasma concentration curve to time zero (Fig. 11.2).

For most drugs, the process of drug elimination is a first-order rate process (monoexponential decline or monophasic), i.e. the process is dependent on the amount or concentration of drug present. So, the rate of change of plasma concentration may be defined as follows:

$$\frac{\mathrm{d}C}{\mathrm{d}t} = k_{\mathrm{e}} \times C \tag{11.3}$$

where dC/dt refers to rate of change of plasma concentration, k_e is elimination rate constant and C is plasma concentration.

The elimination rate constant, k_e, is the second pharmacokinetic parameter in this model, which governs the rate at which the drug concentration in the body declines over time. For most drugs, the process of drug elimination is a first-order rate process, i.e. the process is dependent on the amount or concentration of drug present, and the unit of the elimination rate constant k is time^{-1} (e.g. h^{-1} or 1/h).

Elimination rate constant may be used to compute another key PK parameter called clearance (CL). Drug clearance is defined as the volume of plasma in the vascular compartment cleared of drug per unit time by the processes of metabolism and excretion. Clearance of drug is constant if drug is eliminated by first-order kinetics. Mathematically, clearance is the product of the first-order elimination rate constant (k_e) and the apparent volume of distribution (V_d):

$$\mathrm{CL_{total}} = k_e \times V_d \tag{11.4}$$

where $\mathrm{CL_{total}}$ is the total body clearance. It is sum of all organ clearances ($\mathrm{CL_H}$ = hepatic clearance, $\mathrm{CL_R}$ = renal clearance and $\mathrm{CL_{other}}$ = other less common clearance organs (e.g. biliary excretion)).

$$\mathrm{CL_{total}} = \mathrm{CL_H} + \mathrm{CL_R} + \mathrm{CL_{other}} \tag{11.5}$$

Based on steady-state mass balance considerations, the instantaneous rate of organ elimination is equal to the difference between the rate of drug delivery to the organ in the arterial inflow and its rate of exit in the venous outflow. Therefore, this is equal to the product of the organ's perfusion rate (Q) and the concentration difference between arterial inflow and venous outflow ($C_{\mathrm{art}} - C_{\mathrm{ven}}$). Thus, organ clearance ($\mathrm{CL_{org}}$) may be defined by following equation:

$$\mathrm{CL_{org}} = \frac{Q(C_{\mathrm{art}} - C_{\mathrm{ven}})}{C_{\mathrm{art}}} \tag{11.6}$$

Extraction ratio (E) of organ is essentially the fraction of the drug entering the organ which is removed during transit, and mathematically calculated as the concentration difference between arterial inflow and venous outflow, ($C_{\mathrm{art}} - C_{\mathrm{ven}}$), 'normalised' to the inflow concentration (C_{art}) as follows:

$$E = \frac{(C_{\mathrm{art}} - C_{\mathrm{ven}})}{C_{\mathrm{art}}} \tag{11.7}$$

Thus, Eq. (11.6) can be modified to define organ clearance as equivalent to the product of perfusion rate and extraction ratio:

$$\mathrm{CL_{org}} = \frac{Q(\mathrm{C_{art}} - \mathrm{C_{ven}})}{C_{\mathrm{art}}} = Q \times E \tag{11.8}$$

Two major organ clearances responsible for elimination for most of drugs are hepatic and renal clearances. Their definition, methods of determination and other details are described in Sect. 11.4.3.

Clearance is related to biological half-life ($t_{1/2}$) as follows:

$$t_{½} = \frac{0.693 \times V_{\mathrm{d}}}{\mathrm{CL}} \tag{11.9}$$

Biological half-life (also called as elimination half-life or terminal half-life) of drug is the time necessary to reduce the drug concentration in the plasma or body to one-half or 50%. In simplest interpretation, if a drug had a CL of 4 L/h, this implies that 4 L of the V_{d} is cleared of drug per hour. When the process of absorption is not a limiting factor, half-life is a hybrid parameter controlled by plasma clearance and extent of distribution. In contrast, when the process of absorption is a limiting factor, the terminal half-life reflects rate and extent of absorption and not the elimination process.

11.2.1.2 Intravenous Infusion

Some drugs are administered as an intravenous infusion rather than as an intravenous bolus (Fig. 11.2). Following a continuous infusion, the plasma concentrations will increase with time until the rate of elimination (rate out) equals the rate of infusion (rate in) and will then remain constant. The plateau concentration, i.e. C_{ss}, is the steady-state concentration. Steady state will be achieved in 4–5 times the $t_{1/2}$.

The volume of distribution at steady state (V_{ss}) represents the volume in which a drug would appear to be distributed during steady state as defined by following equation:

$$V_{\mathrm{ss}} = \frac{\text{Amount of drug in body in equilibrium conditions}}{\text{Steady state plasma concentrations } (C_{\mathrm{ss}})} \tag{11.10}$$

V_{ss} is a clearance-independent volume of distribution that is used to calculate the drug amount in the body under equilibrium conditions, i.e. during a drug i.v. infusion and also during multiple drug administration once the steady-state conditions are achieved.

The plasma clearance at steady state after i.v. infusion can be determined as total body clearance:

$$\mathrm{CL}_{\mathrm{total}} = \frac{K_0}{C_{\mathrm{ss}}} \tag{11.11}$$

where K_0 is the infusion rate and C_{ss}, the plasma concentration at steady-state equilibrium.

11.2.1.3 Single Dose Oral Extravascular Administration

The plasma concentration–time profile of a large number of drugs can be described by a one-compartment model with first-order absorption and elimination. Consider the concentration versus time profile following a single oral dose (Fig. 11.2). Assuming first-order absorption and first-order elimination, the rate of change of amount of drug in the body is described by:

$$\frac{dX}{dt} = k_a \times X_a - k_e \times X_t \tag{11.12}$$

where k_a is absorption rate constant; k_e elimination rate constant; X_t is amount of drug in the body at time 't' and X_a amount of drug at the absorption site (X_0 = dose, if all is available). This differential equation may be integrated to get amount X_t in body at time 't' as follows:

$$X_t = \frac{X_0 \times k_a \times \left(e^{k_e t} - e^{k_a t}\right)}{k_a - k_e} \tag{11.13}$$

where X_t can be converted to plasma concentration C_p by dividing it with V_d.

11.2.2 *Non-compartment Analysis*

Non-compartment analysis is model-independent approach where plasma concentration versus time data is analysed by calculating key pharmacokinetic parameters, namely, area under curve (AUC), mean residence time (MRT), C_{max} (peak plasma concentration) and T_{max} (time of peak plasma concentration), V_{ss} and CL. Figure 11.3 shows plasma concentration versus time data after single oral dose and key PK parameters outputted from NCA analysis.

11.2.2.1 Area Under Curve (AUC)

Area under the plasma (or blood) concentration–time curve (AUC) can be calculated from the plasma concentration–time profile after drug administration by equation:

$$\text{AUC} = \int_o^{\infty} C \cdot dt. \tag{11.14}$$

AUC is a primary measure of the extent of drug availability to the systemic circulation (i.e. the total amount of unchanged drug that reaches the systemic circulation following i.v. or extravascular administration). The unit for AUC is concentration per unit time (e.g. nanogram hour per millilitre [ng*h/mL]). AUC is

Fig. 11.3 Key pharmacokinetic parameters determined using non-compartmental analysis of plasma concentration versus time data after single oral dose administration

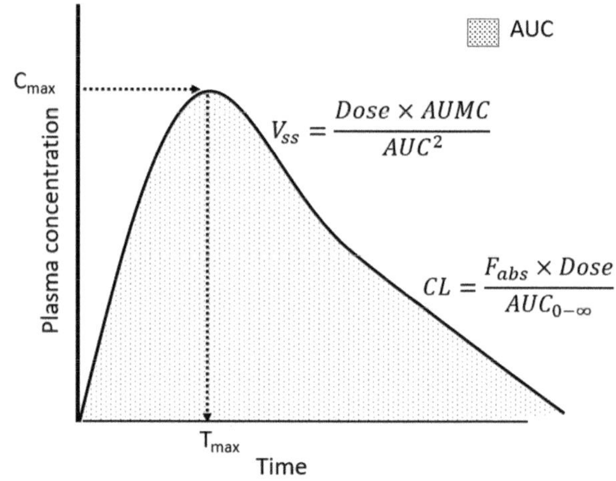

determined using simple integration method as shown in Eq. (11.14) or approximated with a linear trapezoidal method, which is the most widely used approach. The area of each trapezoid is calculated using the following equation:

$$\mathrm{AUC}_{t_1 \to t_2} = \frac{(C_2 + C_1)}{2} \times (t_2 - t_1) \tag{11.15}$$

The extrapolated area from t_{last} to ∞ is estimated as:

$$\mathrm{AUC}_{t_{\text{last}} \to \infty} = C_{\text{last}}/k_{\text{e}} \tag{11.16}$$

where C_{last} is the last observed concentration at t_{last} and k_{e} is the slope obtained from the terminal portion of the curve, representing the terminal elimination rate constant. The total AUC ($\mathrm{AUC}_{0 \to \infty}$) is determined as:

$$\mathrm{AUC}_{0 \to \infty} = \mathrm{AUC}_{0 \to t_{\text{last}}} + \mathrm{AUC}_{t_{\text{last}} \to \infty} \tag{11.17}$$

AUC is used in the calculation of clearance (CL), apparent volume of distribution (V_{d}), and bioavailability and reflects the general extent of exposure over time.

11.2.2.2 Mean Residence Time (MRT)

In NCA, plasma drug concentration is regarded as a random variable. The so-called statistical moments describe the distribution of this random variable. Statistical moments of n-order are defined as follows:

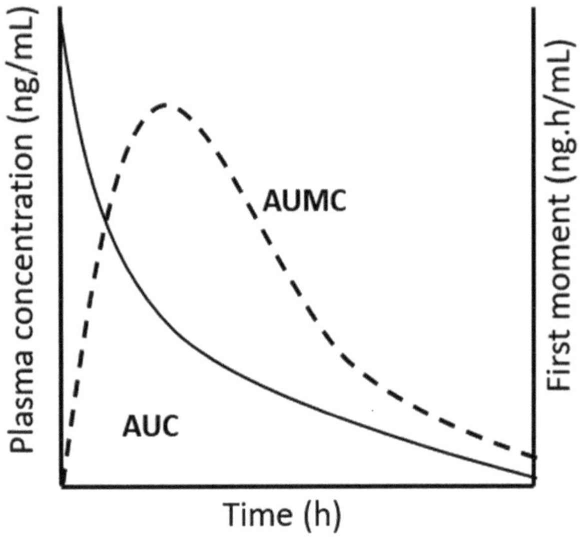

Fig. 11.4 Plasma concentration data versus time after single i.v. bolus dose. AUC is area under curve of plasma concentration versus time curve (continuous curve line) and AUMC is first moment curve (discontinuous curve line) prepared when concentration x time is plotted versus time

$$\int_0^\infty t^n \times C(t) \times \mathrm{d}t \tag{11.18}$$

The area under the plasma concentration ($C(t)$) versus time (t) curve is the area under the zero(-order) moment curve.

$$\mathrm{AUC} = \int_0^\infty t^0 \times Cp \times \mathrm{d}t = \int_0^\infty \mathrm{Cp} \times \mathrm{d}t \tag{11.19}$$

The area under the product of the concentration x time versus time is the area under the first(-order) moment curve denoted as AUMC.

$$\mathrm{AUMC} = \int_0^\infty t \times Cp \times \mathrm{d}t \tag{11.20}$$

The first moment curve is prepared when concentration multiplied with time is plotted versus time. Figure 11.4 shows AUMC and AUC after i.v. bolus administration of drug.

MRT is the average time drug molecules reside in the body before excretion or time when 63.2% of an intravenous dose had been eliminated. MRT is a PK parameter which provides alternative measure of drug elimination and its unit is time (e.g. hour). Following i.v. dosing, MRT is calculated as:

$$\mathrm{MRT} = \frac{\mathrm{AUMC}}{\mathrm{AUC}} = \frac{\int_0^\infty C \cdot t \cdot \mathrm{d}t}{\int_0^\infty C \cdot \mathrm{d}t}, \tag{11.21}$$

where AUMC is the area under the first moment versus time curve from time $t = 0$ to ∞ and calculated using trapezoidal rule similar to AUC.

MRT is used to help interpret the duration of effect for direct-acting molecules (e.g. blood pressure lowering agents), first-in-man dose predictions (Sect. 11.7.3.4) and also used to calculate V_{ss} (Eq. 11.22).

11.2.2.3 C_{max} (Peak Plasma Concentration) and T_{max} (Time of Peak Plasma Concentration)

Peak or maximum plasma concentration (C_{max}) is defined as the maximum observed drug concentration in the plasma concentration–time profile following drug administration. For some drugs, the pharmacological effect is dependent on the C_{max}. For example, aminoglycosides, which are widely used antibiotics, needs to achieve a C_{max} that is at least eight- to tenfold higher than the minimum inhibitory concentration (MIC) to obtain a clinical response ≥90% [2]. Also, the peak plasma concentration is one of the parameters used to determine the comparative bioavailability and/or the bioequivalence between two products (same and or different dosage forms) but containing the same chemical entity or therapeutic agent and may be used to determine the superiority between two different dosage forms or two different routes of administration.

Time of maximum concentration (T_{max}) is the time required to reach C_{max}. The peak time can be used to determine comparative bioavailability and/or bioequivalence, to determine the preferred route of drug administration and the desired dosage form for the patient, to assess the onset of action. Differences in onset and peak time may be observed as a result of administration of the same drug in different dosage forms (tablet, suspension, capsules, etc.) or the administration of the same drug in same dosage forms but different formulations.

Unlike AUC which is robust PK parameter generated by integration, C_{max} and T_{max} are point estimates and their absolute values are susceptible to experimental design and time points of PK sampling. Taken together, C_{max}, AUC and T_{max} provide approximate idea of extent and rate of drug absorption from site of administration.

11.2.2.4 Volume of Distribution (V_{ss}) and Clearance (CL)

NCA of plasma concentration versus time data can be used to determine V_{ss} and CL as follows:

$$V_{ss} = \frac{\text{Dose} \times \text{AUMC}}{\text{AUC}^2} = \text{CL} \times \text{MRT} \tag{11.22}$$

In case of i.v. administration, CL is determined as:

$$\text{CL} = \frac{\text{Dose}}{\text{AUC}_{0-\infty}} \tag{11.23}$$

In case of oral administration, CL_{total} can be estimated by administering a single dose of the drug and calculating the ratio between the available dose and the area under the circulating concentration/time curve (AUC) measured from time zero to infinity as follows:

$$\text{CL} = \frac{\text{F}_{\text{abs}} \times \text{Dose}}{\text{AUC}_{0-\infty}} \tag{11.24}$$

where F_{abs} is absolute bioavailability. Following oral administration, bioavailability of drug must be considered. Bioavailability (BA) is defined as the fraction (percentage) of an administered dose of unchanged drug that reaches the blood stream (systemic circulation). When the systemic availability of an orally administered drug is calculated in comparison to its intravenous (i.v.) administration, it is called as absolute bioavailability (F_{ab}).

$$F_{ab} = \frac{[\text{AUC}_{\text{extravascular}}] \times \text{Dose}_{\text{intravenous}}}{[\text{AUC}_{\text{intravenous}}] \times \text{Dose}_{\text{extravascular}}} \tag{11.25}$$

where AUC is the area under curve of plasma concentration curve after extravascular or intravenous administration of respective Doses ($\text{Dose}_{\text{extravascular}}$) and $\text{Dose}_{\text{intravascular}}$).

Relative bioavailability (F_{rel}) is the systematic availability of the drug from a dosage form as compared to the reference standard given by the same route of administration.

$$F_{rel} = \frac{[\text{AUC}_{\text{extravascular(test)}}] \times \text{Dose}_{\text{intravenous(std)}}}{[\text{AUC}_{\text{intravenous(std)}}] \times \text{Dose}_{\text{extravascular(test)}}} \tag{11.26}$$

where $\text{AUC}_{\text{extravascular(test)}}$ and $\text{AUC}_{\text{extravascular(std)}}$ are the area under curve of plasma concentration curve after test and standard product administration of respective doses ($\text{Dose}_{\text{extravascular(test)}}$ and $\text{Dose}_{\text{extravascular(std)}}$).

F_{ab} is used to characterise drug's inherent absorption properties after extravascular administration whereas F_{rel} is used to characterise absorption of drug from its formulation.

When F_{abs} is unknown, as for most extravascular route of administration, then 'apparent clearance' denoted as 'CL/F' is determined. For example, apparent oral clearance is denoted as follows:

$$CL_{oral} = \frac{CL}{F} = \frac{Dose}{AUC_{0-\infty}} \tag{11.27}$$

11.3 Pharmacokinetics: Qualitative Aspects

As pharmacokinetics constitute various physiological process of absorption, distribution, metabolism and elimination, following sections describe qualitatively each of these processes and how they conjoin in typical plasma concentration–time profile or pharmacokinetics of drug.

11.3.1 Absorption

Drug absorption is process whereby drug enters the systemic circulation from the site of administration, allowing it to be further distributed to other major organs and eventually to its site of action [3]. This is first step in PK for extravascular routes (e.g. peroral and transdermal) of administration. Major factors that influence drug absorption through oral route are:

- Biological factors: membrane permeability, GI environment (volume, pH presence of bile salts, presence of food), transit times, first-pass metabolism, metabolism in the liver, protein binding of drugs.
- Pharmaceutical factors: solubility of the drug, salt form, type of dosage forms, excipients, process of preparation, crystallinity; polymorphism of drug substance; and stereotype and its formation.

In the drug discovery process, some optimal absorption criteria from a biopharmaceutical point of view are shown below [4]:

- High membrane permeability (defined in Sect. 11.4.1) throughout the GI tract.
- Known uptake/influx mechanism for carrier-mediated transport.
- High solubility in aqueous media and over a wide pH range (e.g. pH 1–7).
- No degradation/metabolism in intestinal luminal fluids, intestinal homogenates and/or microsomal preparations from the intestine and liver (i.e. low first-pass metabolism).
- Complete absorption in the GI tract in vivo in several animal species.

Fraction of drug absorbed (f_a) is usually defined as the fraction of the dose that is absorbed from the intestinal lumen into the gut wall. Any possible metabolism of the drug in the enterocytes (gut wall) or in the liver is not included in the absorption process. The fraction absorbed may only be estimated by collecting urine following a radiolabelled dose i.v. and orally.

$$f_a = \frac{C_{po}}{C_{iv}} \times \frac{\mathrm{Dose}_{iv}}{\mathrm{Dose}_{po}} \tag{11.28}$$

where C_{po} and C_{iv} are the total radioactivity (parent compound and metabolites) found in the urine at time infinity after oral and i.v. administration, respectively. It is assumed that the elimination of the compound is the same by two routes of administration, and that the drug is absorbed only as parent compound, not as metabolites.

The fraction absorbed may be approximately calculated from the bioavailability, F_{ab}, and the hepatic extraction (E_H) of the drug, and assuming no extraction within the gut wall.

$$f_a \approx \frac{F_{ab}}{1 - E_H} \tag{11.29}$$

The organ extraction (e.g. hepatic extraction) is estimated from the organ blood clearance (hepatic clearance) divided by the blood flow through that organ. Thus, the hepatic extraction is estimated by comparing the hepatic blood clearance (CL_{BH}) of the drug to the hepatic blood flow (Q_H):

$$E_H = \frac{CL_{BH}}{Q_H} \tag{11.30}$$

If the blood concentration of the drug equals the plasma concentration (or assumed to be equal), the hepatic plasma clearance can be used in the calculation of hepatic extraction. Otherwise the CL_{BH} is calculated from the hepatic plasma clearance (CL_H) and the plasma-to-blood concentration ratio (C_p/C_b) of the drug [5]. Blood-to-plasma ratio of drug is explained in more detail in Sect. 11.4.2.

In general, determinants of oral drug bioavailability include fraction of dose absorbed in the gastrointestinal tract (GIT) and fraction of dose that escapes elimination by the intestinal tract, liver and lung. Thus, oral bioavailability can be defined mathematically by the following equation:

$$F_{ab} = F_a \times F_g \times F_h \tag{11.31}$$

where F_g is the fraction of the dose that escapes pre-systemic intestinal first-pass elimination; and F_h is the fraction of the dose that passes through the liver and escapes pre-systemic liver first-pass elimination. The fraction of the dose that escapes first-pass elimination across the intestine (F_g) and liver (F_h) can be estimated experimentally *via* the comparison of systemic exposures (AUC ratios) where the dosing routes are selected to isolate the contribution by a particular organ.

11.3.2 Distribution

Drug distribution refers to the reversible partitioning of a drug into the various tissues of the body from systemic circulation, driven by blood flow rates or the ability of a drug to cross membrane barrier, leading to defined proportions in the different tissues at steady state, depending on the compound characteristics and composition of the various tissues. Hydrophilic drugs tend to stay in plasma or in the interstitial fluids surrounding the tissue cells, while lipophilic bases have the greatest tendency to cross the membrane barrier and distribute into cells constituting tissues, especially into the fatty tissues. Table 11.1 lists the physiological volumes available for drugs with different physiochemical properties.

Lipid-insoluble drugs are mainly confined to the plasma and interstitial fluid; most do not enter the brain following acute dosing. Lipid soluble drugs reach all compartments and may accumulate in fat-rich tissues. For drugs that accumulate outside the plasma compartment, V_d may exceed the total body volume (Table 11.1). The volume of distribution for digoxin in a healthy volunteer is about 700 L, which is many fold greater than the total body volume (~42 L) of a 70-kg human. Thus, volume of distribution does not represent a real volume but it is an 'apparent' volume that should be considered as the size of the pool of body fluids that would be required if the drug were equally distributed throughout all portions of the body.

Factors involved in drug distribution and diffusion across blood tissue barrier are:

Table 11.1 Physiological volumes in humans (average body weight 70 kg) available for drugs to distribute depending upon their physiochemical properties

Physiological compartment	Distribution volume (L)	Types of drug molecules	Examples
Plasma volume	≤4	Large molecules	Heparin (4 L)
Extracellular volume (plasma and interstitial volume)	4–14	Small, hydrophilic molecules and acids, compounds with high plasma protein binding	Warfarin (8 L), Montelukast (8–11 L)
Total body water (extravascular volume and intracellular volume)	14–42	Lipophilic compounds that diffuse through membranes	Ampicillin (20 L), Ethanol (34–41)
Greater than total body water	>42	Lipophilic bases and drugs that bind strongly to tissues	Fentanyl (280 L), imipramine (1600 L) Chloroquine (12,500 L)

- Membrane permeability and active transport: Distribution into intracellular space is dependent on cell membrane permeability. Cell membranes are made up of phospholipid bilayers that readily allow lipophilic compounds to permeate through them. Due to the negative polar heads of the lipid bilayers, bases have a preferential entry into cell membranes compared to acids and neutrals. For large molecules, the membrane permeability limits the rate of distribution into tissues. Active transport into and out of cells may result in selective accumulation/ distribution of drugs in brain, liver, kidney and other body organs.
- Blood flow: The rate at which a drug is distributed to various organs after a drug dose is administered depends largely on the proportion of cardiac output received by each organ. Drug distribution occurs rapidly into highly perfused tissues with discontinuous capillaries such as liver, spleen and intestine. Tissues with low perfusion rates and with continuous capillaries such as muscle and skin are represented by larger and slowly exchanging extravascular compartments.
- Tissue and plasma protein binding: Various structures are acting as reservoir for storage of drug substance. They are plasma proteins, erythrocytes, and cellular reservoir like muscles, fat tissue, bone, and transcellular compartments. Tissue composition affects the tissue partitioning and thereby the extent of distribution of drugs. Lipophilic drugs distribute into fat-rich organs such as adipose, liver, brain and kidney, while hydrophilic drugs distribute into water-rich organs such as muscle. Drugs bind reversibly to plasma proteins. Only unbound drugs are able to distribute into tissues from systemic circulation. Thus, acids that are extensively and strongly bound to albumin generally have a low fraction unbound in plasma and consequently a low volume of distribution (V_d). Binding to tissue components such as membrane phospholipids, deoxyribonucleic acid and proteins can be reversible or irreversible. Irreversible binding leads to drug accumulation. For example, chloroquine binds to DNA and tends to concentrate in white blood cells and liver cells leading to accumulation.
- pH partitioning: Depending upon drug ionisation potential (pK_a) and pH of body compartment, relative fraction of unionised drug is determined. Since it is unionised drug molecules that can cross biological membranes, pH partitioning play key role in drug distribution process (Fig. 11.5).

11.3.3 Metabolism

Biotransformation or drug metabolism is the enzyme-catalysed conversion of drugs to their metabolites. Metabolism is natural physiological phenomenon in which xenobiotics/drugs are converted to more polar as well as water-soluble products, thus facilitating their excretion by the kidney. Chemical modification of drugs generally results in termination of biological activity through the decreased affinity for receptors. In some instances, metabolism can result in chemically reactive metabolites capable of binding irreversibly to cellular macromolecules such as proteins and nucleic acids leading to toxicological response (e.g. acetaminophen

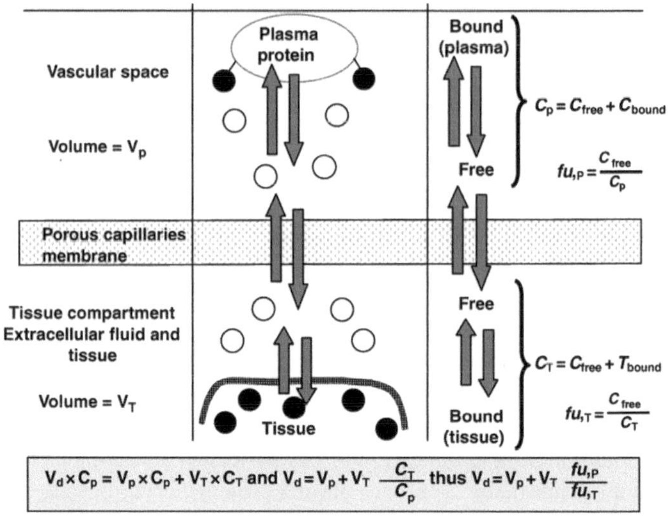

Fig. 11.5 Principles of drug partition between vascular space (plasma) and non-vascular space (tissue), and establishment of a minimal model for volume of distribution. Upper panel: in plasma (volume: V_P), drug is either free (open circle) or bound (black circle) to proteins or red blood cells. The total concentration in plasma (C_P) is the sum of bound (C_{bound}) and free (C_{free}) concentrations. The free fraction in plasma ($f_{u,P}$) is the ratio C_{free}/C_P. Lower panel: only the free drug (open circle) can cross the capillary membrane and gain access to the extracellular fluids outside the plasma. In extravascular space (volume V_T), drug is also either free (open circle) or bound (black circle) to extravascular components; when equilibrium is achieved, the free concentration (C_{free}) in the vascular and extravascular spaces are the same and the free fraction in the extravascular space ($f_{u,T}$) is the ratio C_{free}/C_T; C_T being the total concentration in extracellular space, i.e. $C_{free} + T_{bound}$. (Reprinted from [6] J Vet Pharmacol Ther., 27(6), Toutain PL and Bousquet-Melou A., Volumes of distribution. 441–453, Copyright (2004), with permission from Elsevier)

hepatotoxicity). Liver is the chief organ responsible for biotransformation of most drugs, but drug-metabolising enzymes (DMEs) are found in many other tissues, including the gut, kidneys, brain, lungs and skin.

Most known drugs are substrate of cytochrome P450 (CYP3A4/5) class of enzymes responsible for biotransformation pathways like oxidation, reduction, etc. These are called ‘Phase I’ pathways and often followed up by ‘Phase II’ pathways where metabolite forms adduct with endogenous compounds (e.g. glucuronic acid) to generate larger hydrophilic and less toxic metabolites. Table 11.2 enlists major types of DMEs known to play major role in drug metabolism. In vitro studies involving co-incubation of recombinantly expressed human drug-metabolising enzymes (e.g. rCYP: recombinant CYP) with drug helps to quantitatively assess relative contribution of each type of DME (‘Phenotyping’). In addition, it helps identify structures of likely metabolites in vitro.

Table 11.2 List of major DMEs and clinically relevant transporters routinely screened in drug discovery and development

DMEs/ transporter	Substrate drugs	Inhibitor	Inducer
DME			
CYP1A2	Alosetron, caffeine, duloxetine, melatonin, ramelteon, tacrine, tizanidine, Theophylline, tizanidine	Ciprofloxacin, enoxacin, fluvoxamine	Montelukast, phenytoin
CYP2B6	Bupropion, efavirenz	Clopidogrel, ticlopidine prasugrel	Efavirenz, rifampin
CYP2C8	Repaglinide, Paclitaxel	Fluvoxamine, ketoconazole, trimethoprim	Rifampin
CYP2C9	Celecoxib, Warfarin, phenytoin	Amiodarone, fluconazole, miconazole, oxandrolone	Carbamazepine, rifampin
CYP2C19	Clobazam, lansoprazole, omeprazole, *S*-mephenytoin	Fluconazole, fluvoxamine, ticlopidine	Rifampin
CYP3A	Alfentanil, buspirone, conivaptan, darifenacin, darunavir, dasatinib, eletriptan, eplerenone, felodipine, indinavir, fluticasone, lopinavir, lovastatin	Itraconazole, ketoconazole, posaconazole, ritonavir, saquinavir, telaprevir, telithromycin, voriconazole	Bosentan, efavirenz, etravirine, modafinil, nafcillin
CYP2D6	Atomoxetine, desipramine, dextromethorphan, metoprolol, nebivolol, perphenazine, tolterodine, venlafaxine	Bupropion, fluoxetine, paroxetine, quinidine	None known
Transporters			
P-gp	Aliskiren, ambrisentan, colchicine, dabigatran etexilate, digoxin, everolimus, fexofenadine, imatinib, lapatinib, maraviroc	Amiodarone, azithromycin, captopril, carvedilol, clarithromycin, conivaptan, cyclosporine, diltiazem, dronedarone	Avasimibe, carbamazepine, phenytoin, rifampin, St John's wort, tipranavir/ ritonavir
BCRP	Methotrexate, mitoxantrone, imatinib, irrinotecan, lapatinib, sulfasalazine, topotecan	Cyclosporine, elacridar (GF120918), eltrombopag, gefitinib	Not known
OATP1B1	Atrasentan, atorvastatin, bosentan, ezetimibe, fluvastatin, glyburide, simvastatin acid, pitavastatin, pravastatin, repaglinide, rifampin, valsartan, olmesartan	Atazanavir, cyclosporine, eltrombopag, gemfibrozil, lopinavir, rifampin, ritonavir, saquinavir, tipranavir	Not known

(continued)

Table 11.2 (continued)

DMEs/ transporter	Substrate drugs	Inhibitor	Inducer
OATP1B3	Atorvastatin, pitavastatin, telmisartan, valsartan, olmesartan	Atazanavir, cyclosporine, lopinavir, rifampin, ritonavir, saquinavir	Not known
OCT2	Amantadine, amiloride, cimetidine, dopamine, famotidine, memantine, metformin, pindolol, procainamide, ranitidine, varenicline, oxaliplatin	Cimetidine, quinidine	Not known
OAT1	Adefovir, captopril, furosemide, lamivudine, methotrexate, oseltamivir, tenofovir, zalcitabine, zidovudine	Probenecid	Not known
OAT3	Acyclovir, bumetanide, ciprofloxacin, famotidine, furosemide, methotrexate, zidovudine, oseltamivir acid	Probenecid cimetidine, diclofenac	Not known

11.3.4 Excretion

Excretion of drugs is the final elimination of drug from the body. Table 11.3 tabulates different excretion pathways available to drugs. Drugs and their metabolites may get excreted through biological barriers by specialised types of proteins (called 'transporters') and Table 11.2 enlists few major transporter proteins along with examples of drugs they transport. Further details of transport are provided in subsequent Sects. 11.4.1 and 11.4.3.

11.4 Pharmacokinetic Properties of Drugs

Fundamental drug properties that govern its behaviour in different ADME stages and hence final outlook of plasma concentration profile include membrane permeability, blood-to-plasma ratio, binding to plasma proteins, metabolic stability and carrier-mediated transport. These properties are illustrated below and forms basis for greater understanding in subsequent sections of chapter of how these PK properties play key role in drug discovery and development.

Table 11.3 Major routes of drug excretion in humans

Excreting organ	Excretion medium	Description	Example drugs
Kidney	Urine	Major pathway for parent drug and metabolite through glomerular filtration, secretion, diffusion	Penicillin, ibuprofen, warfarin
Liver	Bile	Major pathway for parent and metabolites through active uptake and/or efflux transport, passive diffusion, pinocytosis	Irinotecan, pitavastatin
Intestine	gastric secretions	Minor pathway by passive diffusion	Ionised organic acids, doxycycline
Salivary glands	Saliva	Minor pathway by passive and active transport	Ethanol, thiamine, propranolol
Lung	exhaled air	Minor pathway by passive diffusion	Camphor, ammonium chloride
Sweat glands	Sweat	Minor pathway by passive diffusion	Opiates, buprenorphine, amphetamines
Mammary gland	Milk	Minor pathway by passive and active transport	Chloramphenicol, tetracyclines, metronidazole

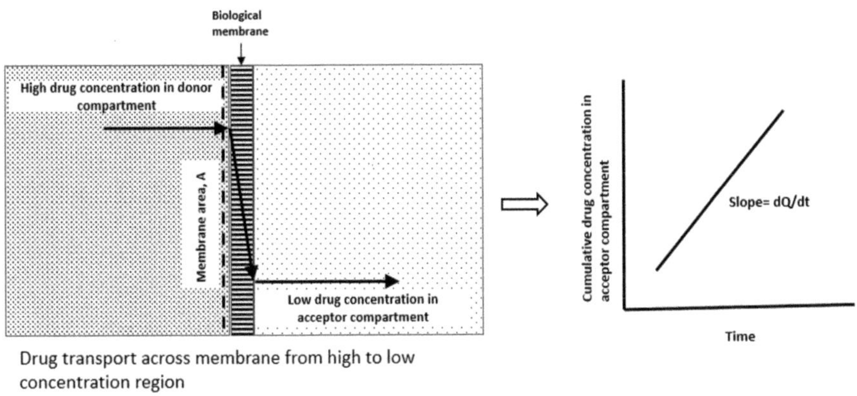

Fig. 11.6 Drug transport across biological membrane or cell monolayers may be used to determine membrane permeability of drugs

11.4.1 Membrane Permeability

The permeability of a compound is the amount of compound that has moved through a membrane in a given time per unit surface area of membrane (Fig. 11.6). It is generally denoted as P_{app} and determined using following equation:

$$P_{\text{app}} = \frac{\mathrm{d}Q/\mathrm{d}t}{C_0 \times A} \tag{11.32}$$

where dQ/dt is the rate of permeation of drug across the membrane, C_0 is the initial concentration at dosing site (called 'donor' compartment) and A is area of membrane available for transport. Membrane permeability of drugs is fundamental intrinsic PK property of drug which controls absorption, distribution and elimination processes of drugs described in preceding sections.

Transcellular permeability is the transport of the compound across the cells through the lipid bilayer membranes, and it is the parameter of most interest at the early screening stage. If compound is transported across biological membranes by carrier protein, it is called carrier/transporter-mediated transport [7]. Since this process mostly involves consumption of energy, it is also termed as active transport. If carrier-mediated proteins transport drugs out of cell, it is called efflux transport, and vice versa is called influx transport. Parallel Artificial Membrane Permeability Assay (PAMPA) utilises artificial membranes designed to 'mimic' biological membranes such as the GI tract and the blood brain barrier (BBB) to measure passive permeability of drugs. Cell lines like Caco-2 express different types of transport proteins (e.g. efflux transporter, P-gp) and are used to characterise carrier-mediated transport. Membrane permeability together with carrier-mediated transport are very important fundamental PK properties of drugs which determine absorption or bioavailability, body bioburden or distribution (specially in physiologically most difficult to access organs like brain) and elimination/clearance pathways. As transporters are differentially expressed in disease states, are genotypically variant, and can be inhibited by other co-dosed drugs, they control drug exposure in clinical settings in patients (pathophysiology) with different ethnic backgrounds (pharmacogenetics) taking multiple drugs (drug–drug interactions) in therapy.

11.4.2 Blood-to-Plasma Ratio and Plasma Protein Binding

The blood-to-plasma ratio (often referred to as $K_{\text{b/p}}$ or B:P ratio) is the ratio of the concentration of drug in whole blood (i.e. red blood cells and plasma) to the concentration of drug in plasma, namely $C_{\text{B}}/C_{\text{P}}$. The blood-to-plasma ratio is determined by following equation:

$$K_{\text{b/p}} = \left(K_{\text{e/p}} \times H\right) + (1 - H) \tag{11.33}$$

where H = haematocrit and $K_{\text{e/p}}$ is the red blood cell to plasma partition coefficient.

The blood-to-plasma ratio which gives indication of drug binding to erythrocytes, in conjunction with other ADME and physicochemical properties, is key determinant for predicting whole body pharmacokinetics. PK parameters are usually determined by analysis of drug concentrations in plasma rather than whole blood.

Therefore, parameters determined using plasma data may be misleading if concentrations of drug differ between plasma and red blood cells as a consequence of differential binding to a specific component in the blood. For example, when $K_{b/p} > 1$, generally due to drug uptake into the erythrocytes, the plasma clearance significantly overestimates blood clearance and could exceed hepatic blood flow.

Plasma protein binding (PPB) normally refers to the reversible binding of a compound with the proteins in the plasma compartment of blood. The amount of unbound compound (f_u) is dependent on its affinity for the various proteins, the binding capacity of those proteins, and the concentration of the compound, which is defined by following equation:

$$\mathrm{fu} = \frac{\text{Free drug in plasma}}{\text{Total drug in plasma}} \tag{11.34}$$

Since it is assumed that it is free drug that is available for distribution in tissues, metabolism and elimination processes in the body, PPB is very important fundamental PK and PD property of drugs. Therefore, an in vitro measure of PPB is used to correct other in vitro parameters, such as volume of distribution and metabolic stability, when making in vivo predictions of PK parameters (e.g. half-life and clearance).

11.4.3 Metabolic Stability, Clearances, Transporters and Induction

Metabolic stability of drugs is resistance to undergo enzymatic conversion (biotransformation) leading to smaller metabolites (catabolism) or adduct (additive reaction). Metabolic stability of drugs determines biological life or duration of availability of drug for pharmacological action after single dosing. Quantitatively metabolic stability is measured as clearance or biological half-life ($t_{1/2}$). In vitro assays measure metabolic stability by incubating the drug with liver components (recombinant enzyme systems, membrane vesicles or microsomes, hepatocytes, liver slices) and then analysing the concentration of substrate drug over a period of time (Fig. 11.7). Drug depletion rate in medium due to metabolism is referred to as intrinsic metabolic clearance ($\mathrm{CL}_{\mathrm{int,vitro}}$) and calculated using following formula:

$$\mathrm{CL}_{\mathrm{int,vitro}} = \frac{0.693 \times V}{t_{½}} \tag{11.35}$$

where $t_{½} = 0.693/k$ and k is elimination rate constant calculated from semi log plot of substrate concentration versus time (Fig. 11.7), as $k = -$ slope and volume V is defined as

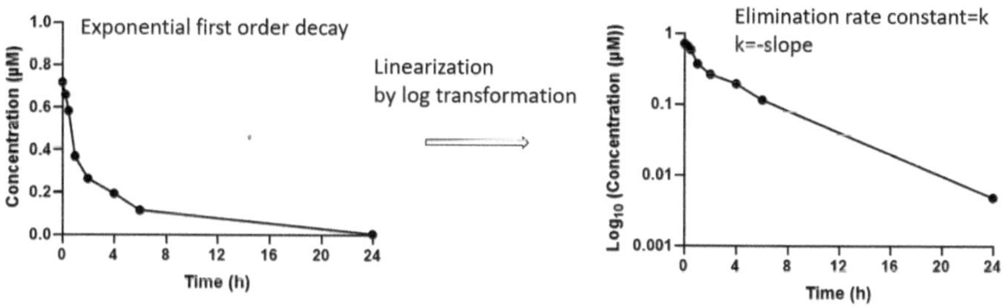

Fig. 11.7 Drug depletion in vitro when incubated with metabolic system (microsomes, recombinant drug-metabolising enzymes, hepatocytes). Drug concentration is shown in linear (**a**) and log scale (**b**), respectively

$$V\frac{\mu\text{L}}{\text{mg of protein or million cells}} = \frac{\text{Volume of incubation medium}}{\text{Protein or cells in incubation medium (mg or million cells)}} \tag{11.36}$$

In case of microsomal incubations, $\text{CL}_{\text{int,vitro}}$ units are expressed as µL/min/mg of microsomal protein, whereas for hepatocytes, $\text{CL}_{\text{int,vitro}}$ units are expressed as µL/min/10^6 cells.

In vitro intrinsic clearance ($\text{CL}_{\text{int,vitro}}$) is scaled to in vivo clearance ($\text{Cl}_{\text{int,vivo}}$; expressed in units mL/min/kg)) using following expression:

$$\begin{aligned}\text{CL}_{\text{int,vivo}} = {} & 10^{-3} \times \text{Cl}_{\text{int,vitro}} \times \text{liver weight} \\ & \times [\text{Cellularity or microsomal protein content}]\end{aligned} \tag{11.37}$$

where liver weight is weight of liver in '*g*'; cellularity is number of cells per gram of liver weight; microsomal protein yield is protein content per gram of liver.

In the last decade, quantitative proteomics has emerged as a promising tool to quantify DMEs and transporters and aid in better translation and in vivo prediction of drug disposition [8].

$$\text{CL}_{\text{invivo}} = \frac{\text{A}_{\text{invivo}}}{\text{A}_{\text{invitro}}} \times \text{CL}_{\text{invitro}} \tag{11.38}$$

where $A_{\text{in vivo}}$ and $A_{\text{in vitro}}$ refer to abundance of DME/transporter protein in vivo and in in vitro assay systems. Also, in simplest form, assuming linear kinetics, i.e. when substrate concentration $\leq 10 \times$ Michaelis-Menten constant (K_{m}), then metabolic or transporter clearance is defined as

$$\mathrm{CL} = \frac{V_{\max}}{K_{\mathrm{m}}} = \frac{[A] \times K_{\mathrm{cat}}}{K_{\mathrm{m}}} \tag{11.39}$$

where V_{max} is maximum velocity of enzyme-catalysed conversion or transporter-mediated transport. Since K_m in the equation indicates affinity of the drug and it is inherent constant, V_{max} is the main variable in the CL equation that determines inter-individual variability primarily because of its dependence on the protein abundance $[A]$. K_{cat} is the turnover number which is defined as the number of substrate molecule each enzyme or transporter site converts to product or transport per unit time. The above model is based on the assumption that protein activity directly correlates with the abundance. In the case of drug metabolism, correlation of activity of DMEs with protein expression is well established [9]. However, whether drug transporter expression correlates with activity depends on the proper localisation of the protein in the plasma membrane.

In vitro intrinsic clearance ($CL_{int,vitro}$) determined using in vitro metabolic clearances using microsomes can be extrapolated to whole body hepatic clearances, CL_H (mL/min/kg). This involves use of well-stirred model [10] according to which liver is assumed as 'well stirred tank', so that drug mixes instantaneously and completely with the organ tissues. Using well-stirred model, CL_H is determined as follows:

$$\mathrm{CL_H} = \frac{Q_{\mathrm{H}} \times \mathrm{CL_H} \times f_{\mathrm{u,blood}}}{Q_{\mathrm{H}} + \mathrm{CL_{int,vivo}} \times f_{\mathrm{u,blood}}} \tag{11.40}$$

where Q_H is liver blood flow in mL/min/kg and $f_{u,blood}$ is blood unbound fraction and expressed as

$$f_{\mathrm{u,blood}} = \frac{f_{\mathrm{u,plasma}}}{\mathrm{B/P}} \tag{11.41}$$

Above expression (Eq. 11.40) can be simplified to

$$\mathrm{CL_{int,vivo}} = \frac{\mathrm{CL_H}}{f_{\mathrm{u,blood}} \times (1 - E)} \tag{11.42}$$

where E (extraction ratio) and $CL_H = E \times Q_H$ as explained in Sect. 11.2.1.1.

Therefore, using in vitro data in conjunction with Eqs. (11.37) and (11.42), it is possible to predict whether drug will be low, medium or high extraction drug. This is very key information to understand if the hepatic clearance of drug is 'flow limited' (i.e. limited by organ blood flow) or 'capacity limited' (i.e. limited by enzyme content).

Intrinsic metabolic clearance ($CL_{int,vitro}$) had multifactorial applications: (1) It can be used to rank the compounds in early discovery with respect their elimination characteristics. Classification bands (Table 11.4) can be used to categorise compounds into low, medium or high clearance drugs and also if give drug clearance will

Table 11.4 Approximate benchmark classification bands that may be used to categorise compounds into low, medium or high clearance in humans based on microsomal clearance values

Metabolic clearance	Extraction ratio (E)	Microsomal CL_{int} (μL/min/mg protein)	Hepatocyte CL_{int} (μL/min/million cells)
Low	0–0.3	≤7	≤ 3
Medium	0.31–0.7	7–36	3–16
High	0.71–1	≥36	≥16

$f_{u,blood}$ assumed to be 1, Q_H: 20.7 mL/kg and liver weight: 25.7 g/kg of body weight [11]; hepatocellularity: 120 million cells/g of liver [12], and 52.5 mg of microsomal protein/g of human liver [10], respectively

be blood flow limited or capacity limited. (2) Interspecies comparison of intrinsic metabolic clearance can be investigated and this may be useful in identifying an appropriate species for preclinical development. (3) Intrinsic metabolic clearance can be scaled to hepatic clearance, which along with any other organ clearance (e.g. renal clearance, CL_R) can add up to provide total body clearance as per Eq. (11.5). The CL_{total}, in conjunction with predicted V_{ss} and k_a, is used in predicting first-in-man dose in late preclinical development stage to achieve target concentrations/profile proposed by PKPD modelling/studies (see details in Sect. 11.7.3).

Polar (low Log$D$7.4 < ~1) compounds are more likely to be eliminated by renal route because of their low PPB, lack of passive permeability (and inability to facilitate reabsorption in the distal tubules of nephron in kidney), and the high expression in the active transporters for anionic/cationic drugs (e.g. OATs and OCTs) in kidney. Renal clearance is determined by glomerular filtration, tubular secretion and reabsorption processes, and can be mathematically described by:

$$CL_R = (f_u \times GFR + CL_{sec}) \times (1 - F_{reabs}) \tag{11.43}$$

where GFR is glomerular filtration rate, CL_{sec} is renal secretory clearance and F_{reabs} is the fraction of filtered and secreted drug that is reabsorbed. Assuming a well-stirred model, CL_{sec} can be expressed as

$$CL_{sec} = Q_R \times \left(\frac{f_u \times CL_{int,sec}}{Q_R + f_u \times CL_{int,sec}} \right) \tag{11.44}$$

where Q_R is the renal blood flow and $CL_{int,sec}$ is the intrinsic secretory clearance that can be described by

$$CL_{int,sec} = \left(\frac{PS_{influx,b} \times PS_{efflux,a}}{PS_{efflux,b} + PS_{efflux,a}} \right) \tag{11.45}$$

$PS_{influx,b}$, $PS_{efflux,b}$, $PS_{influx,a}$ and $PS_{efflux,a}$ are influx and efflux intrinsic transport clearances across the basolateral and apical membranes of proximal tubule cells, respectively [13].

In silico models based on quantitative structure property relationships (QSPRs) can predict human renal clearance [14] and are suitable for use in compound design. Based on a few descriptors (e.g. ion type and $ACDLogD_{6.5}$), a relatively simple algorithm can be used to classify the likelihood of drugs having high (>1 mL/min/kg), medium (>0.1 to <1 mL/min/kg) or low (≤0.1 mL/min/kg) renal clearance.

Since acid and zwitterion compounds are likely to be renally cleared, early determination of shortlisted candidate drugs (CDs) for renal clearance in rats and dogs is recommended. Renal clearance from preclinical species can be used to predict human renal clearance by simple allometric relationship (allometry is described in Sect. 11.7.3.1). If interspecies transability (e.g. rat to dog) is not apparent, then following Eq. (11.46) may be used for prediction of human renal clearance which is based on f_u and renal blood-flow-corrected dog data [15]:

$$\text{Human } CL_R = \text{Dog } CL_R \times \left(\frac{f_{u,\text{human}}}{f_{u,\text{dog}}}\right) \times \left(\frac{Q_{R,\text{human}}}{Q_{R,\text{dog}}}\right) \tag{11.46}$$

where Q_R is renal blood flow and subscript of PK parameters defines the species.

This is based on fact that for acids and zwitterions, human renal clearance is best predicted from the dog [15]. Good prediction of renal clearance of acids and zwitterions is important, because their volumes of distribution tend to be low, and consequently acceptable half-lives are critically dependent upon clearance estimates. In addition, high renal clearance (generally driven by renal transporters) flags likely liability to renal DDI due to potential inhibition of active secretion and/or uptake.

For lipophilic neutral and basic compounds, total clearance is generally expected to be well predicted from in vitro metabolic systems and no specific investigation of renal clearance is recommended. However, urine may be collected during in vivo PK studies and compounds analysed to confirm that renal clearance is low.

Besides quantitative understanding of metabolic clearance, it is important to understand which drug-metabolising enzymes are involved in metabolism of candidate drugs. Information generated by drug metabolism phenotyping is known to have a profound multidimensional impact on drug discovery and development (Fig. 11.8). An interspecies comparison of metabolic profiles in hepatocytes from humans and the animal species is generally helpful in selection of appropriate preclinical animal model for safety studies. This will ensure that there will be exposure to human metabolites in at least one preclinical species in the toxicology studies. Metabolites may not only contribute to the efficacy but may also be responsible for off-target activities (secondary pharmacology). Many drugs are converted by DMEs to chemically unstable species (e.g. reactive metabolites), with a risk of being mutagenic and/or genotoxic.

Knowledge of major metabolic pathway is key information which decides drug liability in clinical drug development to be victim of drug–drug interaction (DDI) when co-dosed with drug (perpetrator) that inhibits this metabolic pathway and hence altering exposure levels in patients. Drugs may inhibit DMEs competitively or non-competitively, time dependently, and induce or supress their expression. In

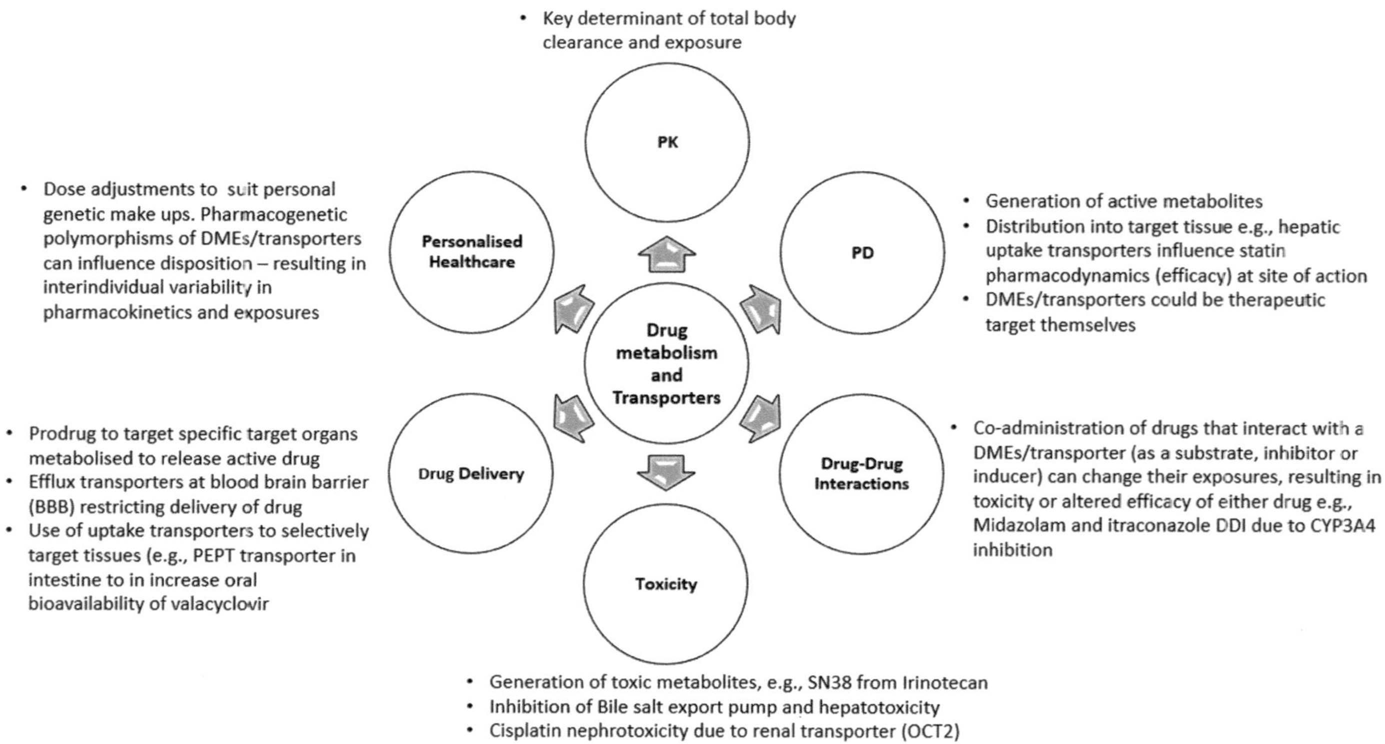

Fig. 11.8 Impact of drug metabolism and transport on drug discovery and development

all these instances, DDIs results due to pharmacokinetic interactions. Table 11.2 enlists examples of drugs which inhibit commonly known DMEs. Various in vitro assays are available which generate in vitro parameters like inhibition constant (IC_{50} or K_i), EC_{50} (50% downregulation of DMEs), K_{inact} (inactivation rate of DMEs by drug), etc. which can be used to predict DDI liability of drug in discovery/preclinical development to understand various potential drug project liabilities. Regulatory guidance from various statutory organisations (United states Food and Drug Administration, US FDA; European Medical agency, EMA) are available which provide detailed information on how to assess DDI liabilities of drugs and this often involve static, semi-mechanistic or dynamic modelling approaches. Since non-CYP metabolic pathways (e.g. Flavin mono-oxygenase or FMOs) are not commonly subjected to inhibition by drugs, they are considered preferred elimination routes for drug projects to minimise clinical PK DDIs. However, scaling methods for in vitro clearance due to drug metabolism by non-CYP pathways are yet not fully validated by in vitro–in vivo correlations (IVIVC) and thus pose a major challenge in terms of predicting metabolic clearances by these routes.

PK of drugs may be influenced considerably if they are substrate of transporters expressed in major ADME organs. Therefore, drugs are screened for their affinity to key clinically relevant transporters (Table 11.2). Interplay of transporters and DMEs may result in complex PK of drug which could be confounding, at least in early stages of drug discovery and development. Recent research focussed on delineating role of transporters and DMEs in drug PK had offered various methods and modelling approaches to provide solution to resolve the PK complexity of drugs eliminated by DMEs and transporters simultaneously. Extended clearance concept [16] expands the hepatic CL_{int} parameter into its component parts to provide a better understanding of the rate-determining mechanisms of clearance as follows:

$$\text{Hepatic } CL_{int} = \frac{\left(PS_{inf,active} + PS_{inf,passive}\right) \times \left(CL_{int,met} + CL_{int,bile}\right)}{\left(PS_{eff,active} + PS_{eff,passive} + CL_{int,met} + CL_{int,bile}\right)} \quad (11.47)$$

where the rate-determining mechanism of clearance is depicted as a complex function of passive clearance rate ($PS_{inf,passive}$), active uptake rate ($PS_{inf,active}$), metabolic intrinsic clearance ($CL_{int,met}$), active sinusoidal clearance ($PS_{eff,active}$), and biliary intrinsic clearance ($CL_{int,bile}$).

As evident from Eq. (11.47), the rate-determining mechanism is not dependent upon the absolute value of any one rate but rather the rates of individual mechanisms in relation to one another. Extended clearance model is essential meta-analysis where clearance parameters are determined experimentally through different assays:

1. $PS_{inf,active}$ and $PS_{inf,passsive}$ in suspension or plated human hepatocytes with and without uptake transporter inhibitors or at 4 °C
2. $CL_{int,met}$ in human liver microsomal or hepatocyte incubations
3. $CL_{int,bile}$ in human sandwich-cultured hepatocyte incubations

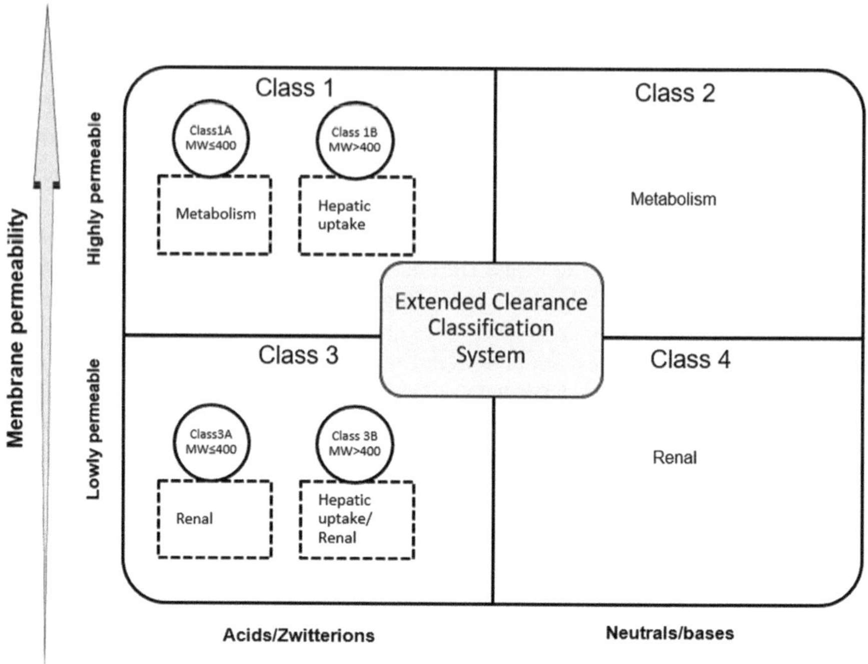

Fig. 11.9 Extended Clearance Classification System (ECCS) as a scientific framework to predict the predominant clearance mechanism (rate-determining process) based on physicochemical properties and passive membrane permeability

Extended Clearance Classification System (ECCS) [13] is such framework which attempts to identify rate-determining clearance mechanism of a drug candidate based on physicochemical properties and passive membrane permeability of drugs (Fig. 11.9). Compounds are classified as: Class 1A—uptake transport-enzyme interplay, with OAT2-mediated uptake and subsequent metabolism as primary systemic clearance mechanism (high permeability acids/zwitterions with molecular weight (MW) ≤400 Da), Class 1B—uptake transport-enzyme interplay, with OATP1B1/1B3-mediated uptake and subsequent metabolism as primary systemic clearance mechanism (high permeability acids/zwitterions with MW >400 Da), Class 2—metabolism as primary clearance mechanism (high permeability bases/neutrals), Class 3A—renal clearance (low permeability acids/zwitterions with MW ≤400 Da), Class 3B—transporter-mediated hepatic uptake or renal clearance (low permeability acids/zwitterions with MW >400 Da), and Class 4—renal clearance (low permeability bases/neutrals).

Similar to DMEs, transporter inhibition may result in clinically relevant DDIs [17]. In vitro assays are routinely run to screen transporter inhibition potential of drugs to flag in advance of likely DDI liabilities to factor in during subsequent clinical pharmacology plans in full development.

11.5 Physiochemical Properties as Determinant of PK

11.5.1 Lipophilicity

Lipophilicity is the measure of a compound's affinity for a lipophilic environment. It can be used, in part, to predict or rationalise a variety of in vitro and in vivo parameters, such as permeability, solubility, tissue distribution, absorption and metabolism. Parameter 'LogP' is a measure of the lipophilicity of a molecule, expressed as the logarithm of the concentration of a molecule in octanol divided by its concentration in water.

$$\log P = \log_{10}(\text{Partition Coefficient}) \quad (11.48)$$

$$\text{Partition Coefficient}, P = \frac{[\text{Compound}]_{\text{octanol}}}{[\text{Compound}]_{\text{water}}} \quad (11.49)$$

For ionisable solutes, the compound may exist as a variety of different ionic species in each phase at any given pH. Therefore, the distribution coefficient (D) is the appropriate descriptor for ionisable compounds since it is a measure of the pH-dependant differential solubility of all species in the octanol/water system (typically used in the logarithmic form log$D$).

$$\log D = \log_{10}(\text{Distribution Coefficient}) \quad (11.50)$$

$$D = \frac{\sum \text{ionised and unionised species}_{\text{octanol}}}{\sum \text{ionised and unionised species}_{\text{water}}} \quad (11.51)$$

Log$D_{\text{octanol7.4}}$ is the log of the distribution coefficient (D) of a compound between octanol and buffer at pH 7.4. Throughout the human body, the pH of the various organs can vary dramatically—from pH 3 in the stomach, to pH 7.4 in the blood. The Log$D$ of an ionisable compound will change with pH depending on whether it is acidic, basic or zwitterionic. This is because the amount of unionised compound that is available for distribution between the two phases is dependent on both the compound's ionisation and the pH of the system.

11.5.2 Ionisation Constant (pK_a)

The pK_{a} of compound is negative value of logarithmic value of its K_{a} (dissociation constant).

$$K_{\text{a}} = -\log_{10}(k_{\text{a}}) \quad (11.52)$$

Table 11.5 pK_a of some typical drugs

Drug	Chemical nature	pK_a
Acetyl salicylic acid	Acid	3.49
Benzyl penicillin	Acid	2.76
Ethosunamide	Acid	9.3
Chlorpropamide	Acid	4.8
Atropine	Base	9.65
Amphetamine	Base	9.8
Lignocaine	Base	7.9
Procaine	Base	8.8

where K_a is equilibrium constant, and for acids (AH) undergoing dissociation reaction as follows:

$$\text{AH (acid)} \rightleftharpoons \text{A}^-(\text{anionic species}) + \text{H}^+ \text{ (hydrogen ions)} = -\log_{10}(k_a)$$

$$K_a = \frac{[\text{A}^-][\text{H}^+]}{\text{AH}} \tag{11.53}$$

Thus, in simplest form, pK_a of compound is pH at which the rate of forward reaction (dissociation) is in equilibrium with the rate of backward reaction. The pK_a of a singly ionising compound is the pH at which the molecule is 50% protonated and it provides an indicator of the extent of ionisation potential of the compound. So, from PK perspective, pK_a of drugs is the pH at which the drug is completely balanced between the uncharged (lipid soluble membrane permeable form) and charged (water-soluble impermeable form). Most drugs are either weak acids or weak bases (Table 11.5). Acids are most highly ionised at a high pH (i.e. in an alkaline environment). Bases are most highly ionised in an acidic environment (low pH). The degree of ionisation of the drug is determined by the pH of the medium (acidic in the stomach and upper intestine, but close to 7.4 systemically) and the pK_a. Therefore, drugs like phenobarbital, a weakly acidic drug with pK_a 7.4, will be in ionised as well as unionised state in equal concentration in blood plasma (pH 7.4). Also, the drug phenobarbital will be in high concentration as unionised in stomach at pH 1 and thus well absorbed through the gut wall.

Lipophilicity and pK_a influence distribution across the lipid bilayer of cells and into tissues, absorption and the binding characteristics of a drug as well as being important factors in determining the solubility of a compound. Many in silico prediction methods rely on these physicochemical properties of a molecule. It is assumed that only the neutral species can cross the lipophilic membrane. Acids are anionic state in the plasma and therefore tend to have a low volume of distribution (although this is also influenced by plasma protein binding) (Fig. 11.5). In contrast, compounds with basic moieties, such as amines, tend to have a high volume of distribution because of sequestration in membranes (due to interaction with anionic phospholipids) and/or trapping in acid organelles such as lysosomes.

Lipinski's rule of 5 ('Ro5') described the physicochemical property space with the highest probability of achieving good oral absorption [18]. Lipinski's 'rule of five,' very well known by medicinal chemists, has been defined as a set of rules, based on the number 5—hence its name, to evaluate the 'drug-likeness' of a given molecule. The rules states that poor absorption or permeation is more likely when there are more than 5 H-bond donors, 10 H-bond acceptors, the molecular weight is greater than 500 and the calculated Log*P* (*C*Log*P*) is greater than 5 (or *M*Log*P* >4.15). Although, generally true several successful drugs violate the 'rule of 5' to some extent: for example, atorvastatin, montelukast, and natural products such as cyclosporine and paclitaxel.

In recent times, there is an increasing focus on the pursuit of less druggable targets, such as the disruption of protein–protein interactions, that offer high potential for the development of new therapeutic agents and may require beyond rule of 5 ('bRo5') chemical space in order to take advantage of these opportunities [18].

11.5.3 Solubility

Solubility is the maximum dissolved concentration of a solute in a particular solvent at a given temperature. Poor solubility can limit the absorption of compounds from the gastrointestinal tract, resulting in reduced oral bioavailability; may necessitate novel formulation strategies and hence increase cost and delays in drug development and can lead to misleading data in in vitro assays.

There appears to be a Gaussian or parabolic relationship between $\mathrm{Log}D_{\mathrm{octanol7.4}}$ and the extent of absorption and bioavailability (Fig. 11.10). For hydrophilic compounds $\mathrm{Log}D_{\mathrm{octanol7.4}} < 0$, solubility is good, but their permeability across membranes is poor, resulting in limited absorption [19]. In addition, their metabolic clearance is usually limited, but clearance by the kidney may be high. In contrast,

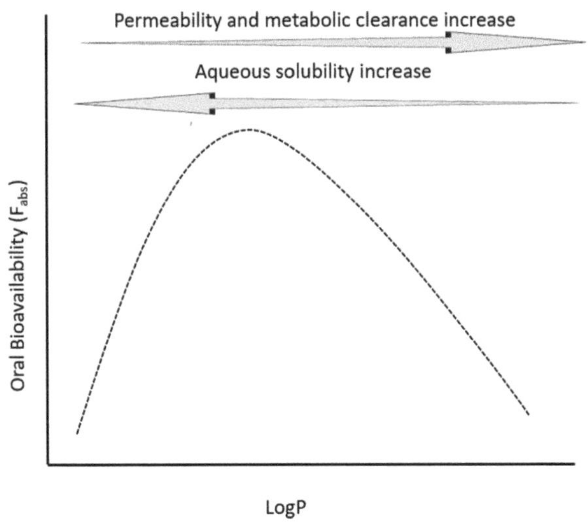

Fig. 11.10 Generic empirical relation between oral bioavailability, permeability, metabolic clearance and log*P*

for compounds with Log$D_{octanol7.4}$ > 5, membrane permeability is good, but the solubility is low, which significantly reduces absorption. In addition, metabolic clearance tends to be higher for compounds with Log$D_{octanol7.4}$ > 5. This effect is enhanced by increased plasma protein binding (i.e. the unbound clearance is much greater at high Log$D_{octanol7.4}$). However, Log*D* range that leads to good absorption and increased metabolic stability is wider at low MW than at high MW [20]. Plasma protein binding and tissue binding increases with increasing Log*P* and Log$D_{octanol7.4}$, and this can lead to a relatively low free drug concentration at the site of action despite reasonable membrane permeability.

11.6 Role of PK in Drug Discovery and Development

Pharmacokinetic science influences each phase of drug discovery and development. Wide range of in silico/in vitro/in vivo and clinical studies are performed for NCE at different stages of drug discovery and development in staged manner. Figure 11.11 is schematic flow diagram illustrating key pieces of information generated during each stage of drug research and development in pharmaceutical industrial set up. Drugs have wide variety of structure and physiochemical properties, and they are being developed for many different types of pharmacological activities. Therefore, PK profiling of each CD is unique and tailor-made, while Fig. 11.11 describes a generic flow of main critical pathway that most drug projects may follow for stepwise progression to support needs of high efficacy/safety and in congruence with regulatory requirements, it is not uncommon to find subtle differences in this workflow among various pharmaceutical companies. As drug project transition its core focus from one stage to another along value chain, there is trail of follow-up/parallel/bespoke studies happening in preceding stages to fill the gaps or troubleshoot specific problem that arise from emerging data.

11.7 Drug Discovery

PK profiling in discovery stage comprises the characterisation of basic physiochemical properties (molecular weight, water solubility, lipophilicity, pK_a and acid/base properties) and ADME properties (permeability, metabolic stability/clearances, transporters, induction) that are important to their function as drugs [21]. Optimisation of these properties by structure activity relationships makes chemical compounds 'druggable' while retaining basic active pharmacophore. Following sections describe the role played by key physiochemical and ADME properties in pharmacokinetic profiling of drugs.

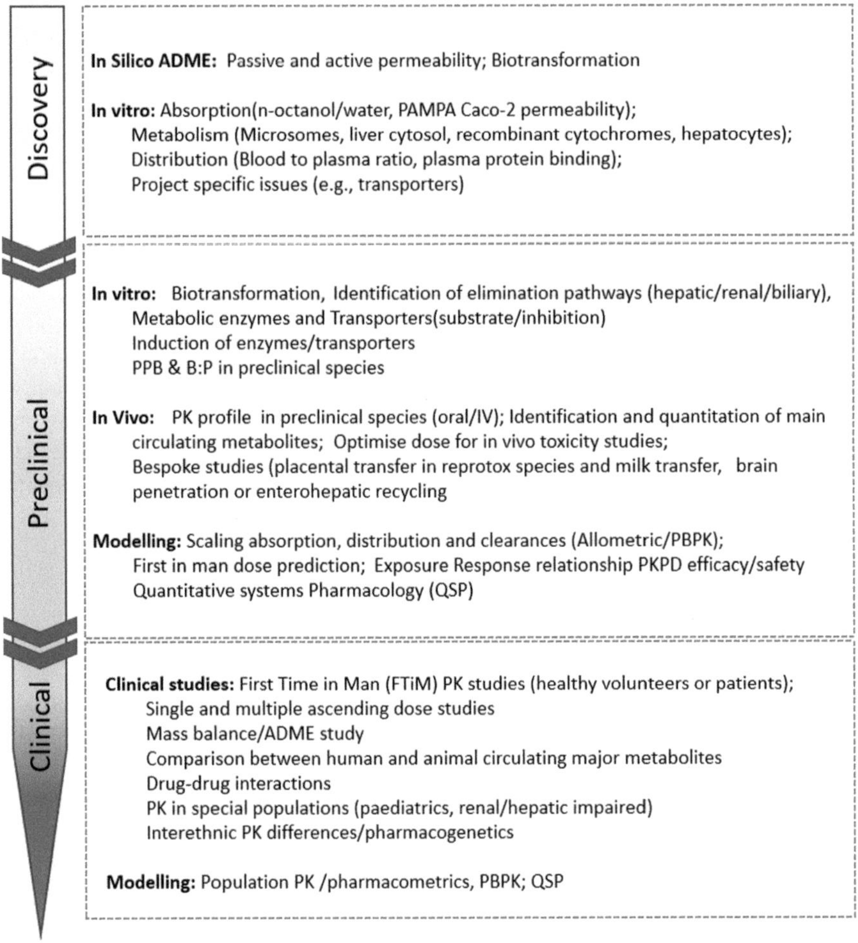

Fig. 11.11 Pharmacokinetic milestones in different stages of drug discovery and development. Key PK studies/parameters/deliverables are listed as drug project moves in value chain up to successful launch in clinic. The shading intensity of arrow is approximate indication of costs, time and labour needed in each stage of discovery and development

11.7.1 Lead Generation/Identification (LG/LI)

Once therapeutic target is identified and validated, the LG/LI process involves identification of major chemical series/groups (clusters or chemical scaffolds) that are suitable for drug discovery. From a pharmacokinetic point of view, the main objective in LG/LI phase is to support medicinal chemistry in assisting hit cluster evaluation and ranking through identification of the DMPK liabilities and assessment of the optimisation potential of the compounds in each cluster.

In addition, pharmacokinetics help to support in vivo pharmacology studies with exposure-based advice on dose and schedule design of pharmacodynamic and efficacy studies. At LI stage, only a few compounds per cluster are tested for in vitro ADME assays, namely, metabolic stability in liver microsomes and/or hepatocytes, Caco-2 permeability, CYP inhibition using human liver microsomes, CYP induction potential based on the PXR assay and efflux transport. Compounds with suitable ADME properties and pharmacologically active are subsequently submitted to rodent PK studies in vivo to examine how the in vitro ADME liabilities translate to in vivo, to determine the PK parameters and to assess their behaviour in the whole organism. Fundamentally the focus here is to gain understanding on what type of DMPK liabilities a hit cluster may carry, assessment of oral bioavailability, how many liabilities are likely and whether all compounds tested show the similar type of liabilities. To mitigate these liabilities, whether there are structural variations which could serve as promising starting points for chemical optimisation (structure property relationships, SPR/structure activity relationships, SAR). Besides the identification of the liabilities of the hit cluster, the project team also defines the desired profile of the drug candidate (sometimes called candidate drug target profile (CDTP) or target product profile (TPP)) as optimisation goal for the LO phase. The key to CDTP/TPP is fundamental understanding during early discovery phases of the relationship between the PK/exposure and the pharmacodynamics/efficacy.

11.7.2 Lead Optimisation (LO)

Once the lead chemical structure cluster is identified, LO is focussed on generating a drug candidate molecule which is efficacious in a well-defined indication and patient population and is able to be administered safely and conveniently to humans ('5R framework: Right target, tissue, safety, patient and commercial potential) [22]. A significant investment into a project on a multidimensional optimisation of the chemical structure to improve the liabilities of the lead structure involves assigning the allocation of large amounts of resources in medicinal chemistry, pharmacology, drug metabolism and pharmacokinetics. Although this exercise needs a significant investment but has succeeded in reducing attrition rates caused by inappropriate human pharmacokinetics at later stage during clinical development.

Drug metabolism and pharmacokinetic (DMPK) role during the LO phase is to optimise the PK profile of the compounds such as to enable efficacy, to mitigate and/or gauge the potential of the compounds to elicit safety risks and drug–drug interactions (DDI) both as victim and perpetrator when given to patients with indication-specific comedications and to ultimately predict first in man efficacious dose and dosing schedule which can be formulated and administered in a way that is convenient for clinical use.

In general, generation of lead molecules follow rigorous 'Design-Make-Test-Analyse (DMTA)' Cycle (Fig. 11.12). This is iterative process of systematically building SAR to select potential pre-CDs, which are optimised for CDTP ensuring

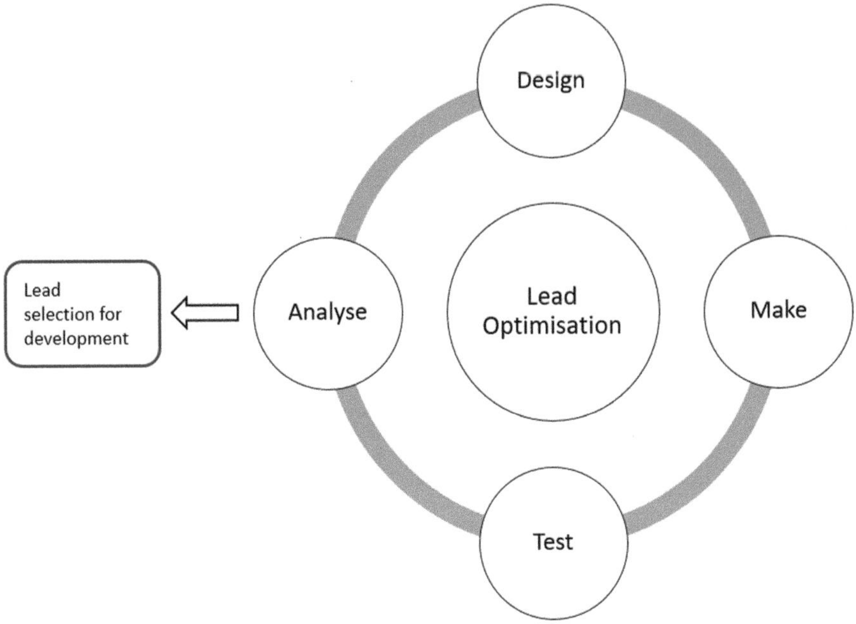

Fig. 11.12 DMTA cycle in lead optimisation at aimed to systematically identifying the best possible candidate drug among thousands of compounds. Rational drug design, followed by high-throughput synthesis and testing of molecules for key parameters is analysed to select CD for more advanced models to identify any risks before proceeding to clinical development

maximum efficacy, minimum toxicity and best pharmacokinetic profile in class. Shortlisted pre-CDs are further characterised in various in vitro and in vivo assays, including disease models, to demonstrate their potential therapeutic use in humans. The scale-up of synthesis of the potential pre-CD and the plan for first-in-human clinical trials are also addressed in the final stages of the LO phase [23]. The experimental data are usually generated with non-validated methods, sometimes from studies in which multiple compounds are administered to the same animal (cassette dosing), with a small number of repetitions and with simplified protocols. Despite these limitations, the results provide a reasonably good knowledge of the pharmacokinetics of the compound already at the beginning of the development process.

'Drug design' is cross-functional multi-disciplinary process involving medicinal, synthetic, computational and physical chemistry, safety and DMPK functions. Some key parameters can be computed or predicted in silico with reasonable reliability before synthesis commences, such as lipophilicity, molecular weight, and structural alerts for reactivity and reactive metabolite generation. In LO stage the target should be to 'make' sufficient material in single batch to serve the requirements of primary in vitro screens (potency, selectivity, physical chemical properties, and an in vitro DMPK panel), an in vivo rat PK (pharmacokinetics) study. Generally, this amounts to 30–35 mg of material for LO studies. 'Testing' of compounds for various assays is

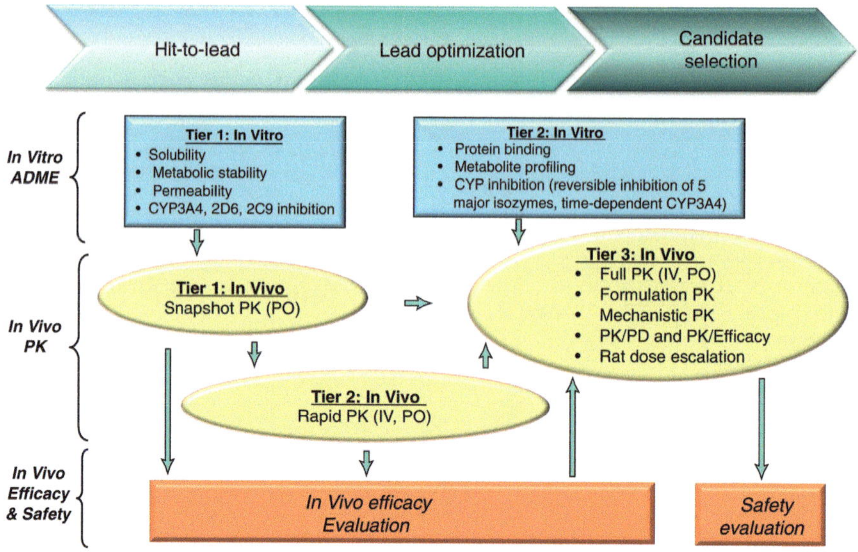

Fig. 11.13 Schematic illustration of how each of the tiered in vivo pharmacokinetic (PK) approaches, snapshot PK, rapid PK and full PK are integrated and applied in the typical drug discovery paradigm. Arrows indicate compound progression. (Reprinted from [24] Drug Discov Today, 18 (1–2), Li C, Liu B, Chang J, Groessl T, Zimmerman M, He YQ, Isbell J, Tuntland T, A modern in vivo pharmacokinetic paradigm: combining snapshot, rapid and full PK approaches to optimize and expedite early drug discovery. 71–78, Copyright (2013), with permission from Elsevier)

done in parallel using high-throughput screening (HTS) technologies which efficiently reduces cost and time while maximising data collection of high quality. 'Analysis' of large amounts of data involves using sophisticated inline data handling tools to visualise, demarcate and rank compounds based on key selection criteria/determinant properties.

Once compounds are shortlisted from the early lead optimisation cycles, animal PK studies are performed to verify the improvements made in vitro in the whole animal in vivo. In vivo rodent PK studies are important to confirm if compounds have appropriate PK properties to be evaluated in preclinical pharmacology and safety studies. In addition, characterisation of in vivo PK of new chemical entities provides insight into complex in vivo biological systems and correlates drug concentration at the site of action with pharmacological response. Various in vivo rodent PK approaches are illustrated in Fig. 11.13 and comprise simple and abbreviated study design of 'snapshot' PK, to the more labour-intensive intravenous/per oral (i.v./po) PK study designs, such as 'rapid PK' and the conventional 'full PK'. Figure 11.14 shows relative placement of these study types in discovery stages and offer a tiered approach to successively profile the potential compounds for screening PK properties.

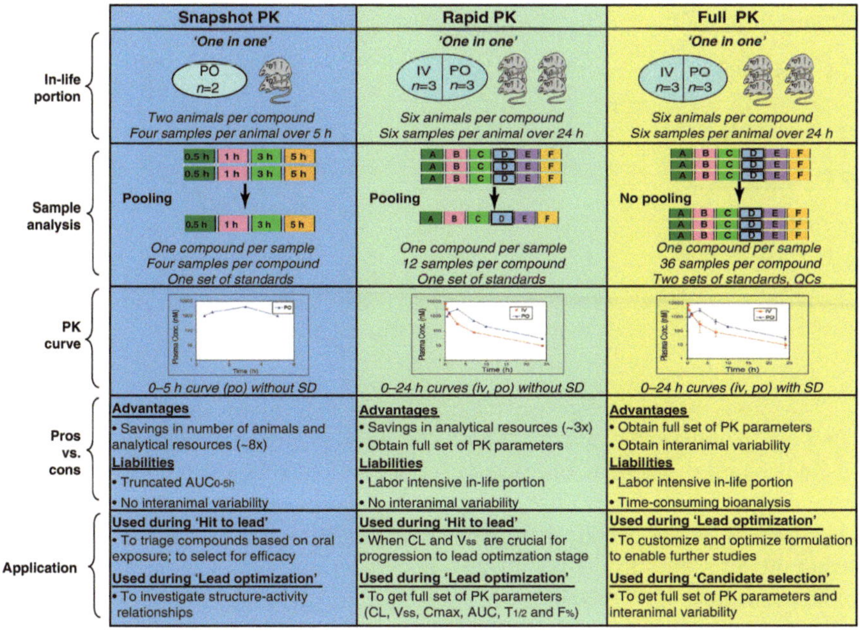

Fig. 11.14 Comparison of tiered in vivo pharmacokinetic (PK) approaches for supporting drug discovery, including in-life study design, sample preparation, data output, pros and cons, as well as their applications. (Reprinted from [24] Drug Discov Today, 18 (1–2), Li C, Liu B, Chang J, Groessl T, Zimmerman M, He YQ, Isbell J, Tuntland T, A modern in vivo pharmacokinetic paradigm: combining snapshot, rapid and full PK approaches to optimize and expedite early drug discovery. 71–78, Copyright (2013), with permission from Elsevier)

Establishing and verifying the link between in vitro and in vivo (IVIVC) is important to establish 'proof of principle' of PK philosophy that the screening assays have the desired impact and that the screening strategy is fit for purpose. IVIVC is achieved by correlation analysis between two or a few more properties and/or by physiologically based PK (PBPK) models, where in vitro ADME data is scaled through to the whole organism which allows simulation of the impact a given property change may have on the concentration–time profile (more details in Sect. 11.7.3.3).

Among the key PK activities, one most important deliverable is to provide quantitative estimate of exposure or plasma concentration profiles in preclinical animal disease model to enable design of exhaustive efficacy studies and conduct of pharmacokinetic and pharmacodynamic (PKPD) analysis. Toxicokinetic is primarily use of PK exposure information applied in toxicology animal species to guide design of safety studies at much higher doses/exposures, to profile any likely adverse safety signals and aid in generation of safety margins/therapeutic index. In these animal safety studies, saturation of PK processes (absorption/clearance) at higher doses may limit the level of high exposure desired to investigate adverse events.

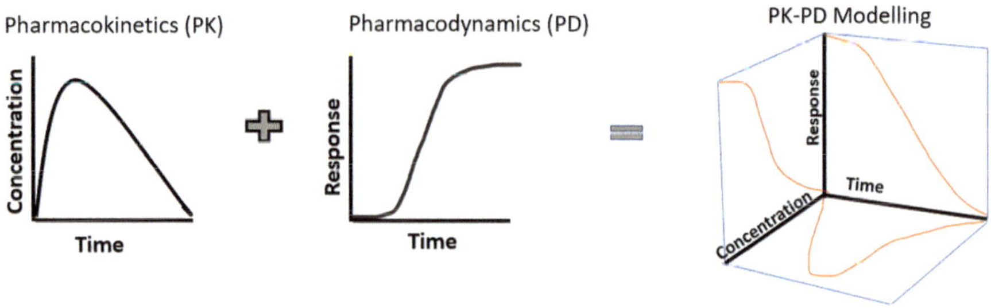

Fig. 11.15 Time–Concentration–Response analysis is important to quantify drug effects (efficacy/ safety) to derive reliable dose recommendations in target population

The dose–exposure–response relationship is defined by the pharmacokinetic (PK) and pharmacodynamic (PD) characteristics of a drug. Pharmacodynamics is pharmacological response (therapeutic effect or toxicological response) resulting from drug concentrations at the effect site. In simplest statement, pharmacokinetics characterises 'what the body does to the drug', whereas pharmacodynamics assesses 'what the drug does to the body' [25]. Integration of both pharmacological disciplines fundamentally driven by mathematical relationships (PK/PD modelling) allows conceptualisation of the effect–time course resulting directly from the administration of a certain dose (Fig. 11.15). Figure 11.16 gives systematic evolution of PK-PD modelling in drug project and deliverables to support various stage of discovery and development. Preclinical pharmacokinetic/pharmacodynamic (PK/PD) analysis is an efficient tool for the translational research and proof of mechanism/concept in animals.

After multiple DMTA LO cycles, the lead compounds will have significant improvements on the liabilities seen in the original lead structures. Late LO phase continues screening efforts on a broader examination of the most promising 'druglike' compounds in vivo, e.g. more sophisticated animal PD and efficacy models, non-rodent PK studies as well as pilot toxicology and safety pharmacology studies. These activities gradually lead into the phase of candidate selection and profiling.

11.7.3 *Candidate Selection and Profiling*

Candidate selection and profiling comprises comprehensive PK characterisation to examine whether promising molecules qualify as potential drug candidate. Efforts are largely focussed on predicting the PK in human, estimating the human therapeutic exposure, predicting the therapeutic dose in patients and estimating the therapeutic window in human. This phase will also identify the potential DMPK-related risks and likely probability of success, which have to be addressed specifically during preclinical and clinical development to de-risk the programme and to

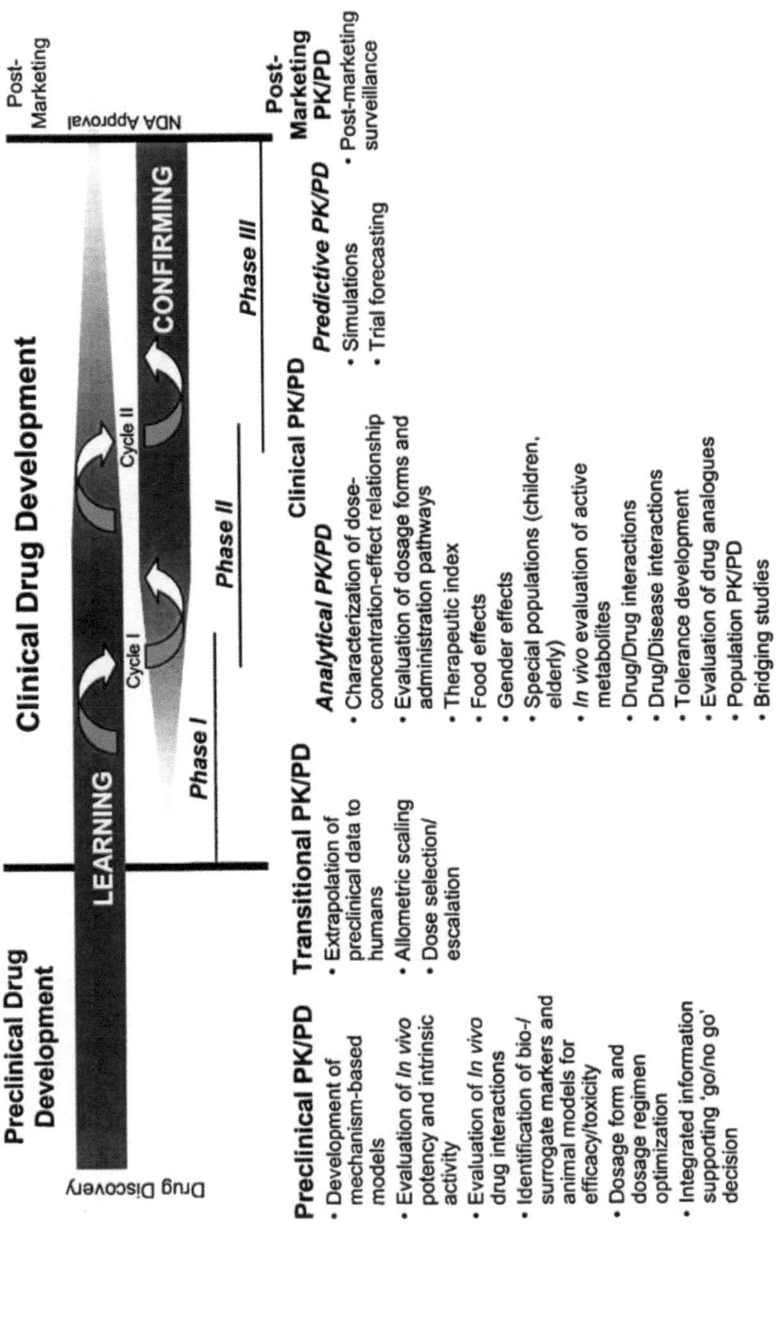

Fig. 11.16 Potential applications of PK/PD concepts during preclinical and clinical drug product development. (Reprinted from [26] J Pharm Sci., 91(1), Meibohm B and Derendorf H, Pharmacokinetic/pharmacodynamic studies in drug product development. 18–31, Copyright (2002), with permission from Elsevier)

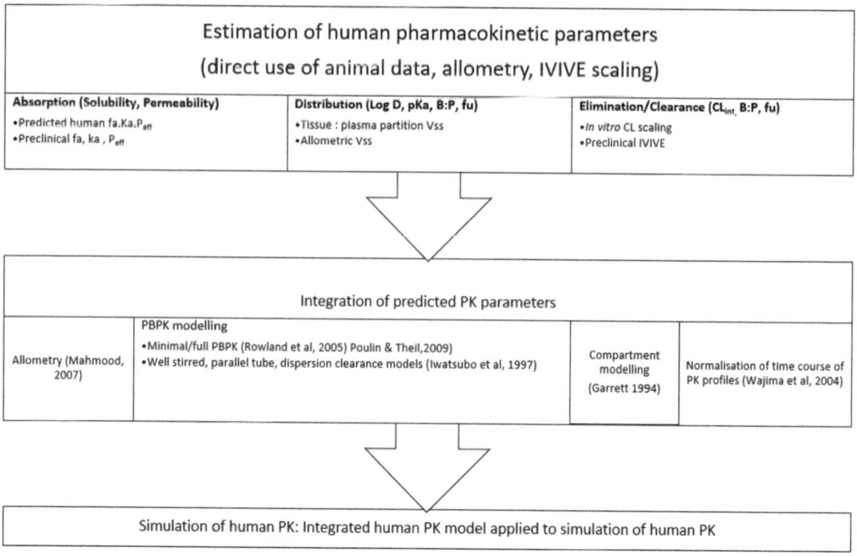

Fig. 11.17 Systematic steps towards prospective prediction of human PK involves two distinct stages: first, the estimation of individual kinetic parameters, and second, the integration of these processes into a model to simulate a concentration–time profile

allow nomination of CDs for the future investments by way internal governance structure for project funding.

The prediction of human PK involves two distinct stages: first, the estimation of individual kinetic parameters, and second, the integration of these processes into a model to simulate a concentration–time profile (Fig. 11.17). In practice, there are four options for modelling the data: allometry, compartmental models, PBPK models and Wajima et al. method [27].

11.7.3.1 Allometry

Allometry is based on principle that physiological variables such as clearance, volume of distribution, blood flow, heart rate or other biochemical processes from lower to higher species are related to body weight or surface area [28]. When a PK parameter is available from one species, the allometric equation can be used to predict the value of the parameter in humans:

$$y_{\text{human}} = y_{\text{animal}} \times \left(\frac{W_{\text{human}}}{W_{\text{animal}}}\right)^{\text{b}} \tag{11.54}$$

The value of '*b*' has been observed to often be related to the type of parameter that is being measured: (1) biological time ($t_{1/2}$ or MRT) $= 0.25$, (2) distribution volume = 1.0, and biological rate (blood flow, renal and hepatic CL) = 0.75 [29]. Empirical observations indicate that many physiological parameters change as a function of size and the relationship can be described as

$$y = a \times W^{\mathrm{b}} \tag{11.55}$$

where 'y' is the variable of interest (e.g. CL, V_{ss}), 'a' is the allometric coefficient, 'W' is the body weight of a given species and 'b' is the allometric exponent. If PK data in multiple species is available, a plot of log$Y$ versus log$W$ results in an intercept of log (a) and a slope of b, as shown in the equation below:

$$\log\ y = \log(a) + b \times \log(W) \tag{11.56}$$

In this case, the value of b is derived from compound specific data as opposed to the predefined values of b when scaling is based on data from a single species. By plotting the physiologic variable (e.g. CL, V_{ss}) on the y-axis versus the body weight for a range of species on a log–log plot, a linear relationship can be achieved. Performing linear regression using the following power equation allows for the estimation of an unknown variable for a given species by plugging in the desired body weight (e.g. human BW = 70 kg).

11.7.3.2 Garrett Method

For compounds that are expected to show monoexponential kinetics, one-compartment models (Sect. 11.2.1 and Fig. 11.2) may be useful in simulating human concentration–time profiles as well as for obtaining an early integrative view of human PK and dose [30]. In simplest form, one-compartment model to predict the dose required to maintain trough concentrations at a minimum effective concentration (MEC) over a given dose interval (τ) is shown in the following equation:

$$\text{Dose} = \frac{\text{MEC} \times (k_{\mathrm{a}} - k_{\mathrm{e}}) \times V_{\mathrm{ss}}}{k_{\mathrm{a}} \times F \times \left(\mathrm{e}^{-ke\tau} - \mathrm{e}^{-ka\tau}\right)} \tag{11.57}$$

where F_{abs} = bioavailability and k_e is elimination rate constant. This requires predicting relatively few parameters, including the absorption rate constant k_a, CL and V_{ss}. For compounds with multiexponential model, PBPK models are preferred, because the scalability between species is more manageable.

11.7.3.3 Physiologically Based Pharmacokinetic (PBPK) Modelling

PBPK models have a unique advantage compared to other in silico approaches in that they consider not only the drug and the formulation characteristics but also the underlying physiology of the individual subject and its variability within a population in prediction of drug absorption, distribution and elimination [31]. These processes can be explained with a system of coupled differential equations that can be parameterised from knowledge of human physiology and drug properties characterised in vitro. Mechanistic understanding of processes involved in the drug liberation from formulation, dissolution and absorption of the drug, distribution and elimination does not only depend on the drug but also the physiology, for example, when the same drug formulation is given to different human subjects; it may result in different pharmacokinetic profiles.

Components of PBPK Modelling

As depicted in Fig. 11.18, the overall PBPK Modelling Process can be divided into four main components: (1) drug formulation parameters that can be obtained in vitro; (2) in vitro to in vitro extrapolation of in vitro drug parameters to in vivo situations; (3) physiology or systems data; (4) PBPK structural model; and (5) study design and drug administration.

Drug and Formulation Parameters

This includes in vitro for physicochemical parameters described earlier, namely solubility, permeability, metabolism, binding and partitioning. Dissolution rates

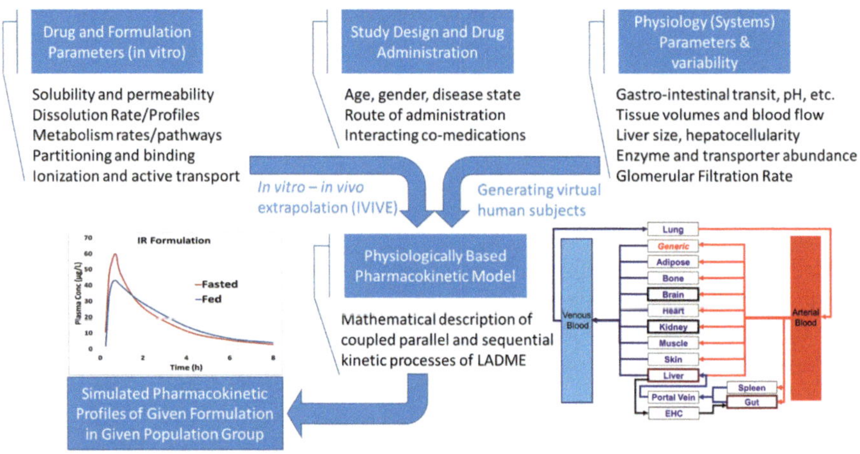

Fig. 11.18 PBPK modelling Process can be divided into four main components: (*1*) Drug formulation parameters that can be obtained in vitro; (*2*) In vitro *to* in vitro extrapolation of in vitro drug parameters to in vivo situations; (*3*) Physiology or systems data; (*4*) PBPK Structural Model; and (*5*) Study design and drug administration

from the formulation are obtained from suitable in vitro apparatus with appropriate fluid media for dissolution.

In Vitro and In Vivo Extrapolation (IVIVE)

Solubility of the drug can be measured in vitro at several specified pH and/or bile salt concentrations such as FaSSIF (Fasted State Simulated Intestinal Fluid, pH 6.5 and [BS] 3 mM) or FeSSIF (Fed State Simulated Intestinal Fluid, pH 5 and [BS] 15 mM). pH is known to vary significantly along the GI tract (~1 in fasted stomach, ~6.5 in proximal small intestine and ~7.5 in colon), with prandial state (fasted stomach pH ~1 while fed stomach pH ~5) and between individuals. Telemetric capsule and aspiration studies from human GI tract also suggest that the pH can vary along GI tract and between individuals as well as vary over time in a given segment of the GI tract. Similarly, the bile salt concentrations is known to vary significantly along the GI tract, between subjects and strongly depends on prandial state [32]. Nonetheless, it is challenging to measure all permutations and combinations of pH and bile salt concentration likely to occur in vivo for various individuals along the GI tract. Thus, suitable IVIVE approach is needed that can be parameterised with minimum standard in vitro experiments and then used to estimate solubility with various permutations and combinations of pH and bile salt concentrations likely to occur in vivo *via* PBPK model simulations. Equation (11.58) is widely used for IVIVE of solubility where pK_a can be obtained from the solubility measured in three or more pH media and $K_{m:w}$ (bile micelle to water partition coefficient) from FaSSIF (fasted state) and FeSSIF (fed state) solubility measurements [32]. Once parameterised, the equation can be used to estimate solubility at any pH and bile salt concentration. Equation (11.58) is linear in nature; hence if there is saturation of solubility with increasing bile salt concentration or a salt limited solubility is observed or expected, further refinement might be warranted during modelling.

$$S_{(\mathrm{BS})_{\mathrm{Tot}}} = \left[\begin{array}{l} \left([\mathrm{BS}] \cdot \dfrac{S_0}{C_{\mathrm{H_20}}} \cdot 10^{\log K\mathrm{m:w,unionised}} + S_0 \right) + \\ \left([\mathrm{BS}] \cdot \dfrac{S_{\mathrm{ionised}}}{C_{\mathrm{H_20}}} \cdot 10^{\log K\mathrm{m:w,ionised}} + S_{\mathrm{ionised}} \right) \end{array} \right] \tag{11.58}$$

where [BS] is the concentration of bile salt (the BS:lecithin molar ratio in the study used to generate the model is 4:1); S_0 is the aqueous intrinsic solubility; S_{ionised} refers to the aqueous solubility of the ionised form of the drug at a given pH; $S_{(\mathrm{BS})\mathrm{Tot}}$ is the total solubility (aqueous at given pH + micellised) at a stated concentration of [BS]; C_{H20} is the concentration of water; Log$K_{\mathrm{m:w,unionised|ionised}}$ is respectively the micellar partition coefficient defined for neutral (or ionised) molecular species.

IVIVE (in vitro-in vivo extrapolation) is a method to predict in vivo PK based on in vitro dissolution data. IVIVE of dissolution can be obtained *via* diffusion layer model such as Eq. (11.59) used in the Advanced dissolution, absorption and metabolism (ADAM) model of proprietary PBPK modelling software platform Simcyp® or other variants of Noyes-Whitney equation:

$$\mathrm{DR}(t) = -N\frac{D_{\mathrm{eff}}}{h_{\mathrm{eff}}(t)}4\pi r(t)\ (r(t) + h_{\mathrm{eff}}(t))\ \left(C_{\mathrm{surface}} - C_{\mathrm{bulk}}(t)\right) \tag{11.59}$$

where DR(t) is the dissolution rate at given time t, N is the total number of particles available for dissolution, $r(t)$ is the radius of dissolving particle at time t, $h_{\mathrm{eff}}(t)$ is the thickness of unstinted boundary layer around dissolving particle, C_{surface} is the solubility of the drug at surface of dissolving particles which commonly is assumed as the thermodynamic solubility of the drug, $C_{\mathrm{bulk}}(t)$ is the concentration of the drug at time t which depends on amount dissolved as well as volume of fluid available for dissolution.

Equation (11.58) offers translation of in vitro solubility and formulation characterisation (e.g. particle size) to in vivo dissolution considering the pH, bile salt and fluid volumes present at various segments of GI tract while the formulation transits through the GI tract in PBPK Simulation [33].

IVIVE of metabolism is one of the most studied and established area of PK modelling. IVIVE scaling for liver metabolic clearance was described in Sect. 11.4.3. Similar approach can be applied for gut wall metabolism; however, permeation rate should be accounted for especially for poorly permeable compounds.

Physiology Data

This includes parameters such as size and volume of various tissues, cardiac output, tissue blood flows and their covariations with age, gender, ethnicity and disease conditions. For example, the variability in GI tract physiology in fasted and fed states and enzyme abundance and its covariations is well documented in literature and used to parameterise the absorption and metabolism processes in the PBPK model [33]. By changing the physiology database once can simulate the PK of particular drug from healthy subjects to diseased population or from healthy adult population to paediatric or geriatric population groups (Fig. 11.19). Such prospective simulations could be highly valuable to support drug development decision-making and identify any population specific risks in exposure levels [34].

PBPK Structural Models

PBPK model can have multiple organs (e.g. 15-organ fully body PBPK model reported by Jamei et al. [35] or it can be a so-called minimal PBPK model where majority of organs are lumped into one or two compartments and separate liver compartment that accounts for metabolic processes. Further mechanistic details for specific organ or tissue of interest can be added, for example, to model mechanistic processes in drug absorption; multi-compartment GI tract models are routinely used

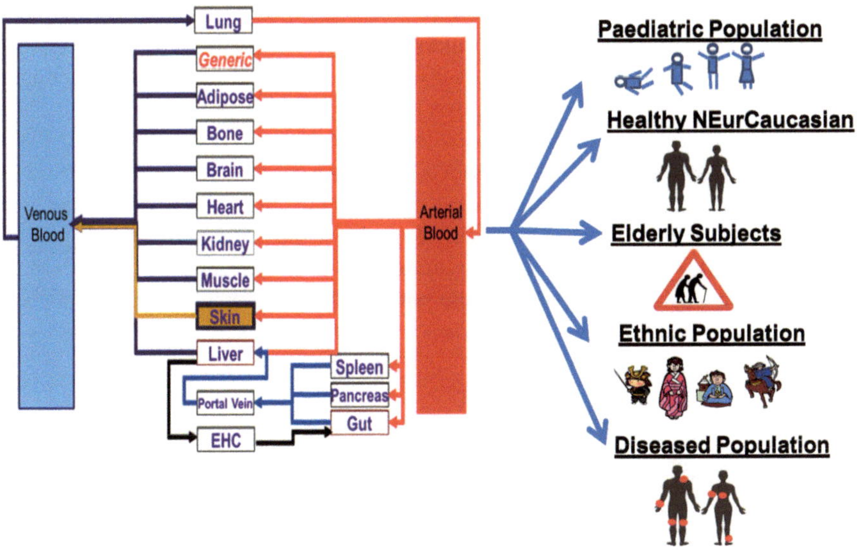

Fig. 11.19 Population-based PBPK modelling involves use of demographic and physiological parameters from specific population groups (e.g. paediatrics) to scale PK from typical 'individual PK' to 'population PK' where intra- and inter-individual variabilities are driven by physiology

such as ADAM or Advanced compartment absorption and transit (ACAT) model. Brain, kidney and liver can be further defined as multi-compartmental system to model the specific processes in particular organ in more details.

General form of coupled differential equations is shown in Eqs. (11.60) and (11.61).

$$\frac{dC_{\mathrm{vb}}}{dt} = \frac{1}{V_{\mathrm{vb}}}\left[\left(Q_{\mathrm{C}} - \sum\nolimits_{\mathrm{Tissues}} Q_{\mathrm{Tissue}}\right) C_{\mathrm{ab}} + \sum\nolimits_{\mathrm{Tissues}} \left(Q_{\mathrm{Tissue}} \frac{C_{\mathrm{Tissue}}}{P_{\mathrm{Tissue}:p}/B:P}\right)\right.$$
$$\left. - Q_{\mathrm{C}} C_{\mathrm{vb}} - \frac{\mathrm{CL_R}}{B:P} C_{\mathrm{vb}}\right] \tag{11.60}$$

$$\frac{dC_{\mathrm{Tissue},i}}{dt} = \frac{Q_{\mathrm{Tissue},i}}{V_{\mathrm{Tissue},i}}\left(C_{\mathrm{ab}} - \frac{C_{\mathrm{Tissue},i}}{P_{\mathrm{Tissue}:p}/B:P}\right) - \mathrm{CL}_{\mathrm{Tissue},i} * C_{\mathrm{Tissue},i} \tag{11.61}$$

where C_{vb} is the concentration in venous blood; C_{ab} is the concentration in arterial blood; $C_{\mathrm{Tissue},i}$ is the concentration in the ith tissue; $Q_{\mathrm{Tissue},i}$ is the blood flow to ith tissue; B:P is the blood-to-plasma ratio; $P_{\mathrm{Tissue},i:p}$ is the partition ratio of ith tissue and plasma; $V_{\mathrm{Tissue},i}$ is the volume of ith tissue; $\mathrm{CL}_{\mathrm{Tissue},i}$ is the clearance of drug from ith tissue.

$P_{\mathrm{Tissue},i:p}$ can be determined from lipophilicity of the drug and it's ionization at physiological pH and known information of tissue composition. There are two widely used methods and their variants for prediction of $P_{t:p}$, one reported by Poulin and Theil [36] as corrected by Berezhkovskiy [37] and the other by Rodgers and Rowland [38]. The latter splits the tissue water volume into intra- and extracellular components, along with the addition of a tissue acidic phospholipid fraction. These equations take explicit account of the extent of ionisation of a compound at the pH of the compartment concerned and have been shown to improve the prediction of tissue:plasma partition coefficients, and consequently volume of distribution, for strong bases. Clearance can be estimated as shown in Eq. (11.37) by scaling suitable in vitro intrinsic clearance to whole organ level and applied into the tissue disposition equation of particular tissue. These equations can account for inter-individual variability in blood flows, tissue volumes and compositions.

Study Design and Drug Administration

The design of the clinical study to be simulated and the way the drug is administered determines the setting of the simulation to be run and allows framing the scientific question that needs to be addressed *via* simulations. For example, study design determines adult or paediatric population, healthy or hepatic impaired population, age and gender of the subjects to be simulated virtually. Similarly, route of administration (oral, topical, intravenous, etc.), type of formulation (immediate release, controlled release, gel, cream ointment, injection, etc.), prandial state (fasted or fed), any comedications that can interact directly with the drug or physiology that can affect the kinetics of the drug of interest (e.g. metabolic inhibitors, inducers, proton pump inhibitors) can be defined to understand the impact of these differences in kinetics of the drug of interest in population of interest.

Qualification of PBPK Modelling

Recently various regulatory guidance by US FDA [39] and EMA [40] have emerged on recommendations on qualification of PBPK models. The documentation needed to support the qualification and verification of PBPK platforms should cover three components of the platform: software, systems parameters and drug model [41]. The software qualification is intended to ensure that the software does what it is intended to do from a computational perspective. Qualification of the system-dependent components involves documentation of the physiological framework, the equations used to describe the system, as well as the physiological parameters feeding it. The drug model verification documents consistency between the input parameters and underlying mechanisms and assumptions within the related physiological system and the ability of the model to successfully simulate sets of observed data, sometimes following several iterations of a learn and confirm process.

Application of PBPK Modelling

PBPK modelling allows realistic interaction between drug and physiology making it possible to ask various 'what if' questions before conducting a clinical trial based on prior knowledge of drug and human physiology. Hence, it serves a powerful tool to reduce, refine or avoid unnecessary clinical studies and make well-informed decisions early on drug development. It allows assessment of first-in-man dose by preclinical data, formulation on bioavailability, drug–drug interactions, exposure in special populations such as paediatric or organ impairment, virtual bioequivalence, product specification settings, biopharmaceutics risk associated with scale-up and post-approval changes (SUPAC), personalised dosing or personalised safety assessment *via* simulating physiology of a particular subject in silico 'virtual twin' of the subject, etc. There are number of publications from consortium of pharmaceutical companies, regulatory bodies such as US FDA, EMA and PMDA highlighting the use of PBPK modelling in drug development and regulatory assessment [42].

11.7.3.4 Wajima Method

Wajima et al. (2004) [27] proposed method which is based on the assumptions that concentration–time profiles of a drug can be superimposed among species by normalisation of time axis with the MRT and the concentration axis with dose/V_{ss}. Dose normalised curves are obtained by dividing the time axis with mean residence time (MRT set equals to V_{ss}/CL) and the concentration axis with dose/V_{ss} (C_{ss}) of the respective preclinical species (Fig. 11.20). The concentration–time profile in humans after intravenous injection can be simulated using the normalised curve for an animal and the predicted values of CL and V_{ss} for humans. In other words, this step is accomplished by multiplying the concentration and time scales of the normalised curve of each preclinical species by the predicted human C_{ss} and MRT, respectively. The concentration–time profile in humans after intravenous injection can be simulated using the normalised curve for an animal and the predicted values of CL and V_{ss} for humans. To simulate the plasma concentration–time profiles following oral administration in humans, the additional parameters, namely percentage absorbed (F_{abs}) and rate constant of absorption (K_a), are used in combination with microconstants of the predicted IV profiles from compartmental analyses. The F_{abs} and K_a used here are average from preclinical species PK data.

Subsequent to predicting human PK, key decision-making question is—what exposure is needed to have therapeutic benefit (efficacy), and also what exposure is likely to result adverse effects (safety). The predicted PK along with dose–exposure relationship in humans is used to estimate a dose/dosing schedule for humans which is anticipated to mimic the unbound concentration–time profile that is expected to be efficacious as well as safe in humans. Lowe et al. [44] had described a general basic four-step projection process for the anticipation of human doses (AHD), which includes:

Step 1: characterisation of preclinical exposure–response or effect relationships

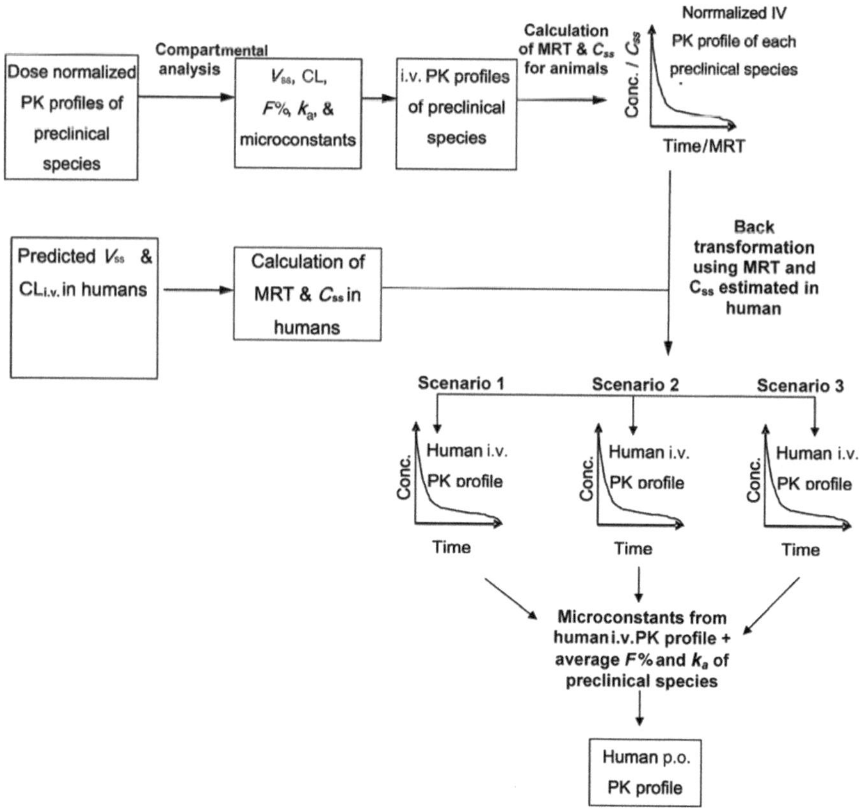

Fig. 11.20 Schematic presentation of the C_{ss}-MRT approach developed by Wajima et al. (2004). MRT and C_{ss} correspond to V_{ss}/CL and dose/V_{ss}, respectively. (Reprinted from [43] J Pharm Sci., 100(10), Vuppugalla R, Marathe P, He H, Jones RD, Yates JW, Jones HM, Gibson CR, Chien JY, Ring BJ, Adkison KK, Ku MS, Fischer V, Dutta S, Sinha VK, Björnsson T, Lavé T, Poulin P., PhRMA CPCDC initiative on predictive models of human pharmacokinetics, part 4: prediction of plasma concentration–time profiles in human from in vivo preclinical data by using the Wajima approach. 4111–4126, Copyright (2016), with permission from Elsevier)

Step 2: correction for interspecies differences, such as differences in tissue and plasma protein binding, blood cell binding, and target receptor binding (potency)

Step 3: prediction of human pharmacokinetic parameters

Step 4: prediction of human dose–responses for dose selection in phase I protocols

11.8 Preclinical or Non-clinical Development

Whether drug project is follow-up of an advanced preclinical or clinical compound (hunt of back-ups or 'Best in class') or entirely new project ('First in class'), different stages of drug discovery are in continuum and often parallel in vitro and in vivo activities are ongoing simultaneously (Fig. 11.13). Data generated in early discovery (LI and LO) is intended for screening best compounds for preclinical development, and on that basis, shortlisted compounds are then profiled for detailed PK characterisation which often involve bespoke in vitro/in vivo studies (e.g. bile duct cannulation studies in rodents). Preclinical or non-clinical development majorly constitutes the activities that bridge drug discovery in the laboratory to initiation of human clinical trials, defined by the filing of an Investigational New Drug (IND) application to regulatory authorities. Therefore, preclinical development is final round of preparations to meet all regulatory requirements of drug project to go first-in-human (FIH) studies. Preclinical studies can be designed to develop the best procedure for new drug scale-up; select the best formulation; determine the route, frequency and duration of exposure; and ultimately support the intended clinical trial design. Toxicology and safety studies identify potential target organs for adverse effects and define the safety margin/therapeutic index to set the initial starting doses in clinical trials. The non-clinical data in PD, PK and toxicology and their translation to human are important basis for planning and conduct of a FIH early clinical trial. Once a lead candidate is nominated as CD for intent of full development, a typical preclinical development plan is charted out and generally consists of six major efforts: (1) manufacture of drug substance (DS)/active pharmaceutical ingredient (API), (2) pre-formulation and formulation (dosage design), (3) analytical and bioanalytical methods development and validation, (4) metabolism and pharmacokinetics, (5) toxicology (both safety and genetic toxicology), and (6) good manufacturing practice (GMP) production and documentation of drug product for use in clinical trials.

Since this is final stage prior to human dosing, studies (e.g. pivotal preclinical safety studies) are stringently controlled by regulatory requirements of Good Laboratory Practices (GLP) and GMP as defined by FDA, International Conference on Harmonisation (ICH) and EMA.

At the preclinical development phase, the primary methods used to assess safety include single- and repeat-dose toxicology studies conducted in rodent and non-rodent species. Definitive animal studies establish the safety characteristics, including the no observable adverse effect level (NOAEL), of the CD. Contrary to the desired pharmacological effect, the mechanism of toxic effects and their relation to exposure is often poorly understood at this stage. Due to this reason, toxicology studies often provide a qualitative or semi-quantitative description of toxicity, i.e. which organs are affected, severity and reversibility of the effect with thresholds such as no observable adverse effect (NOAEL) or lowest observable adverse effect levels (LOAEL). Therefore, the human equivalent dose for safety (HED) is considered to be the dose that leads to human exposure similar to that at the NOAEL in the most sensitive animal species. Development of empirical, mechanistic PKPD and quantitative systems toxicology models based on target organs and anticipated toxicities commensurate to specific therapy areas is useful to derive more reliable and quantitative estimates of therapeutic margins [45].

11.9 Clinical Development and Beyond (Phase I, Phase II, Phase III, and Phase IV/Post-marketing Surveillance

11.9.1 Phase I

Drug development in humans follows standardised pathway to approval for human use, mostly in the form of a checklist laid out in regulatory guidance's that form basis of approval for use in man. Pharmacokinetics continue to play significant role and considerable studies in clinical pharmacology package of drug project are designed to further refine understanding of drug exposure in humans in different clinical scenarios (Fig. 11.11).

Phase I studies are designed mainly to investigate the safety/tolerability (if possible, identify maximum tolerated dose, MTD), pharmacokinetics and pharmacodynamics of an investigational drug in humans. Single-ascending-dose studies (SAD) comprise randomised, placebo-controlled study with small numbers of subjects (healthy volunteers or patients) dosed with either the drug or placebo, and safety is monitored by recording adverse events, clinical laboratory measurements, vital signs, electrocardiograms, and additional tests depending on concerns raised in the animal studies or from the known pharmacology. Sometimes upper limit on PK parameters of exposure is pre-set and escalation schedule or planned doses may be revised depending upon the outcome of the previously dosed groups. To characterise a PK profile, sampling is optimised to capture absorption, peak concentrations, distribution and elimination phases, and would minimally be 2–3 times the elimination half-life, preferably 4–5 times the elimination half-life. Emerging data from study is evaluated concurrently, for example, once at least two dose levels have been administered, it may be checked if the increase in exposure is proportional to the increase in dose (dose proportionality) and if so to predict what exposures might be at the next dose level, given that dose proportionality continues to the next dose.

In multiple ascending dose (MAD) study the drug is administered repeatedly as per predefined schedule (dosage/regimen) for the number of doses. Aim is to reach steady-state levels for the highest exposure a given drug regimen will achieve. This is done at increasing dose levels up to maximum possible dose without any clinical safety issue. Again, the main purpose of the study is to determine safety at maximum exposures, but PK at these exposures is applicable to the design of the next study in the drug development programme. If PK properties after a single dose are known, then the number of repeated doses given at equal intervals for a duration of approximately 5 times the half-life will reach predictable steady-state levels.

Confirmation of steady state in this type of study is usually assessed by determining if trough (pre-dose) concentrations for the last few doses are approaching a constant value. Taken together, SAD and MAD PK studies provide information on general PK characteristics, variability, linearity, dose proportionality, steady-state parameters (accumulation, time-dependency) and preliminary exploration of drug elimination (urine PK, metabolite identification).

Human ADME study or mass balance study is conducted to understand the full clearance mechanisms of the drug and its metabolites in humans. This study is typically single dose PK study with healthy males (n = 4–6) radiolabelled (C14) drug is given by intended route of administration. Generally, plasma, urine and faecal samples are analysed to measure parent and metabolite concentrations. Study is aimed to confirm primary mechanism(s) of elimination/excretion from the body and proportion of parent drug that is converted to metabolite(s). Exploratory human ADME studies can be conducted using micro dosing method. Microdosing is an approach to early drug development where exploratory pharmacokinetic data are acquired in humans using inherently safe sub-pharmacologic doses of drug. The data from a microdose study are exploratory and need to be carefully contextualised when scaling results from a sub-pharmacologically active dose to the higher doses relevant to clinical use.

Bioavailability (BA) and Bioequivalence (BE) studies are key PK studies to evaluate the rate (C_{max}, T_{max}) and extent (AUC) of absorption of drug from a test formulation (vs. reference formulation). BE means that the identical active pharmaceutical ingredient of two medicines have the same rate and extent of absorption. BA study is typically, crossover, single dose (if linear PK) study in healthy subjects where concentration of parent drug and major active metabolites is measured in blood/plasma for ≥ 3 $t½$. BE studies are crossover study in which fasted healthy subjects are given single doses of test & reference products administered at same molar doses and blood/plasma concentration of only parent drug is measured. ‘Pivotal’ BE study is required to bridge the to-be-marketed formulation (test formulation) to that used in Phase 3 clinical trials (reference formulation). These studies help understanding relative BA, absolute BA of drug from a formulation and BE (no significant difference in BA) of test *versus* reference drug products.

Clinical pharmacology package normally accounts for impact of all extrinsic (food effect, smoking, comedications) and intrinsic factors (age, race, organ dysfunction, disease state, gender, genetics, pregnancy/lactation) that can impact PK in humans [46]. Information from all these studies is important to characterise the PK of drug in various routine clinical settings and forms important part of drug dossier to regulatory authorities as well as end up on drug label of final medicine.

During Phase 1, sufficient information about the drug’s pharmacokinetics and pharmacological effects should be obtained to permit the design of well-controlled, scientifically valid, Phase 2 studies. The total number of subjects and patients included in Phase 1 studies varies with the drug, but is generally in the range of 20 to 80. Data from Phase 1 studies is also used for ‘back translation’ from humans to animals as a ‘learn and confirm’ exercise to update/refine preclinical strategies that enable improved translation of back up lead molecules in drug project.

Data from all these PK studies is extensively analysed primarily by population pharmacokinetic (PPK or PopPK) and PBPK approaches. Population PK analysis is a well-established, quantitative method that can quantify and explain the variability in drug concentrations among individuals. Population parameters were originally estimated either by fitting the combined data from all the individuals, ignoring individual differences (the 'naive pooled approach'), or by fitting each individual's data separately and combining individual parameter estimates to generate mean (population) parameters (the 'two-stage approach'). However, in more recent times, sophisticated statistical tools are employed to do population modelling where non-linear mixed-effect (NLME) approaches are used. In this approach of PopPK, PK is studied at the population level while data from all individuals in a population are evaluated simultaneously using a non-linear mixed-effects model. 'Non-linear' refers to the fact that the dependent variable (e.g. concentration) is non-linearly related to the model parameters and independent variable(s). 'Mixed-effects' refers to the fact that model is built in using combination of parameters that do not vary across individuals ('fixed effects') and parameters that vary across individuals ('random effects'). Population models comprise several components: structural models, stochastic models and covariate models. Structural models are functions that describe the time course of a measured response and can be represented as algebraic or differential equations. Stochastic models describe the variability or random effects in the observed data, and covariate models describe the influence of factors such as demographics or disease on the individual time course of the response.

PopPK can be applied to various stages in clinical development (Fig. 11.21). The population-based PKPD analysis may be used to estimate population parameters of a therapeutic response of a drug in phases I and IIb, and the simulations from developed model may increase the efficiency and specificity of drug development by suggesting more informative designs and analyses of experiments. The population approach can also be applied to phases IIa and III of drug development to gain information on drug safety/efficacy and to gather additional information on drug pharmacokinetics (and pharmacodynamics) in special populations, such as the elderly; it is also useful in post-marketing surveillance (phase IV) studies.

11.9.2 Phase II, III and IV

Phase II includes the well-controlled, closely monitored clinical studies conducted to evaluate the effectiveness of the drug for a particular indication or indications in patients with the disease or condition and to determine the common short-term side effects/risks associated with the drug. Phase II studies are typically conducted in a relatively small number of patients, usually involving no more than several hundred subjects. Phase II studies are sometimes divided into Phase IIA and Phase IIB, where Phase IIA trials are specifically designed to assess optimum dosing ('definite dose-finding' studies) at which the drug shows biological activity with minimal side

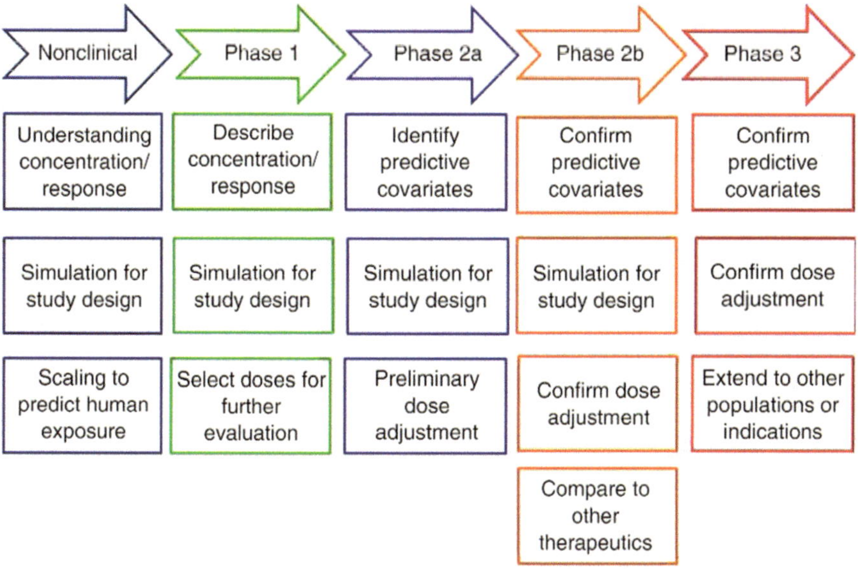

Fig. 11.21 Deliverables of modelling and simulation at different stages of development. (Reprinted from [47] Clin Pharmacol Ther. Pharmacometrics Syst Pharmacol (2012), 1(9), Mould DR, Upton RN, Basic concepts in population modeling, simulation, and model-based drug development. 1–14. Creative Commons Attribution License (Attribution 3.0 Unported), Wiley)

effects, and Phase IIB studies are usually pilot studies designed to demonstrate clinical efficacy or biological activity ('proof of concept' studies).

Phase III studies are randomised multi-centre, controlled and uncontrolled trials in large patient population. These trials compare new treatments with the best currently available treatment (the standard treatment). PopPK-PD models are further evolved on emerging Phase II and Phase III data to support modelling and simulation activities (Fig. 11.21). Phase IV trials (Post-marketing surveillance) involve the safety surveillance (pharmacovigilance) and ongoing technical support of a drug after it is approved.

11.10 Summary and Conclusions

As evident from aforementioned sections, one may appreciate that understanding of PK plays a very significant role in selection and development of new chemical entity (NCE) as prospective drug. Recently published work analyses successes and failures of oncology drug projects and based on learnings, developed a question-based approach called Pharmacological Audit Trail (PhAT) to identify viable drug projects (Fig. 11.22) [48]. At the centre of PhAT philosophy is patient population followed

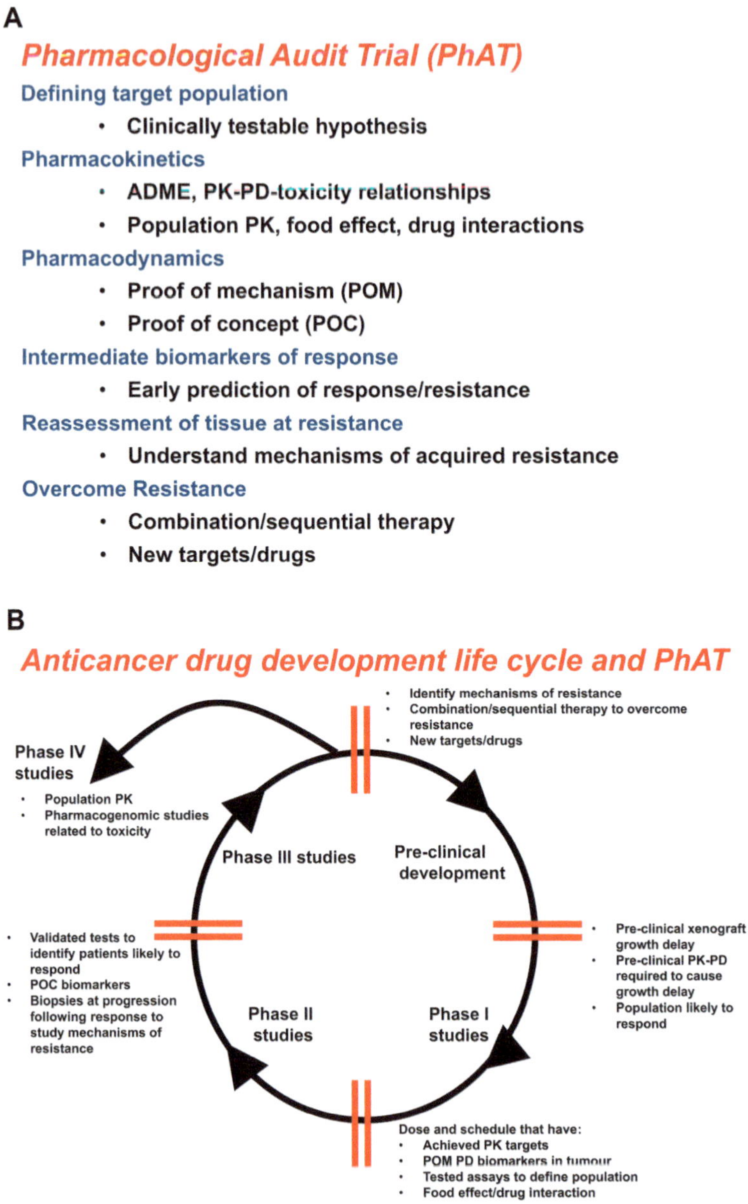

Fig. 11.22 The pharmacological audit trail (PhAT). (**a**) The six crucial aspects of the pharmacological audit trail. (**b**) The relationship of the PhAT to the various phases of the life cycle of an anticancer drug. The red lines indicate 'checkpoints' between different phases of drug development where go-no-go decisions are made. (Reprinted from [48] Semin Oncol, 43(4), Banerji U and Workman P, Critical parameters in targeted drug development: the pharmacological audit trail. 436–445, Copyright (2016), with permission from Elsevier)

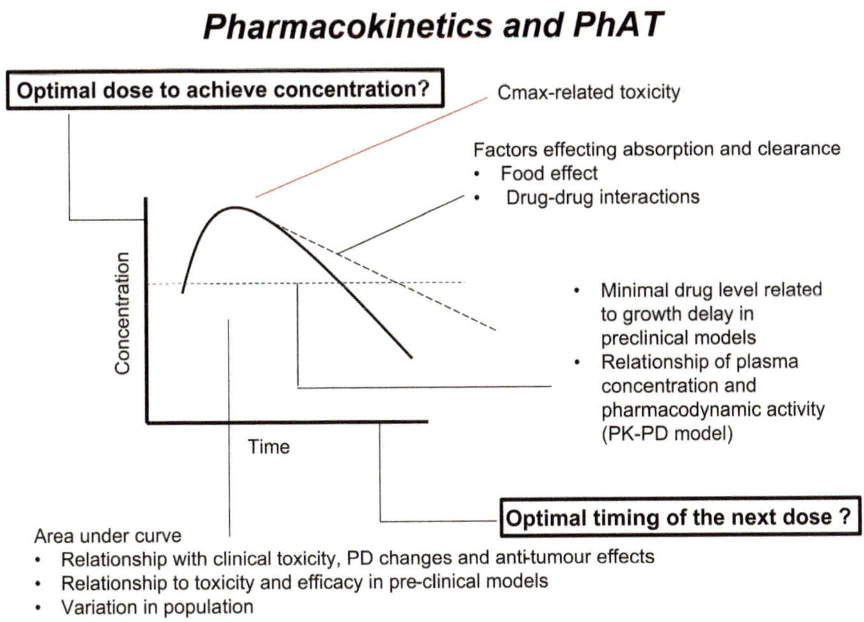

Fig. 11.23 Pharmacokinetics in the PhAT. The two main questions asked using PK analysis in early phase drug development are highlighted. Multiple parameters studied within the umbrella of pharmacokinetics and their clinical importance are also shown. (Reprinted from [48] Semin Oncol, 43(4), Banerji U and Workman P, Critical parameters in targeted drug development: the pharmacological audit trail. 436–445, Copyright (2016), with permission from Elsevier)

by pharmacokinetics, pharmacodynamics and biomarkers. Although, work focuses on oncology portfolio, the core philosophy stands equally viable to any other disease area. Figure 11.23 shows typical plasma concentration graph of drug after oral administration and how different parts of this graph play role in answering PhAT questions that are key determinant to efficacy and safety. In general, following is multidimensional impact of drug PK in discovery and development:

- Optimising the exposure of drug at target site to enable therapeutic efficacy. This is generally two-step process. First is to ensure appropriate concentration levels are achieved in preclinical species to drive measurable and meaningful pharmacological effect. Second, using translation approaches to predict exposure in humans that causes therapeutic activity in preclinical models. This two-step process is very central to drug projects and plays an important part in go-no-go decisions during phase I studies.
- Understanding of the drug exposure at unintended targets (secondary pharmacology receptors) eliciting undesirable side effects/safety issues. Toxicokinetics (TK) is use of PK studies to help clarify toxicity, especially by correlating PK profiles with dose-limiting toxicities. Some drugs have toxicities related to C_{max} (the maximal concentration achieved in plasma) and adjusting the dose of the

drug such that the C_{max} falls below levels that cause unacceptable toxicity is a critical aspect of Phase I studies.

Recommending appropriate dosing schedule (dose amount and time interval between successive doses) in preclinical and clinical studies. This involves quantifying rate of input (absorption) and output (elimination/clearance) in preclinical species and in humans.

- Identifying effect of extrinsic environmental factors (e.g. comedications and food) and intrinsic factors (e.g. disease state and genotype) on PK of drug, which may have cascading impact on efficacy and/or safety.
- Selection of most promising drug-like compound by ranking/comparing their ADME properties that are desired to be key to reaching the therapeutic target and then maintaining effective concentrations for reasonable period of time.

In conclusion, pharmacokinetics is key scientific discipline supporting drug discovery and development throughout various stages of drug project. It contributes significantly to understanding extent and duration of drug exposure, which in turn determines efficacy and safety, helps to decide optimum dose and schedule in patients, adaptation of these doses in clinical settings for diseased patients of different ethnicity and age and in presence of food and/or comedications and/or food.

Disclosures and Conflict of Interest Pradeep Sharma, Nikunjkumar Patel, Bhagwat Prasad and Manthena V.S. Varma are employees and/or shareholders of AstraZeneca, Simcyp Division (Certara UK Limited), Washington State University and Pfizer respectively. Simcyp® is properitary PBPK modelling software platform of Simcyp division of Certara UK Ltd.

References

1. Gabrielsson J, Weiner D. Pharmacokinetic & pharmacodynamic data analysis: concepts and applications. 4th ed. Stockholm: Swedish Pharmaceutical Press; 2017.
2. MacGowan AP. Role of pharmacokinetics and pharmacodynamics: does the dose matter? Clin Infect Dis. 2001;33(Suppl 3):S238–9.
3. Sharma P, Chawla HPS, Panchagnula R. The role of sorption promoters in increasing the bioavailability of drugs in oral preparations. Drugs Future. 1999;24(11):1221–40.
4. Varma MV, Khandavilli S, Ashokraj Y, Jain A, Dhanikula A, Sood A, et al. Biopharmaceutic classification system: a scientific framework for pharmacokinetic optimization in drug research. Curr Drug Metab. 2004;5:375–88.
5. Yang J, Jamei M, Yeo KR, Rostami-Hodjegan A, Tucker GT. Misuse of the well-stirred model of hepatic drug clearance. Drug Metab Dispos. 2007;35(3):501–2.
6. Toutain PL, Bousquet-Mélou A. Volumes of distribution. J Vet Pharmacol Ther. 2004;27 (6):441–53.
7. Sharma P, Atari M, Elsby R, Stahl S, Hilgendorf C, Fenner K. Review of characteristics, validation and applications of in silico and in vitro models of drug transporters. In: Nicholls G, Youdim K, editors. Drug transporters: role and importance in ADME and drug development. Cambridge: Royal Society of Chemistry; 2016. p. 231–97.

8. Prasad B, Achour B, Artursson P, Hop CECA, Lai Y, Smith PC, et al. Toward a consensus on applying quantitative liquid chromatography-tandem mass spectrometry proteomics in translational pharmacology research: a white paper. Clin Pharmacol Ther. 2019;106(3):525–43.
9. Ohtsuki S, Schaefer O, Kawakami H, Inoue T, Liehner S, Saito A, et al. Simultaneous absolute protein quantification of transporters, cytochromes P450, and UDP-glucuronosyltransferases as a novel approach for the characterization of individual human liver: comparison with mRNA levels and activities. Drug Metab Dispos. 2012;40(1):83–92.
10. Iwatsubo T, Hirota N, Ooie T, Suzuki H, Shimada N, Chiba K, et al. Prediction of in vivo drug metabolism in the human liver from in vitro metabolism data. Pharmacol Ther. 1997;73 (2):147–71.
11. Davies B, Morris T. Physiological parameters in laboratory animals and humans. Pharm Res. 1993;10(7):1093–5.
12. Bayliss MK, Bell JA, Jenner WN, Wilson K. Prediction of intrinsic clearance of loxtidine from kinetic studies in rat, dog and human hepatocytes. Biochem Sot Trans. 1990;18:1198–9.
13. Varma MV, Steyn SJ, Allerton C, El-Kattan AF. Predicting clearance mechanism in drug discovery: extended clearance classification system (ECCS). Pharm Res. 2015;32 (12):3785–802.
14. Paine SW, Barton P, Bird J, Denton R, Menochet K, Smith A, et al. A rapid computational filter for predicting the rate of human renal clearance. J Mol Graph Model. 2010;29(4):529–37.
15. Paine SW, Ménochet K, Denton R, McGinnity DF, Riley RJ. Prediction of human renal clearance from preclinical species for a diverse set of drugs that exhibit both active secretion and net reabsorption. Drug Metab Dispos. 2011;39(6):1008–13.
16. Izumi S, Nozaki Y, Komori T, Takenaka O, Maeda K, Kusuhara H, et al. Comparison of the predictability of human hepatic clearance for organic anion transporting polypeptide substrate drugs between different in vitro-in vivo extrapolation approaches. J Pharm Sci. 2017;106 (9):2678–87.
17. Sharma P, Butters CJ, Smith V, Elsby R, Surry D. Prediction of the in vivo OATP1B1-mediated drug-drug interaction potential of an investigational drug against a range of statins. Eur J Pharm Sci. 2012;47(1):244–55.
18. DeGoey DA, Chen HJ, Cox PB, Wendt MD. Beyond the rule of 5: lessons learned from AbbVie's drugs and compound collection. J Med Chem. 2018;61(7):2636–51.
19. Varma MV, Gardner I, Steyn SJ, Nkansah P, Rotter CJ, Whitney-Pickett C, et al. pH-Dependent solubility and permeability criteria for provisional biopharmaceutics classification (BCS and BDDCS) in early drug discovery. Mol Pharm. 2012;9(5):1199–212.
20. Johnson TW, Dress KR, Edwards M. Using the golden triangle to optimize clearance and oral absorption. Bioorg Med Chem Lett. 2009;19:5560–4.
21. Reichel A, Lienau P. Pharmacokinetics in drug discovery: an exposure-centred approach to optimising and predicting drug efficacy and safety. Handb Exp Pharmacol. 2016;232:235–60.
22. Morgan P, Brown DG, Lennard S, Anderton MJ, Barrett JC, Eriksson U, et al. Impact of a five-dimensional framework on R&D productivity at AstraZeneca. Nat Rev Drug Discov. 2018;17 (3):167–81.
23. Andersson S, Armstrong A, Björe A, Bowker S, Chapman S, Davies R, et al. Making medicinal chemistry more effective—application of Lean Sigma to improve processes, speed and quality. Drug Discov Today. 2009;14(11–12):598–604.
24. Li C, Liu B, Chang J, Groessl T, Zimmerman M, He YQ, et al. A modern in vivo pharmacokinetic paradigm: combining snapshot, rapid and full PK approaches to optimize and expedite early drug discovery. Drug Discov Today. 2013;18(1–2):71–8.
25. Derendorf H, Meibohm B. Modeling of pharmacokinetic/pharmacodynamic (PK/PD) relationships: concepts and perspectives. Pharm Res. 1999;16:176–85.
26. Meibohm B, Derendorf H. Pharmacokinetic/pharmacodynamic studies in drug product development. J Pharm Sci. 2002;91(1):18–31.
27. Wajima T, Yano Y, Fukumura K, Oguma T. Prediction of human pharmacokinetic profile in animal scale up based on normalizing time course profiles. J Pharm Sci. 2004;93(7):1890–900.

28. Mahmood I. Application of allometric principles for the prediction of pharmacokinetics in human and veterinary drug development. Adv Drug Deliv Rev. 2007;59(11):1177–92.
29. Jang GR, Harris RZ, Lau DT. Pharmacokinetics and its role in small molecule drug discovery research. Med Res Rev. 2001;21:382–96.
30. Garrett ER. The Bateman function revisited: a critical re-evaluation of the quantitative expressions to characterize concentrations in the one compartment body model as a function of time with first-order invasion and first-order elimination. J Pharmacokinet Biopharm. 1994;22 (2):103–28.
31. Rostami-Hodjegan A, Tucker GT. Simulation and prediction of in vivo drug metabolism in human populations from in vitro data. Nat Rev Drug Discov. 2007;6(2):140–8.
32. Patel N, Polak S, Jamei M, Rostami-Hodjegan A, Turner DB. Quantitative prediction of formulation-specific food effects and their population variability from in vitro data with the physiologically-based ADAM model: a case study using the BCS/BDDCS Class II drug nifedipine. Eur J Pharm Sci. 2014;57:240–9.
33. Pathak SM, Schaefer KJ, Jamei M, Turner DB. Biopharmaceutic IVIVE-mechanistic modeling of single- and two-phase in vitro experiments to obtain drug-specific parameters for incorporation into PBPK models. J Pharm Sci. 2019;108(4):1604–18.
34. Schwenger E, Reddy VP, Moorthy G, Sharma P, Tomkinson H, Masson E, et al. Harnessing meta-analysis to refine an oncology patient population for physiology-based pharmacokinetic modeling of drugs. Clin Pharmacol Ther. 2018;103(2):271–80.
35. Jamei M, Bajot F, Neuhoff S, Barter Z, Yang J, Rostami-Hodjegan A, et al. A mechanistic framework for in vitro-in vivo extrapolation of liver membrane transporters: prediction of drug-drug interaction between rosuvastatin and cyclosporine. Clin Pharmacokinet. 2014;53 (1):73–87.
36. Poulin P, Theil FP. Development of a novel method for predicting human volume of distribution at steady-state of basic drugs and comparative assessment with existing methods. J Pharm Sci. 2009;98(12):4941–61.
37. Berezhkovskiy LM. Volume of distribution at steady state for a linear pharmacokinetic system with peripheral elimination. J Pharm Sci. 2004;93(6):1628–40.
38. Rodgers T, Rowland M. Mechanistic approaches to volume of distribution predictions: understanding the processes. Pharm Res. 2007;24(5):918–33.
39. United States Food and Drug Administration. Guidance for Industry: physiologically Based Pharmacokinetic Analyses—Format and Content. December, 2016. https://www.fda.gov/downloads/Drugs/GuidanceComplianceRegulatoryInformation/Guidances/UCM531207.pdf. Accessed 28 Nov 2019.
40. European Medicines Agency. Guideline on the qualification and reporting of physiologically based pharmacokinetic (PBPK) modelling and simulation. 2016. http://www.ema.europa.eu/docs/en_GB/doc. Accessed 28 Nov 2019.
41. Shebley M, Sandhu P, Emami Riedmaier A, Jamei M, Narayanan R, Patel A, et al. Physiologically based pharmacokinetic model qualification and reporting procedures for regulatory submissions: a consortium perspective. Clin Pharmacol Ther. 2018;104(1):88–110.
42. Grimstein M, Yang Y, Zhang X, Grillo J, Huang SM, Zineh I, et al. Physiologically based pharmacokinetic modeling in regulatory science: an update from the U.S. Food and Drug Administration's Office of Clinical Pharmacology. J Pharm Sci. 2019;108(1):21–5.
43. Vuppugalla R, Marathe P, He H, Jones RD, Yates JW, Jones HM, et al. PhRMA CPCDC initiative on predictive models of human pharmacokinetics, part 4: prediction of plasma concentration-time profiles in human from in vivo preclinical data by using the Wajima approach. J Pharm Sci. 2011;100(10):4111–26.
44. Lowe PJ, Hijazi Y, Luttringer O, Yin H, Sarangapani R, Howard D. On the anticipation of the human dose in first-in-man trials from preclinical and prior clinical information in early drug development. Xenobiotica. 2007;37(10-11):1331–54.
45. Shankaran H, Cronin A, Barnes J, Sharma P, Tolsma J, Jasper P, Mettetal JT. Systems pharmacology model of gastrointestinal damage predicts species differences and optimizes

clinical dosing schedules. Clin Pharmacol Ther Pharmacometr Syst Pharmacol. 2018;7 (1):26–33.
46. Huang SM, Rowland M. The role of physiologically based pharmacokinetic modeling in regulatory review. Clin Pharmacol Ther. 2012;91(3):542–9.
47. Mould DR, Upton RN. Basic concepts in population modeling, simulation, and model-based drug development. Clin Pharmacol Ther Pharmacometr Syst Pharmacol. 2012;1(9):1–14.
48. Banerji U, Workman P. Critical parameters in targeted drug development: the pharmacological audit trail. Semin Oncol. 2016;43(4):436–45.

Chapter 12
Regulatory Toxicology Testing of Pharmaceuticals

Venkatesha Udupa and K. S. Rao

12.1 Introduction

This chapter on regulatory toxicology testing in pharmaceuticals is intended for a beginner in pharmaceutical drug development to get familiar with regulatory aspects of drug safety testing. New molecular entities have to be subjected to battery of nonclinical safety assessment before first dose in human, and prior to their marketing approval. Safety testing of drugs and the accompanying regulations have been one of evolution due to demand by the public following, in some cases, unintended catastrophic toxic episodes of human suffering due to drug exposure. It is important for a newcomer in drug development to understand the conceptual basis of safety testing. It is important to be flexible to tweak the development program based on the type of the drug and its use pattern and to some extent consider any potential abuse or misuse of the drug. Unified regulatory approaches for the nonclinical studies to support the drug development were put together by continued effort by three largest regions (United States, Europe and Japan), which was collectively published through the International Conference for Harmonization (ICH) compendium of guidance. Much of the following discussion relates to small molecules, although in many cases the same concepts are applied to biotechnology-derived products.

V. Udupa (✉)
Glenmark Pharmaceuticals Limited, Mumbai, India
e-mail: venkatesha.udupa@glenmarkpharma.com

K. S. Rao
Eurofins Advinus Limited, Bengaluru, India

R. Poduri (ed.), *Drug Discovery and Development*,
https://doi.org/10.1007/978-981-15-5534-3_12

12.2 Early Years of Drug Development and Marketing

Due to a lack of regulatory laws, before federal legislation, products were being sold over-the-counter (OTC) to buyers, with assurance to cure diseases like asthma, diabetes, cancer, etc. There were limited laws prescribing safety and efficacy testing, the content, labeling, or advertising. These "patented medicines" were prevalent and included many products that are yet to be recognized currently, including Ex-Lax, Listerine, Sominex, Vaseline, etc.

The first significant legislation passed by the US congress was the Pure Food and Drug Act in 1906, which had more to do with tainted and adulterated food than drug regulation. This legislation consisted strictly of a labeling law prohibiting the marketing of processed foods and drugs that were misbranded. No approval process by the FDA was involved, and enforcement relied on post-marketing surveillance and prosecution. The 1906 Food and Drugs Act, the first primary federal law governing therapeutic drugs, gave the FDA the authority to intercede in marketing drugs that were adulterated or misbranded concerning the article's identity and its strength centration thereof. This act required the FDA to pursue and prove fraudulent claims before a product could be taken off the market. There were no requirements for the manufacturer to prove the statements of ingredients before using them. The burden of proof was on the government to prove the claim was false or misleading after the product was already on the market being sold. Efficacy was not a consideration until 1911 when the Sherley Amendment outlawed fraudulent therapeutic claims.

12.3 Good Manufacturing Practice (GMP)

Manufacturing of any drugs must be compliant to "good manufacturing practice" [1]; otherwise, it is considered misbranded. FDA laid down the standards of "good manufacturing requirement" for drug manufacturers to avoid selling substandard products to consumers. To ensure such practices being followed, FDA routinely inspects the facility and issues warning or abandon the companies if they do not meet the compliance. Agency made it mandatory to comply to GMP inspection for product approval or supplemental new drug applications (NDA) since 1991.

12.4 Alternative Means of Accelerating Registration of Drugs

A decade after the 1962 amendments, it was apparent that the effectiveness standard, coupled with the requirement of affirmative FDA approval, cost patients more (by slowing the introduction of new drugs) than it saved them (by preventing other thalidomide incidents). Defenders of the proof-of-effectiveness standard have sought to justify both Congress's central choice and the FDA's implementation [2]. This debate provides the background for examining two efforts by Congress to compensate for the new drug approval system's effects. A series of means were implemented, allowing faster access of drugs to the market and helping patients. These will be discussed in this section.

12.4.1 Drug Efficacy Study Implementation (DESI) Program

Under the DESI program, FDA would review the effectiveness of drugs approved between 1938 and 1962. Prior to Drug Amendments of 1962, sponsors of NDAs had to prove their products' safety only. The proof of effectiveness and safety was needed to all drugs approved based on safety alone. Thus, when the 1962 amendments came into effect, a number of drugs were on the market whose effectiveness was far from satisfactory or not known. The regulators acted on with the legislation: §102(e) of the legislation amended FDCA §505(e) to require the FDA to withdraw prior approval of a drug if it found: *"based on new information before [it] concerning such drug, evaluated together with the evidence available to [it] when the application was approved, that there is a lack of substantial evidence that the drug will have the effect it purports or is represented to have."*

The DESI review began in 1966 in association with the National Academy of Sciences/National Research Council (NAS-NRC). See 31 Fed. Reg. 9425 (Jul. 9, 1966). While DESI review has been a long, slow process. Most of the NDAs approved in the period 1938 to 1962 have been reviewed till now. The DESI program was intended to classify all pre-1962 drugs that were already on the market as either effective, ineffective, or needing further study. However, the drugs marketed prior to 1938 were allowed to be sold because they were Generally Recognized As Safe [3] and effective.

12.4.2 First Generic

The history of generic drugs started nearly hundred years back. Bayer filed a court case to keep generic versions off the shelves. The company lost in court, and consumers had choices in plethora of generic aspirin. Today, generic drugs are both widely available and carefully regulated as per *Generic Drugs 1996* [4]. First generics are important to public health, and regulatory authorities in general prioritizes review of these submissions.

First-Time Generics are the very first generic versions of marketed brand-name drug products to be approved by the FDA. The generic version is developed to produce identical effect and therapeutic benefit as the brand-name product.

Generic versions are introduced in the market after the patent on a brand-name drug expires. The implementation of the Drug Price Competition and Patent Term Restoration Act[6], commonly known and much acclaimed as the Hatch-Waxman Act the era of generic drugs started.

As a resultant of the act it is easier and affordable to introduce a new generic drug to market. This is mainly due to not to undergo time-consuming and cost-prohibitive human trials. Manufacturers merely had to prove that their drug had the same active pharmaceutical ingredients, and they produce identical response as that of brand-name drug. The implementation of the Act paved way to new generic drug industry. Approximately one half of the prescribed drugs are generic drugs which are affordable, and it is a great relief to the patients who do not have third party payment facility.

12.4.3 Prescription Drug User Fee Act (PDUFA)

The PDUFA was established and enacted in 1992 to streamline the process and reduce the time to approve the new drugs by introducing the new fee structure. The funds collected were used for engaging more reviewers, which resulted in a drop-in approval time of the new drugs.

12.5 Toxicological Testing for Typical Drug Registration

In the early days of drug (and chemical warfare agents) research, it was common practice to use humans (prisoners, etc.) to understand the harmful effects of drugs. It is unethical to use humans as the first species to test drugs; hence it has become routine to use laboratory animals for testing (both for pharmacology and toxicology) drugs. Nonclinical safety testing is a critical aspect of new molecular entity development before it is first used in humans. The conventional wisdom in toxicology for registration of a drug depends on the type of the drug, therapeutic area, population to be treated, and treatment duration. Nevertheless, preclinical toxicology studies design and execution to determine safety before start of a clinical trial is axiomatic. It has become a standard practice to use laboratory animals for toxicology studies.

12.5.1 Why Use Animals for Drug Toxicology Studies?

The results obtained in laboratory animals, which are acclimatized in controlled environment and monitoring the physiological, biochemical, histological, and genetic characteristics before and after the drug administration, can be extrapolated to humans fairly well. Besides, due to the widespread use of animals in drug toxicology over several years, sizeable historical control data have been generated on various drugs that can be used for the design of clinical trials. Hence, in vivo models are commonly used in the preclinical development of new drugs to predict its pharmacological and pharmacokinetic profile in humans. The essence of toxicity testing is not just to check how safe a test substance is, but to characterize the possible toxic effects it can produce. The guiding principles of toxicity testing are to check the effect of the test substances on laboratory animals and anticipate toxic effect on humans. Further, high doses of test drug are administered to laboratory animals to evaluate its possible hazards on humans, who are normally exposed at much lower dose.

12.5.2 Which Animal Species to Use for Drug Toxicology Studies?

It is fair to say that no single species is ideal for all drugs, which gives relevant information that can be used with certainty for predicting human adverse effects. The selection of right species for toxicological studies of a new drug depends on ADME profile, species tolerance, pharmacological activity, and relevance to humans. Regulatory agencies generally recommend rats for repeated dose studies, carcinogenicity, and reproductive toxicology while mice for carcinogenicity studies. On the other hand, for short-term and chronic studies dogs are preferred. Similarly, route of administration also determines the species selected for the toxicological studies. Rats are used for peroral and inhalation studies. To study the toxic effects of the test substance on embryofetal development and local tolerance, rabbits are used. Guinea pigs are used mainly for sensitization testing and are rarely used for embryofetal toxicology studies. Rationale must be provided for the use of species other than those outlined above. Transgenic mice (e.g., rasH2) are the preferred strain for conducting carcinogenicity testing with 6 months of test article administration.

Retrospective analysis of more than 150 drugs has shown that the concordance rate for true human toxicity would be approximately 71% and with rodent and non-rodent species, 43 and 63%, respectively. Hence, the present safety testing is required in two mammalian species, one rodent (rat or mouse) and one non-rodent (dog, minipig, or nonhuman primate), before administration of potential new medicines to man. However, there are incidences where only one species may be acceptable like monkeys in case the drug does not cross-react (e.g., antibodies) with rodent species. Although animal species will be selected based on scientific rationale with the data available before first-in-man trial Criteria for Dose Selection and Route of Administration [5].

The highest dose selected in general toxicology studies should produce observable toxicity as defined by decrements either in weight gain or food consumption and/or alterations in clinical observations, clinical pathology, or organ weight-related alterations. The low dose which is equivalent to proposed human therapeutic dose should ideally demonstrate NOEL. However, depending on the level of off-target engagement, it is not uncommon to find other effects at the low dose that are related to the pharmacology of the drug candidate, but which are deemed to be not adverse. This would be an example of a dose that can be considered as NOAEL.

The five general criteria for defining the high dose are:

1. Maximum tolerated dose
   (a) A dose where clinical signs of toxicity is seen and target-organ toxicity is anticipated without mortality.
2. Maximum feasible dose (MFD)
   (a) A dose that cannot be administered at a high level due to formulation limitation reasons or physicochemical properties of the active pharmaceutical ingredient (API).
3. 50-fold margin dose
   (a) Maximum safety margin based on mean area under the curve (AUC) values of the API or the pharmacologically active moiety of a pro-drug in systemic exposure clinical conditions.
4. Limit dose
   (a) In general, the limit dose does not provide information to predict safety in human. Limit doses for acute and repeat dose studies of 1000 milligrams (mg)/kilogram (kg)/day for rodents and non-rodents are recommended. *However, "if the abovementioned dose does not attain the mean exposure margin of tenfold to the therapeutic condition and the dose exceeds 1 gram (g) per day, then the doses in the toxicity studies should be limited by a tenfold exposure margin or a dose of 2000 mg/kg/day or the MFD, whichever is lower"* [5].
5. Saturation of exposure
   (a) Saturation of exposure indicates the limitation to absorption of the drug, under study, wherein the lowest dose that shows the maximum systemic exposure should be taken as the high dose.

Three doses are required to prove a dose-response relationship, which is essential in determining the optimum dose and in determining the margins of safety.

The route of administration of the new drug should be same as that of intended clinical use, and the dosing regimen will depend on the Pk/Pd properties of the test substance. In case the drug is intended to be administered by more than one route in clinical set up, a bridging study also may be designed.

12.5.3 *What Type of Tests to be Conducted?*

There is no standard type or list of toxicology studies to be conducted which is applicable to all drugs. Toxicity studies must be designed such that to obtain a correct risk assessment. Toxicity of substance depends on two factors: the effect and the level of exposure (dose) at which the effect is observed. Some tests are designed specifically to detect a particular effect (such as skin and eye irritancy, skin sensitization, mutagenicity). Other tests are designed to detect a wider range of less-specific effects on organs or body systems and the dose range over which the effect develops.

A successful toxicity testing and selection of doses are made based on intended duration and frequency of administration of the test item in human. Understanding of stability of test item and nature of its breakdown are helpful in designing the types of toxicity tests and for selection of the dosing regimen. The toxicity tests like acute, subchronic, and chronic are characterized by exposure duration; these tests are designed to evaluate the systemic effects, when the test items are administered for single to several days, and are used to assess the human health risk for those exposure durations.

Regulatory toxicology studies for pharmaceuticals are generally performed in rats, mice, rabbits, dogs, and monkeys, with rats and dogs being used at greater frequency. Testing guidelines generally require that common laboratory strains and species are used. Routine toxicology study design includes three dose groups and a control group. For the general toxicology studies, agencies require that the selected high dose exhibit signs of toxicity with no deaths and the lowest dose without any evidence of toxicity, thus a NOEL. The numbers of animals required are defined in each study protocol and range from five rats or 3 dogs/monkeys per sex per dose in 28-day toxicity studies to 10 rats or 4 dogs/monkeys per sex per dose in subchronic studies to 50–60 rats or mice per sex per dose group in carcinogenicity studies. For embryofetal developmental studies, a minimum of 20 L per dose is required. The statistical power of a study is calculated on the basis of total number of animals used and the sensitivity of the end point being evaluated.

12.6 Basic Necessities for Toxicity Testing for Drugs

The adverse effects of a new drug in humans are predicted for the results of animal experimentation either in vitro or in vivo testing that includes but not limited to acute toxicity, genotoxicity, repeat dose toxicity, carcinogenicity, developmental and reproductive toxicity, and local and special toxicity.

The following set of principles enunciated by [6] are accepted by toxicologists. It is important to adopt these principles before commencing a toxicity tests:

1. Use of one/more species which are physiologically similar in responding to the test substance qualitatively and/or quantitatively as similarly as possible to that of humans. For this, absorption, distribution, metabolism, excretion, and other physiological effects might be contemplated [7].
2. Use several doses in the experiments as all toxicologic and pharmacologic actions in human and animals are dose dependent. The use of a single maximum dose which is relatively nontoxic and sufficiently large multiple of that is attainable and practicable [7].
3. Effects produced at higher dose levels that are useful for delineating mechanism of action. For any substance, a dose level exists below which no adverse effects will appear, which will be considered as NOAEL [7].
4. Statistical significance are only valid on the experimental sample are randomized control and treated groups [7].
5. The drug administered by two routes may produce different toxicity in a species; the toxic effects by one route of administration to animal species are not applicable to effects by another route of administration to humans. The variabilities in physicochemical properties of the test substance the ADME profile may vary and it may low or high compared to that of normal profile. The possibility of adverse/side effects is more if the absorption is on higher side.

12.6.1 Drug Substance Purity, Stability, and Storage Conditions

The purity, solubility in aqueous and nonaqueous medium, and stability of drug substance to be used in GLP complied studies should be fully characterized and certified by authorized person. This information is useful in experimental design.

12.6.2 Vehicle, Drug Formulation, and Analysis of the Drug in the Formulation

Before one starts any toxicology studies, in particular pivotal repeat dose toxicity studies, it is critical to decide on the vehicle's biological suitability [8] and the vehicle stability of the drug formulation. It is essential to develop a method of analysis under GLP, which will be used in all GLP toxicology studies to determine the drug concentration in the formulation used in toxicology studies. The latter is essential for all GLP studies of any duration. In all GLP toxicology studies, one must monitor the drug concentration in the actual formulation used in the study. The frequency of such monitoring does vary with the study duration.

12.6.3 Bioanalytical Method for Analysis of the Drug Concentration in Plasma

Before starting any repeat dose GLP toxicology study, it is important to develop a suitable analytical method in plasma (or suitable matrix) under GLP. In all drug toxicology studies, the concentration of the drug in the central compartment plasma as an index of systemic exposure should be estimated.

12.6.4 Toxicokinetic (TK) Assay

Monitoring the drug concentration in the plasma (or blood in some cases) compartment is a standard procedure in all pivotal GLP drug toxicology studies as an index of exposure [9]. Toxicokinetics (TK) is to determine the effect of a drug to test animals at doses that elicit a toxic response. Incorporation of TK in a testing program can play a vital role in toxicity studies, ensuring dosage profile relevant to humans, minimize the generation of unnecessary findings at very high dose and reduce animal welfare concerns. Toxicokinetic data, NOEL/NOAEL can form the basis for clinical investigator to decide an optimum starting and upper doses in the initial Phase I clinical study.

In a typical repeat dose toxicity study, TK is an integral part of the study. In rodents, TK animals are a satellite (subset) group or animals that are distinct from the accompanying main toxicology animals in the repeat dose toxicity study protocol.

Animals: The same animal species should be used in TK studies that are used in Toxicity studies. Sufficient animals should be included in the rodents TK study design to guarantee a minimum of three data samples per time point of analysis.

Chemical: Radiolabeled chemical can be used, and it is optional.

Route of exposure: The route of administration should be same as that in human.

Dose levels: Three doses of the new drug which are used in the toxicity studies are recommended to show dose-response relationship. In case of the nonlinear kinetics observed, more doses may be needed for accurate conclusion.

Samples: The drug under study is administered intravenously, blood samples should be collected at multiple (7–8) time points. The plasma concentration of the drug reaches the LOD of the analytical method at approximately 24 h samples might be collected at 5, 15, 30, and 60 min and 2, 4, 8, and 24 h after dosing. In case the drug is administered other than intravenous route, blood samples should be collected at eight to ten time points sufficient to characterize the drug concentration in plasma or other tissues versus time profiles. However, sampling intervals and frequency will depend on the rate of absorption and clearance of the drug.

Following is a typical study design (Table 12.1) of a repeat dose toxicity study with the inclusion of satellite TK rats:

Table 12.1 Repeat dose toxicity and toxicokinetic (TK) study in rats

Dose group	Dose levels (mg/kg/day)	Number of animals		
				Toxicokinetic
		Main study group	Recovery group	Satellite group[a]
Vehicle control	0	10♂/10♀	5♂/5♀	3♂/♀
Low	TBD[b]	10♂/10♀	–	9♂/9♀
Mid	TBD[b]	10♂/10♀	–	9♂/9♀
High	TBD[b]	10♂/10♀	5♂/5♀	9♂/9♀

[a]Toxicokinetic animals will be used for blood sample collection and analysis, and will not be evaluated with the main study animals
[b]*TBD* To be determined

Table 12.2 Rat blood samples for bioanalysis (BA) towards toxicokinetic (TK) assessment

Treatment group	No. of rats/group/sex	Total no. of rats per group	No. of rats bleeding per time point (♂ + ♀ combined)								
			Sample collection times (hours)								
			T1	T2	T3	T4	T5	T6	T7	T8	Total samples
Control	3	6	6			6		–	–	–	12
Low dose	9	18	–	6	6	6	6	6	6	6	42
Middle dose	9	18	–	6	6	6	6	6	6	6	42
High dose	9	18	–	6	6	6	6	6	6	6	42
Total samples		–	6	18	18	24	18	18	18	18	138

Table 12.3 Rat blood samples for bioanalysis (BA) towards toxicokinetic (TK) assessment on last treatment day

Treatment group	No. of rats/group/sex	Total no. of rats per group	No. of bleeding per time point (♂ + ♀combined)								
			sample collection times (hours)								
			T1	T2	T3	T4	T5	T6	T7	T8	Total samples
Control	3	6	6	–	–	6	–	–	–	–	12
Low dose	9	18	6	6	6	6	6	6	6	6	48
Middle dose	9	18	6	6	6	6	6	6	6	6	48
High dose	9	18	6	6	6	6	6	6	6	6	48
Total samples	–	–	24	18	18	24	18	18	18	18	156

Toxicokinetics: 8 sampling time points (sparse sampling on rotation) on Day 1 (Table 12.2) and last treatment day (Table 12.3) from 3 animals/sex/group for each time point. Parameters: C_{max}, T_{max}, $T_{1/2}$, $AUC_{0\text{-}24h}$, and $AUC_{0\_infinity}$.

TK in Non-rodent (dogs, monkeys, etc.): In non-rodent studies, generally, main toxicology animals are used for TK blood collection. All other features remain the same with the exception of the number of animals which will be in the range of 3–5 per group per sex.

TK Parameters: The following parameters are usually measured using Phoenix WinNonlin, a software:

(a) *Maximum plasma concentration (C_{max})*
(b) *Area Under Curve (AUC) of plasma concentration versus time*
(c) *Time to reach maximum plasma concentration (T_{max})*
(d) *Time to 50% plasma level ($t_{1/2}$; often, this parameter is not measured in routine toxicity studies owing to limited datasets)*

The important outcome of preclinical TK data is the interspecies comparison of test drug toxicity. The toxic effects can be extrapolated from animals to humans when these comparisons are based on TK and PK [10]. The safety margin is the ratio of animal AUC at NOAEL to human AUC at efficacious dose and is an indicator of toxicity risk in humans. The safety margin is large, the expected risk of toxicity is low.

The determination of PK profile of a substance using several doses leads to good interpretation of the toxicological findings. The TK study also evaluates whether a drug under study accumulates in a specific organ and/or tissue. The TK assessment is part of the nonclinical test battery recommended by the regulatory agencies [11].

12.7 Toxicology Study Guidelines

Up until the 1980s, no formalized guidelines existed for toxicology testing. However, each region or country has developed its own set of practices for testing drugs, requiring additional testing for the same effect if the drug company wished to register their drug across different regions or countries. This resulted in unnecessary use of animals and wastage of limited resources, as most drugs (particularly patented drugs) were registered in more than one country or region.

To avoid such duplication of testing, efforts were made by

12.7.1 OECD Guidelines

Scientists have developed SOPs for determining the potentially toxic effects of chemicals and established scientific methods to obtain the high-quality data which is essential for assessing human hazards and risks. Test Guidelines Program of the OECD developed testing methods that are adopted by all OECD Member Countries [12] through an *agreement "the mutual acceptance of data"* [13]. The OECD Guidelines for the Testing of Chemicals [14] is a collection of testing methods used by government, industry, and independent laboratories to identify and characterize potential hazards of chemicals. These tests also cover health effects. A list of OECD guidelines can be viewed on OECD Website, https://www.oecd.org/chemicalsafety/testing/ oecdguidelinesforthetestingofchemicals.htm

Though these tests are used for testing of chemicals and pesticides, they are also used to pharmaceutical testing. The OECD methods have removed the duplication of testing to comply regulatory authorities in different countries and substantially reduced the number of animals used for toxicity testing of new drugs. It encourages on alternative methods to test the toxicity as per 3 R principles (replace, reduce, or refine).

12.7.2 ICH Guidelines

"The International Council for Harmonization [15] of Technical Requirements for Pharmaceuticals for Human Use was created in 1990 and brought together regulatory and pharmaceutical industry authorities to discuss technical aspects of drug registration and development."

The objective of ICH is better standardization worldwide to ensure that safe, efficacious, and quality medicines are developed and marketed in the most resource-efficient manner. Greater uniformity was achieved through the creation of ICH Guidelines. The ICH Steering Committee (SC) consists of representatives from six parties, which is informally referred to as the "Six Pack," presented in Table 12.4 below.

The International Conference on Harmonization of Technical Requirements for Registration of Pharmaceuticals for Human Use (ICH) to standardize testing of medicines.

A series of guidelines (listed below in Tables 12.5 and 12.6) which have been published and accepted by the three regions:

Table 12.4 ICH six pack

Regulatory agency		Industrial associations	
1	European Union (including European Medicines Agency and Committee for Medicinal Products for Human Use (CHMP)	4	European Federation of Pharmaceutical Associations (EFPIA)
2	US Food and Drug Administration	5	Pharmaceutical Research and Manufacturers Association (PhRMA)
3	Japanese	6	Japanese Pharmaceutical Manufacturers Association (JPMA)

Table 12.5 ICH guidelines

M3(R2): Guidance on Nonclinical Safety Studies for the Conduct of Human Clinical Trials and Marketing Authorization for Pharmaceuticals, 2009	S4: Duration of Chronic Toxicity Testing in Animals (Rodent and Nonrodent Toxicity Testing), 1998
M4: Organization of the Common Technical Document for the Registration of Pharmaceuticals for Human Use, 2003	S5(R2): Harmonized Tripartite Guideline—Detection of Toxicity to Reproduction for Medicinal Products and Toxicity to Male Fertility, 2005
S1A: The Need for Carcinogenicity Studies of Pharmaceuticals, 1995	S6(R1): Preclinical Safety Evaluation of Biotechnology Derived Pharmaceuticals, 2011
S1B: Testing for Carcinogenicity of Pharmaceuticals, 1997	S7A: Safety Pharmacology Studies for Human Pharmaceuticals, 2000
S1C(R2): Dose Selection for Carcinogenicity Studies of Pharmaceuticals, 2008	S7B: The Preclinical Evaluation of the Potential for Delayed Ventricular Repolarization (QT Interval Prolongation), 2000
S2(R1): Guidance on Genotoxicity Testing and Data Interpretation for Pharmaceuticals Intended for Human Use, 2011	S8: Immunotoxicity Studies for Human Pharmaceuticals, 2005
S3B: Pharmacokinetics: Guidance for Repeated Dose Tissue Distribution Studies, 1994	S9: Nonclinical Evaluation for Anticancer Pharmaceuticals, 2010
S3A: Toxicokinetics; the Assessment of Systemic Exposure Studies in Toxicity Studies, 1995	S10: Photosafety Evaluation of Pharmaceuticals, 2013
Q3B(R2): Impurities in new drug products	Q3A(R) Impurities in New Drug Substances, 2008

Table 12.6 FDA guidance for industry related to nonclinical studies

Oncology Therapeutic Radiopharmaceuticals: Nonclinical Studies and Labeling Recommendations, 2019	Investigational Enzyme Replacement Therapy Products: Nonclinical Assessment, 2019
Oncology Pharmaceuticals: Reproductive Toxicity Testing and Labeling Recommendations, 2019	Testicular Toxicity: Evaluation During Drug Development, 2018
Q3D: Elemental Impurities in Drug Products, 2018	Assessment of Abuse Potential of Drugs, 2017
Safety Testing of Drug Metabolites, 2016	Nonclinical Evaluation of Endocrine-Related Drug Toxicity, 2015
Nonclinical Safety Evaluation of Pediatric Drug Products, 2006	Nonclinical Safety Evaluation of Drug or Biologic Combinations, 2006
Exploratory IND Studies, 2006	Estimating the Maximum Safety Starting Dose in Initial Clinical Trials for Therapeutics in Adult Healthy Volunteers, 2005
Code of Federal Regulations: 21CFR58: Good Laboratory Practice for Nonclinical Laboratory Studies, 2004	S2B Genotoxicity: A Standard Battery for Genotoxicity Testing of Pharmaceuticals, 1997
S2A Specific Aspects of Regulatory Genotoxicity Tests for Pharmaceuticals, 1996	

12.7.3 Good Laboratory Practice (GLP) Regulation

The GLPs were initiated in the 1970s in the USA due to lack of the validity of nonclinical safety data submitted to the FDA. The audit report of the testing facilities made public in the Kennedy-Hearings of the US Congress resulted in publication of *Proposed Regulations on GLP in 1976, with the establishment of the Final Rule [16] 1 in June 1979 (21 CFR 58).* Since that time, many regions have developed GLP regulations; they are the OECD GLP [17]. Toxicology studies submitted to regulatory agencies to support clinical trials should be conducted according to GLPs.

GLP is a quality system which deals the organizational procedures and the conditions under which nonclinical studies are designed, executed under continuous monitoring, documented, archived, and reported [17]. The purpose of GLP is to promote the quality of testing data and provide a tool to ensure a sound approach to managing the laboratory studies end to end. The Principles may be considered as a set of SOPs for ensuring specifications for test system facilities and chemicals and reagents and record keeping which can be verifiable and can be retrieved at any stage. The institutions should assign the roles and responsibilities to staff to ensure good management of each study and focus on all aspects of study execution that are important for the reconstruction of the whole study [17]. Since all these aspects are of equal importance for compliance with GLP principles, it is not permissible to partially implement GLP requirements and still claim GLP compliance. No test facility may rightfully claim GLP compliance if it has not been implemented, and does not comply with, the full array of the GLP rules. GLP is not directly concerned with the scientific design of studies.

In vitro or in vivo studies are carried out to determine the toxicological profile prior to clinical trial initiation. All the studies that are required for IND application need to be conducted in compliance with GLP regulations. The GLP principles were the basis for ensuring that the reports submitted to any regulatory agencies are fully validated and reliably reflected in the experimental work.

12.8 Description of Toxicity Tests

The main objective of preclinical toxicological testing in drug development is to characterize potential dose-related side/adverse effects and its potential reversibility. The findings of preclinical studies are useful in determining the initial safe dose to be used in the clinical trials and identifying parameters for clinical monitoring of the possible adverse effects in humans [18]. The preclinical toxicology program is designed to characterize potential adverse effects that might occur under the clinical trial conditions.

The toxicology assessment means that the doses used for preclinical exposure will be multiple times higher than that of humans. Thus, the characterization of side/adverse effects can be identified in the preclinical study. Due to this fact, regulatory documents are written to inform about the general "stepwise" process of safety assessments, while also alluding to the concept of the case-by-case approach. For small molecules, the case-by-case consideration may be associated with a combination of side effects and pharmacodynamics effects, but for biologically derived molecules, most case-by-case relations are associated with pharmacodynamics effects (or exaggerated pharmacology).

The guidelines cited above under OECD and ICH sections offer a detailed description of tests, which will not be repeated in the following sections. The M3 (R2) "Nonclinical Safety Studies for the Conduct of Human Clinical Trials and Marketing Authorization for Pharmaceuticals" of ICH are for sponsoring the drug development studies. The harmonization is a process to provide consistency and streamline the procedures to enhance the safety and effectiveness of the regulatory assessment process for new drug applications and reducing the times and resources needed for drug development. The registration and approval of a new drug can differ in countries, and the responsibility is of the sponsor to find out the specific regional registration or marketing requirements.

In order to facilitate the reader of this chapter to be self-contained, an effort is made in this section to offer an abbreviated version of each of the major types of toxicity tests.

12.8.1 Acute Toxicity

In acute toxicity studies adverse effects if any are observed on administration of single dose of a new drug. Separate tests are required to detect the effect(s) of the application of the drug to skin and eye and the effects on internal organs of a substance on administration of the drug by various routes.

12.8.1.1 Dermal Irritation

Dermal irritancy is assessed by applying the test substance to shaved areas of the backs of rabbits and observing adverse effects, viz., redness, swelling, and ulceration over a period of 72 h. Recently, such in vivo tests are replaced by alternatives to animals.

12.8.1.2 Eye Irritation

In the eye irritation testing, test drug is administered directly into the rabbit's eye and signs of irritation, corneal opacity, and reddening are observed. Recently, such in vivo tests are replaced by alternatives to animals.

12.8.1.3 Skin Sensitization

In the case of skin-sensitization testing, multiple doses of the drug are applied to guinea pigs' skin and if a later dose will cause a strong immune reaction, indicating sensitization to the chemical. The local lymph node assay using mice is being used.

12.8.1.4 Local Lymph Node Assay

The drug is applied to the ears of the mice and after an interval, are euthanized. The sensitization is determined by estimating the level of induction of DNA synthesis in the lymph nodes. The advantage of the test is the use of fewer animals with little pain and distress compared to the skin sensitization test.

The primary purpose of skin and eye testing is to classify the test substance as corrosive, irritant, and sensitizing chemicals. The current classification of chemicals as corrosive or irritant to the skin can be predicted by studying physicochemical properties, such as pH value. Using isolated human/animal tissue cells or ex vivo tissues or organs to identify chemicals that produce irritation or corrosion is known as non-animal pre-screens. SAR (structure-activity relationships) studies can predict many potential sensitizers and can replace animal tests.

12.8.1.5 Acute Systemic Toxicity

Acute toxicity is tested by administering a single dose of test drug by the proposed route to rodents. The objective of this test is to determine the nature of acute toxic response and its duration. The information regarding the maximum nonlethal dose is also obtained which is helpful in design of clinical trials.

12.8.1.6 Maximum Tolerated Dose (MTD)

MTD studies are carried out to aid the later process of dose selection. These tests are conducted in larger species such as the dog and primates. The design of the study is to steadily increase the dose (single or several consecutive doses) until adverse effects, viz., vomiting and convulsions, occur. This is normally determined by careful observation of the animals. However, there is no accepted definition of the MTD and adverse effects.

12.8.2 Subchronic Toxicity Studies (up to 90-days)

These studies are carried out in two species, rodent and non-rodent. The duration of these studies will depend on the protocol of clinical trials. The drug should be administered 7 days/week by the proposed route as per clinical use. Generally, mortality, body weight, food consumption, and biochemical and histopathological parameters are assessed. Biomarkers of anticipated target organ, cytokines, and immunophenotyping are also evaluated on a case-by-case basis. In addition, for the non-rodents, other parameters such as electrocardiogram, blood pressure, and heart rate were also recorded in Week 1 and at the end of the study period.

The recovery group should be planned to evaluate any adverse effects' reversibility. To estimate the index of systemic exposure, the drug concentration in plasma is determined. Due to the limited amount of the blood available in rodents, it is a standard practice to employ sparse blood collection, wherein no rat is bled more than three times in a 24-time period with a maximum collection of 2-ml in 24 h. Hence, maintaining a satellite group of rats for toxicokinetics will not be subjected to monitoring other toxicologic parameters as the main toxicology rats. However, in the case of non-rodents, the same animals are used for blood collection for TK.

All the precautions as given in guidelines regarding the blood volume collected/ 24-h period should be followed to avoid physiological complications and thereby changes in bioanalytical data.

A typical study design for repeat dose rodent and non-rodent studies is depicted in Tables 12.7 and 12.8.

Subchronic tests provide data which help in identification of ideal dose(s) for chronic and carcinogenicity studies. Clinical or behavioral signs of toxicity (daily), gross morphological, biochemical, and histological parameters are monitored in these studies. They can consist of activity, gait, excreta, hair coat, and feeding and drinking patterns. Bodyweight and food consumption data are recorded weekly throughout the study. Ophthalmoscopy is conducted at the beginning and end of the study.

The results of hematology testing indicate any abnormality in the function of hematopoietic system and clinical chemistry and urinalysis results indicate potential adverse effect on functions of kidneys, liver, and heart.

A complete necropsy is performed after sacrifice/death of the test animal. Generally, all tissues (about 40^{+}) are examined from control and high dose. If the test drug is cytotoxic, dosing and study design should be as per the [19] guidelines.

12.8.3 Chronic Toxicity Studies

Chronic toxicity studies are conducted for 6 and 9 months in rodents and non-rodents, respectively. The results of these studies provide inferences about the long-term effect of a test substance in animals, which may be extrapolated to the

Table 12.7 Repeat dose toxicity study and TK in rats with recovery

Regulatory status	GLP				
Guideline	Mutually agreed guideline				
Objective	The objective of this study is to determine the toxicity potential and toxicokinetic profile of the test item following 28-day repeated dose administration to rats. This study is designed to provide information on major toxic effects, kinetics, target organs, and determination of NOAEL				
Test system	Rats				
Age	6–9 weeks at the start of dosing				
No. of doses	+ 3 doses: Control, Low, Mid, and High dose				
Experimental design	**Dose group**	**Dose levels (mg/kg)**	**Number of animals**		
			Main study group	**Recovery group**	**Toxicokinetics Satellite group**[a]
	Vehicle control	0	10♂/10♀	5♂/5♀	3♂ or ♀
	Low	TBD[b]	10♂/10♀	—	9♂/9♀
	Mid	TBD[b]	10♂/10♀	—	9♂/9♀
	High	TBD[b]	10♂/10♀	5♂/5♀	9♂/9♀
Route	Intended human route				
Dose volume	To be determined				
Dose formulation analysis	Formulation analysis for active ingredients (A.I) and Homogeneity—Two times (at the start of treatment and in the fourth week)				
Treatment	Animals will be dosed for 2 weeks to 90 consecutive days (or as per study plan) with necropsy of main study animals 24 h following the last dose and necropsy of recovery animals following 14–28 days of treatment-free days				
Observations:					
Ophthalmology	Pre-dose and at termination				
Mortality/morbidity	Daily				
Daily clinical observation	Daily, or more frequently based on the requirement				
Detailed clinical examination	Weekly				
Functional observational battery	One week prior to sacrifice of respective groups				
Body weight	Once a week unless there are any specific requirements				
Feed consumption	Weekly and presented as average consumption per day				
Toxicokinetics	As per study plan				
Clinical lab workup	At sacrifice				
Necropsy and organ weights	As planned				
Histopathology	As per study plan				

[a]Toxicokinetic animals will be used for blood sample analysis and will not be evaluated with the main study animals

[b]*TBD* To be determined

Table 12.8 Repeat dose toxicity study in dogs with toxicokinetics and recovery

<table>
<tr><td>Regulatory status</td><td colspan="4">GLP</td></tr>
<tr><td>Guideline</td><td colspan="4">OECD and ICH</td></tr>
<tr><td>Objective</td><td colspan="4">To determine the toxicity potential and toxicokinetic profile of the test item following 3-month repeated dose administration by oral gavage to dogs. This study is designed to provide information on major toxic effects, kinetics, target organs, and determination of NOAEL</td></tr>
<tr><td>Test system</td><td colspan="4">Beagle dogs</td></tr>
<tr><td>Age</td><td colspan="4">5–9 months at the start of dosing</td></tr>
<tr><td>Treatment groups</td><td colspan="4">1 Control and 3 treatment groups</td></tr>
<tr><td>Group</td><td>Treatment</td><td>Dose (mg/kg/day)</td><td>Male</td><td>Female</td></tr>
<tr><td>G1</td><td>Vehicle control</td><td>0</td><td>4</td><td>4</td></tr>
<tr><td>G2</td><td>Low dose</td><td>X</td><td>4</td><td>4</td></tr>
<tr><td>G3</td><td>Mid dose</td><td>XX</td><td>4</td><td>4</td></tr>
<tr><td>G4</td><td>High dose</td><td>XXX</td><td>4</td><td>4</td></tr>
<tr><td>G1R</td><td>Vehicle control recovery</td><td>0</td><td>2</td><td>2</td></tr>
<tr><td>G4R</td><td>High dose recovery</td><td>XXX</td><td>2</td><td>2</td></tr>
<tr><td>Route of exposure</td><td colspan="4">Intended human route</td></tr>
<tr><td>Volume of administration</td><td colspan="4">To be determined</td></tr>
<tr><td colspan="5">• Dose formulation concentration analysis: Periodically during the treatment period
• Formulation analytical method to be validated
• Vehicle: To be determined</td></tr>
<tr><td colspan="5">1. Main toxicology dogs will be treated daily for the duration of the study
2. Recovery designated dogs will be treated as planned followed by 2–4 weeks of no treatment prior to sacrifice
3. Control dogs will receive an equivalent amount of the vehicle daily till they are sacrificed</td></tr>
<tr><td>Mortality/morbidity</td><td colspan="4">Twice a day</td></tr>
<tr><td>Clinical signs</td><td colspan="4">As per study plan</td></tr>
<tr><td>Food consumption</td><td colspan="4">Measured daily (≈ 300 g input per each dog offered ≈ 2 h post-dose and retained in the kennel for further 4–5 h and then withdrawn).</td></tr>
<tr><td>Body weight</td><td colspan="4">Weekly</td></tr>
<tr><td>Detailed clinical examination</td><td colspan="4">Weekly</td></tr>
<tr><td>ECG</td><td colspan="4">Pre-dose (basal) and at post-dose T_{max} at periodic intervals</td></tr>
<tr><td>Ophthalmological examination</td><td colspan="4">Beginning of the treatment, end of the treatment, and recovery period.</td></tr>
<tr><td>Neurological evaluation</td><td colspan="4">Neurological examination prior to the initiation of treatment, and at periodic intervals. General attitude and behavior, gait, postural reactions, cranial nerve function such as tone, eye reaction, nasal reflex, tongue, and pharynx; spinal nerve function such as joint reflexes.</td></tr>
<tr><td>Blood collection for toxicokinetics</td><td colspan="4">Pre-dose (basal) and at periodic intervals. Blood from all dogs will be collected for TK at 8 time points</td></tr>
<tr><td>Clinical lab</td><td colspan="4">Pretreatment and at periodic intervals</td></tr>
<tr><td>Necropsy and organ weights</td><td colspan="4">As per study plan</td></tr>
<tr><td>Histopathology</td><td colspan="4">All organs from all dogs on the study</td></tr>
</table>

safety of the new drug in humans. The observations of various parameter as described in the previous para are studied. A satellite TK group should be included in the rodent study.

12.8.4 Carcinogenicity Studies

12.8.4.1 Two-year Carcinogenicity Study

The carcinogenicity testing (ICH Guideline S1A) object is to identify a carcinogenic potential in animals and assess the relevant risk to humans. The design of these studies depends on the duration and exposure of the drug for human treatment, information regarding the drug/related drugs, genotoxicity, the indication, patient population, and route of and extent of systemic exposure. A carcinogenicity study is required when the drug is indicated for chronic use (daily/6 months).

The ICH S1B guideline entitled "Guideline on the Need for Carcinogenicity Studies of Pharmaceuticals" provides basis for evaluation of carcinogenicity.

The ICH S1C(R2) guideline entitled "Dose Selection for Carcinogenicity Studies of Pharmaceuticals" offers guidance for dose selection.

The ICH M3(R2) provides the guidelines regarding the time to initiate the study, and the results of the carcinogenicity study are needed to support market authorization.

Under the Prescription Drug Act of 1992, the FDA should evaluate and comment on carcinogenicity protocol. It is a standard practice for the industry to present to the FDA, a summary of all available short-term toxicology data to make a better judgment on the selection of dose(s) for the study.

Typically, rodents (usually HanWistar rats; CD—1, Swiss or B6C3F1 mice) are used in carcinogenicity testing. The study needs to be carried out in the same strain used in toxicology studies. The tests are performed for major portion of an animal's lifespan. It is typical to have three dose groups. Each group should have at least fifty animals/sex. The highest dose should not induce mortality or reduce animals' life span by more than 10% of normal. The historical tumor pathology data, the survival of strains, and spontaneous/background incidence of pathology are taken into consideration for selection of animal model and comparison of tumor incidences in treated groups versus contemporary and historical controls.

The test may be terminated after 18 months in mice and hamsters and after 24 months with rats if the animals do not develop tumor and toxicity. The animals are sacrificed, and gross pathological changes are noted, and histopathological studies are carried out on all the tissues. Histopathology is the basis for determining the carcinogenicity of a molecule.

12.8.4.2 Alternative Models to Evaluate Potential Carcinogenicity in Lieu of a 2-Year Study

Since the 1990s when several transgenic mouse models became available, these are becoming popular to use in short-term (6-month) carcinogenicity study with drugs.

At present, about 20% of mouse cancer bioassays use transgenic mice. The insertion of oncogene or deletion of tumor suppressor gene results in the increased sensitivity of these animals to neoplastic development and leads to decrease in the duration of the test.

Among the transgenic models available, the rasH2 and hemizygous p53 knockout mice model are currently the most frequently used.

The rasH2 model carries multiple copies of the human homolog of the Harvey rat sarcoma virus (c-Ha-ras, commonly known as HRAS) under the control of its own promoter and enhancer elements. This model has a low incidence of spontaneous tumor formation up to 6 months of age, and the c-Ha-ras oncogene renders these mice highly susceptible to tumor development following carcinogen exposure with less false positives and assesses both genotoxic and nongenotoxic mechanisms of carcinogenesis [20–22].

In the transgenic model study, a 4-week dose range finding in wild-type mouse (rasH2 or p53 knockout) with all routine parameters evaluation including histopathology to select the doses for 6-month pivotal carcinogenicity bioassay in transgenic animals. A positive control (e.g., N-methyl N-nitrosourea, Urethane, etc.) is usually included. A group size of 25 transgenic mice per sex per group for the main study and 12 wild-type mice per sex as a satellite group for TK is generally followed. At the end of the study, a complete necropsy and histopathology assessment is conducted by experienced pathologist.

12.8.5 Reproductive and Developmental Toxicity Studies

In the reproductive and developmental toxicology studies, the test system is exposed at multiple stages leading to the variability of possible effects. Sterility or decreased fertility or affecting ova or sperm or affecting endocrine functions of organs involved in reproduction is expected when the test system is exposed to toxic substances in sexually mature animals. Depending upon the stage of reproductive cycle and development, the toxic substance exposure results in fetal death, congenital malformations, reversible or irreversible growth retardation, or premature or delayed parturition and delayed postnatal effects such as cancer, neurological effects, growth retardation, and death.

The general purpose of reproductive and developmental toxicity assays is to produce phenotypically normal offspring. All or most of the reproductive cycle is evaluated.

Table 12.9 Standard testing for reproductive and developmental toxicity assessment any

No.	Reproductive cycle covered	Species	ICH guideline
1	The fertility and general reproductive performance study (Segment I)	Rodents	ICH 4.1.1
2	The prenatal and postnatal study (Segment III)	Rodents	ICH 4.1.2
3	Embryofetal development (Segment II)	Rodents and rabbits	ICH 4.1.3

According to the ICH S5(R2) document, Guideline for industry, Detection of Toxicity to Reproduction for Medicinal Products, a reproductive and developmental toxicity testing regimen should be selected to allow adults' exposure and all stages of development from conception to maturity.

To determine the immediate and latent effects of any of drug under study, observations should be continued for one life cycle. Hence, in drug development, the common practice for integrated reproductive cycle assessment is to conduct the following three types of studies listed in Table 12.9:

1. Fertility and General Reproduction (Segment I):

   The aim of this study is to determine the adverse effects of the administration of test drug to males and females before mating and through mating. In females, this should detect effects on the stages of reproduction cycle. In males, it detects functional effects on epididymal and sperm maturation that might be observed by histological examinations of the male reproductive organs.
2. Prenatal and Postnatal Study (Segment III):

   This study aims to detect adverse effects on the pregnant/lactating female and the development of the conceptus and the offspring following exposure of the test drug on the female from implantation through weaning.
3. Embryofetal Development Study (Segment II):

   Based on historical precedence, it is customary to conduct this test in two species: rat (or mice) and rabbits, due to the extreme sensitivity of rabbits to thalidomide. The study is carried out to assess any maternal and/or embryofetal toxicity of the test drug in rat and rabbits, following daily administration during the period of major organogenesis, from implantation to the approximate day of closure of the hard palate which vary depending on animal species and followed by caesarian sections.

   Developmental toxicity studies give information on the potential hazards to the unborn that may arise when the mother is administered to test substances during pregnancy. The pregnant rats or rabbits which are administered the drug during organogenesis are sacrificed before delivery. Maternal toxicity, embryo/fetal death, defects in growth, structural abnormalities in the fetus, etc. are evaluated in this study. Rabbits are used in addition to rats or mice to improve the predictability of risk in humans. The results of both tests are used in classification by hazard and risk assessment.

12.8.6 Genetic Toxicology—Mutagenicity Studies

Genotoxicology studies are generally performed at the early stage of discovering compounds to minimize the risk of attrition of the compounds at a later stage of development. The genotoxic potential is also used during the qualification of major drug metabolites, impurities, and excipients. ICH S2(R1) guidelines describe the genotoxicity testing for small molecule drugs, which was revised in 2011.

Unlike conventional toxicology, where each study by itself offers a conclusive evidence of toxicity in that species, in mutagenicity there is no single test that can be recommended to determine the mutagenic potential of a drug. Hence, it is a standard practice to use several tests to assess the potential of test drug to affect genetic material. The test selection depends on its ability to detect with assay methods, the capacity of the chemical to alter genetic material in cells.

It is a standard practice to conduct initial mutagenicity studies in vitro using either bacterial (Salmonella) or mammalian (*Chinese Hamster Ovary-CHO or Human Peripheral Blood Lymphocytes-HPBL*) cell systems.

Genetic toxicology assays conducted to support drug development are intended to evaluate their ability to induce in vitro *gene* mutations (point mutation) and clastogenicity (chromosomal structural aberration and numerical alteration). Because genotoxicity could be attributable to a metabolite and/or the parent compound, in vitro assays are conducted with and without metabolic activation. Subcellular liver fractions (S9) from rats treated with agents that induce enzymatic activity are routinely used in the in vitro medium.

Mutagenic potential of a test substance is heritable in humans based on the weight of the evidence, when it is demonstrated in an in vitro mutagenic assay.

The current guidelines would require an initial battery of tests consisting of:

1. Ames Salmonella typhimurium reverse mutation assay
2. Chromosomal aberration or Micronucleus assay in Mammalian cells (either CHO or Peripheral Human Blood Lymphocytes)

Results derived from abovementioned studies and other toxicity studies would mandate further mutagenicity testing involving in vivo assessment too. However, it is a standard practice to conduct one in vivo chromosomal assay in rodents at late stage of drug development before registering a new drug. Table 12.10 lists standard in vitro and in vivo tests that are currently used for drugs.

1. Ames Test (reverse mutation assay):

   Salmonella typhimurium strains such as *TA 98, TA 100, TA 102, TA 1535, TA 97*, or *Escherichia coli* WP2 *uvr*A or *Escherichia coli* WP2 *uvr*A (pKM101) are used. In vitro exposure (with and without metabolic activation, S9 mix) should be done at a minimum of five log dose levels. Solvent and positive controls are employed. The positive control could include 9-amino-acridine, 2-nitrofluorine, sodium azide, and mitomycin, respectively, in the Ames' tester strains mentioned above. Three replicates in each set are the minimum requirement. Two-and-a-

Table 12.10 Mutagenicity tests use drug induced genotoxicity

No	Assay	Damage detected	Guideline
In vitro assays			
1	Ames—Salmonella bacterial reverse mutation assay	Gene mutations	OECD 471
2	Chromosomal aberration assay (mammalian cells)	Chromosomal damage—Structural and/or numerical	OECD 473
3	Micronucleus assay (mammalian cells)	Chromosomal damage	OECD 487
In vivo			
1	Chromosomal aberration assay (rodents)	Chromosomal damage—Structural and/or numerical	OECD 475

half-fold (or more) increase in the number of revertants than that of spontaneous revertants would be considered +ve.

2. In vitro Chromosome Aberration Assay in Mammalian Cells:

   For this assay, either CHO or HPBL cells are used. The desired level of toxicity for in vitro clastogenicity test should be no greater than fifty per cent reduction in cell number or confluency. For lymphocyte cultures, inhibition of mitotic index by greater than fifty per cent is considered sufficient. In vitro exposure (with and without metabolic activation, S9 mix) should be done using a minimum of three log doses. Solvent and positive controls are included along with test doses. Cyclophosphamide is used as +ve control with metabolic activation and mitomycin C without metabolic activation should be used to give a reproducible and measurable increase in clastogenic effect over the background, which demonstrate the test system's sensitivity. An increased number of aberrations in metaphase chromosomes should be used as the criteria for evaluation.

3. In vivo Micronucleus Assay:

   One rodent species (preferably mouse) is used. Route of administration of the test drug should be the same as that proposed for humans. Five animals/sex/dose group should be used. A minimum of three dose levels, a solvent, and positive controls should be tested. A positive control like mitomycin C or cyclophosphamide should be used. Dosing should be done on two consecutive days, and 6 h after the administration of the last dose of the drug the animals are sacrificed. Bone marrow from both the femora should be taken out, flushed with fetal bovine serum (20 ml), pelleted, and smeared on glass slides. Giemsa-MayGruenwald staining should be used for staining slides. An increase in number of micronuclei in polychromatic erythrocytes (minimum 2000) should be used as the evaluation criteria.

Mutagenicity study (the in vitro genotoxicity tests) results are to be submitted in support of Phase I clinical trial. At the same time, the in vivo assay results are generally submitted in support of the New Drug Application (NDA).

12.8.7 Special Toxicology Studies

12.8.7.1 Phototoxicity

Phototoxicity, photoallergy, photogenotoxicity, and photocarcinogenicity are essential components of photosafety testing, although the latter two tests are not considered vital for pharmaceuticals. Phototoxicity assessment is a critical test in the safety assessment of new APIs and excipients for systemic administration, dermal products (application or patches), ocular products, and photodynamic therapy products.

Phototoxicity (photoirritation) is a light-induced tissue response to API or excipient [23].

Photoallergy is developed as an immunologically mediated reaction to a drug, initiated by the formation of photoproducts (e.g., protein adducts) following a photochemical reaction [23].

Issuance of the FDA and EMA photosafety guidance provided some structure to the approach to photosafety evaluation. The guidance indicates that for a chemical to elicit phototoxicity, the following any one or more of the criteria are critical [23]:

1. *Absorbs in the range of natural sunlight (290–700 nm).*
2. *Generates a reactive species following absorption of UV or visible light.*
3. *Distributes sufficiently in light exposed tissues (e.g., skin or eye).*
4. *A molar absorption characteristic/coefficient greater than 1000 L/mol.cm at any wavelength between 290 and 700 nm is recognized as cutoff.*

Assessment of photosafety testing is not dependent alone on photostability testing, but the above criteria should be considered.

The complicated nature of plasma/tissue and its interactions with the drug, coupled with the effect of light on the test drug to propose minimal test material exposure to trigger or prevent photosafety testing. Its photoreactive character of the test drug is more important rather than its interaction with the tissue and its concentration. Metabolites do not require additional phototoxicity testing.

Photosafety evaluation is carried out in vitro and in vivo tests. The guideline for in vitro assay for phototoxicity using 3 T3 Neutral Red Uptake (3 T3 NRU-PT) is available [24]. This is a reliable procedure for testing the phototoxic potential of a test compound. A phototoxicity study in animals should be conducted if the result is positive in the above protocol.

No standard design of phototoxicity testing in animals is available. Pigmented and nonpigmented animals like a guinea pig, mice, or rats are used and selected based on irradiation sensitivity, heat tolerance, and control substance response. However, nonpigmented skin tends to be more sensitive than pigmented skin for detecting phototoxicity due to melanin-binding effect. Usually, a single dose of test article administration is followed to evaluate phototoxicity. Based on the author's experience, at least 3 days of repeat dose administration using intended clinical route of administration would better characterize the test article for any phototoxicity risk potential.

In case of dermal formulation, based upon the PK characteristics the irradiation of the animals is done at the approximate Tmax. Systemic drug levels are not needed in dermal phototoxicity studies [23].

For all animal models, following UVR exposure, evaluation of the skin responses are performed at least once on the day of UVR exposure, and then 24 and 48 h after exposure. The skin sites are assessed using a Draize-type scale [25]. Histopathological evaluation is not necessary as the skin's repair capacity is rapid enough.

ICH M3(R2) suggests a stepwise approach to photosafety assessment. An initial assessment of phototoxic potential based on photochemical properties and pharmacological/chemical class should be undertaken before outpatient studies.

Juvenile Toxicology

Juvenile is the time between birth and sexual maturation. Age comparison of various species is given in Table 12.11. Developmental processes in pediatric patients may differentially affect drug ADME and its effects compared to adult therapeutic use. Some of the adverse effects may be difficult to detect in clinical trials or in post-marketing surveillance. Literature indicates ineffective dosing/unnecessary exposure to ineffective therapies, overdosing of effective drugs and identification of novel pediatric adverse effects/events. Juvenile animal studies result in identifying post-natal developmental toxicities that are inadequately assessed in reproductive toxicity studies and may not be adequately determined in pediatric clinical trials [18].

The decision to conduct juvenile animal toxicity studies to evaluate the test article's safety is driven by whether pediatric populations would be included in the clinical trials and what safety data from previous adult human experience is available. Before the initiation of trials in pediatric populations, results from repeated-dose toxicity studies of appropriate duration in adult animals, the core safety pharmacology, and battery of genotoxicity studies results should be available. If all these studies data are insufficient to support pediatric studies, a juvenile study in animals should be conducted. If a study is warranted, one relevant species, preferably rodent, is generally considered adequate. A study in a non-rodent species can be appropriate when scientifically justified [15]. Single animal species are generally considered sufficient for juvenile animal toxicity testing to support pediatric drug development [18].

Table 12.11 Age comparison of various species

Species	Neonate	Infant	Child	Adolescent
Rat	0–1 (weeks)	1–3 (weeks)	3–9 (weeks)	9+ (weeks)
Dog	0–0.75 (month)	0.75–1.5 (month)	1.5–5 (month)	5+ (month)
Primate	0–0.05 (year)	0.05–0.5 (year)	0.5–3 (year)	3+ (year)
Man	0–0.1 (year)	0.1–2 (year)	2–12 (year)	12+ (year)

Note: Does not account for organ maturation

The following are important in determining the suitability and design of juvenile animal studies:

1. Targeted use of the drug in children
2. The timing of dosing concerning phases of growth and development in pediatric populations and juvenile animals
3. The potential similarities or differences in pharmacological and toxicological profiles between mature and immature systems
4. Any established temporal developmental differences in animals relative to pediatric populations

The frequency of dosing administration should be relevant to the intended clinical use of the drug. The treatment duration in animals should include at least the significant periods of appropriate postnatal development for the selected species.

Selection of doses in juvenile studies should follow standard practice in any toxicology study to establish a clear dose-response by choosing appropriate low, mid, and high doses. Study parameters additional to those used in adult work can include:

- Neurobehavioral testing
- Specific growth measurement
- Reproductive development
- Developmental landmarks in rats in pre-weaned rats
- Physical development
- Sensory and reflex development

Modifications of existing toxicity designs or *de novo* juvenile studies should be used depending on the concerns to be addressed.

12.8.8 Abuse Potential

Drug abuse is defined as the intentional, non-therapeutic use of a drug product or substance, even once, to achieve a desired psychological or physiological effect. Therefore, abuse potential refers to the likelihood that abuse will occur with a particular drug product or substance with CNS activity. Desired psychological effects can include euphoria, hallucinations, perceptual distortions, alterations in cognition, and mood [26].

Evaluation of abuse potential may be needed for new drugs (including the drugs derived from biological origin). For the test drug that produce CNS activity, regardless of its use, it should be considered for abuse liability evaluation. Preclinical studies should support the design of the clinical assessment of abuse potential, classification/scheduling by regulatory agencies, and product information.

A variety of approaches that can be used to assess a drug product's abuse potential are discussed in the FDA's Guidance on the Assessment of Abuse Potential of Drugs. In vitro receptor binding, ion channels, and association with some of the neurotransmitter systems are important to abuse potential evaluation. Preclinical abuse liability evaluations could be conducted in rodents if the metabolite profile and target for drug activity in rodents are similar to that of humans.

Three types of study are often conducted: (1) drug discrimination, (2) self-administration of the compound, and (3) an assessment of withdrawal.

In general, drug discrimination and self-administration studies are performed standalone. Assessments of withdrawal can sometimes be incorporated in designing the reversibility arm of a repeated-dose toxicity study. Abuse-related studies in animals typically use test doses that produce Cmax levels, as well as at least 2–3 times greater than the Cmax produced by the highest proposed therapeutic dose.

The studies should include a positive control drug group and vehicle group for comparison with the test drug group. Data collection should be at Tmax, with additional measurements made before and afterward to ensure full characterization of the test drug.

12.8.9 Metabolite Testing

The majority of the oral drugs are cleared via biotransformation from the body generating circulating and/or excreted drug metabolites. Metabolites need to be quantified in toxicology studies and adequately characterized through their exposure in animals when the parent drug is administered directly.

Metabolic profiles can vary across species both quantitatively and qualitatively. Clinically relevant metabolites of a test drug may not be identified/characterized during nonclinical safety studies if the metabolite is formed only in humans/ more in humans than that of animal species. Drug metabolites in human can raise a safety concern are those formed greater than 10% of total drug-related exposure at steady-state and significantly greater levels in humans than the maximum exposure seen in the toxicity studies [5]. Structural evaluation of the metabolite for its potential toxicity should be evaluated first. Definitive data on the total metabolite profile of a novel drug candidate in humans can only be obtained from in vivo studies in which radiolabeled compound is administered and blood/plasma and excreta samples are collected to analyze qualitatively and quantitatively the radioactive components.

It is important to consider the following factors for the nonclinical study design for a disproportionate metabolite [27]:

- Structural similarity of the metabolite to the parent molecule
- Pharmacological or chemical class
- Solubility
- Stability in stomach pH
- Phase I versus Phase II metabolite
- Relative amounts detected in humans versus the quantities seen in animals

The following studies may need to be conducted to assess the safety of the disproportionate drug metabolite.

- In vitro receptor profiling and safety pharmacology (e.g., hERG)
- General toxicity study
  - The toxicity of the drug metabolite should be carried out as per GLP study at multiples of the human dose or at least at levels comparable to those measured in humans at the intended clinical route of administration as per ICH M3 (R2). Other routes, if administered, have to be scientifically justified. The duration of the study is subjected to case by case, a minimum of 3 months in single species is desirable.
- Genotoxicity study
  - In vitro assay that detects point mutation and chromosomal aberrations be conducted according to the recommendations in the ICH S2 (R1). If one or both of the in vitro tests are equivocal and/or positive, results from a complete standard battery of genotoxicity studies may be warranted.
- Embryofetal development study
  - Embryofetal development in single species should be evaluated if the intended use in a population includes women of childbearing potential.
- Reproductive toxicity study
  - This is requested by the regulators on a case-by-case basis, depending on the general toxicity and embryofetal development studies results.
- Carcinogenicity study
  - Disproportionate metabolites should be subjected to carcinogenicity testing if the drug is administered for at least 6 months, or that are used intermittently in the treatment of chronic or recurrent conditions when the carcinogenic potential of the metabolite cannot be adequately evaluated from carcinogenicity studies conducted with the parent drug. A single carcinogenicity study or an alternative bioassay should be conducted and the studies should be conducted in accordance with the ICH S1A and S1C(R2).

In general, toxicity studies are done for the disproportionate metabolites; the same has to be submitted to the Agency before initiating large-scale clinical trials. However, drugs for cancer or serious life-threatening diseases, the metabolite qualification studies are either not required or modified on a case-by-case basis. It is preferred for Sponsors to contact the appropriate review division of the Agency to discuss such situations [27].

12.9 Investigational New Drug Application (IND)

It is important for sponsor during preclinical development to evaluate safety of the product for use in humans and that drug shows therapeutic effect that is required for its intended use. Information should be collected to ensure that the drug is safe with use in humans when used in early studies in humans.

12.9.1 Parts of an IND

The IND application should focus on three major areas [28]:

1. Pharmacology and Toxicology Studies in Animals—Generation of safety data in preclinical species to allow an evaluation to know if the drug is safe for first dose in humans. The earlier knowledge with the drug in humans are also included.
2. Information on manufacturing—Information related to the content, producer, stability, and controls used for production of the drug substance and the drug product. This ensures that the pharmaceutical company can manufacture the drug adequately with consistency and supply to meet the demand.
3. Clinical Protocols and Investigator Information—Study protocol for clinical trials to evaluate whether humans will be exposed to adverse effects. Protocols for proposed clinical studies should be submitted in detail to evaluate if the initial-phase studies pose any risks to subjects. The qualifications of principal investigators, generally physicians, who monitor the delivery of the trial drug to determine whether they are eligible to achieve the clinical study responsibilities should also be included.

In this section, we will concentrate on the development of preclinical studies towards an IND filing.

Ensure there is an adequate amount of the test material (API) with certificate of analysis. Just to give a ballpark estimate for a newcomer to drug development, on average we need 3.5 Kg of the test material for a compound that is not acutely toxic to complete all preclinical studies.

Embryofetal developmental studies in animals are mandatory post toxicity observed with Thalidomide use during pregnancy where children were born with limb deficiency.

Based on these information, the FDA required an application of IND for all new drugs that require studies in humans. The application of IND must enclose the data on safety and efficacy of the drug before the first exposure to humans [29].

12.9.2 Regulatory Toxicology Studies for an IND

In the drug development process, regulatory toxicology studies are compulsory with the main objective to assess toxicity of a drug. The study procedures used will be based on the guidance meant for the conduct of studies in preclinical species. It is imperative to highlight that these studies should be conducted meeting GLP compliance. The planning of these studies could be based on the data obtained by the exploratory studies of both efficacy and toxicity. These findings could help to define doses, the duration of study, and any side effects that could require special attention.

The duration of toxicity study with repeat doses is vital and is decided based on the duration and indication of use in the clinical trial.

In general, the length of toxicity studies should not be greater than the maximum time outlined in M3 (R2) guidance [15] for each tested species (see more details in Table 12.12). It is required to be conducted in one rodent and one non-rodent mammalian species and the period should be same or longer than the clinical studies. Table 12.12 outlines the suggested duration of repeated dose toxicity studies. A requirement of single and or repeat dose toxicity study to support phase 1 study is dependent on the target population, therapeutic indication of use, desired number of cycles of treatment and risk-benefit in humans.

For short-term toxicity studies with repeat doses, the duration of 1–3 months or single dose toxicity studies may be enough to support a short duration phase 1 study

Table 12.12 Recommended duration of repeated-dose toxicity studies to support the conduct of clinical trials

Maximum duration of clinical trial	Recommended minimum duration of repeated-dose toxicity studies to support clinical trials	
	Rodents	Non-rodents
Up to 2 weeks	2 weeks[a]	2 weeks[a]
Between 2 weeks and 6 months	Same as clinical trial[b]	Same as clinical trial[b]
>6 months	6 months[b,c]	6 months[b,c,d]

[a]In the United States, as an alternative to 2-week studies, extended single-dose toxicity studies (see footnote c in Table 12.12) can support single-dose human trials. Clinical studies of less than 14 days can be supported with toxicity studies of the same duration as the proposed clinical study

[b]In some circumstances clinical trials of longer duration than 3 months can be initiated, provided that the data are available from a 3-month rodent and a 3-month non-rodent study, and that complete data from the chronic rodent and non-rodent study is made available, consistent with local clinical trial regulatory procedures, before extending dosing beyond 3 months in the clinical trial. For serious or life-threatening indications or on a case-by-case basis, this extension can be supported by complete chronic rodent data and in-life and necropsy data for the non-rodent study. Complete histopathology data from the non-rodent should be available within an additional 3 months

[c]There can be cases where a pediatric population is the primary population, and existing animal studies (toxicology or pharmacology) have identified potential developmental concerns for target organs. In these cases, long-term toxicity testing starting in juvenile animals can be appropriate in some circumstances

[d]In the EU, studies of 6 months duration in non-rodents are considered acceptable. However, where studies with a longer duration have been conducted, it is not appropriate to conduct an additional study of 6 months

for the treatment of cancer. In the treatment of life-threatening diseases during phase 1 trial, additional doses other than tested in toxicity studies may be administered to subjects without the instant requirement for additional preclinical evaluation. This is an exception to the standard practice. Based on expectation from regulatory agencies, an amendment to the clinical dosing regimen is expected to be followed by preclinical studies of adequate length and dosage levels to ensure safety to humans [30].

The nature of the drug, therapeutic use, duration, and type of clinical trial guides the type of preclinical studies needed for a drug. In fact, for the IND of small molecule there is requirement of three different types of studies and they are:

1. Repeat dose (duration can vary from 2 weeks to 4-weeks or beyond) toxicology by the intended human route in two species.
   (a) As per the latest regulations testing is conducted in two species, a rodent species (usually rat or mouse) and a non-rodent species (usually the dog).
2. Mutagenicity studies based on the nature of the test material. Typically, minimum of the following two studies are required for an IND:
   (a) Ames *Salmonella typhimurium* reverse mutation assay
   (b) Chromosomal aberration or Micronucleus assay in Mammalian cells (either CHO or Peripheral Human Blood Lymphocytes).
3. Safety Pharmacology studies, which requires assessment of three vital organs, namely:
   (a) Central Nervous System—Functional observation battery (FOB).
   (b) Respiratory plethysmography.
   (c) Cardiovascular.
      (i) hERG (the human Ether-à-go-go-Related Gene) using CHO-hERG cells.
      (ii) Cardiovascular Telemetry Study in conscious Beagle Dogs.

All the above tests have been described in Sect. 12.1. The most important information obtained is characterization of the test drug toxicity and dose-response relationship to determine the NOAEL.

Additional toxicology data may be required based on the results of standard toxicological studies. For example, if the test item induces alterations in the immunological cells or the lymphoid system tissues, immunogenicity studies could be necessary. These can be predicted from hematological, clinical chemistry, and histopathological changes seen from previous studies. Immunophenotyping, T-cell dependent antibody response (TDAR), NK cell activity, organ weights, and histological changes are evaluated further to understand ongoing pathology [31]. CHMP/SWP/2145/2000 Rev. 1 [32] contains details about local tolerance tests and systemic toxicity testing.

Although the regulatory agencies suggest an essential battery of tests to be performed, it is fundamental that the developmental team anticipates possible additional side effects to elaborate a complete clinical development plan with important information and avoiding unnecessary studies. Thus, the previous and direct contact between the pharmaceutical industry and the regulatory agencies is highly recommended, aiming to establish the most appropriate tests for each drug candidate.

12.9.3 Drug Metabolism and Pharmacokinetics (DMPK) of New Drugs

The success of a drug candidate is also related to its kinetic properties. Based on preclinical studies, one should characterize the PK profile of a test item to establish an adequate dosage that enables patient adherence to treatment and the correct interpretation of results obtained from efficacy and safety studies. The toxic effects of a drug leading to the discontinuation of treatment and withdrawal or their progression to develop are associated with prolonged systemic drug exposure, the formation of toxic/reactive metabolites, and/or possible drug-drug interactions [33].

Data about the pharmacokinetic properties of a test item and preliminary studies about its safety and efficacy are critical to deciding to continue or not (go/no-go decision) the development process of a new drug [34].

12.9.4 Determination of the First-in-Human (FIH) Dose

Calculation of starting dose without any risk for humans is the most important step before any new product undergoes clinical testing for the first time. The starting dose should produce any harm in humans and expected show efficacy resulting in reducing the number of volunteers exposed to ineffective dose in the FIH trials [35]. The regulatory agencies such as the US FDA and the EMA have published guidance documents to select the maximum recommended starting dose (MRSD) in the FIH study [30, 36]. The FDA guidance emphasis is placed on the NOAEL assessed in preclinical toxicology studies. Based upon NOAEL data the human equivalence dose is computed by using scaling factor for accounting the body surface area variations in different species. However, the EMA guidance give importance to the minimal anticipated biological effect level (MABEL) approach. The NOAEL- or the MABEL-derived human equivalence dose can be reduced by applying the safety factor, a ratio between calculated human equivalence dose and the first dose in humans.

Biologics require TK studies and anti-drug antibody (ADA) analyses to confirm systemic exposure of drug and correlation of exposure levels with observed toxicities. According to ICH S6, both TK and ADA sampling should be incorporated into study designs for easy interpreting results. This is essential due to development of immunogenicity following administration of the biologic. PD data inclusion, if available, may also be useful to confirm expected on-target biological activity and assess the suitability of the rodent model.

12.10 Conclusion

Nonclinical toxicity assessment plays an essential role in drug development in evaluating the new drug safety and predicting adverse effects in humans through in vitro and in vivo studies. Toxicology studies use common laboratory animals, viz., rodents and non-rodents, and nonhuman primates. Results of toxicological studies are dependent on the dose, duration of the treatment, and availability of the test substance at the target site.

Nonclinical and clinical studies constitute major part of the drug development process. Nonclinical studies are mainly conducted in animals adopting Good Laboratory Practice (GLP) except early studies. During the early nonclinical studies, a drug candidate needs to pass through several steps, such as determining bioavailability, ADME, preliminary dose range finding, mutagenicity, hERG assessment, etc. The purpose of these studies is to ascertain the drug safety and to obtain its tolerability in the test system. The repeat dose toxicity, genotoxicity, and safety pharmacology studies have to be performed in compliance to GLP norms. Subsequent to the IND approval, evaluating long-term safety, carcinogenicity, reproductive and developmental toxicity are carried out in parallel to clinical studies and are performed depending on the new drug clinical application.

There is no single sequence of nonclinical studies in drug development. These studies may be performed in parallel or in the sequence which can vary based up on therapeutic needs. GLP standards are necessary, especially for the evaluation of safety studies, and is a decisive factor for accepting nonclinical studies.

References

1. Good Manufacturing Practice. FD&C Act, Section 501(a)(2)(B). 1976.
2. Merrill R. Regulation of drugs and devices, an evolution. Health Aff. 1994;13(3):47–69.
3. GRAS. U.S. Food and Drug Administration. 1972.
4. Generic Drugs. www.fda.gov/fdac/special/newdrug/benlaw.html. Justina A. Molzon, The Generic Drug Approval Process, 5 J. PHARMACY & LAW 275, 276 (1996) (discussing the steps generic drugs go through to be approved by the FDA and highlighting the passage of the Drug Price Competition and Patent Term Restoration Act). 1996.
5. ICH M3 (R2). Guidance on Nonclinical Safety Studies for the Conduct of Human Clinical Trials and Marketing Authorization for Pharmaceuticals. 2009.
6. Weil CS. Guidelines for experiments to predict the degree of safety of a material for man. Toxicol Appl Pharmacol. 1972;21:194–9.
7. National Research Council. Pesticides in the Diets of Infants and Children. Washington, DC: The National Academies Press; 1993. https://doi.org/10.17226/2126.

8. Gad SC, Cassidy CD, Aubert N, Spainhour B, Robbe H. Nonclinical vehicle use in studies by multiple routes in multiple species. Int J Toxicol. 2006;25:499–521.
9. Federal Register. Vol. 60, No. 40, Wednesday, March 1, pp. 11264–11268. 1995.
10. Nyman A-M, Schirmer K, Ashauer R. Importance of toxicokinetics for interspecies variation in sensitivity to chemicals. Environ Sci Technol. 2014;48:5946–54.
11. ICH S3A. International Conference on Harmonization (ICH). Note for guidance on toxicokinetics: the assessment of systemic exposure in toxicity studies. 1994.
12. OECD. Member Countries include the UK and other European Countries, Japan and the USA. 1983. http://www.oecd.org/.
13. OECD. Mutual Acceptance of Data (MAD). 2001. http://www.oecd.org/document/.
14. OECD. Test Guidelines. 1982. https://www.oecd.org/chemicalsafety/testing/oecdguidelinesforthetestingofchemicals.htm.
15. ICH. Guidelines. 2009. https://www.ich.org/home.html.
16. FDA GLP. Code of Federal Regulations (CFR) Anonymous, Food and Drugs Chapter I. Food and Drug Administration Department of Health and Human Services Subchapter A-General; 2015. Title 21. Part 58. Good Laboratory Practice for Nonclinical Laboratory Studies. 1979.
17. OECD. OECD Principles of Good Laboratory Practice (as revised in 1997): Council Decision [C (97) 186/Final]26th November 1997. 1997.
18. FDA. US Food Drug Administration. Guidance for Industry: Nonclinical safety evaluation of pediatric drug products. 2006.
19. ICH S9. Nonclinical Evalaution for Anticancer Pharmaceuticals. 2010.
20. Morton D, Alden CL, Roth AJ, Usui T. The Tg rasH2 mouse in cancer hazard identification. Toxicol Pathol. 2002;30:139–46.
21. Nambiar PR, Morton D. The rasH2 mouse model for assessing carcinogenic potential of pharmaceuticals. Toxicol Pathol. 2013;41:1058–67.
22. Tamaoki N. The rasH2 transgenic mouse: nature of the model and mechanistic studies on tumorigenesis. Toxicol Pathol. 2001;29(Suppl):81–9.
23. ICH S10. Photosafety evaluation of pharmaceuticals. 2013.
24. OECD. Test No. 432, In vitro 3T3 NRU Phototoxicity Test, OECD Guidelines for the Testing of Chemicals, Section 4, OECD Publishing. 2004.
25. Draize JH, Woodard G, Calvery HO. Methods for the study of irritation and toxicity of substances applied topically to the skin and mucous membranes. J Pharmacol Exp Ther. 1944;82:377–90.
26. FDA. US Food Drug Administration. Guidance for Industry: Assessment of abuse potential of drugs. 2007.
27. FDA (2016). US Food Drug Administration. Guidance for Industry: Safety testing of drug metabolites.
28. FDA. US Food Drug Administration. Guidance for Industry: Investigational New Drug (IND) Application; 2017. https://www.fda.gov/drugs/types-applications/investigational-new-drug-ind-application.
29. FDA. US Food Drug Administration. Guidance for Industry: content and format of investigational new drug applications (INDs) for phase 1 studies of drugs, including well-characterized, therapeutic, biotechnology-derived products. 1995.
30. FDA. US Food Drug Administration. Guidance for Industry: Estimating the Maximum Safe Starting Dose in Initial Clinical Trials for Therapeutics in Adult Healthy Volunteers. Rockville, MD: Food and Drug Administration; 2005.
31. ICH S8. Immunotoxicity studies for human pharmaceuticals. 2005.
32. EMA. European Medicines Agency (EMA) Guideline on non-clinical local tolerance testing of medicinal products, draft EMA/CHMP/SWP/2145/2000. 2000.
33. Singh SS. Preclinical pharmacokinetics: an approach towards safer and efficacious drugs. Curr Drug Metab. 2006;7:165–82.

34. Ducharme J, Dudley AJ, Thompson RA. Pharmacokinetic issue in drug discovery. In: Rang HP, editor. Drug discovery and development. Philadelphia, PA: Churchill Livingstone Elsevier; 2006. p. 141–61.
35. Ivy SP, Siu LL, Garrett-Mayer E, Rubinstein L. Approaches to phase 1 clinical trial design focused on safety, efficiency, and selected patient populations: a report from the clinical trial design task force of the national cancer institute investigational drug steering committee. Clin Cancer Res. 2010;16:1726–36.
36. EMA. Committee for Medicinal Products for Human Use. In: Guideline on Strategies to Identify and Mitigate Risks for First-in-Human Clinical Trials with Investigational Medical Products. London, UK: European Medicines Agency; 2007; http://www.ema.europa.eu/docs/en_GB/document_library/Scientific_guideline/2009/09/WC500002988.pdf.

Chapter 13
Nanomedicine: Implications of Nanotoxicology

Mohd Aslam Saifi, Ramarao Poduri, and Chandraiah Godugu

13.1 An Introduction to Nanotechnology and Nanomedicine

The advancements in science and technology have rewarded the humankind with a number of breakthroughs and nanotechnology is one of them. Nanotechnology is a branch of science dealing with the study and manipulations of the materials at the level of nanoscale. The history of nanotechnology goes back to 1959 when Richard Feynman laid the conceptual foundation of nanotechnology in his lecture "There's plenty of room at the bottom: An invitation to enter new field of physics" at annual meeting of American Physical Society at Caltech. Nanotechnology utilizes the improved characteristic properties of materials once they transform their size to the nanometre level. There is no clear-cut definition of nanoparticles (NPs); however, broadly speaking, a nanoparticle has its, at least one of its, dimensions in the range of 1–100 nm. The nanotechnology has impacted almost every aspect of our life including general household items to high-end smartphones. The particle size influences the physicochemical properties of materials, and with decrease in size, the surface area/volume ratio (also called aspect ratio) and surface reactivity of the material increases. Once the particle size approaches nanoscale, the fundamental properties of material change surprisingly due to gain and dominancy of quantum properties at this scale. So, a material at nanometre scale behaves quite differently

M. A. Saifi · C. Godugu (✉)
Department of Regulatory Toxicology, National Institute of Pharmaceutical Education and Research (NIPER), Hyderabad, Telangana, India
e-mail: chandra.niperhyd@gov.in

R. Poduri
Academic Affairs, School of Basic and Applied Sciences Professor, Department of Pharmaceutical Sciences and Natural Products, Central University of Punjab, Bathinda, Punjab, India

R. Poduri (ed.), *Drug Discovery and Development*,
https://doi.org/10.1007/978-981-15-5534-3_13

from its corresponding bulk material due to changes in properties such as melting point, fluorescence, magnetism, chemical reactivity and electrical conductivity. For example, the bulk gold is yellow in colour while nano-gold shows completely different red or purple colour. These altered properties of material at nanoscale have been harnessed by humans for specific purpose in different fields by utilizing the concept of tunability. The result is that the scientists can modify these properties as per their needs by altering the size and other physicochemical parameters of the material. The flexibility to modify the pharmacokinetic and pharmacodynamic properties has ushered the area of nanotechnology for amelioration of certain disorders such as cancer. The field of cancer research specifically gets benefits from the advancements in the nanotechnology as the NPs originated from various materials with optimum sizes demonstrate passive targeting of tumour by involving enhanced permeation and retention (EPR) effect which favours their leakage from damage blood vessels and subsequent retention in the tumour tissues. On the other hand, the flexibility to actively target the NPs by making use of ligands and other recognizing agents on the surface of NPs helps to selectively target the tumour cells. Nanotechnology has even made it possible to efficiently deliver poorly water-soluble drugs, co-delivery of two or more drugs and delivery of drugs across the difficult-to-cross physiological barriers. In numerous occasions NPs were made for drugs to reduce drug-induced toxic effects. Moreover, the ability to finely tune the surface characteristics of NPs makes them potential candidates for the drug delivery and targeting purpose. Due to all these versatile applications offered different types of nanoparticles for various medical applications, the new discipline Nanomedicine has been envisaged. Nanomedicine is the branch of medicine that employs the concepts, knowledge and tools of nanotechnology by making use of nanoscale materials for the prevention, diagnosis, therapy or a combination, thereof. Nanomedicine offers certain advantages as compared to conventional therapy such as improved biological properties, improved delivery, selectivity, specificity, targeted delivery and therapy while decreasing the toxic effects. In contrast, the use of large particle sized materials or small molecules is associated with poor solubility, low bioavailability and instability; all these problems ultimately affect the pharmacokinetic and pharmacodynamic properties of therapeutic agents. However, the application of nanotechnology concepts helped us to use the small molecules by improving their properties. For example, the use of doxorubicin is associated with dose-dependent irreversible cardiomyopathy. On the other hand, the nanotechnology-based liposomal formulation of doxorubicin (Doxil®) shows better safety profile and exhibits less cardiotoxic potential [1]. The harmonious integration of nanotechnology with drug discovery and development led to an exponential growth in the number of nanotechnology-based therapeutic agents. Inspired by the promising benefits of nanomedicine, the US Food and Drug Administration (FDA) has already approved certain nanomedicine-based therapeutic agents including well-known Doxil® (liposomal doxorubicin), Abraxane (nanoparticles of albumin bound (NAB) Paclitaxel) and AmBisome® (liposomal amphotericin B). Further, there are more than 50 nanomedicine-based drug candidates in the developmental pipeline. Table 13.1 summarizes the current status of novel drug development based on nanomedicine

Table 13.1 Approved and investigational nanomedicine-based drugs

S. No.	Name	Type	Drug	Indication	Status
1.	Doxil	Liposome	Doxorubicin	Kaposi's sarcoma, metastatic breast cancer, multiple myeloma	Approved
2.	AmBisome	Liposome	Amphotericin B	Fungal infection	Approved
3.	DaunoXome	Liposome	Daunorubicin	HIV-related Kaposi's sarcoma	Approved
4.	DepoCyt	Liposome	Cytarabine	Lymphomatous meningitis	Approved
5.	DepoDur	Liposome	Morphine	Postoperative analgesia	Approved
6.	Marqibo	Liposome	Vincristine	Acute lymphatic leukaemia, acute myeloid leukaemia	Approved
7.	Onivyde	Liposome	Irinotecan	Pancreatic cancer	Approved
8.	Visudyne	Liposome	Verteporfin	Wet age-related macular degeneration	Approved
9.	Arikayce	Liposome	Amikacin	*Mycobacterium avium* complex lung disease	Approved
10.	Estrasorb	Micelle	Estradiol	Vasomotor symptoms in menopause	Approved
11.	Genexol PM	Micelle	Paclitaxel	Metastatic breast cancer, advanced lung cancer	Approved
12.	Avinza	Nanocrystal	Morphine sulphate	Psychostimulant	Approved
13.	Emend	Nanocrystal	Aprepitant	Antiemetic	Approved
14.	Ryanodex	Nanocrystal	Dantrolene sodium	Malignant hyperthermia	Approved
15.	Abraxane	Protein nanoparticle	Albumin bound paclitaxel	Breast cancer, non-small cell lung cancer, pancreatic cancer	Approved
16.	LiPlaCis	Liposome	Cisplatin	Advanced or refractory tumours, metastatic breast cancer	Phase I/II
17.	ThermoDox	Liposome	Doxorubicin	Breast cancer, hepatocellular carcinoma,	Phase I/II completed
18.	L9NC	Liposome	Camptothecin	Ewing's sarcoma, lung and endometrial cancer	Phase I/II completed
19.	SPI-077	Liposome	Cisplatin	Ovarian cancer, malignant	Phase I/II/III

(continued)

Table 13.1 (continued)

S. No.	Name	Type	Drug	Indication	Status
				mesothelioma and osteosarcoma	
20.	Lipoxal	Liposome	Oxaliplatin	Advanced or metastatic solid tumour	Phase I completed
21.	EndoTAG-1	Liposome	Paclitaxel	Pancreatic, liver, HER2 and TNBC	Phase II completed
22.	OSI-211	Liposome	Lurtotecan	Ovarian neoplasm, small cell lung cancer (SCLC)	Phase I/II completed
23.	LE-DT	Liposome	Docetaxel	Solid tumours, pancreatic cancer	Phase I/II completed
24.	LEP-ETU	Liposome	Paclitaxel	Breast cancer, neoplasm, gastric carcinoma	Phase I/II/IV
25.	TKM-080301	Lipid nanoparticle	siRNA against PLK1	Advanced hepatocellular carcinoma, solid tumours or lymphomas that are refractory to conventional therapies; colorectal, gastric, breast and ovarian cancers with hepatic metastases	Phase I/II completed
26.	Atu027	Liposome	siRNA against PKN3	Advanced solid tumours, pancreatic cancer	Phase I/II completed
27.	2B3–101	Glutathione PEGylated liposome	Doxorubicin	Advanced solid tumours, brain metastases, lung and breast cancers, melanoma, malignant glioma	Phase I/II completed
28.	MTL-CEBPA	Liposome	Small activating RNA	Liver cancer	Phase I
29.	TLI	Liposome	Topotecan	SCLC, ovarian cancer, solid tumours	Phase I
30.	MM-398 Onivyde	Liposome	Irinotecan	Solid tumours, ER/PR positive and triple negative breast cancer, metastatic breast cancer with active brain metastasis, SCLC, metastatic pancreatic cancer	Phase I/II/III
31.	MM-302	Liposome	Doxorubicin	Breast cancer	Phase I

(continued)

Table 13.1 (continued)

S. No.	Name	Type	Drug	Indication	Status
32.	ATI-1123	Liposome	Docetaxel	Advanced solid tumours	Phase I completed
33.	SGT-53	Liposome	p53 pDNA	Solid tumours, recurrent glioblastoma	Phase I/II
34.	SGT-94	Liposome	RB94 pDNA	Solid tumours, recurrent glioblastoma	Phase I/II
35.	Anti-EGFR-IL-DOX	Immunoliposome	Doxorubicin	Solid tumours	Phase II
36.	RNL	Radioactive liposome	Rhenium-186	Glioblastoma and astrocytoma	Phase I/II
37.	Paclical	Polymeric nanoparticles	Paclitaxel	Ovarian cancer	Phase III completed
38.	NK105	Micelle	Paclitaxel	Gastric cancer	Phase III completed
39.	BIND-014	Polymeric nanoparticles	Docetaxel	NSCLC, solid tumours	Phase II completed
40.	CALAA-01 RRM2	Polymeric nanoparticles	siRNA	Solid tumours	Phase II terminated
41.	CRLX101	Polymeric nanoparticles	Camptothecin	NSCLC	Phase II completed

approach. In addition, the commercialization of nanotechnology has also translated to billion-dollar business worldwide. The President's 2020 Budget has requested over $1.4 billion for the National Nanotechnology Initiative (NNI) totalling almost $29 billion since the start of the NNI in 2001 (https://www.nano.gov/about-nni/what/funding). This suggests that there is going to be more and more use of nano-based materials in future which also increases the chances of human and environmental exposure to nano-based products during various phases of their product cycle such as manufacture, utility and disposal.

13.2 Different Types of Nanoparticles and their Classification

NPs can be broadly classified into carbon based, inorganic and organic. The carbon-based NPs are completely made up of carbon and can be found in different shapes such as sheets, rods and tubes. Fullerenes and carbon nanotubes (CNTs) are two of the most important types of carbon-based NPs. Fullerenes are hollow spherical structures made up of pentagon or hexagon carbon units, while the CNTs can be single-walled (SWCNTs), double-walled (DWCNTs) or multi-walled CNTs (MWCNTs). Fullerenes and CNTs have been extensively used as drug delivery vehicles. In addition, graphene quantum dots (GQDs), nanodiamonds, etc. are other types of carbon-based NPs. On the other hand, inorganic NPs cover a broad range of different substances including metals, metal salts and metal oxides and generally have smaller size. These NPs have greater thermal and chemical stability. In addition, these NPs have biocompatibility and hydrophilic nature, which make them suitable as drug delivery carriers. The organic NPs include liposomes, micelles, dendrimers and polymeric nanoparticles. Polymeric nanoparticles (PNPs) are mostly biodegradable in nature, which give them an upper hand for preclinical and clinical uses. They are generally nanosphere or nanocapsular in shape made up of polymeric matrix with a hollow central compartment. The central compartment (core) as well as polymeric matrix (coat) can encapsulate or adsorb different drugs on their surfaces. Further, the outer polymeric matrix can also be functionalized with certain functional moieties to modify their surface characteristics and improve delivery and targeting capabilities. The most commonly used polymers for NPs include polyesters such as polycaprolactone (PCL), polylactic acid (PLA) and poly(lactide-co-glycolic) acid (PLGA). The biggest advantage of using these polymers is their biodegradable nature. In addition to that, there are many types of NPs that are being used in different fields of day-to-day life.

13.3 Nanotoxicology: Toxicology at the Nano Level

The possibilities of nanotechnology are endless due to its application in wide variety of fields. However, it goes without saying that "too much of everything is bad", and nanomedicine and nanotechnology are no different from it. With ever-increasing use of nanotechnology-based products, there also come the concerns of their toxicity. The exceptionally high reactivity and the ability of the NPs to cross different physiological barriers due to their small size draws the attention of toxicologists to avoid any undesirable outcome from use of these nanomaterials. Our limited knowledge of nanomaterial properties and their interaction with living matter raised concerns on the use of nanomaterials. This led to the foundation of the branch of science dealing with the toxicological aspects of nanomaterials called nanotoxicology [2, 3]. This branch is relatively younger than toxicology branch as it primarily deals with possible hazards of the nanomaterials on human health and environment. However, the most striking feature of the nanotoxicology is that the material which is not toxic at bulk level could be toxic in its NPs form. Humans are exposed to NPs every now and then, unintentionally or deliberately for various biomedical applications. Through air, humans are exposed to natural NPs such as ultrafine dust. Moreover, NPs easily contaminate the air and are available in the air to be respirated into lungs. On the other hand, ingestion of food or water can also be a source of NPs exposure to humans. Once the NPs come in contact with living tissue, they can directly penetrate the cell membrane due to their small size and consequently interfere with a number of cell organelles. A wide variety of NPs can produce toxicity at the cellular level primarily by binding to cell surface, dissolution

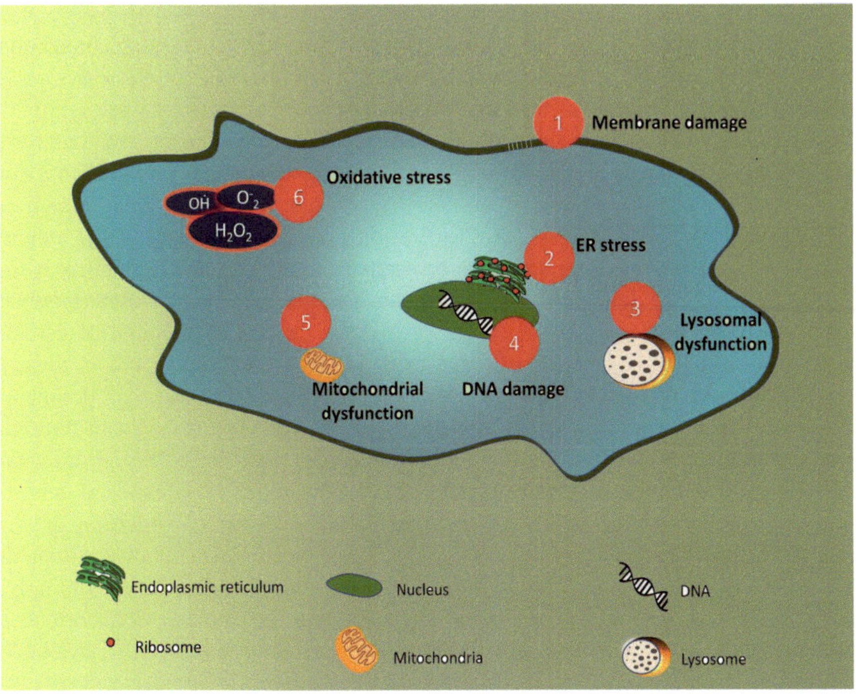

Fig. 13.1 Mechanism of NPs cellular toxicity. Once the NPs come in contact with cell, they can cause membrane disruption and enter the cell. The NPs then interfere with a number of organelles such as endoplasmic reticulum, lysosome and mitochondria and alter their functions. Further, NPs can directly cause DNA damage upon gaining entry into the nucleus. Furthermore, the mitochondrial dysfunction generates free radicals and causes oxidative stress which is centrally involved in NPs toxicity

of toxic ions or inducing oxidative stress [4]. NPs can directly induce cellular death, apoptosis and necrosis after exposure. NPs such as silver, titanium, gold, SWCNTs, MWCNTs and fullerenes are reported to directly induce cellular death of different cells at variable concentrations [5–10]. However, it is not the cellular death which is the end point of NPs toxicity. A number of NPs can interfere and disturb the physiological functioning of the cellular systems such as mitochondrial dysfunction, lysosomal dysfunction, DNA damage or cytoskeletal disorganization without even inducing cellular death. Some of the most important mechanisms by which NPs produce toxicity are shown in the Fig. 13.1. The NPs can affect the membrane integrity directly, or by inducing oxidative stress by oxidizing membrane lipids [11, 12]. Sun et al. used silica at varying concentration of 0, 25, 50, 100 and 200 μg/ml for 3 and 24 h and found that NPs induced oxidative stress which damaged mitochondrial membrane and subsequently caused cellular apoptosis [13]. On the other hand, Athinarayanan et al. used nanosilica which is present in commercial food and investigated the toxic effects on human lung fibroblasts [14]. The authors reported that nanosilica was able to induce oxidative stress, cellular death and mitochondrial membrane potential depletion. Further, the nanosilica also altered the cell cycle distribution and changed the expression levels of metabolic stress responsive genes. Zn NPs which are widely used in cosmetics, biosensors and food additives are reported to interfere with mitochondrial function by releasing the Zn^{++} ions [15]. The authors showed that overexpression of microsomal glutathione transferase I was able to protect the cells from toxicity induced by silica NPs but not by ZnO NPs. They also reported that the cytotoxicity of silica NPs was reduced in the presence of serum while the ZnO were toxic even in the presence of serum. Moreover, ZnO NPs also interfere with lysosomal functioning and destabilize them by release of Zn^{++} ions [16]. On the other hand, a number of NPs produce genotoxicity by damaging DNA of the cell. DNA damaging potential of certain NPs including gold [17], silver [18], zinc oxide [19], iron oxide [20], etc. is already well reported. Moreover, NPs are toxic not only to humans but also to environment. The natural or synthesized NPs contaminate the air, soil or water ecosystems through which a number of living organisms come in contact with these NPs. Studies on aquatic organisms have shown that C_{60} fullerenes were toxic to *Daphnia* and fathead minnow [21, 22]. Similarly, NPs have also been found to be toxic to microorganisms such as bacteria and algae. Sadiq et al. reported that 20 nm TiO_2 NPs were toxic to *E. coli*, *P. aeruginosa* and *B. subtilis* by damaging the membrane [23]. Similarly, Xiu et al. demonstrated that Ag nanoparticle toxicity towards *E. coli* correlates directly with Ag^+ concentration [24]. They used commercially available polyvinylpyrrolidone coated (18, 51, and 72 nm) as well as synthesized glycol-thiol coated Ag NPs (3, 5 and 11 nm) and suggested that Ag NPs properties shown elsewhere to influence antimicrobial activity, including size, surface charge and surface coating, influence toxicity only by changing the dissolution properties. Further, ZnO and CuO NPs are also reported to show toxicity towards microorganisms [25]. In relation to algal toxicity, Hund-Rinke and Simon tested toxicity of Ti NPs to *Desmodesmus subspicatus* and found EC_{50} of 44 mg/L [26]. In addition, there is ample literature stating NPs toxicity to microorganism along with plants (reviewed in [27, 28]). On the other hand, most of the polymeric NPs have been reported to be safer as compared to metal-based NPs. However, there are chances that even these polymeric NPs could be toxic in certain conditions. For example, most of the tumour studies carried out with these NPs focused on tumour regression and/or reduced burden of tumour in the animals. However, tumour is not the only site that receives NPs and they can also go to other tissues such as liver, spleen and kidney which can translate to higher levels of the payload delivered by NPs in these organs. Despite this fact, most of the studies have neglected this aspect of suspected toxicity of polymeric NPs. However, we observed significantly higher levels of paclitaxel-loaded PLGA NPs in the RES organs (Unpublished data). Biodistribution study of NPs showed significantly higher levels of paclitaxel in the liver, spleen, lungs and kidneys after 48 h. Although a significant portion of paclitaxel reached the tumour, the unwanted high levels of paclitaxel in different organs raised the concerns on the use of PLGA NPs. In addition, the Tween 80 coating of NPs significantly increased the brain paclitaxel concentration. Our results suggested that the bare NPs (220 nm) may not cross the BBB while their surface modification with Tween

80 may enhance the brain penetration by increasing their endocytosis. Similarly, Singh and Poduri also reported the toxicity of accumulated degraded polymeric products of different polymeric NPs including PLA and PLGA in RAW 264.7 cells [29]. Hence, there is a great need to focus on comprehensive toxicological profile of the polymeric NPs even if they have been found to be quite safer.

However, there is no general trend for the cytotoxic potential of the NPs and there is lot of contradiction between results of different studies possibly due to varying experimental conditions or interference of NPs with toxicity assays. NPs prepared/synthesized from various methods also indicated contrasting toxicological outcomes. Further, the lack of suitable preclinical models is one of the biggest concerns. The in vitro toxicity testing of NPs has mostly been done on a single type of cellular system which does not recapitulate the ideal conditions needed to carry out toxicity testing. On the other hand, in vivo toxicity studies also show their concerns due to unrealistically high doses of NPs used. Moreover, the low sensitivity of bioanalytical methods used for in vivo toxicity testing also affects the outcome and results of the studies. However, this does not invalidate the fact that NPs are toxic at a certain dose level and there is a greater need to focus on this aspect of nanomedicine. Based on these evidences, it is believed that smaller may be not always better and nanotechnology may result in nanotoxicology concerns. Nevertheless, it is apt to say that nanotechnology and nanotoxicology are two sides of the same coin.

13.4 Factors Influencing Nanoparticles Toxicity

Nanotechnology, despite its wide application in different fields, can also pose serious threats to the environment or human health. The toxicological profile of NPs primarily depends on different factors including, but not limited to, material type, composition, size, shape and surface chemistry.

13.4.1 Size

Size is one of the most important determinants for the observed toxicity of NPs. As the size of particle decreases, the surface area/volume ratio increases, which in turn increases surface reactivity due to higher number of molecules exposed on the surface. Further, the nanometre size of NPs helps in increasing their uptake and increases the chance of their interactions with biological components as they can be internalized in the cell easily. Unlike bulk material, at this nanoscale level, the effect of dose becomes redundant and the effect of surface area increases in determining toxicity. For example, the low dose exposure of 20 nm TiO_2 NPs caused more damage as compared to high dose of 300 nm TiO_2 NPs. There is a size-dependent change in the toxicity profile of the nanoparticles. For example, gold NPs with a size of 1.4 nm were found to be toxic to fibroblasts, epithelial cells, macrophages and melanoma cells, while the NPs with size of 15 nm were found to be non-toxic at up to 60-fold higher concentrations [30]. In addition, there are reports available demonstrating the size-dependent toxicity of NPs in different scenarios. Even a fine scale variation of 1 nm resulted in significantly improved antimicrobial activity of palladium NPs [31]. Similarly, the silver NPs with a size of 10 nm were found to be more toxic as compared to 80 nm, further supporting the dependency of size on the toxicity of NPs [32]. In addition, the SWCNTs are more toxic to bacteria as compared to MWCNTs [33]. However, there are deviations observed from this trend as one study reported that NPs with a size of 60 nm were found to be most toxic to the cells from a bunch of other NPs ranging from 20 to 200 nm [34]. In addition, there are multiple examples of NPs which do not show size-dependent toxicity profiles. However, the absence of a strong correlation between size and toxicity profile suggest that there is more to it. Therefore, detailed and reliable characterization of NPs is required to reach possible conclusions on relative toxicity of NPs of different sizes.

13.4.2 Shape

In addition to the size, shape of NPs is also one important factor which influences the toxicity. The factor of shape comes into picture because the NPs can be synthesized in a number of shapes and different shapes interact and behave differently with living systems. The sphere-shaped NPs exhibit better endocytosis and transport properties as compared to non-spherical shape NPs such as ellipsoid, cylinder, cubes and rods, which help to increase their interactions with biological systems. Further, SWCNTs block ion channel more effectively as compared to spherical fullerenes [35]. In addition, the rod-shaped iron NPs are reported to be more toxic to the macrophages as compared to the sphere-shaped NPs. Lee et al. demonstrated that rod-shaped NPs produced higher degree of LDH leakage, membrane damage and necrosis as compared to sphere-shaped NPs [36]. Similarly, wire-shaped silver NPs showed greater injurious effect than sphere-shaped silver NPs [37]. The spherical-shaped ZnO NPs which bound more proteins showed stronger immunogenic response as compared to sheet-shaped NPs [38]. These available reports suggest a direct role of shape of NPs in imparting toxicity profiles.

13.4.3 Material Type and Composition

Another important feature to affect the NPs toxicity is the material type and composition. The material used in the NPs synthesis affects NPs toxicity as the inherent properties of material largely determine the toxicological potential of the NPs. The PNPs are generally safer as compared to the inorganic NPs as the inorganic NPs are primarily non-degradable. On the contrary, most of the PNPs are biodegradable and have less toxicity potential. For example, the NPs made from biodegradable polymers such as PLGA and PLA have wider safety profile. However, most of the inorganic NPs such as gold silver and zinc iron are reported to produce toxicity as discussed in previous headings. NPs toxicity could also be due to the presence of impurities. For example, cetyltrimethylammonium bromide (CTAB), which is widely used as a surfactant in the NP synthesis, imparted toxicity potential to NPs [39]. On the other hand, the 20 nm NPs of SiO_2 and ZnO showed different mechanisms of toxicity on mouse fibroblasts due to difference in composition [40]. In addition, Fe and Ni impurities found in MWCNTs also produced oxidative stress and inflammatory reactions [41]. Quantum dots made of different types may produce varying toxicity profiles based on their chemical composition. However, when compared among different metal NPs some of them are highly safe like cerium oxide and yttrium oxide NPs, whereas other NPs made up of different metal type exhibited potential toxic effects.

13.4.4 Surface Chemistry

The nature of surface of NPs defines some of the critical properties of NPs along with toxicity profile. In general, the NPs with neutral surface are more biocompatible as compared to the NPs with cationic or anionic surface. Further, the cationic NPs are found to be more cytotoxic as compared to anionic NPs. The higher cytotoxic potential of cationic NPs could be attributed to their strong interactions with negatively charged cellular components. Further, NPs with highly negative or positive charge can readily form "protein corona" consequently changing their in vivo distribution and clearance characteristics. It has been reported that protein-coated NPs can elicit a greater immunological response as compared to bare NPs due to their uptake by phagocytes [42]. However, the adsorption of proteins on the surface of NPs also render them biocompatible and less toxic [43, 44]. Along the similar lines, it was demonstrated that coating of SiO_2 NPs with serum decreased cytokine secretion [45]. In addition, the surface charge can affect the agglomeration properties of NPs consequently affecting the toxicity of the same. It was found that micron-sized CNTs agglomerates were more toxic as compared to well-dispersed CNTs [46]. Additionally, cationic NPs used for gene delivery such as polyethylene amine (PEI) reported to be toxic in nature. Moghimi et al. reported that linear and branched chain PEI can induce membrane damage and initiate apoptosis in human cell lines [47]. A pictorial representation of different factors affecting NPs toxicity is given in Fig. 13.2.

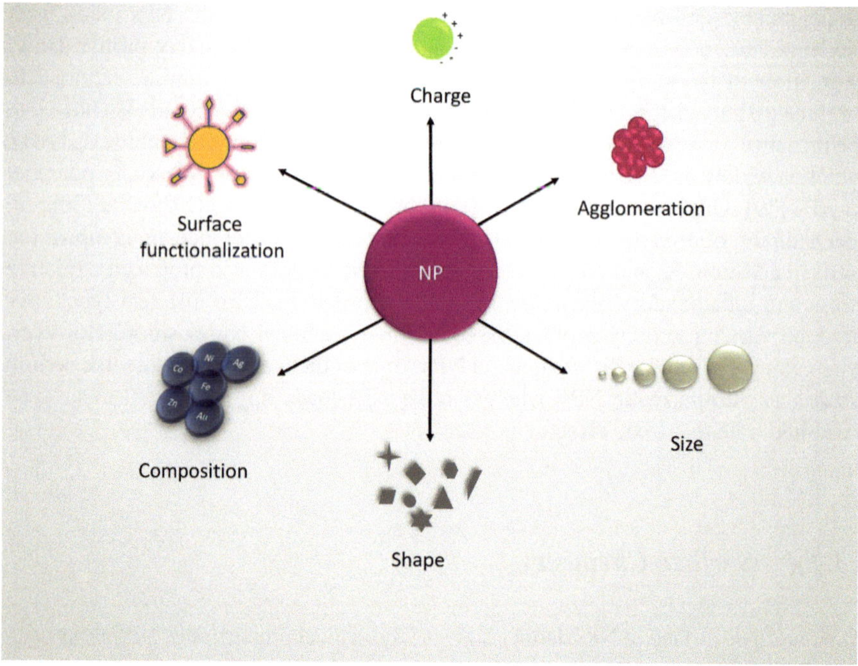

Fig. 13.2 Factors affecting nanoparticles toxicity

13.5 Toxicological Outcomes of Nanoparticles

The toxicological effects produced by NPs cover a range of effects including increased oxidative stress, immunogenic and inflammatory response, misfolding of proteins, damage of different cellular components such as DNA, mitochondria and membrane, immunoreactivity, autophagic changes, complement activation, blood coagulation and accumulation in organs. There are three main mechanisms responsible for NPs toxicity, viz. NPs binding to cell surface, release of toxic ions from NPs and generation of harmful ROS. Generation of reactive oxygen species (ROS) by NPs is probably most important contributors to toxicological outcomes of the NPs. One of the best examples is iron-based nanoparticles which produce free radicals by Fenton reaction. The ROS induction with NPs is such a common phenomenon that most of the NPs including fullerenes, CNTs and metal oxides have been reported to increase ROS in one or the other way. Although ROS are necessary for the proper physiological functioning of the cell, the higher levels of ROS can cause lipid peroxidation, membrane damage, protein alterations, altered signalling and gene transcription in a cell. Moreover, the antioxidants including N-acetylcysteine provided protection from NP-induced oxidative stress highlighting the importance of generation of ROS in NP-induced toxicity. The higher levels of ROS activate inflammatory cascades by involving redox sensitive nuclear factor kappa-light-chain-enhancer of activated B cells (NF-κB) or mitogen-activated protein kinase (MAPK) pathway. Silver NPs toxicity studies on *Caenorhabditis elegans* demonstrated increased levels of ROS, p38, MAPK and hypoxia-inducible factor (HIF)-1 [48]. In addition, silica NPs toxicity was also attributed to p38 and MAPK pathways [49]. The oxidative stress also increases the activation of transcription factors responsible for the inflammatory signalling and induces release of inflammatory mediators such as IL-1β, IL-6, TGF-β, TNF-α and others. Silica NPs are reported to induce the expression of proinflammatory cytokines such as IL-6, IL-8 and MCP-1 [49]. Further, NPs can also induce granuloma formation and fibrosis in different organs. Inhalation of MWCNTs showed granulomatous inflammation and fibrosis with increase in TGF-β and PDGF expression [50, 51]. Similarly, SWCNTs exposure resulted in increased levels of TGF-β, a profibrotic cytokine [52]. Moreover, the increase in oxidative stress by NPs can also induce cellular death by inducing apoptosis. Wilhelmi et al. demonstrated that zinc quartz and silica NPs induced apoptosis in macrophages [53]. However, NPs produce a variety of cellular effects not mentioned here and the readers are requested to read excellent reviews mentioned here [54, 55].

13.6 Biodistribution across Biological Barriers and Toxicological Outcomes

One of the greatest implications of wide distribution characteristics of NPs is the crossing of NPs across the crucial barriers of the body. The small size of NPs acts as a double-edged sword as it also favours the transport of NPs across the biological barriers such as blood brain barrier (BBB), blood placental barrier (BPB) and blood testis barrier (BTB). The BBB is a protective barrier made from endothelial cells, astrocyte foot processes and pericytes which protects the brain from xenobiotics. However, certain NPs can cross this barrier and reach the brain. There are two pathways responsible for the entry of NPs into the brain. The first pathway involves uptake of sensory nerve endings followed by axonal transport. NPs are taken up by nerve endings of olfactory bulb which forms the basis of nose-to-brain delivery. A number of NPs have been reported to follow this pathway. Manganese oxide NPs were observed in different parts of the brain following the olfactory route. However, the second pathway, which is more important from toxicity point of view, is the uptake through the BBB via systemic route. Though this pathway has been exploited by delivery experts as an approach to deliver drugs into the brain, this also raises the concern of neurotoxicity after systemic administration of the NPs. Sharma and Sharma provided the evidence that Ag, Cu or Al NPs disrupted the BBB after systemic administration and altered the brain physiology [56]. Similarly, Ma et al. demonstrated that abdominal cavity administration of TiO_2 NPs for 14 days induced the oxidative stress in the brain of mice [57]. The generation of oxidative stress is believed to be responsible behind the observed neurotoxicity of NPs as brain is highly susceptible to oxidative stress. The placenta constitutes the primary barrier between maternal and foetal tissue which performs the function of selective transport across it. The literature provides evidences that NPs can cross the BPB to reach the foetal circulation and accumulate in foetal organs which is a serious concern associated with use of NPs [58, 59]. Titanium oxide NPs exposed animals showed neurobehavioural abnormalities and altered gene expression in offspring [60, 61]. Similarly, the maternal exposure of silica and titanium dioxide NPs caused pregnancy complications in mice [62]. The NPs were found in foetal liver and brain and the mice treated with NPs had smaller foetuses. On the other hand, there are contradictory reports demonstrating no foetal uptake for PEGylated SWCNTs [63] and gold NPs [64]. These incidences raise the concerns of use of NPs in pregnancy and demands the comprehensive knowledge of transport systems of placenta and the NPs properties affecting this transport. However, there is no general trend that can be used to speculate the transport of the NPs across the placenta. Further, the lack of suitable animal model makes it more difficult to extrapolate the results of these studies for humans. Nevertheless, the toxicity of NPs in pregnancy cannot be ruled out and needs more comprehensive studies.

Another important barrier is blood testis barrier which restricts the transport of xenobiotics to testis. However, a variety of NPs are known to cross BTB as shown in different reports. Park et al. observed significant accumulation of 22 and 42 nm Ag NPs in testis after 2 weeks [65]. Similarly, Bai et al. injected 20–30 nm carbon NPs intravenously in mice and observed accumulation of carbon NPs in the testes after 10 min, 60 min and 24 h [66]. However, it was suggested that penetration of BTB may not be dependent on particle size solely and the penetration of BTB could be a result of increased inflammatory response that weakens the integrity of BTB [67]. Although the NPs can be deliberately directed towards brain, placenta or testis for therapeutic purpose, the unintentional exposure of NPs to these organs warrants the need of toxicological studies to explore the mechanisms involved and ultimately avoid toxicity of these organs (Fig. 13.3).

13.7 Reticuloendothelial Uptake of Nanoparticles and their Safety Profiles

One of the major concerns associated with the use of NPs is their clearance and retention by reticuloendothelial system (RES) of the body. RES is a part of the immune system of the body which removes immune complexes from circulation and the tissue. The RES involves blood monocytes, macrophages, lymph organs, bone, liver and lungs. Particles with more hydrophobic surface are preferentially taken up by the liver, followed by the spleen and lungs [68]. The opsonins present in the blood bind to surface of the NPs and make them prone to recognition by macrophages. As a result of it, RES of the body clears NPs rapidly from blood and directs them to the organs with high RES activity such as liver and spleen. However, this also results in

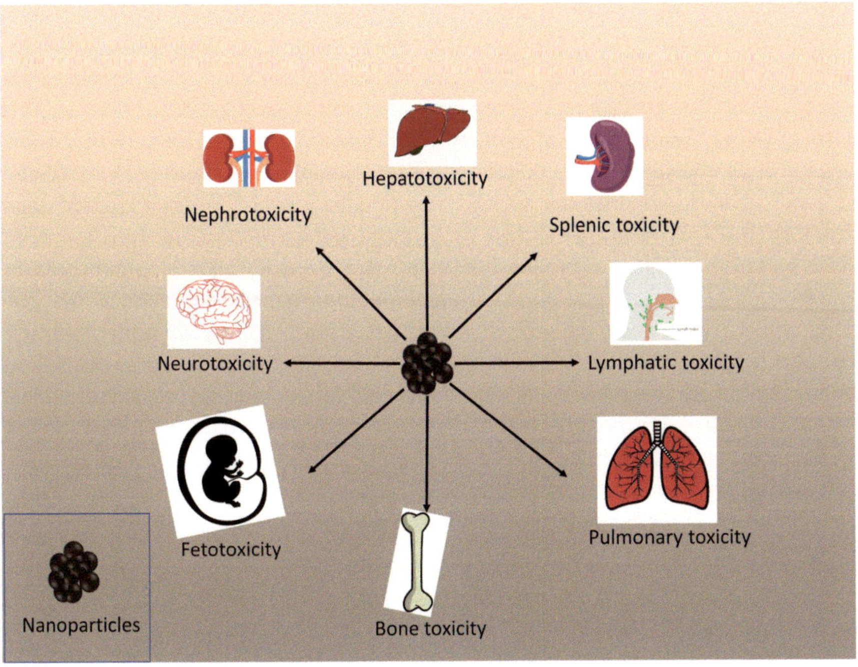

Fig. 13.3 Distribution and toxicity of nanoparticles in different organs. The NPs can either show toxicity of organs with high RES activity or even show toxicity of organs protected by physiological barriers such as brain and placenta primarily due to their small size

unintended delivery of NPs to these organs compromising delivery purpose and also toxicity concerns for these organs. For example, Xu et al. showed that after intravenous administration, MWCNTs aggregated into liver Kupffer cells and aggravated non-alcoholic steatohepatitis in high fat-fed rats [69]. Even one of the highly versatile drug delivery carrier systems, liposomes, are reported to be rapidly cleared from systemic circulation due to their high uptake by RES [70]. We reported the biopersistence of CNTs in lungs even after 3 months of intratracheal administration of 5 mg/kg CNTs as evaluated by UV and SEM (Unpublished data). These results were also correlated with histopathological technique where we observed significant lung damage evident by interstitial fibrosis, inflammatory macrophages, neutrophils and small abscesses. Further, 10 mg/kg of intravenous CNTs administration showed significant persistence of CNT in RES organs including lungs as demonstrated by black appearance of liver, spleen and lungs after 1 month of administration. In addition, there was a significant increase in the oxidative stress, inflammatory cytokines and reduced antioxidant status of these organs. These results suggested that the CNTs have potential safety concerns due to their persistence and low elimination profile. On the other hand, PNPs which have wide safe margin could also be toxic in certain conditions. Although most of the studies carried out with PNPs were oriented towards activity profile of the PNPs neglecting toxicity spectrum, there are chances that the PNPs may also produce toxicity. The lack of toxicity testing studies with PNPs could be a new avenue to explore for the suspected toxicity of the PNPs ultimately leading to safe and efficacious nanomedicine therapy. However, the surface modification of NPs with an aim to reduce opsonization helps to avoid the clearance of NPs by RES organs. Jain et al. demonstrated that toxicity of MWCNTs critically depends on functionalization status [71]. Similarly, we have also shown that the COOH functionalization of CNTs favoured their clearance from the body as shown by reduced persistence of the functionalized CNTs in various organs (Unpublished data). However, the most common surface modification of NPs is done by using amphiphilic polymer chain such as polyethylene glycol (PEG) and evade the RES clearance. Further, the reduction of particle size also helps in reducing the chances of clearance by RES. These modifications minimize the RES clearance of the NPs to provide longer blood levels and improved pharmacokinetics. Further, this also reduces the chances of toxicity of RES organs. On the other hand, RES uptake of NPs can also be utilized to target RES organs such as liver and spleen. Nevertheless, the RES uptake is one of the important mechanisms to affect the NPs toxicity of major organs.

13.8 Strategies to Reduce Nanoparticles Toxicity

The incidences of NPs toxicity in preclinical models led us to strive for the strategies to reduce NP toxicity. However, these strategies primarily depend on the mode of toxicological effects the NPs produce. Hence, the modification of NPs to reduce their toxicity include the modification of NPs surface to reduce their interaction with cellular surface, reduce the toxic metal ion release, reduce the free radical generation, increase susceptibility to biodegradation, alter their biodistribution characteristic, increase the clearance and use of scavenger receptor inhibitors. The first and foremost strategy to reduce NPs toxicity is the coating of NPs surface by PEG or "PEGylation". The PEGylation process also provides "stealth" characteristics to NPs and improves the drug delivery aspects such as longer circulation time [72]. The PEGylation of lipoplexes reduced their haemotoxicity and aggregation of erythrocytes as demonstrated by Eliyahu et al. [73]. Similarly, PEG conjugation of PAMAM dendrimers reduced their hemotoxic potential [74]. Further, the PEGylation reduced the toxicity and improved activity of Mitomycin C liposomes in a tumour model [75]. Furthermore, PEGylation of ZnO NPs reduced their cytotoxicity as a result of decreased cellular uptake [76]. Similarly, surface modification of NPs by materials other than PEG also reduced their toxicity as shown in literature [77, 78]. In addition to that, the NPs toxicity can also be reduced by reducing the release of metal ions from them. The release of metal ions can be reduced by stabilizing the NPs. For example, the release of toxic Zn^{++} ions and eventual toxicity was reduced after stabilizing the ZnO NPs by iron doping [79]. On the other hand, as oxidative stress is one of the most important mechanisms behind NPs toxicity, reducing the generation of free radicals by NPs will minimize the toxicity potential. Functionalization of an antioxidant to NPs surface is one of the most common strategies to reduce the oxidative stress produced by NPs. Nie et al. functionalized gold NPs with vitamin E analogue, Trolox and suggested the low cytotoxic potential of the nano-conjugate [80]. Other than that, replacing the highly redox active material with less active one or doping with another material to change its electronic structure are used to reduce the oxidative stress inducing potential of NPs [4]. However, there are few NPs that are inherently antioxidant such as cerium, selenium and yttrium and their use could be superior to other NPs [81–83]. Since metal NPs produce varying degree of toxicity profiles, the selection of metal NPs should be done based on their toxicity data reported in the literature. The toxic metal can be replaced with non-toxic metal for Nanomedical applications without compromising the pharmacological benefits.

Scavenger receptors are expressed by different cells including macrophages, dendritic, endothelial and epithelial cells. They are receiving attention due to their role in NPs recognition and uptake and toxicity. Singh and Ramarao demonstrated that silver NPs induced toxicity by involving internalization of NPs by scavenger receptors [84]. Similarly, Wang et al. also demonstrated the involvement of scavenger receptor in endocytosis of silver NPs in macrophages [85]. Further, MWCNTs are known to induce apoptosis in macrophages by mitochondrial pathway and scavenger receptors [86]. Due to potential role of scavenger receptors in NPs-mediated toxicity, the inhibition of scavenger receptors has been considered as a potential target to reduce the toxicological outcomes of NPs. The 2-(2-butoxyethyl)-1-cyclopentanone thiosemicarbazone (Blt-2) mediated inhibition of scavenger receptor type B1 has shown reduction in uptake of silver NPs [87]. Similarly, dextran sulphate, a scavenger receptor inhibitor, reduced NPs uptake by immune cells and increased delivery to target tissue, thereby reducing the toxicity of NPs [88]. Moreover, the silencing of scavenger receptor type A on macrophages reduced uptake and inflammatory responses to silica NPs [89].

13.9 Nanotoxicological and Conventional Toxicity Testing Procedure: Similar or Dissimilar?

New chemical entities (NCEs) and medical products come under the purview of toxicological testing suggested and governed by certain agencies including Organization for Economic Cooperation and Development (OECD), International Council for Harmonisation of Technical Requirements for Pharmaceuticals for Human Use (ICH) & World Health Organization (WHO) and Food and Drug Administration (FDA). The in vitro tests suggested by these agencies have narrow coverage and do not consider the wide variety of toxicological effects produced by NPs. The array of toxicological mechanisms shown by NPs are quite unique and complex which are generally not observed with conventional toxicological studies. On the other hand, the conventional toxicity testing procedures involve analysis of toxicity parameters such as single dose acute toxicity studies, genotoxicity, reproductive toxicity, neurotoxicity and carcinogenicity studies, all these studies are neglecting the toxic effects on molecular level. However, the NPs often produce sublethal toxicity which does not directly cause organ toxicity but alter the physiological functions of the cells. For example, adsorption of NPs on cell membrane is one example of NP toxicity that does not directly induce cellular death but affects the cell in a negative way. Moreover, conventional toxicity testing does not keep into account the biopersistence and pharmacokinetic properties of the NPs. As the NPs show pharmacokinetic properties different from their bulk counterparts, the conventional toxicity testing procedure fails to fulfil ideal requirement for toxicity testing of NPs. Hence, the toxicity testing procedures used for conventional compound is inadequate, if same is applied to the toxicity testing of these NPs as the application of these conventional tests cannot give the accurate results. Further, the common mechanisms through which NPs produce toxicological effects such as increased oxidative stress, activation of Inflammatory pathway and induction of proinflammatory cytokines are not at all covered or evaluated during recommended conventional toxicological studies required for NCEs testing. Therefore, we need to have additional toxicological methods to evaluate toxicological effects of various NPs.

Further, there are some inherent unique properties of NPs that interfere with testing procedures and affect the NPs toxicity testing. For example, CNTs are well known to interfere with 3-(4,5-dimethylthiazol-2-yl)-2,5-diphenyltetrazolium bromide (MTT) assay, an assay used to assess metabolic activity of the cell based on the reduction of yellow MTT to purple formazan. CNTs adsorb the reduced formazan dye resulting in false results [90]. Further, liposomes are also reported to interfere with MTT assay due to their lipidic nature compromising the validity of MTT assay for the in vitro toxicity testing of liposomes [91]. NPs may also adsorb cell culture media components and deplete vital requirements for cells, which may result in false positive toxic effects [92]. In addition to that, the agglomeration effects of NPs affect their toxicity profiles as shown by silica, titanium dioxide and zinc oxide NPs [93]. Further, the conventional in vitro testing procedures also fall short of the mimicking of ideal in vivo conditions as most of the in vitro testing is performed on 2-D culture. On the other hand, the cells in the body are embedded in the extracellular matrix in a 3-D fashion and the conventional in vitro conditions do not mimic it. These discrepancies bring us to need of the alternative or advanced approaches to avoid the limitations associated with conventional toxicity testing. The alternative models to address these issues include in silico model, 3-D culture, microfluidic device and organ culture. In addition, the different "Omics" approaches such as proteomics, genomics, transcriptomics and metabolomics offer certain advantages for the nanotoxicology (reviewed in [94]). One of the advantages is the identification of novel targets and biomarkers. In addition, low interference, information about post-translational modification, protein corona effect, high sensitivity and reduced duration are few of the advantages with use of Omics approaches. However, these advanced or alternative techniques are in initial phases that need a lot of improvement and their proper implementation for the nanotoxicological testing.

13.10 Regulatory Guidelines: A Dire Need

Based on available data on NPs toxicological effects, it is clear that these materials behave differently compared to their bulk counterparts. Extrapolating existing toxicity guidelines which are meant for new drugs to nanomaterial safety evaluation may result in inadequate safety outcomes. By just adopting the conventional toxicity methods we may end up in underestimation of toxicological outcomes of NPs. Unfortunately, we do not have any dedicated regulatory guidelines for NPs toxicological and safety evaluations. In that case for regulatory approval of nanomedical products, we are still depending on conventional toxicity guidelines. Though there are regulatory guidelines for nanomaterial safe handling, they deal with safe handling of nanomaterials during manufacture, disposal and occupational exposure. This clearly suggests that there is a need for framing new regulatory guidelines for safety evaluation for nanomaterials. Once the new regulatory guidelines are available, we will be able to efficiently demonstrate their toxicity profiles so that approval for clinical trials and drug approvals can be achieved. The lack of strict guidelines for nanomaterials was one of the biggest hurdles in controlling the toxicity of NPs. Further, these NPs were considered same as their bulk counterparts even after demonstrating the different activity/toxicity profiles. However, the regulatory bodies such as OECD have already started working on the regulatory control on the use of NPs in different fields which led to establishment of OECD's Working Party on Manufactured Nanomaterials (WPMN) in 2006 to assess the safety implications of nanomaterials. Quite recently, Government of India has also released guidelines for the evaluation of nanopharmaceuticals in India. According to the guidelines, the toxicological studies for nanopharmaceuticals should follow the general guidelines according to the route of administration as specified in clause 2 of the Second Schedule of the New Drugs and Clinical Trials Rules, 2019. It is advised that toxicological studies should be carried out in the most clinically relevant animal model. Rodent and non-rodent species, generally rats and dogs of both sexes, should be used for the toxicological studies. Similarly, the most relevant species should be selected for the biodistribution, PK and RES accumulation of nanopharmaceuticals by keeping in mind the inter-species variations. If there is no data on nanomaterial toxicity, it is important to include blank NPs in the toxicity studies. On the other hand, if the active pharmaceutical ingredient (API) is novel, toxicity study of API should be presented with that of nanoformulation. Hence, the harmonious communication and integration of regulatory guidelines along with the toxicity testing data will help to regulate the safe use of NPs.

References

1. Chatterjee K, Zhang J, Honbo N, Karliner JS. Doxorubicin cardiomyopathy. Cardiology. 2010;115:155–62.

2. Saifi MA, Khan W, Godugu C. Cytotoxicity of nanomaterials: using nanotoxicology to address the safety concerns of nanoparticles. Pharm Nanotechnol. 2018;6:3–16.
3. Saifi MA, Khurana A, Godugu C. Nanotoxicology: toxicity and risk assessment of nanomaterials. In: Nanomaterials in chromatography: Elsevier; 2018. p. 437–65.
4. Buchman JT, Hudson-Smith NV, Landy KM, Haynes CL. Understanding nanoparticle toxicity mechanisms to inform redesign strategies to reduce environmental impact. Acc Chem Res. 2019;52:1632.
5. Asare N, et al. Cytotoxic and genotoxic effects of silver nanoparticles in testicular cells. Toxicology. 2012;291:65–72.
6. Eom H-J, Choi J. p38 MAPK activation, DNA damage, cell cycle arrest and apoptosis as mechanisms of toxicity of silver nanoparticles in Jurkat T cells. Environ Sci Technol. 2010;44:8337–42.
7. Chueh PJ, Liang R-Y, Lee Y-H, Zeng Z-M, Chuang S-M. Differential cytotoxic effects of gold nanoparticles in different mammalian cell lines. J Hazard Mater. 2014;264:303–12.
8. Cui W, et al. Effects of aggregation and the surface properties of gold nanoparticles on cytotoxicity and cell growth. Nanomedicine. 2012;8:46–53.
9. Coradeghini R, et al. Size-dependent toxicity and cell interaction mechanisms of gold nanoparticles on mouse fibroblasts. Toxicol Lett. 2013;217:205–16.
10. Jia G, et al. Cytotoxicity of carbon nanomaterials: single-wall nanotube, multi-wall nanotube, and fullerene. Environ Sci Technol. 2005;39:1378–83.
11. Lin J, Zhang H, Chen Z, Zheng Y. Penetration of lipid membranes by gold nanoparticles: insights into cellular uptake, cytotoxicity, and their relationship. ACS Nano. 2010;4:5421–9.
12. Voinov MA, Pagán JOS, Morrison E, Smirnova TI, Smirnov AI. Surface-mediated production of hydroxyl radicals as a mechanism of iron oxide nanoparticle biotoxicity. J Am Chem Soc. 2010;133:35–41.
13. Sun L, et al. Cytotoxicity and mitochondrial damage caused by silica nanoparticles. Toxicol In Vitro. 2011;25:1619–29.
14. Athinarayanan J, Periasamy VS, Alsaif MA, Al-Warthan AA, Alshatwi AA. Presence of nanosilica (E551) in commercial food products: TNF-mediated oxidative stress and altered cell cycle progression in human lung fibroblast cells. Cell Biol Toxicol. 2014;30:89–100.
15. Shi J, et al. Microsomal glutathione transferase 1 protects against toxicity induced by silica nanoparticles but not by zinc oxide nanoparticles. ACS Nano. 2012;6:1925–38.
16. Vandebriel RJ, De Jong WH. A review of mammalian toxicity of ZnO nanoparticles. Nanotechnol Sci Appl. 2012;5:61.
17. Li JJ, et al. Gold nanoparticles induce oxidative damage in lung fibroblasts in vitro. Adv Mater. 2008;20:138–42.
18. Ahamed M, et al. DNA damage response to different surface chemistry of silver nanoparticles in mammalian cells. Toxicol Appl Pharmacol. 2008;233:404–10.
19. Dufour EK, Kumaravel T, Nohynek GJ, Kirkland D, Toutain H. Clastogenicity, photo-clastogenicity or pseudo-photo-clastogenicity: genotoxic effects of zinc oxide in the dark, in pre-irradiated or simultaneously irradiated Chinese hamster ovary cells. Mutat Res Toxicol Environ Mutagen. 2006;607:215–24.
20. Sadeghiani N, et al. Genotoxicity and inflammatory investigation in mice treated with magnetite nanoparticles surface coated with polyaspartic acid. J Magn Magn Mater. 2005;289:466–8.
21. Zhu S, Oberdörster E, Haasch ML. Toxicity of an engineered nanoparticle (fullerene, C60) in two aquatic species Daphnia and fathead minnow. Mar Environ Res. 2006;62:S5–9.
22. Lovern SB, Klaper R. Daphnia magna mortality when exposed to titanium dioxide and fullerene (C60) nanoparticles. Environ Toxicol Chem. 2006;25:1132–7.
23. Mohammed Sadiq I, Chandrasekaran N, Mukherjee A. Studies on effect of TiO2 nanoparticles on growth and membrane permeability of Escherichia coli, Pseudomonas aeruginosa, and Bacillus subtilis. Curr Nanosci. 2010;6:381–7.
24. Xiu Z, Zhang Q, Puppala HL, Colvin VL, Alvarez PJJ. Negligible particle-specific antibacterial activity of silver nanoparticles. Nano Lett. 2012;12:4271–5.

25. Li Y, Zhang W, Niu J, Chen Y. Mechanism of photogenerated reactive oxygen species and correlation with the antibacterial properties of engineered metal-oxide nanoparticles. ACS Nano. 2012;6:5164–73.
26. Hund-Rinke K, Simon M. Ecotoxic effect of photocatalytic active nanoparticles (TiO2) on algae and daphnids (8 pp). Environ Sci Pollut Res. 2006;13:225–32.
27. Navarro E, et al. Environmental behavior and ecotoxicity of engineered nanoparticles to algae, plants, and fungi. Ecotoxicology. 2008;17:372–86.
28. Maurer-Jones MA, Gunsolus IL, Murphy CJ, Haynes CL. Toxicity of engineered nanoparticles in the environment. Anal Chem. 2013;85:3036–49.
29. Singh RP, Ramarao P. Accumulated polymer degradation products as effector molecules in cytotoxicity of polymeric nanoparticles. Toxicol Sci. 2013;136:131–43.
30. Pan Y, et al. Size-dependent cytotoxicity of gold nanoparticles. Small. 2007;3:1941–9.
31. Adams CP, Walker KA, Obare SO, Docherty KM. Size-dependent antimicrobial effects of novel palladium nanoparticles. PLoS One. 2014;9:e85981.
32. Ivask A, et al. Size-dependent toxicity of silver nanoparticles to bacteria, yeast, algae, crustaceans and mammalian cells in vitro. PLoS One. 2014;9:e102108.
33. Kang S, Herzberg M, Rodrigues DF, Elimelech M. Antibacterial effects of carbon nanotubes: size does matter! Langmuir. 2008;24:6409–13.
34. Kim I-Y, Joachim E, Choi H, Kim K. Toxicity of silica nanoparticles depends on size, dose, and cell type. Nanomedicine. 2015;11:1407–16.
35. Park KH, Chhowalla M, Iqbal Z, Sesti F. Single-walled carbon nanotubes are a new class of ion channel blockers. J Biol Chem. 2003;278:50212–6.
36. Lee JH, et al. Rod-shaped iron oxide nanoparticles are more toxic than sphere-shaped nanoparticles to murine macrophage cells. Environ Toxicol Chem. 2014;33:2759–66.
37. Stoehr LC, et al. Shape matters: effects of silver nanospheres and wires on human alveolar epithelial cells. Part Fibre Toxicol. 2011;8:36.
38. Heng BC, et al. Evaluation of the cytotoxic and inflammatory potential of differentially shaped zinc oxide nanoparticles. Arch Toxicol. 2011;85:1517–28.
39. Takahashi H, et al. Modification of gold nanorods using phosphatidylcholine to reduce cytotoxicity. Langmuir. 2006;22:2–5.
40. Yang H, Liu C, Yang D, Zhang H, Xi Z. Comparative study of cytotoxicity, oxidative stress and genotoxicity induced by four typical nanomaterials: the role of particle size, shape and composition. J Appl Toxicol. 2009;29:69–78.
41. Firme CP III, Bandaru PR. Toxicity issues in the application of carbon nanotubes to biological systems. Nanomedicine. 2010;6:245–56.
42. Ruge CA, et al. Uptake of nanoparticles by alveolar macrophages is triggered by surfactant protein A. Nanomedicine. 2011;7:690–3.
43. Ge C, et al. Binding of blood proteins to carbon nanotubes reduces cytotoxicity. Proc Natl Acad Sci. 2011;108:16968–73.
44. Hu W, et al. Protein corona-mediated mitigation of cytotoxicity of graphene oxide. ACS Nano. 2011;5:3693–700.
45. Panas A, et al. Screening of different metal oxide nanoparticles reveals selective toxicity and inflammatory potential of silica nanoparticles in lung epithelial cells and macrophages. Nanotoxicology. 2012;7:259–73.
46. Sayes CM, et al. Functionalization density dependence of single-walled carbon nanotubes cytotoxicity in vitro. Toxicol Lett. 2006;161:135–42.
47. Moghimi SM, et al. A two-stage poly (ethylenimine)-mediated cytotoxicity: implications for gene transfer/therapy. Mol Ther. 2005;11:990–5.
48. Lim D, et al. Oxidative stress-related PMK-1 P38 MAPK activation as a mechanism for toxicity of silver nanoparticles to reproduction in the nematode Caenorhabditis elegans. Environ Toxicol Chem. 2012;31:585–92.
49. Liu X, Sun J. Endothelial cells dysfunction induced by silica nanoparticles through oxidative stress via JNK/P53 and NF-κB pathways. Biomaterials. 2010;31:8198–209.

50. Ryman-Rasmussen JP, et al. Inhaled multiwalled carbon nanotubes potentiate airway fibrosis in murine allergic asthma. Am J Respir Cell Mol Biol. 2009;40:349–58.
51. Porter DW, et al. Mouse pulmonary dose-and time course-responses induced by exposure to multi-walled carbon nanotubes. Toxicology. 2010;269:136–47.
52. Shvedova AA, et al. Vitamin E deficiency enhances pulmonary inflammatory response and oxidative stress induced by single-walled carbon nanotubes in C57BL/6 mice. Toxicol Appl Pharmacol. 2007;221:339–48.
53. Wilhelmi V, et al. Evaluation of apoptosis induced by nanoparticles and fine particles in RAW 264.7 macrophages: facts and artefacts. Toxicol In Vitro. 2012;26:323–34.
54. Khlebtsov N, Dykman L. Biodistribution and toxicity of engineered gold nanoparticles: a review of in vitro and in vivo studies. Chem Soc Rev. 2011;40:1647–71.
55. Elsaesser A, Howard CV. Toxicology of nanoparticles. Adv Drug Deliv Rev. 2012;64:129–37.
56. Sharma HS, Sharma A. Nanoparticles aggravate heat stress induced cognitive deficits, blood–brain barrier disruption, edema formation and brain pathology. Prog Brain Res. 2007;162:245–73.
57. Ma L, et al. Oxidative stress in the brain of mice caused by translocated nanoparticulate TiO2 delivered to the abdominal cavity. Biomaterials. 2010;31:99–105.
58. Buerki-Thurnherr T, von Mandach U, Wick P. Knocking at the door of the unborn child: engineered nanoparticles at the human placental barrier. Swiss Med Wkly. 2012;142:w13559.
59. Hougaard KS, et al. A perspective on the developmental toxicity of inhaled nanoparticles. Reprod Toxicol. 2015;56:118–40.
60. Hougaard KS, et al. Effects of prenatal exposure to surface-coated nanosized titanium dioxide (UV-titan). A study in mice. Part Fibre Toxicol. 2010;7:16.
61. Shimizu M, et al. Maternal exposure to nanoparticulate titanium dioxide during the prenatal period alters gene expression related to brain development in the mouse. Part Fibre Toxicol. 2009;6:20.
62. Yamashita K, et al. Silica and titanium dioxide nanoparticles cause pregnancy complications in mice. Nat Nanotechnol. 2011;6:321.
63. Campagnolo L, et al. Biodistribution and toxicity of pegylated single wall carbon nanotubes in pregnant mice. Part Fibre Toxicol. 2013;10:21.
64. Rattanapinyopituk K, et al. Demonstration of the clathrin-and caveolin-mediated endocytosis at the maternal–fetal barrier in mouse placenta after intravenous administration of gold nanoparticles. J Vet Med Sci. 2013:13–512.
65. Park E-J, et al. Repeated-dose toxicity and inflammatory responses in mice by oral administration of silver nanoparticles. Environ Toxicol Pharmacol. 2010;30:162–8.
66. Bai Y, et al. Repeated administrations of carbon nanotubes in male mice cause reversible testis damage without affecting fertility. Nat Nanotechnol. 2010;5:683.
67. Lan Z, Yang W-X. Nanoparticles and spermatogenesis: how do nanoparticles affect spermatogenesis and penetrate the blood–testis barrier. Nanomedicine. 2012;7:579–96.
68. Brannon-Peppas L, Blanchette JO. Nanoparticle and targeted systems for cancer therapy. Adv Drug Deliv Rev. 2004;56:1649–59.
69. Xu Y-Y, et al. Intravenous administration of multiwalled carbon nanotubes aggravates high-fat diet-induced nonalcoholic steatohepatitis in Sprague Dawley rats. Int J Toxicol. 2016;35:634–43.
70. Senior JH. Fate and behavior of liposomes in vivo: a review of controlling factors. Crit Rev Ther Drug Carrier Syst. 1987;3:123–93.
71. Jain S, et al. Toxicity of multiwalled carbon nanotubes with end defects critically depends on their functionalization density. Chem Res Toxicol. 2011;24:2028–39.
72. Suk JS, Xu Q, Kim N, Hanes J, Ensign LM. PEGylation as a strategy for improving nanoparticle-based drug and gene delivery. Adv Drug Deliv Rev. 2016;99:28–51.
73. Eliyahu H, Servel N, Domb AJ, Barenholz Y. Lipoplex-induced hemagglutination: potential involvement in intravenous gene delivery. Gene Ther. 2002;9:850.

74. Qi R, et al. PEG-conjugated PAMAM dendrimers mediate efficient intramuscular gene expression. AAPS J. 2009;11:395.
75. Gabizon AA, et al. Reduced toxicity and superior therapeutic activity of a mitomycin C lipid-based prodrug incorporated in pegylated liposomes. Clin Cancer Res. 2006;12:1913–20.
76. Luo M, et al. Reducing ZnO nanoparticle cytotoxicity by surface modification. Nanoscale. 2014;6:5791–8.
77. Lankoff A, et al. Effect of surface modification of silica nanoparticles on toxicity and cellular uptake by human peripheral blood lymphocytes in vitro. Nanotoxicology. 2012;7:235–50.
78. Yung MMN, et al. Physicochemical characteristics and toxicity of surface-modified zinc oxide nanoparticles to freshwater and marine microalgae. Sci Rep. 2017;7:15909.
79. Xia T, et al. Decreased dissolution of ZnO by iron doping yields nanoparticles with reduced toxicity in the rodent lung and zebrafish embryos. ACS Nano. 2011;5:1223–35.
80. Nie Z, et al. Enhanced radical scavenging activity by antioxidant-functionalized gold nanoparticles: a novel inspiration for development of new artificial antioxidants. Free Radic Biol Med. 2007;43:1243–54.
81. Khurana A, Tekula S, Saifi MA, Venkatesh P, Godugu C. Therapeutic applications of selenium nanoparticles. Biomed Pharmacother. 2019;111:802–12.
82. Sangomla S, Saifi MA, Khurana A, Godugu C. Nanoceria ameliorates doxorubicin induced cardiotoxicity: possible mitigation via reduction of oxidative stress and inflammation. J Trace Elem Med Biol. 2018;47:53–62.
83. Khurana A, et al. Yttrium oxide nanoparticles reduce the severity of acute pancreatitis caused by cerulein hyperstimulation. Nanomedicine. 2019;18:54–65.
84. Singh RP, Ramarao P. Cellular uptake, intracellular trafficking and cytotoxicity of silver nanoparticles. Toxicol Lett. 2012;213:249–59.
85. Wang H, Wu L, Reinhard BM. Scavenger receptor mediated endocytosis of silver nanoparticles into J774A. 1 macrophages is heterogeneous. ACS Nano. 2012;6:7122–32.
86. Wang X, et al. Multi-walled carbon nanotubes induce apoptosis via mitochondrial pathway and scavenger receptor. Toxicol In Vitro. 2012;26:799–806.
87. Shannahan JH, et al. Formation of a protein corona on silver nanoparticles mediates cellular toxicity via scavenger receptors. Toxicol Sci. 2014;143:136–46.
88. Shannahan JH, Bai W, Brown JM. Implications of scavenger receptors in the safe development of nanotherapeutics. Receptors Clin Investig. 2015;2:e811.
89. Orr GA, et al. Cellular recognition and trafficking of amorphous silica nanoparticles by macrophage scavenger receptor a. Nanotoxicology. 2011;5:296–311.
90. Wörle-Knirsch JM, Pulskamp K, Krug HF. Oops they did it again! Carbon nanotubes hoax scientists in viability assays. Nano Lett. 2006;6:1261–8.
91. Angius F, Floris A. Liposomes and MTT cell viability assay: an incompatible affair. Toxicol In Vitro. 2015;29:314–9.
92. Guo L, et al. Adsorption of essential micronutrients by carbon nanotubes and the implications for nanotoxicity testing. Small. 2008;4:721–7.
93. Adams LK, Lyon DY, Alvarez PJJ. Comparative eco-toxicity of nanoscale TiO2, SiO2, and ZnO water suspensions. Water Res. 2006;40:3527–32.
94. Fröhlich E. Role of omics techniques in the toxicity testing of nanoparticles. J Nanobiotechnol. 2017;15:84.

Chapter 14
Clinical Research in Pharmaceutical Drug Development

Srinivas Ghatta and Michelle Niewood

14.1 Introduction

Clinical research is the science of designing, conducting, analyzing, and interpreting the results of clinical trials. The main aim is to understand whether a test article, such as a drug or a device or a procedure, has an effect on selected endpoints. Clinical research provides data or evidence based on a group of study participants rather than a single study participant.

By implementing methodical science and operations together, clinical research provides effective and/or improved therapies for patients. Most of the traditional medicines came into existence based on historical knowledge, personal experiences, and observational evidences, but they were not based on any formal validation. However, evolving times called for more rigorous and robust ways to evaluate the safety and efficacy of potential new therapies.

The main rationale behind clinical research is to find whether an intervention is efficacious and/or safe and that a response is not by chance. Utilization of proper statistics can help reduce bias. Clinical trials and studies need to prespecify endpoints and use sound statistical methods to analyze results. Like in any research, clinical trials may have limitations. For example, the study sample may not be representative of the population or the results are not generalizable.

Broadly, clinical trials are divided into Descriptive and Hypothesis trials. Descriptive trials do not test hypothesis and do not provide relation between the test article and clinical outcome. Most of the Phase I trials are Descriptive type. To evaluate whether a test article improves outcome, hypothesis testing trials are performed. Most of the hypothesis testing uses randomization.

S. Ghatta (✉) · M. Niewood
Genmab US, Inc, Plainsboro, NJ, USA
e-mail: sgh@genmab.com; mnd@genmab.com

R. Poduri (ed.), *Drug Discovery and Development*,
https://doi.org/10.1007/978-981-15-5534-3_14

Designing a trial plays a key role in clinical research. Well-designed trials yield interpretable results, whereas poorly designed trials yield uninterpretable results. Some of the questions that need to consider before and during study design and execution planning are as follows:

1. How to reduce bias?
2. What are the aims?
3. Do those aims address the appropriate issues?
4. What procedures are being followed to make sure the patient's safety and rights are protected?
5. Is this trial feasible and practicable?
6. What is the target population for the study?
7. What kind of trial design is needed to address the study aims?
8. Is this trial ethically correct?
9. Do we have enough trial participants available to enroll?
10. Do we have the correct dose(s) to conduct study?

14.2 Ethical Issues: History of Clinical Research

There are unfortunately, several past experiences where study participants were subjected to unethical treatment and research procedures. As a result, clinical research is now a highly regulated field and regulations continue to evolve to protect human subjects. Before planning or conducting a clinical trial, clinical researchers must carefully consider if the study's projected value outweighs the projected risks for study participants.

During World War II, German Nazi scientists performed horrific and dangerous experiments on concentration camp prisoners without any scientific rationale. Because of this, Nuremberg code emerged in 1946 with the following outcomes: (1) subject participation in clinical trial is voluntary (2) the researcher has to be qualified and should make every effort to protect the subject (3) the study design should be justifiable and scientifically sound. The Nuremberg code serves as a foundational document in clinical research, however its only focus was on human rights and did not address the questionable practices by investigators.

In 1937, an elixir, Sulfanilamide, which contained diethylene glycol, killed 107 people, many of whom were children. Even though FDA had a presence in the US before 1937, it was the Sulfanilamide case that provided the need to establish drug safety before marketing and made the passing of the 1938 Federal Food, Drug, and Cosmetic Act possible.

In post-World War II era, the use of sleep aids was widespread in the US and Europe. Thalidomide, a non-barbiturate sedative, entered into German market in 1957 and was marketed in 40+ countries by 1960. This was prescribed to pregnant women to reduce morning sickness without any studies supporting this specific use. Several kids were born with phocomelia: shortened, absent, or flipper-like limbs. In the US, FDA inspector Frances Kelsey prevented the Thalidomide's approval despite pressure from the pharmaceutical company and from FDA supervisors. Kelsey felt the application for thalidomide contained incomplete and insufficient data on its safety and effectiveness. This motivated profound changes in the FDA. By passing the Kefauver-Harris Drug Amendments Act in 1962, legislators tightened restrictions surrounding the surveillance and approval process for drugs to be sold in the US which requires manufacturers to prove the safety and efficacy before they are marketed.

Tuskegee experiment: One of the most notorious cases which supported the development of the informed consent process and the Informed Consent Form (ICF) was the Tuskegee experiment. Researchers followed 400 African American men with syphilis for the natural course of disease however, researchers denied those men the available treatment for syphilis during the trial period. In addition, there was no scientific need for including only African Americans. This led the National Institutes of Health requirement for institutions conducting clinical research to have an Institutional review board (IRB) or Ethics committee (EC) to assess and approve clinical trial protocols. The primary purpose of an IRB or EC is to protect study participants.

As a result of all these historical tragedies, Good Clinical Practice (GCP) guidelines came into existence in clinical research. Furthermore, the International Conference on Harmonization (ICH) requirements for registration of pharmaceuticals for human use created standards for the design, conduct, performance, monitoring, auditing, recording, analyses, and reporting of clinical trials that provides assurance that the data and reported results are credible and accurate, and that the rights, integrity, and confidentiality of trial subjects are protected.

The main ethical principles for clinical research are (1) a clinical study's potential benefits should always outweigh the risks, (2) informed consent from study participants, (3) fairness towards all study participants, and (4) study participant's rights to keep their information confidential.

Two pioneering regulatory authorities, the US Food and Drug Administration (FDA) and the European Medicines Agency (EMA), have created the Investigational New Drug Application (IND) and the Clinical Trial Application (CTA) process, respectively, before granting the approval to conduct clinical trials in humans. For an IND application, all nonclinical study reports, nonclinical summaries, detailed chemistry, manufacturing, and control (CMC) information, as well as the protocol and Investigator Brochure (IB) (summary of the known nonclinical and clinical safety and efficacy information of the test article) are submitted. FDA has a 30 day review period for an initial IND. For a CTA, the protocol, informed consent form (ICF), IB, and Investigational Medicinal Product Dossier (IMPD) are required for submission. The average timelines for review of a national CTA is ~60 days [1].

14.3 Phases of Clinical Trials

Demarcation of Clinical trial Phases (I, II, III, and IV) are not clearly defined as it depends on therapeutic areas. Clinical development success rates are dependent on several factors and vary from one phase to another phase and from one therapeutic area to another. BIO, which is a trade association of biotechnology companies, has published the success rates of Phase 1 to regulatory approval to be around 10%. This was higher in rare diseases and lower in chronic diseases. With respect to success in different phases, lowest success rates were observed in the transition from Phase 2 to Phase 3 [2].

Phase 1 studies: Phase 1 studies are also addressed as First-in-Human (FIH) studies where adequate scientific justification and well-designed methodology is an essential part to move the test drug to advanced phases. Phase Is in most therapeutic areas (exceptions are Oncology and rare diseases) involve healthy volunteers. Typically, these studies are smaller with around ~20–80 study participants. These studies aim to establish a safe dose. Starting dose selection should be conservative in the approach so that the first dose tested does not produce unmangeable adverse events. Phase I is a very exciting time for drug development where researchers work to understand whether the drug is safe; researchers may also begin to explore whether the drug has direct or indirect efficacy. In addition to the safety and efficacy, Phase Is may also provide information about tolerability and pharmacological effects of the test drug.

The majority of the phase 1 studies are conducted in Phase 1 units where study participants are in an inpatient setting with close monitoring. Starting dose generally determined in first-in-human studies is based on mostly either No Observable Adverse Effect Level (NOAEL) that involves applying appropriate scaling factors to adjust for body surface area among different species or based on minimal anticipated biological effect level (MABEL) approach where all in vitro and in vivo information is taken into consideration. Dose escalation is conducted in different ways in these studies, i.e., single ascending dose (SAD) and multiple ascending dose (MAD) studies (Fig. 14.1). In SAD studies, drug is administered to a small group of study participants. Starting dose of SAD may be multiple orders lower than the expected efficacious dose. After completion of dose limiting toxicity (DLT) period, a new group of study participants will be exposed to the next dose level of drug. Based on route of administration, regulatory authorities may also require drug–drug or drug–food interaction studies. In addition, in order to rule out cardiovascular liabilities, thorough QT studies are also requested by regulators. Some key considerations that sponsors need to concentrate are (1) selection of correct doses, (2) paying attention to any adverse events of special interest that may emerge, (3) any preliminary correlation of biomarkers to safety and/or efficacy, and (4) selection of the countries/sites follow ICH guidelines. From this stage, approximately 70% of test drugs move to the next phase.

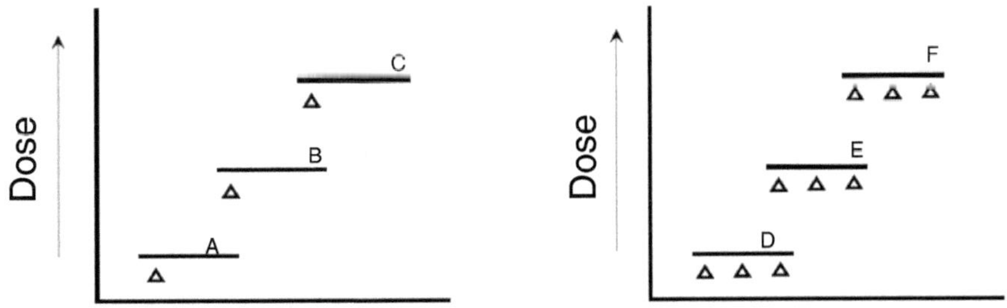

Fig. 14.1 Graphical representation of Single Ascending Dose (SAD) (Left) and Multiple Ascending Dose (MAD) (Right) schema. Each triangle represents dosing and each alphabet represents different cohorts

14.3.1 Phase I Lessons Learned

(a) TeGenero Immuno Therapeutics was developing TGN1412 for B cell lymphoma or rheumatoid arthritis. TGN1412 is a humanized monoclonal antibody which binds and causes agonism to the T cell's CD28 receptor. In preclinical experiments, TGN1412 did not cause any pro-inflammatory responses, however TGN1412 led to T cell expansion. TeGenero also conducted toxicological studies using rhesus and cynomolgus monkeys because they have similar affinity for TGN1412 because of 100% sequence homology of extracellular domain of CD28 receptor. They also conducted a repeat dose pilot study in cynomolgus and rhesus monkeys with dose range from 5 to 50 mg/kg; however they did not observe any systemic immune system dysregulation or hypersensitive reactions even with the highest dose tested. Based on the animal data, TeGeneroTGN1412's phase I was initiated in 2006 by PAREXEL in London. Starting dose was selected based on a NOAEL approach. Six volunteers who were administered a subclinical dose of 0.1 mg/kg were within a short span of time admitted to the intensive care unit. Although the dose was 500 times lower than in animal studies, all six volunteers had multi-organ failure from rapid release of cytokines by activated T cells, i.e., cytokine release syndrome (CRS) minutes after infusion. One of the participants experienced a balloon head similar to elephant man. United Kingdom's Medicines and Healthcare Products Regulatory Agency (MHRA) initiated an investigation on the trial procedures and ethics and they did not find any flaws with procedures associated with preclinical to clinical study transition; however there were several deficiencies in trial conduct, such as inadequate maintenance of medical records, improper qualification of trial staff, inadequacy in ensuring insurance protection of the sponsor, and failure in arranging early medical coverage.

An important reason for the lack of CRS in animal models compared to humans was laboratory animals may not have the similar density of memory T cells when compared to humans and that these memory cells could have been activated by the TGN1412 in the trial. Some lessons learned from the TGN1412 trial are (1) prior to clinical trials, more in vitro human tissue experiments are needed, such as immobilized mAb-based assay or endothelial cell co-culture assays which may predict pro-inflammatory response in humans, (2) NOAEL alone may not be enough for starting dose selection, (3) all the volunteers of same dose cohort should not be dosed at the same time, and (4) prepare for unexpected adverse events based on preclinical data and the mechanism of action of the drug.

(b) In 2016, BialPortela, a Portuguese pharmaceutical company, had started a Phase I with BIA 10–2474, a fatty acid amide hydroxylase (FAAH) inhibitor which led to one death of a subject and caused serious neurological damage in few other healthy volunteers. This trial was done by a contract research organization, Biotrial in France on behalf of the BialPortela. The trial was approved by the French National Agency for Medicines and Health Products Safety (ANSM) and a local institutional review board. Endpoints for the study were safety, tolerability, pharmacokinetic, and pharmacodynamic profile of BIA 10–2474 with single-ascending dose (SAD) and multiple-ascending doses (MAD) in healthy volunteers. A few other companies had taken similar molecules through phase I and II trials without any incidents. Based on NOAEL data in the rat, the human equivalent dose was determined to be 100 mg. Planned starting dose in SAD was 0.25 mg/day. After 0.25 mg/kg, other dosing cohorts were planned with 1.25 mg, 2.5 mg, 5 mg, 10 mg, 20 mg, 40 mg, and 100 mg/day. A total of 90 subjects completed treatment without incident in the SAD and in the first four MAD cohorts. One subject became ill after the fifth dose in the fifth MAD cohort, and was admitted to the hospital with symptoms similar to stroke. Despite this, remaining subjects in the cohort continued to be dosed until the study was suspended later that day. Four of the five subjects who were dosed were eventually hospitalized, and the first subject incurred brain injury that resulted in death. Hemorrhagic and necrotic lesions were seen on brain magnetic resonance imaging (MRI) of the subjects. One probable reason for this serious adverse event was BIA-102474-101's non-selectivity to FAAH at the higher doses and can bind other molecular targets. Doses administered in the affected cohort of the BIA-102474-101 study were several-fold higher than required to fully inhibit FAAH. Following this incident, some companies had voluntarily suspended the development of FAAH inhibitors. Some of those were even in Phase II stage, such as JNJ-42165279.

Some lessons learned from this trial are (1) incorporate detailed stopping rules at the subject, cohort, and study levels, (2) ensure integration of the pharmacokinetic/pharmacodynamic data and modeling to determine the appropriate doses and schedule, (3) screen the off-target effects of test drugs with higher dose ranges as MADs can be higher than SADs, and (4) NOAEL may not be suitable for all test drugs in selecting the starting dose.

Phase II studies: The purpose of this phase is to understand efficacy and side effects of test article. In general, Phase II studies enroll patients (n = 100–300) and generate early efficacy and additional safety data. This phase can be divided into IIa and IIb. Phase IIa is the first trial in patients with several doses and phase IIb is to find the efficacy with the selected dose of IIa. After Phase II, the sponsor can meet with

FDA to obtain guidance on Phase III design and studies (aka "end of Phase II meeting"). Consulting FDA can be informative however, an end of Phase II meeting will not guarantee regulatory approval. From Phase II, approximately 33% of test drugs move to the Phase III.

14.3.2 Phase II Lessons Learned

(a) Olmutinib, by Hanmi Pharmaceuticals, is a third-generation epidermal growth factor receptor (EGFR)-targeted therapy for EGFR mutation-positive lung cancer and was approved in Korea. FDA had granted breakthrough therapy designation for Olmutinib's development in the US; however, South Korea's Board of Audit and Inspection found significant deviations in Hanmi's monitoring and reporting of the Phase II study where patients had developed toxic Stevens-Johnson syndrome, one of which was fatal. Hanmi was cited by regulators for not reporting the death until 14 months later.

Some lessons learned from Olmutinib's clinical development are (1) transparency with regulatory authorities is critical in development and (2) timely reporting of any adverse events and deaths is required.

Fraud in clinical research is rare but sometimes occurs with individual clinical investigators at the site of clinical trial. Fraud can tarnish science, an institution's reputation, and patients beliefs. Any fraud found should be reported to the IRB/EC and regulatory authorities immediately. Several countries have legal protection for whistle-blowers. Some examples of investigator fraud are as follows:

1. Professor Werner Bezwoda of the University of the Witwatersrand, Johannesburg, South Africa, reported in conferences in the 1990s the beneficial effect of autologous stem cell transplantation with chemotherapy in solid tumors. He showed a striking benefit of high-dose chemotherapy for both lymph node-positive and metastatic breast cancer. As there were striking differences with the results from other investigators, NCI physician auditors conducted an on-site audit and found fraud. It is known that high-dose chemotherapy is toxic and autologous stem cell transplantation is very expensive, but Bezwoda's study required patients to be exposed unnecessarily. In addition, inspectors were not able to find all patients' consent forms and not all the patients were eligible for the study. Bezwoda only provided data from the high-dose chemotherapy patients of the trial and did not share the control group data. He later admitted to falsifying the data.
2. Dr. Robert Fiddes was the director of the Southern California Research Institute in the 1990s and was the lead clinical investigator for a large number of clinical trials conducted for various sponsors. Dr. Fiddes was famous for rapid recruitment of patients into clinical trials with a low dropout rate. A whistle-blower contacted the FDA about the enrollment of ineligible patients and fictitious patients. Several laboratory data were altered and fabricated. For example, in

order to obtain proteinuria urine samples, he allegedly paid an employee who was proteinuric for urine sample. After thorough investigation, Dr. Fiddes pled guilty to fraud in 1997 and was sentenced to 15 months in prison.

Phase III studies: Phase III trials are generally large global trials with many study participants ($n = 300$–20,000 or more) and are very expensive to conduct. At the end of these Phase IIIs, marketing approval will typically be submitted and reviewed by regulatory authorities. There is no specific rule however, at least one successful Phase III trial is needed to demonstrate a drug's safety and efficacy, in order to obtain full marketing approval from regulatory agencies such as FDA or the EMA. Most of the time, FDA requires large, confirmatory or registrational, phase 3 studies to ensure the safety and efficacy of test drugs. One research from 359 studies showed the main reasons for Phase III suspensions are ranked in order of efficacy>commercial>safety [3]. Some efficacy reasons for Phase III failures are (1) high placebo response, (2) biomarker-based efficacy prediction failures, and (3) different mechanism of action. From Phase III, approximately 25–30% of test drugs move to the next phase.

14.3.3 Phase III Lessons Learned

(a) Most Phase III studies utilize global multicenter randomized study designs. Randomization is considered the gold standard of clinical trials, however it is important to note that not all regions perform the studies with the same quality standards. NIH initiated the Treatment of Preserved Cardiac Function Heart Failure with an Aldosterone Antagonist (TOPCAT) trial to determine whether treatment with spironolactone would improve clinical outcomes in patients with symptomatic heart failure and a relatively preserved ejection fraction. There was a benefit with the treatment in South and North America, however no change in Russia and Georgia. To assess adherence in different regions as post hoc analysis, the spironolactone metabolite canrenone was measured in that study. Canrenone levels were undetectable in 30% of participants from Russia, as compared with only 3% from the United States and Canada. One important lesson learned from TOPCAT is even though global clinical development is less expensive in certain countries, differences in the level of quality should be carefully considered.

(b) *Phase IV studies:* Phase IV trials focus the side effects caused over time by an approved marketed drug. These studies specifically look for adverse effects that were not seen in earlier trials and may also study how well a new treatment works over a long period of time. Phase IV clinical trials may include thousands of people. Sometimes these are also called as post-marketing surveillance trials. Most of the time, Phase IV drugs are withdrawn from the market because of either any safety finding or commercial viability.

14.3.4 *Investigator-Initiated Studies (IISs)*

Investigator-Initiated Studies are research studies typically sponosored by investigators with support from pharmaceutical companies. IISs generate additional efficacy data and often aim to address clinical questions observed in everyday practice. Most pharmaceutical companies who establish IIS programs require Investigators to submit a proposal that must be reviewed and accepted by supporting company prior to conducting the trial. Some advantages of IIS are (1) reduced cost and complexity compared to larger multi-center trials, (2) reduce off-label use of drugs, and (3) generate data which may lead to supplemental approvals.

14.4 Clinical Trial Protocol Contents

The clinical trial protocol is a key document that describes how a clinical trial will be conducted and ensures the safety of the trial subjects and integrity of the data collected. Below are some of the sections included in a clinical trial protocol (not all protocols have the same sections or same order).

14.4.1 *Trial Synopsis*

This section provides brief summary of the whole protocol with rationale, objectives, endpoints, trial design, etc.

14.4.1.1 Visit Assessment Schedule or Schedule of Activities or Time and Events Table

This is an important section where information on various study procedures and timing of study assessments are outlined.

14.4.1.2 Introduction

This section provides information on the disease(s) that the protocol is evaluating and also information about the investigational agent. In addition, a summary of preclinical and clinical studies (both efficacy and safety) will also be present. This section will include key information about rationale and benefit-risk assessment of the study.

14.4.1.3 Objectives and Endpoints

This section outlines the aims of the protocol. The endpoints (outcomes) are divided mainly into primary, secondary, and exploratory categories and assess efficacy, safety, pharmacokinetic, and pharmacodynamic measurements. Endpoints should be reliable and robust. Endpoints will differ based on the phase of the study, objectives, and the indication that is being evaluated. In some cases where clinical outcome endpoints are not feasible or practical, surrogate endpoints can be used.

14.4.1.4 Trial Design

In this section, the type of design used for the study will be described. The designs can be either a traditional design or flexible design. Traditional design has no flexibility, i.e., they are rigid to changes. Flexible designs can be altered based on safety and efficacy. Even though flexible designs need more time and effort, they can move faster than traditional designs.

14.4.1.5 Trial Population

Type of trial population needed for the study is described in this section. The criteria should be described properly without any confusion or openness. In this section, both inclusion and exclusion criteria will be described. These criteria need to have practically feasible requirements, to avoid slow or limited recruitment. In addition to study entry, information on rescreening will be elaborated here for screen failure subjects.

14.4.1.6 Treatment

In this section, the manner in which treatment is assigned to the study subjects will be described. The type of design plays an important role here because treatment assignment can be either open or blinded. The protocol describes how the subjects are numbered and how subjects get their test article assignment. This section has more information related with Interactive Response Technology (IRT). IRT system can be either Interactive Voice Response or Interactive Web Response systems or both.

The other section in treatment includes information on the preparation and administration of the test drug. This may be included in a document called pharmacy manual. In some studies, premedication and supportive care administration are also needed and those will be described here as well. Permitted and prohibitive concomitant medication information are also added in this section.

If there is any issue with the preparation of test article, such as presence of particles in prepared product and discolorations, site staff must notify the sponsor.

14.4.1.7 Dose Modifications and Safety Management Guidelines

In this section, safety management information from preclinical studies and previous clinical studies are summarized. Dose limiting toxicity periods are discussed and also the type of adverse events which can lead to study discontinuation are described here. For some adverse events, the dose can be reduced with proposed reductions or the dose can be delayed. To prevent some specific adverse events, mitigation plans can be described in this section.

14.4.1.8 Discontinuation, Follow-up, and Completion

This section provides information on the reasons behind discontinuation of the treatment and trial. It also has information on follow-up of safety evaluations. In addition, the protocol describes how follow-up is conducted and with what frequency. The protocol also provides information on lost to follow-up subjects.

14.4.1.9 Trial Assessments

In this section, the protocol describes the parameters to be collected from participants, such as Demography, Baseline Assessments, Medical History, Concomitant Medication, Physical Examination, Body Measurements, Vital Signs, Electrocardiograms (ECG) and Prior Therapy and Surgery. This section also describes the methods that will be used to measure Efficacy Assessments, Pharmacokinetics, and Biomarkers.

14.4.1.10 Safety Monitoring and Adverse Event Reporting

This section has several safety aspects of protocol, including method to report adverse events. If there are any adverse events of special interest including special scenarios, such as pregnancy of a trial participant, they are described here. This section also has information about intensities and relatedness of the adverse events. Some studies utilize Data Monitoring Board (DMB) to get unbiased view of safety data. If study uses DMB, this section delineates as to when they investigate that data.

14.4.1.11 Statistics

This is a crucial section for any study. This section describes how analysis will be conducted and what sets of data (safety, efficacy, pharmacokinetic, biomarker, vital signs, laboratory abnormalities, and ECGs) will be used. Sample size selection details are also described. Missing values are very common in clinical trials and this section describes the information on using missing values here.

14.4.1.12 Data Handling and Record Keeping

This section provides the information associated with data entry, recording, and retention policies. It also has information related to Investigator Responsibilities, EC or IRBs, Informed Consent, and Privacy of Personal Data.

14.4.1.13 Administrative Procedures

This section has all administrative procedures, such as Contractual and Financial Details and Insurance, Indemnity and Compensation.

14.5 Clinical Trial Designs

As not all clinical trials are the same, clinical design selection depends on aspects, such as randomization (process of assigning the treatment to subjects in a random manner may be by a permutation of a sequence) and stratification, blinding (concealment of treatment group allocation from one or more individuals involved), placebos, sample size, selection of a control group, target population, and endpoints. The main objective of a clinical trial is to demonstrate the effect of test article.

14.5.1 Parallel Design

There are two types of parallel designs: (a) single arm and (b) multiple arm. In single-arm design, there are no placebo or comparison interventions. Even though technically single-arm design is not parallel design, it is still considered as parallel design. One of the issues with single-arm design is determining whether the change is due to other factors or the intervention. Even though this design is not the same level of rigor as a randomized study, single-arm designs are easier and less costly to conduct. These designs are appropriate when the efficacy will not occur by itself generally, when the efficacy is obvious, and when the patients are rare and homogenous. Multiple arm parallel design is the common clinical study design where it has less complex underlying assumptions when compared to other designs. In multiple arms, each group receives one type of treatment and this design has least risk for bias. Compared to other clinical designs, these design studies do not take long time to complete. One main issue with this design is variability in groups may have an impact. To reduce this variability, within-patient or between-patient designs can be used.

14.5.2 Within-Patient Designs

Each patient receives both treatments, i.e., one treatment in one period, and different treatment in another period. This switch can also be with placebo to active treatment. This comparison in each subject with themselves minimizes the variability among different subjects. Because of the same subjects' utilization for different treatments, this design needs fewer subjects than other designs; hence it may not be as expensive to run as parallel designs. Some disadvantages with this design are temporal effects (disease improvement or worsening in one direction), carryover effects, in between dropouts of subjects, and longer time to complete the studies.

Common within patient designs are crossover trial design and Latin square design (Fig. 14.2). Crossover trials are commonly used in Bioequivalence studies, pharmacokinetic studies, and drug–food interaction studies. Latin square designs are common with orphan diseases where the subject availability is difficult.

14.5.3 Factorial Designs

Factorial design contains two or more interventions with two or more quantities of the interventions; hence 2 × 2 design has two levels and two factors (Fig. 14.2).This design is widely used for very large mortality studies. Greco-Latin square Factorial design is with a Latin square and it is uncommonly used in clinical studies. Some drawbacks of this design are (1) possibility for interaction among different interventions and (2) this design is incomplete if it misses any possible combination.

14.5.4 Group Sequential Designs

Group sequential design is a type of adaptive design where the number of patients is not set in advance. Stopping rules specify when and why a trial might be halted. This design facilitates interim analysis and positive data allows to continue the study and negative data makes the study to stop. Patients are divided into an equal number of groups and data is analyzed at predetermined points in the trial. Basic requirements

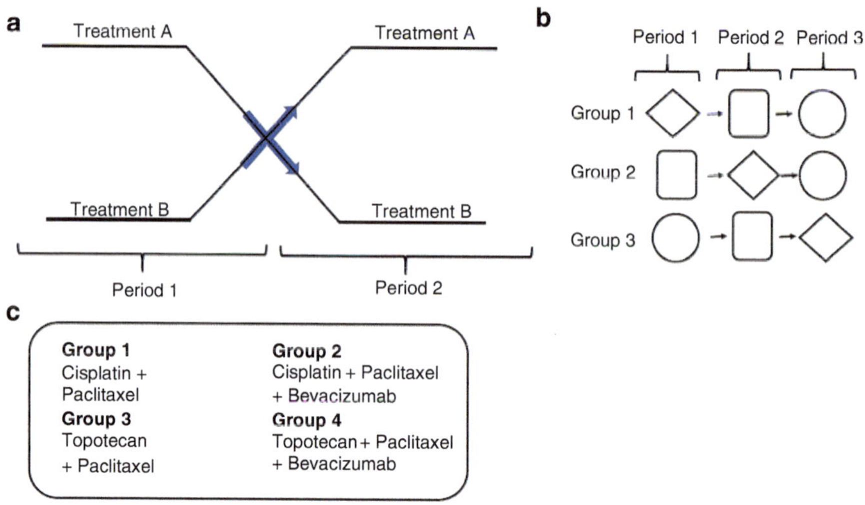

Fig. 14.2 (**a**) Crossover study design. (**b**) Latin square design. (**c**) Factorial design (e.g., GOG240 2 × 2 design)

for this design are (1) intervention's effects occur rapidly and (2) modification of trial is possible. Oncology trials often use this type of trials where they have preset arbitrary efficacy bars with a different bar in each stage of the sequential trial.

Precision Medicine Trials: Based on precision medicine premise, biomarker-positive patients get treatment benefit and negative patients may not get benefit. Two types of trials where biomarker-positive patients are specifically enrolled to test the efficacy especially in oncology and they are

1. Basket trials: If the plan is to find the effect of a specific treatment within a biomarker-positive subgroup, a basket trial will be good way to find out the effect in different subtypes of disease. Example: In BRAF V600 mutation biomarker patients were tested in different cancers (5 baskets = 5 subtypes of cancers) by the drug Vemurafenib [3].
2. Umbrella trials: In contrast to basket trials, the umbrella trial evaluates many treatments within a single histology. Trial participants are assigned to a specific treatment arm of the trial based on their specific molecular makeup of their cancer. Example: In Lung-MAP, patients get a genomic profile to determine the genomic alterations, or mutations, which may drive the growth of their cancer and are treated based on the results [4].

14.6 Key Staff at Investigative Sites (at Clinic)

14.6.1 Principal Investigator (PI)

The PI is the lead individual at the site that is responsible for leading the clinical trial. This individual is often a medically qualified doctor. The PI's job is to protect the rights, safety of study participants. PIs typically have several Sub-Investigators (Sub-I) who are also medically qualified doctors. The PIs and Sub-Is, ensure the protocol is executed exactly as written, oversee the activities of the research team, and oversee all protocol amendments, regulatory compliance paperwork, and reviewing adverse events. PIs also supervise data collection, analysis, interpretation, and presentation.

14.6.2 Research Nurse/Site Coordinator

Research nurses hold an active registered nurse (RN) license and handling some of the medical duties, such as administering drugs or performing exams. They communicate regularly with the principal investigator and train staff and educate patients about the trial. They play a crucial role in study monitoring, quality assurance, and data entry management.

Site coordinators who do not have RN license or medically qualified degree will not perform any medical duties, however these individuals handle the daily conduct of study such as screening participants, obtaining informed consent, checking eligibility, collecting, and entering data.

14.6.3 Data Manager (DM)

Data managers manage the entry of data and provide data output to the required parties throughout the course of a clinical trial. DMs also prepare summaries for interim and final data analysis.

14.7 Key Staff at Sponsor or CRO

Safety Physician*, Safety Scientist*, Data programmer*, Stats programmer*, Pharmacometrician*, Medical writer*, CRA*, CMC lead*, Project manager*, Translational medicine lead*. (*—Description for these roles are not provided here).

14.7.1 Medical Monitor (MM)

Medical monitors are mostly physicians. MMs provide medical expertise and oversight for the entire clinical trial and provide guidance of the medical aspects of the protocol. MMs closely work with other team members in designing the clinical trial design. They ensure the clinical integrity of the trial subjects and provide safety accountability across the duration of the study with great knowledge on good clinical practice (GCP). MMs may also present and discuss data findings at advisory committee meetings. They also provide medical input to relevant pharmacovigilance activities throughout the trial.

14.7.2 Clinical Research Scientist (CRS or CS)

CRSs are mostly qualified scientifically or clinical trianed professionals with degrees ranging from BS, MSN, to MD/PhD. An important aspect is that CRSs need to have knowledge about disease(s) included in the protocol, the test article, GCP, and organizational management skills. CRSs work very closely with MMs and Clinical Operations Lead. Some of the activities that CRSs spearhead are:

1. Develops relationships with appropriate KOLs to obtain feedback on protocol design and strategy.
2. Provide input on study design and author protocols. Provides protocol training to internal/external team members.
3. Ongoing medical review of data (issuing/closing queries as applicable).
4. Assist in the review of clinical data for key deliverables.
5. Tracks, review, and/or summarizes safety and efficacy data for ongoing studies (as needed).
6. Reviews and summarizes safety data for safety calls (Phase 1 studies) or safety review meetings.
7. Respond to questions about the protocol by health authority, ethics committees, and investigational sites.
8. Assists in the development and review of abstracts, posters/presentations, and manuscripts.

14.7.3 Clinical Operations Lead or Clinical Management Team (CMT) Lead

The Clinical Operations Lead plans, directs, and coordinates operational activities of clinical studies. The CMT lead works very closely with the sites, CRAs, and other functional area team members. Some of the activities of CMT lead, but not limited to, are

1. Drive vendor selection and vendor management.
2. Day-to-Day management of the trial operational activities.
3. Work closely with MM, CSR, Data Manager, and Biomarker operations management team.
4. Master service agreement (MSA) and Clinical Trial Agreements setup.
5. Responsible for developing recruitment plans.
6. Perform Sponsor oversight/Booster visits.
7. Work closely with project manager that study is progressing according to agreed to timelines and budget.
8. CMT lead provides oversight on CRO management, if there is a CRO involvement.

14.7.4 Clinical Data Manager (CDM) or Data management Lead (DML)

CDMs are responsible for collecting data from various sources (both internal systems and external systems) in the clinical study. They work collaboratively with data programmers and statisticians to make sure data is collected, managed, and reported clearly, accurately, and securely. Clinical data managers have a variety of educational backgrounds and professional experience. Most have degree in life sciences, computer science, or information technology. Some may have graduate certifications in areas such as clinical data management, health informatics, or biometrics. Some of the activities of CDM, but not limited to, are

1. Electronic case report form (Where the trial participant's data is entered) (eCRF)development.
2. Data quality review.
3. Data delivery for documents and presentations.

14.7.5 Clinical Study Statistician

Statisticians can work on study design, randomization, sample size estimation, statistical methodology, and analysis and reporting. Their main role is to ensure that study's questions will be answered while satisfying regulatory requirements. Statisticians work very closely with statistics programmers who uses programming software such as SAS® to deliver quality data. Statisticians have Masters or a PhD in Statistics or Mathematics. Some of the activities of statisticians, but not limited to, are:

1. Contribution to trial design.
2. Contributor to development of protocol.

3. Perform vendor oversight related to key deliverables, e.g., SAP, Statistical QC plan, interim analysis.
4. Perform exploratory analysis, ad hoc analyses, and modeling of data.
5. Review and approve the tables, figures, and listings of clinical study.
6. Presenting study data in presentations and publications.

14.8 Summary

Clinical research is highly regulated field as it supports the evaluation of experimental therapeutics in human subjects. Ethical principles and historical lessons led to the development of guidance documents that serve to protect human subjects from wrong doing and establish standards that must be adhered to when conducting clinical research. A brief history of clinical research lessons, trial phases, clinical protocol contents, trial designs, and clinical research team compositions has been shared to support your quest to learn more about this exciting field. To increase the success rate from Phase I to marketing approvals, the drug development industry is working on novel study designs and biomarker-driven approaches. As Science and Regulations around clinical research are evolving rapidly, current and prospective researchers should keep abreast of current regulations mainly through the FDA, EMEA, and ICH websites.

References

1. Chiodin D, Cox EM, Edmund AV, Kratz E, Lockwood SH. Regulatory affairs 101: introduction to investigational new drug applications and clinical trial applications. Clin Transl Sci. 2019;12:334–42.
2. Bioteschnology Innovation organization, Biomedtracker, Amplion Clinical Development Success Rates 2006-2015 (BIO, Washington, DC, BioMedTracker, CA, Ampion, OR, 2016). https://www.bio.org/sites/default/files/Clinical%20Development%20Success%20Rates%202006-2015%20-%20BIO,%20Biomedtracker,%20Amplion%202016.pdf.
3. Hyman DM, et al. Vemurafenib in multiple nonmelanoma cancers with BRAF V600 mutations. N Engl J Med. 2015;373:726–36.
4. Herbst RS, Gandara DR, Hirsch FR, Redman MW, LeBlanc M, Mack PC, Schwartz LH, Vokes E, Ramalingam SS, Bradley JD, Sparks D, Zhou Y, Miwa C, Miller VA, Yelensky R, Li Y, Allen JD, Sigal EV, Wholley D, Sigman CC, Blumenthal GM, Malik S, Kelloff GJ, Abrams JS, Blanke CD, Papadimitrakopoulou VA. Lung master protocol (Lung-MAP)—A biomarker-driven protocol for accelerating development of therapies for squamous cell lung cancer: SWOG S1400. Clin Cancer Res. 2015;21(7):1514–24.

Chapter 15
Pharmacovigilance

Pramil Tiwari and Prity Rani Deshwal

15.1 Introduction

In the past fifty plus years, several safety issues on drugs have caught the attention of the regulators, prescribers, users, and the pharmaceutical industry alike. The need of safe and efficacious drug therapy has been captured at the scientific meetings, workshops, conferences and routine meetings of various associations.

All pharmacotherapy are associated with both desirable and undesirable effects. There is no drug which is completely safe under all situations. Even right drug given at the right time and correctly dosed may also cause either an adverse drug reaction (ADR) or side effect(s). ADRs pose a high level of variability in the individuals who receive medications. Adverse drug reactions are rated as the fifth leading cause of death among all diseases. It is well known that all drugs carry the potential to produce both desirable and undesirable effects. No drug is absolutely safe under all circumstances of use or in all patients and ADRs may occur even if a drug is correctly selected and dosed.

In 1975, Karch and Lasagna observed that the data on adverse drug reactions are incomplete, unrepresentative, uncontrolled, and lacking in operational criteria for identifying ADRs. In light of this, no quantitative conclusions can be drawn from the reported data in regard to morbidity and mortality. They had suggested that to evaluate the impact as well as the causes of ADRs, representative populations, including general hospital and ambulatory patients of all medical specialties, must be studied, and operationally defined criteria must be used to establish the presence of an ADR in a prospective study that incorporates appropriate control populations. They had underlined the value of risk–benefit assessment in pharmacotherapy. The

P. Tiwari (✉) · P. R. Deshwal
Department of Pharmacy Practice, National Institute of Pharmaceutical Education and Research (NIPER), S.A.S. Nagar, Punjab, India
e-mail: ptiwari@niper.ac.in

R. Poduri (ed.), *Drug Discovery and Development*,
https://doi.org/10.1007/978-981-15-5534-3_15

researchers, rightly, concluded that until such studies are performed, estimates of the problems posed by the ADR can be only guesses [1].

Lazarou et al., in 1998, published their findings of a meta-analysis in an attempt to understand the incidence of adverse drug reactions in the hospitals of the United States. Of 39 prospective studies analysed, they reported that the overall incidence of serious adverse drug reactions was 6.7%. While cautioning to use their findings, they concluded that the incidence of serious and fatal ADRs in the US hospitals was extremely high [2].

Rightly, the 'thalidomide disaster' continues to be a starting point for a discussion on drug safety issue, which caused limb reduction malformations (phocomelia) in approximately 10,000 children [3]. Wiedemann was the first to report on a series of children with malformations in September 1961 [4].Therefore, before 1960s, drug safety was not the matter of concern; drug's efficacy was dominant priority instead of drug's safety. However, the 'Thalidomide disaster' changed the perspective of looking at safe use of drugs.

Pharmacovigilance is the science and activities relating to the detection, assessment, understanding and prevention of adverse effects or any other drug-related problem [5].The reporting of ADRs has become an important component of the monitoring and the evaluation activities which are performed in hospitals. Although many of the ADRs are relatively mild and disappear when drug is stopped or dose is reduced, others are more serious and last longer. The ADRs produced by a certain new drug are often recognized when the medication is undergoing its phase three clinical trials. A clinician may have problems recognizing the scenario as an ADR, because of the background symptoms of the patient's original illness.

At this point, some relevant definitions need to be touched upon. A side effect, adverse event and adverse drug reactions are commonly understood to be synonymous by the novice. In fact, they are not.

Side effect: Any unintended effect of a pharmaceutical product occurring at doses normally used in humans which is related to the pharmacological properties of the medicine. Such effect may be either positive or negative. Such effects may be well known and even expected and may require little or no change in patient management.

Adverse event: Medical occurrence temporally associated with the use of a medicinal product, but not necessarily causally related.

Adverse drug reaction: An appreciably harmful or unpleasant reaction, resulting from an intervention related to the use of a medicinal product; adverse effects usually predict hazard from future administration and warrant prevention, or specific treatment, or alteration of the dosage regimen, or withdrawal of the product [6].

According to WHO [5], an ADR is 'a response to a drug which is noxious and unintended, and which occurs at doses normally used in man for the prophylaxis, diagnosis, or therapy of disease, or for the modifications of physiological function'.

The American Society of Health-System Pharmacists [7] provides another definition of ADR. It describes an ADR as any unexpected, unintended, undesirable, or excessive response to a medicine that

1. Requires discontinuing the medicine (therapeutic or diagnostic).
2. Requires changing the pharmaceutical therapy.
3. Requires modifying the dose (except for minor dosage adjustments) • Necessitates admission to a hospital.
4. Prolongs the patient's stay in a healthcare facility.
5. Necessitates supportive treatment.
6. Significantly complicates diagnosis.
7. Negatively affects prognosis.
8. Results in temporary or permanent harm or disability, or in death.

Causality: The probability that a particular medicine or substance is responsible for an isolated effect or ADR.

Signal: Reported information on a possible causal relationship between an adverse event and a medicine, the relationship being previously unknown or incompletely documented. Usually more than one signal report is required to generate a signal, depending on the seriousness of the event and the quality of the information.

Serious adverse drug event or reaction: Any untoward medical occurence that at any dose leads to death, life-threatening situation, requires inpatient hospitalization, or prolongation of existing hospitalization, results in persistent of significant disability or incapacity.

15.2 Drug Safety Frameworks

The WHO Programme for International Drug Monitoring is a group of more countries that share the vision of safer and more effective use of medicines. Currently, 142 countries are members of the WHO Programme for International Drug Monitoring, and 29 associate member countries, in the early stages of establishing their pharmacovigilance systems, are preparing themselves for full membership. They work nationally and collaborate internationally to monitor and identify the harm caused by medicines, to reduce the risks to patients and to establish worldwide pharmacovigilance standards and systems.

Uppsala Monitoring Centre (UMC) has been responsible for the technical and operational aspects of the programme. It is an independent, not-for-profit foundation, a centre for international scientific research, based in Sweden—closely associated with the World Health Organization (WHO), since 1978. The UMC sells products related to pharmacovigilance to finance the support to WHO while still maintaining its intellectual and scientific independence.

The WHO Programme, created in 1968, to ensure that evidence about harm to patients was collected from as many sources as possible has enabled individual countries to be alerted to patterns of harm that were emerging across the world and which might not be evident from their local data alone.

In the past few years, considerable effort has gone into improving the adverse drug reaction reporting system in the United States, which is now called

MEDWATCH [8]. Still, the voluntary reporting system in the United States is deficient compared with the legally mandated systems of the United Kingdom, Canada, New Zealand, Denmark and Sweden [9] (Table 15.1).

The growth and development of the pharmacovigilance programme in India has been well documented [10–14]. Though not exhaustive, the interested reader may refer to these publications to understand the development of the subject in India.

The government of India started the Pharmacovigilance Programme of India (PvPI) on 14th July 2010 with the All India Institute of Medical Sciences (AIIMS), New Delhi, as the National Coordination Centre for monitoring Adverse Drug reactions and promoting the safe use of medicines in India. This programme led to setting up of 22 ADR monitoring centres including AIIMS in the year 2010.

In April 2011, the National Coordination Centre was shifted from the AIIMS, New Delhi, to the Indian Pharmacopoeia Commission, Ghaziabad, India, for effective implementation of this programme. Ever since then, the commission is responsible for achieving the objectives of the Pharmacovigilance Program of India.

Pharmacovigilance Program of India (PvPI) works with international organizations and authorities to promote pharmacovigilance activities, to establish high-quality pharmacovigilance systems for medical products and to upgrade ADRs reporting cultures. The program also interfaces with the Council for International Organisations of Medical Sciences (CIOMS).

In addition to healthcare providers reporting the ADRs, the patients can report an ADR through a toll-free telephone helpline number (1800–180–3024); through email; by submitting the blue form titled "Medicines Side Effect Reporting Form for Consumers" that may be downloaded from the official website of the commission. This form is now available in 10 vernacular languages [11].

In 2017, the National Coordination Centre-Pharmacovigilance Programme of India, Indian Pharmacopoeia Commission, Ministry of Health and Family Welfare, Government of India was launched as a WHO Collaborating Centre for Pharmacovigilance in Public Health Programmes and Regulatory Services [14].

In 2015, India had become the first country to report over one lakh individual case safety reports (ICSRs) to Vigiflow, Uppsala Monitoring Centre's (UMC) web-based pharmacovigilance system. Further, India is currently the seventh largest contributor to the UMC's international drug safety database (Vigibase). Adding another feather to India's cap is the UMC's completeness score of 0.94 out of 1 assessed for Indian ICSR's, positioning India among the top-rankers in the completeness score criterion. It is worth mentioning that the pharmacovigilance outsourcing industry in India has grown by leaps and bounds in the past few years, with the number of pharmacovigilance professionals in the country amounting to almost 15,000 people. Ranging from basic case processing activities to complex functions such as signal detection and analysis, the spectrum of pharmacovigilance capabilities available in India has been expanding. With India now being looked upon as the preferred destination for global pharmacovigilance services, it is quite logical to expect this knowledge to be utilized for safer pharmacotherapy [15].

Table 15.1 Countries and agencies responsible for pharmacovigilance

#	Country	Agency	Website
1	India	Central Drug Standard Control Organization (CDSCO) and NCC-PvPI	cdsco.gov.in/opencms/opencms/en/PvPI/
2	Australia	Therapeutic Goods Administration (TGA)	www.tga.gov.au/
3	Brazil	Agencia Nacional de Vigiloncia Sanitaria (ANVISA)	portal.anvisa.gov.br/english
4	Canada	Health Canada	www.canada.ca/en/health-canada/services/drugs-health-products.html
5	China	State Food and Drug Administration	www.sfdachina.com/
6	Denmark	Danish Medicines Agency	laegemiddelstyrelsen.dk/en/#
7	Europe	European Medicines Agency (EMEA)	www.ema.europa.eu/en
8	Germany	Federal Institute for Drugs and Medical Devices	www.bfarm.de/EN/Home
9	Ireland	Health Products Regulatory Authority	www.hpra.ie/
10	Italy	Italian Pharmaceutical Agency	www.agenziafarmaco.com/en
11	Japan	Ministry of Health, Labour & Welfare (MHLW)	www.mhlw.go.jp/english/policy/health-medical/pharmaceuticals/index.html
12	Malaysia	National Pharmaceutical Control Bureau	npra.gov.my/index.php/en/
13	Nigeria	National Agency for Food and Drug Administration and Control (NAFDAC)	www.nafdac.gov.ng/
14	New Zealand	Medsafe—Medicines and Medical Devices Safety Authority	www.medsafe.govt.nz/
15	Netherlands	Medicines Evaluation Board	english.cbg-meb.nl/
16	Pakistan	Drugs Control Organization	www.dra.gov.pk/
17	Philippines	Food and Drug Administration (FDA)	www.fda.gov.ph/pharmacovigilance/
18	South Africa	Medicines Control Council	www.sahpra.org.za/Publications/Index/15
19	Sri Lanka	SPC, Ministry of Health	www.spc.lk/
20	Sweden	Medical Products Agency (MPA)	lakemedelsverket.se/english/
21	Switzerland	Swissmedic, Swiss Agency for Therapeutic Products	www.swissmedic.ch/swissmedic/en/home.html
22	UK	Medicines and Healthcare Products Regulatory Agency (MHRA)	www.gov.uk/government/organisations/medicines-and-healthcare-products-regulatory-agency
23	USA	Food and Drug Administration (FDA)	www.fda.gov/drugs/drug-safety-and-availability

15.3 Process of Pharmacovigilance

From execution point of view, the process involves five steps as outlined below:

1. Collect and record ADRs.
2. Causality assessment and analysis of ADRs.
3. Collate and code in database.
4. Compute risk–benefit and suggest regulatory action, if needed.
5. Communicate for safe use of drugs among stakeholders, as appropriate.

An alternative and perhaps more comprehensive classification scheme is 'DoTS,' which classifies reactions dependent on the

1. **D**ose of the drug
2. The **T**ime course of the reaction
3. Relevant **S**usceptibility factors (such as genetic, pathological and other biological differences)

DoTS has the advantage of being helpful to consider the diagnosis and prevention of ADRs in practice [16].

15.3.1 Types of Adverse Drug Reaction

Adverse drug reactions are classified based on the frequency, severity and onset of action. Based upon onset of action and severity, ADRs can be classified into 5 types [17].

1. **Type—A (Augmented):** It is the commonest type of adverse drug reaction. This type of adverse drug reaction is dose dependent, and severity increases with dose. It is preventable in most part by slow introduction of low dosages. It is predictable by the pharmacological mechanisms. For example—hypotension caused by beta-adrenergic blockers, hypoglycaemia caused by insulin or oral hypoglycaemic agents, or NSAID-induced gastric ulcers.
2. **Type—B (Bizarre):** It is a rare and unpredictable. It is an idiosyncratic type of adverse drug reaction. It is genetically determined. Mechanisms of this type of adverse drug reactions are unknown. This type of adverse drug reaction is serious and it can be fatal also. It is unrelated to the dose. For example—hepatitis caused by halothane, neuroleptic malignant syndrome caused by some anaesthetics and antipsychotics.
3. **Type—C (Chronic):** These reactions are associated with long-term drug therapy. For example—Benzodiazepine dependence and Analgesic nephropathy. They are well known and can be anticipated.
4. **Type—D (Delayed):** These reactions refer to carcinogenic and teratogenic effects. These reactions are delayed in onset and are very rare since extensive mutagenicity and carcinogenicity studies are done before drug is licensed.

Table 15.2 Frequency-based category of ADRs

Category	No. of incidence	% of incidence
Very common	≥1/10	≥10%
Common (frequent)	≥1/100 and <1/10	≥1% and <10%
Uncommon (infrequent)	≥1/1000 and <1/100	≥0.1% and <1%
Rare	≥1/10,000 and <1000	≥0.01% and <0.1%)
Very rare	<1/10,000	<0.01%

5. **Type—E (End of Treatment):** These reactions occur on withdrawal especially when drug is stopped abruptly. For example—withdrawal seizures on stopping phenytoin, adrenocortical insufficiency on withdrawal of steroids.

Based upon the severity of reaction, adverse drug reactions can be classified into three categories—i.e. mild, moderate and severe.

Mild: A reaction that does not required treatment or prolongation of hospital stay.

Moderate: A reaction that requires treatment and/or prolongs hospitalization by at least 1 day.

Severe: A reaction that is potentially life threatening or contributes to the death of patient is permanently disabling, requires intensive medical care or results in a congenital anomaly cancer or unintentional overdose.

Further, the adverse drug reactions are classified into five categories based on the frequency of occurrence (Table 15.2).

Adverse drug reactions are classified into three categories based upon the onset of reaction.

1. **Acute adverse drug reaction:** Acute adverse drug reactions occur within 60 min of the drug administration.
2. **Sub-acute adverse drug reaction:** Sub-acute adverse drug reactions occur between 1 and 24 h of the drug administration.
3. **Latent adverse drug reaction:** Latent adverse drug reactions take more than 2 days of the drug administration.

15.4 Characterization of Adverse Drug Reactions

The most challenging task in the pharmacovigilance is identification of an adverse drug reaction.

The characterization has three parts, namely—causality, severity and preventability assessment. To achieve these, various tools have been developed, standardized and are used globally as a gold standard.

Adverse drug reaction can be assessed by causality, severity, and preventability.

Causality assessment: Causality assessment is the method by which the extent of relationship between a drug and a suspected reaction is established. Causality assessment can highlight individual case reports, identify areas of uncertainty, reduce disagreement between assessors, and help regulatory decision-making. Performing causality assessment in pharmacovigilance may involve making a decision based on the information in a single adverse event or suspected ADR report (or a series of such reports) on the relationship between exposure to a drug and the reported adverse event or the suspected ADR. While this process is generally referred to as causality assessment, it is in fact an assessment of possible relatedness between exposure to a drug and an adverse reaction. This evaluation, which is useful in pharmacovigilance and essential in clinical medicine, is different from the broader process of assessing causality of adverse reactions on the basis of data from multiple sources.

There are many similarities between the clinical diagnosis of suspected ADRs in individual patients and causality assessment of case reports in pharmacovigilance. The following issues need to be considered:

1. The temporal relationship (time to onset): The temporal relationship is perhaps the most important point to consider in assessing the relationship between exposure to a drug and a subsequent adverse event. Since most ADRs are 'Type A' reactions, a plausible temporal relationship between exposure and the onset of the suspected ADR, taking into consideration the pharmacological characteristics of the drug, is normally the first point to consider in the assessment of causality.
2. The clinical and pathological characteristics of the events: Since ADRs are caused by definite pathophysiological mechanisms (although many are unclear at the time of assessment), conformity with the recognized clinical and pathophysiological patterns of the ADR is supportive of a causal relationship. For example, the clinical manifestations and laboratory abnormalities of patients with anaemia following the use of alpha-methyldopa are expected to be consistent with the diagnosis of haemolytic anaemia (a recognized association) and doubt about relatedness should be raised when other types of anaemia are attributed to the use of this product.
3. Pharmacological plausibility: Pharmacological plausibility is based on previous knowledge of the drug, including its pharmacodynamic and pharmacokinetic characteristics. It is self-evident that causality is supported for events that can be explained on the basis of the pharmacological characteristics of the suspect drug, e.g. when the occurrence of the event coincides with the expected peak concentration in the affected organ.
4. Existing information: Whether the event has been previously reported as an adverse reaction (in clinical trials and post-marketing) needs examination. Although the Summary of Product Characteristics (SmPC), or the Investigator Brochure for products in development, includes a list of adverse events or suspected ADRs that have been previously reported, the search for important events may need to include prescribing and pharmacological information in other sources.

5. Concomitant medication: The role of concomitant medications and medications taken prior to the event that could have caused the event is an important factor. In patients who are taking more than one drug, it is frequently difficult to decide which one is the more likely cause of the suspected ADR. All the points that need to be considered for causality assessment for the suspect drug apply to concomitant medications or drugs to which the patient was exposed.
6. Underlying and concurrent illnesses: Some events which are attributed to drug exposure may be simply manifestations of preexisting conditions. Furthermore, patients with some diseases, e.g. AIDS, respond differently to drugs, with qualitatively and quantitatively different ADR profiles. Patients with some conditions may experience altered responses to medications, e.g. patients with heart failure may experience augmented responses to cardiovascular medications.
7. Dechallenge or dose reduction: Recovery after stopping the drug or dose reduction is important, particularly when the timing of the full recovery or improvement is consistent with the pharmacological characteristics of the drug. However, it is noteworthy that some events, e.g. deafness with aminoglycosides, are irreversible.
8. Rechallenge or dose increase: The occurrence of ADRs after dose increases or rechallenge is a strong indicator of causality, particularly if the onset of the suspected ADR is pharmacologically plausible. A note of caution is to resist temptation for using rechallenge as a diagnostic tool unless clinically warranted an informed consent has been obtained. There must be a valid reason for the rechallenge and the practitioner and the institute must be trained and prepared for all eventualities, e.g. severe anaphylaxis and cardiac arrest.
9. Patients' characteristics and previous medical history: Previous medical history including previous history of drug allergies and the presence of renal or hepatic impairment can be very relevant in the diagnosis of suspected ADRs. For example, patients with a history of allergy to drugs are more likely to experience allergic ADRs.
10. Drug interactions: A plausible temporal relationship with the introduction or cessation of a concomitant drug with potential for interaction is an important consideration in causality assessment.
11. The quality of information in the report: Whereas minimal information is required for valid regulatory reports, causality assessment requires all the necessary data elements. In many cases in pharmacovigilance practice, this can be obtained only after seeking detailed follow-up information from the reporter [18].

15.4.1 Methods for Causality Assessment

Many systems for causality assessment have been proposed. These have been mostly used during regulatory reporting by pharmaceutical companies, assessment by regulatory authorities or in academic research.

These systems can broadly be divided into three categories [19].

1. Unrestricted clinical evaluation/global introspection.
2. Algorithms, with or without scoring.
3. Bayesian probabilistic methods.

Unrestricted clinical evaluation/global introspection: This is the most commonly used method by clinicians and by many pharmaceutical companies and regulatory authorities. Although the method is easy to apply and resembles the day-to-day medical practice with its 'common-sense' approach, it lacks transparency and is subject to the fallibility of human judgement. The method permits inconsistency, as well as missing or misinterpreting important points.

Structured algorithms with or without scoring: There are many structured or semi-structured algorithms for causality assessment. These range from simple flow charts (with or without scoring) to lengthy questionnaires with tens of questions that need to be analysed by powerful computers. The advantages of structured assessments include transparency and consistency. When a structured method for causality assessment is applied properly, no points are missed.

A consistent method of assessment facilitates communication between healthcare professionals and aids documentation and audit. The disadvantages include reduced ability to apply clinical judgement and, with algorithms that include scoring, adherence to the scores in place of clinical judgement. With regard to algorithms, one that has gained popularity in pharmacovigilance assessment is the system proposed by Naranjo et al. [20], perhaps because of its simplicity.

Bayesian probabilistic methods: This approach is based on assigning a prior probability to the event under investigation. In drug safety, the prior probability of an adverse reaction is based on information obtained from pre-marketing clinical trials and epidemiological studies for patients with the underlying illness.

There has been long-standing academic interest in the application of Bayesian approaches to causality assessment, but the progress has not been as fast as the proponents of the methods would have hoped, nor have the methods been generally embraced by regulatory authorities or pharmaceutical companies. There are two main reasons for this:

1. While the Bayesian approach to statistics resembles, to some extent, human intelligence, the apparent complexity of its mathematics deters those unfamiliar with statistics.
2. Even with increasing availability of epidemiological and pharmacoepidemiological databases, background information for calculating prior probabilities is often either sparse or unavailable. Furthermore, pre-marketing data from clinical trials are mostly confidential between the

pharmaceutical companies and regulatory authorities and are not available for the scientific community to be used for estimating the prior probability. Even when they are available, they may not be in a format that could be used for Bayesian evaluation of the causality of ADRs [18].

Naranjo scale: The Naranjo ADR Probability Scale was developed to help standardize assessment of causality for all adverse drug reactions. The scale was also designed for use in controlled trials and registration studies of new medications, rather than in routine clinical practice [20].

There are 10 questions on the scale that need to be answered and assigned a score. Based upon a total of the score obtained, an ADR is classified as 'doubtful,' 'possible,' 'probable' or 'definite.' The explanations are described

- ***Doubtful*** *ADR (<2): The reaction was likely related to factors other than a drug.*
- ***Possible*** *ADR* (2–4)*: The reaction followed a temporal sequence after a drug, possibly followed a recognized pattern to the suspected drug and could be explained by characteristics of the patient's disease.*
- ***Probable*** *ADR* (5–8)*: The reaction followed a reasonable temporal sequence after a drug, followed a recognized response to the suspected drug, was confirmed by withdrawal but not by exposure to the drug, and could not be reasonably explained by the known characteristics of the patient's clinical state.*
- ***Definite*** *ADR (≥9): The reaction followed a reasonable temporal sequence after a drug or in which a toxic drug level had been established in body fluids or tissues, followed a recognized response to the suspected drug and was confirmed by improvement on withdrawing the drug and reappeared on re-exposure.*

The WHO-UMC scale [21] and the RUCAM (Roussel Uclaf Causality Assessment Method) are two more tools for the causality assessment.

15.4.1.1 Severity Assessment

The term severity is often used to describe the intensity of a medical event, as in grading 'mild,' 'moderate' and 'severe.' Severity assessment categorizes the ADRs as mild, moderate, or severe based on the steps taken for the management of the ADRs. The severity of an ADR is assessed by Modified Hartwig's severity scale given below [22] (Table 15.3).

15.4.1.2 Preventability Assessment

Preventability of ADRs is assessed by using Modified Schumock and Thornton scale. This scale categorizes an ADR as 'definitely preventable,' 'probably preventable' and 'not preventable' [23]. It may be interesting to note that very many ADRs reported are 'preventable.'

Table 15.3 The modified Hartwig's severity scale

Severity level	Description
1	An ADR occurred but required no change in treatment with the suspected drug
2	The ADR required that treatment with the suspected drug be held, discontinued, or otherwise changed. No antidote or other treatment requirement was required. No increase of length of stay
3	The ADR required that treatment with the suspected drug be held, discontinued, or otherwise changed. AND/OR An antidote or other treatment was required. No increase in length of stay
4	Any level 3 ADR which increases length of stay by at least 1 day. OR The ADR was the reason for the admission
5	Any level 4 ADR which requires intensive medical care
6	The adverse reaction caused permanent harm to the patient
7	The adverse reaction either directly or indirectly led to the death of the patient

Mild = Level 1 and 2
Moderate = Level 3 and 4
Severe = Level 5, 6 and 7

15.4.1.3 Management of ADRs

Rapid action is sometimes important because of the serious nature of a suspected adverse drug reaction, for example, anaphylactic shock. Emergency treatment and withdrawal of all medicines is occasionally essential, in which case cautious reintroduction of essential medicines should be considered. Otherwise, using clinical benefit–risk judgement, together with help from investigations, one decides which medicine or medicines should be withdrawn as a trial.

A problem immediately arises if one or more of the medicines are essential to the patient. If the culprit is fairly clear, a benefit–risk decision needs to be taken about the need for the drug (are there equally effective substitutes that are unlikely to produce the same adverse drug reaction?), the severity of the reaction, and its potential for treatment. If several medicines could be causative, the non-essential medicines should be withdrawn first, preferably one at a time, depending on the severity of the reaction. If the reaction is likely to be dose related, dose reduction should be considered. Many prescribers unnecessarily withhold a drug when interactions are suspected, rather than adjusting the dose. The patient should be observed during withdrawal. The waiting period will vary, depending on the rate of elimination of the drug from the body and the type of pathology. For example, urticaria usually disappears quickly when the drug is eliminated, whereas fixed psoriatic skin reactions can take weeks to resolve. If the patient is clearly getting better, in keeping with the prediction, alternative medicines for the basic disease can be introduced if necessary. If the patient is not doing well after withdrawal of the first drug, the next most likely culprit should be considered, and the process repeated.

On the other hand, the patient may be suffering through being deprived of the medicine withheld. In that case, either another suitable drug should be substituted (remembering the possibility of cross-sensitivity), or the same drug should be tried at a lower dosage (for a dose-related reaction). The latter approach should be tried if more than one drug was withheld, for instance if an interaction was suspected or if the seriousness of the reaction made it wise to withhold several possible drugs. Reintroduce apparently essential medicines one at a time, starting with the one least likely to be the culprit. If the patient cannot manage without a medicine that has caused an adverse reaction, provide symptomatic relief while continuing the essential treatment. For example, severe nausea and vomiting are routinely treated symptomatically in patients receiving anti-cancer drugs. However, when treating an adverse drug reaction, it is important not to introduce more medicine than are essential. For better management of patient, a clear therapeutic objective should be kept in mind and patient should be reviewed regularly.

Onakpoya et al. [24] have reported a systematic review of post-marketing withdrawal of 462 medicinal products because of adverse drug reactions. The interested reader may peruse the publication that is reasonably comprehensive.

15.5 Roles of Pharmacists in the Management of ADRs

Pharmacists worldwide, by virtue of their education and training, are equipped with the knowledge of the pharmacology of drugs. This empowers them to connect the real patient situations and any side effect or an ADR. An intelligent pharmacist can always ensure safer drug therapies for the patients using his/her knowledge and training.

The pharmacists, especially the practicing ones, have a definite role in the following domains for ensuring safer pharmacotherapy.

1. Monitoring the patients who are at a greater risk of developing ADRs.
2. Monitoring the patients who are prescribed with drugs highly susceptible to cause ADRs.
3. Assessing and documenting the patient's previous allergic status.
4. Assessing the patient's drug therapy for its appropriateness.
5. Assessing possible drug interactions in case of multiple therapies.
6. Assisting healthcare professionals in detection and assessment of ADRs.
7. Encouraging/stimulating healthcare professionals in reporting an ADR.
8. Documentation of suspected reported reactions for future reference.
9. Follow-up of patients to assess the outcome of the reaction and management.
10. Obtaining feedback about the reported reaction.
11. Educating healthcare professionals about the importance of reporting of an ADR.
12. Educating patients.
13. Creating awareness about ADRs among healthcare professionals, patients and public.
14. Preparation and utilization of promotional materials.

15. Communication with other healthcare professionals such as nurses and community pharmacist.
16. Presentation of reports in meetings and conferences.
17. Conducting workshops/conferences/seminars on ADRs for healthcare professionals.
18. Dissemination of signals generated through publication of reports in bulletins or journals [25].

In 2020, Xue et al. reported a method for the prediction of potential ADRs by combining Biomedical Tripartite Network and Graph Representation Model. They have reported a novel strategy to predict the drug-ADR associations by combining deep learning and the biomedical tripartite network. This heterogeneous network contains biomedical linked data of three entities, for example, drugs, targets, and ADRs. For the first time, GraRep, a deep learning method for distributed representations, is introduced to learn graph representations and identify hidden features from the tripartite network that is further used for ADR prediction. Through this approach, drug-ADR associations could possibly be discovered from a systemic perspective [26].

The authors suggest that the novice can surf the WHO-UMC website and reach the 'global pharmacovigilance' tab. This tab blows up ample learning resources under the heads—UMC materials, WHO materials and other scientific publications.

Additionally, the knowledge of pharmacovigilance to a student of pharmacy helps to enhance the employability. Companies like Parexel, Novartis, Tata consultancy, APCER, Continuum, IQVIA, Wipro, Kinapse, and Cognizant hire pharmacy students for safety-based work.

References

1. Karch FE, Lasagna L. Adverse drug reactions. A critical review. JAMA. 1975;234:1236–41.
2. Lazarou J, Pomeranz BH, Corey PN. Incidence of adverse drug reactions in hospitalized patients: a meta-analysis of prospective studies. JAMA. 1998;279:1200–5.
3. Kim JH, Scialli AR. Thalidomide: the tragedy of birth defects and the effective treatment of disease. Toxicol Sci. 2011;122(1):1–6.
4. Peters P, Miller RK, Schaefer C. General commentary on drug therapy and drug risks in pregnancy. In: Drugs during pregnancy and lactation. 3rd ed. Amsterdam: Elsevier; 2014. p. 1–23.
5. The importance of pharmacovigilance. http://apps.who.int/medicinedocs/pdf/s4893e/s4893e. Accessed 15 Oct 2019.
6. Aronson JK, Ferner RE. Clarification of terminology in drug safety. Drug Saf. 2005;28 (10):851–70.
7. ASHP. Guidelines on adverse drug reaction monitoring and reporting. AJHP. 1995;52:417–9.
8. Brewer T, Colditz GA. Postmarketing surveillance and adverse drug reactions: current perspectives and future needs. JAMA. 1999;281:824–9.
9. Oates JA. The science of drug therapy. In: Brunton LL, Lazo JS, Parker KL (eds.) Goodman and Gilman's: the pharmacological basis of therapeutics. 11th ed. McGraw Hill; 2006. p.135.

10. Kalaiselvan V, Prasad T, Singh A. Current status of adverse drug reactions monitoring centres under pharmacovigilance program of India. Indian J Pharm Pract. 2014;7:19–22.
11. Kalaiselvan V, Prasad T, Singh GN. Pharmacovigilance programme of India: recent developments and future perspectives. Indian J Pharmacol. 2016;48:624–8.
12. Kalaiselvan V, Srivastava S, Singh A, Gupta SK. Pharmacovigilance in India: present scenario and future challenges. Drug Saf. 2019;42(3):339–46. https://doi.org/10.1007/s40264-018-0730-7.
13. Mandal SC. Evolution of pharmacovigilance programme: present status in India. Pharma Times. 2017;49:31–6.
14. Singh KM, Kanase HR. Pharmacovigilance programme of India: the beginning, current status and recent progress. Adv Pharmacoepidemiol Drug Saf. 2017;6:219.
15. Venkatraman V. Current status of pharmacovigilance in India. Pharmabiz 2015. Accessed 29 Dec 2019.
16. Aronson JK, Ferner RE. Joining the DoTS: new approach to classifying adverse drug reactions. BMJ. 2003;327:1222–5.
17. Rawlins MD. Clinical pharmacology: adverse reactions to drugs. BMJ. 1981;282:974–6.
18. Shakir SAW. Causality and correlation in pharmacovigilance. In: Talbot J, Waller P, editors. Stephens' detection of new adverse drug reactions. 5th ed. Hoboken, NJ: John Wiley & Sons, Ltd.; 2004. p. 329–43.
19. Davern TJ. Drug-induced liver disease. Clin Liver Dis. 2012;16:231–45.
20. Naranjo CA, Busto U, Sellers EM, Sandor P, Ruiz I, Roberts EA, et al. A method for estimating the probability of adverse drug reactions. Clin PharmacolTher. 1981;30:239–45.
21. The use of the WHO-UMC System for Standardised Case Causality Assessment. http://www.who-umc.org/Graphics/24734.pdf. Accessed 15 Oct 2019.
22. Hartwig SC, Siegel J, Schneider PJ. Preventability and severity assessment in reporting adverse drug reactions. Am J Hosp Pharm. 1992;49(9):2229–32.
23. Schumock GT, Thornton JP. Focusing on the preventability of adverse drug reactions. Hosp Pharm. 1992;27(6):538.
24. Onakpoya IJ, Heneghan CJ, Aronson JK. Post-marketing withdrawal of 462 medicinal products because of adverse drug reactions: a systematic review of the world literature. BMC Med. 2016;14:10. https://doi.org/10.1186/s12916-016-0553-2.
25. Parthasarathi G, Hansen K, Nahata M. A textbook of clinical pharmacy practice: essential concepts and skills. 4th ed. Annasalai, Chennai: Orient Longman; 2004. p. 84–101.
26. Xue R, Liao J, Shao X, Han K, Long J, Shao L, et al. Prediction of adverse drug reactions by combining biomedical tripartite network and graph representation model. Chem Res Toxicol. 2020;33(1):202–10. https://doi.org/10.1021/acs.chemrestox.9b00238.

Chapter 16
Regulatory Process for New Drug Approval in India

K. Bangarurajan

Abbreviations

BA/BE	Bioavailability/Bioequivalence
CDSCO	Central Drugs Standard Control Organization
CMC	Chemistry & Manufacturing Control
CT	Clinical Trial
IND	Investigational New Drug
IPC	Indian Pharmacopeia Commission
ND	New Drug
NOC	No Objection Certificate
SEC	Subject Expert Committee

16.1 Introduction

The drug approval is a regulatory procedure, by which pharmaceutical unit get the permission to market the new drug from the Regulator, Drugs Controller General (India), Central Drugs Standard Control Organization, Ministry of Health and

General Comment: For the details related to forms and fees for new drug approval processing in India refer the Eighth Schedule and Sixth Schedule of New Drugs and Clinical Trials Rule, 2019 [G.S.R. 227(E)].
Web Page Link: https://cdsco.gov.in/opencms/opencms/system/modules/CDSCO.WEB/elements/download_file_division.jsp?num_id=NDU0Mg==

K. Bangarurajan (✉)
Central Drugs Standard Control Organization, Directorate General of Health Services, Ministry of Health & Family Welfare, Government of India, FDA Bhawan, New Delhi, India
e-mail: bangaru@cdsco.nic.in; http://www.cdsco.gov.in

R. Poduri (ed.), *Drug Discovery and Development*,
https://doi.org/10.1007/978-981-15-5534-3_16

Family Welfare, Government of India. The purpose of this approval process is to ensure the availability of safe and efficacious drugs to the patients. This organization is having authority for approvals of New Drugs/Investigational New Drug:

The New Drugs and Clinical Trial Rules 2019 define the New Drug (ND) or Investigational New Drug as

(i) An active pharmaceutical ingredient or phytopharmaceutical drug, not approved as safe and efficacious by the Drugs Controller General (India) for its claims

(ii) A drug marketed in India for certain claims, now to be marketed with certain new claims, indication, route of administration, dosage, and dosage form

(iii) Two or more drugs in a fixed combination [1–3]

New Chemical Entity (NCE) is the one which has not been approved by any drug regulators of any country as a drug and for first time it is to be developed as a Drug.

16.2 Clinical Trials

For approval of a New Drug, the safety and efficacy of the molecule must be proved in human beings in four stages (phases)

(a) **First Stage or Phase I:** The objective to conduct the Phase I clinical trial is to prove the safety of drugs in healthy volunteers. Further the maximum dose tolerated that can be administered in the next phase is determined. In addition, rate of ADME (Pharmacokinetics), Pharmacodynamic and Pharmacokinetic parameters in healthy volunteer/patients with the target disease and the secondary pharmacological activities (off target), if any, are studied.

(b) **Stage Two or Phase II:** The effectiveness of a drug for treating the disease is evaluated, the side effects and adverse effects of the test drug in homogeneous group is determined in small group of subjects. From this study the dose and treatment schedule for the next phase is decided.

(c) **Stage Three or Phase III.**

  (i) This involves large number of patients in multiple centres. This phase is to demonstrate the therapeutic effects of the drug. These trials are to confirm the results of the Phase II that the drug is safe and efficacious. Marketing approval is given on the positive outcome of these studies by the regulators.

  (ii) If the drug under study is to be used for longer period, protocols should be designed accordingly.

  (iii) Further for the drugs already approved by the other country regulators, Phase III studies to be undertaken in Indian population taking into consideration the genetic variations.

(d) **Stage Four or Phase IV:** This study is carried out after the drug approval as post-marketing survey. It is done in a larger population and less frequently occurring side/adverse effects are monitored and recorded for necessary action.

Studies in special populations: As the physiological processes vary depending upon the factors like age, pregnancy, nursing mothers and renal/other organ failure, the study protocol needs to consider these factors. Special consideration to be emphasized for drugs purported to be used mainly in Geriatrics, Paediatrics, Pregnant or nursing women, etc. The requirements are elaborately given in the New Drugs and Clinical Trial Rules, 2019 [3].

16.2.1 Conduct of Clinical Trial

These studies are to be conducted only after the approval from regulator, i.e. the Licensing authority as per New Drugs and Clinical Trial Rules, 2019 [3]. Moreover, the necessary approval from the Central Licensing Authority or intimating to him about any amendment of the protocol and the rationale for it to be obtained by the applicant. In addition, the serious adverse events occurred shall be reported to the regulator.

16.2.2 Analysis

The results of the study should be tabulated. The data should be subjected to statistical analysis and inference of the study need to be documented and certified by the Principal investigator/authorized person.

16.3 New Drug Approval Process

Any person to manufacture or to import any new drug for the purpose of conducting clinical trials should obtain approvals from the Regulator, Central Licensing Authority. This process involved is known as New Drug approval process and are detailed in the New Drugs and Clinical Trial Rules, 2019 [3].

16.3.1 Approval Process for New Drug

Permission to carry out Clinical testing for marketing of New Drugs involves submission of an application to the regulator with certain information like:

1. Chemical and pharmaceutical information
2. Non-clinical Study Data, i.e. toxicological and pharmacological data in animals
3. Human clinical pharmacology data

(a) The submission of Human Clinical Pharmacological data depends upon whether the new drug substance discovered in India or other countries

If developed in India Clinical Studies to be conducted from phase I and if developed in other countries, based on the phase I data of country of origin permission to be granted to either repeat phase I or to proceed with Phase II trials and then Phase III trials.

(b) Phase III trials are to be conducted if the drug is already approved and to be introduced for the first time in India.

(c) Data like, the regulatory status of the drug in other country, full prescribing information, all package insert, promotional literature and approved prescribing information complete testing protocol along with impurity profile are also to be submitted along with the application.

Exemptions for the submissions of certain data for consideration of New Drug in certain cases are also given in the New Drugs and Clinical Trial Rules, 2019 [3] which is based on the class of New Drug.

Further expedited approval process is also allowed if therapeutic outcome is superior to the existing drugs.

Accelerated approvals also be granted in case of pandemic situation in the country.

16.3.1.1 Chemical and Pharmaceutical Information

The chemical and pharmaceutical information required to be submitted to the Regulatory Authority includes:

1. The details of the drug and the pharmacological class.
2. Information on active pharmaceutical ingredients—Drug information (Common Name, Chemical Name/International Non-proprietary Names (INN)).

   Further the data should include

(A) Physicochemical data like chemical name, structure, molecular weight and physical properties like solubility, partition coefficient, etc.

(B) Analytical data with elemental analysis UV, IR, NMR, Mass spectra and polymorphic identification.

(C) Complete standards like Identification, Identity of impurities and assay, etc.

(D) Validations of assay method, Impurities estimation method and residual solvent.

(E) Stability studies

(F) Formulation data like Dosage form, Composition, Master manufacturing formula, etc. [3–5]

16.3.1.2 Non-clinical Study Data

The animal toxicological studies are to be designed based on the characteristics of the drug under investigation, characteristics of disease, total period and exposure in study subject and route of administration of the drug.

(A) Animal Toxicology

Animal toxicological studies are to be carried out as per Good Laboratory Practices (GLP) by trained toxicologist. It should be done with proper written protocols, proper SOPs with calibrated and standard equipments. The validated test substances and system to be used and the documents preserved as record for a minimum of 5 years after marketing of the drug.

1. Systemic toxicity studies

   Systemic toxicity conducted as single-dose and repeated-dose toxicity studies

   Single-dose toxicity studies: Single-dose toxicity study is performed in rodents and route administration is same as intended for humans. One more route of administration also be used in one more species to make sure the test drug is absorbed systemically. Maximum Tolerated Dose (MTD) and Minimum Lethal Dose (MLD) are to be found out after specified time. LD_{10} and LD_{50} are reported by 95 percent Confidence Interval.

   Repeated-dose systemic toxicity studies: Repeated-dose systemic toxicity studies are conducted in two mammalian species of which one is a non-rodent. The experimental design is as specified in New Drugs and Clinical Trial Rule, 2019 [3].
2. **Male fertility study:** The experiments are conducted using one rodent species (preferably rat) and the test drug is administered by the same route as to be used clinically for a period specified [3] before they are paired with female animals. The fertility index calculated after day 13 of the gestation.
3. **Female reproduction and developmental toxicity studies:** If the drug is to be used in females of the reproductive age, it should undergo the female reproduction and developmental toxicity studies. This is conducted in mice/ rats and rabbits.

   (3.1) *Female fertility study:* Conducted in one rodent species. In male the drug is administered for 28 days and in female for 14 days before mating. Three doses including the MTD should be administered and pups of treated and control groups are to be observed for any morphological changes.

   (3.2) *Teratogenicity study:* In one rodent and one non-rodent species the drug is administered during the period of development of organs, using three dose levels. All foetuses are subjected for gross examination, fifty per cent for skeletal abnormalities and other fifty percent for visceral abnormalities.

   (3.3) *Perinatal study:* If the drug is to be given to pregnant or nursing mothers for prolonged periods, then study to be conducted in one rodent species as specified [3].

4. **Local toxicity:** It is to be carried out if the drug is to be administered by some special route. These studies are performed by using three dose levels and vehicle. If the drug is absorbed from the site of application, appropriate systemic toxicity studies also to be carried out.

   Local toxicity studies like dermal toxicity study, photoallergy or dermal phototoxicity study, vaginal toxicity test, and rectal tolerance test are to be conducted as specified in the New Drugs and Clinical Trial Rules, 2019 [3].

   Further Allergenicity or Hypersensitivity studies, Genotoxicity studies and Carcinogenicity studies are to be carried out as specified in the rules [3].

   The New Drugs and Clinical Trials Rules, 2019 specifies the animal toxicity testing requirements for clinical trials and marketing of a new drug in a comprehensive table [3] which is annexed for a ready reference **(annexure attached in the end of chapter).**

   To conduct the different stages of Clinical studies of an Investigational new Drug, following safety evaluation data of non-clinical toxicity testing is required

   For Stage I/Phase I Studies: Single-dose, Dose-Ranging and Repeat-dose systemic toxicity studies data along with data of *male fertility study, in vitro* genotoxicity tests, Allergenicity and Photoallergy test.

   For Stage II/Phase II Study: In addition to data required for stage I/Phase I study, in vivo genotoxicity tests and Segment II reproductive or developmental toxicity study.

   For Stage III/Phase III Study: All the summary data submitted for permissions for Stage I/Phase I and Stage II/Phase II studies, with appropriate references.

   If the permission is for conducting Stage III/Phase III study straight away, complete details of the non-clinical safety data resubmitted to obtain Stage I/Phase I and Stage II/Phase II study data.

   For Stage IV/Phase IV Study: Summary data of all the non-clinical safety data as listed above with appropriate references.

(B) Animal Pharmacology

Animal pharmacological studies to be conducted to support use in humans. This involves both general and specific pharmacological action.

1. **Specific pharmacological actions**—This is required to demonstrate the therapeutic potential for humans and to be designed based on the individual properties and intended uses of the drug.
2. **General pharmacological actions**

   (2.1) Essential safety pharmacology—This study is required to find out the potential undesirable pharmacodynamic effects of the drug on physiological functions in relation to exposure within the therapeutic range and above. The study design is to be such that identify undesirable pharmacodynamic properties of the drug in human safety. This study is to find out the effects of the test drug safety on vital functions and vital organ like cardiovascular, respiratory, and central nervous systems.

(2.1.1) Cardiovascular system (CVS): The study of the drug in CVS includes the effect on blood pressure, heart rate and ECG.
(2.1.2) Central nervous system (CNS): The study of the drug in CNS includes the effects on motor activity, behavioural changes, coordination, sensory and motor reflex responses, and body temperature.
(2.1.3) Respiratory system: The study of the drug in respiratory system includes the effects on respiratory rate, tidal volume, and haemoglobin oxygen saturation.

3. **Follow-up and supplemental safety pharmacology studies:** These studies are to be conducted, if needed based on the pharmacological properties of the new drug/NCE.

(3.1) Follow-up studies for essential safety pharmacology: Follow-up studies provide additional information on safety on CVS, CNS, and respiratory system by doing certain additional studies.
(3.2) Supplemental safety pharmacology studies: It is to determine the possible adverse pharmacological effects which are not performed in the essential safety pharmacological studies. These includes study in urinary system, autonomic nervous system, gastrointestinal system and other organ systems like skeletal muscle, immune and endocrine functions.

4. **Timing of safety pharmacology studies in relation to clinical development:**

(4.1) Prior to first administration in humans: The essential safety pharmacology study to be carried out. Further if any supplemental studies are identified, these studies are also to be performed.
(4.2) During clinical development: Additional investigations are required due to suspected adverse effects in animals and humans during clinical study.
(4.3) Before applying for marketing approval: Follow-up and supplemental safety pharmacological and toxicological data or toxicological studies addressing safety pharmacological endpoints.

16.4 Post-marketing Assessment of New Drug

Based on the limited number of patient's data in stage III/Phase III Clinical Studies the Regulatory Authorities issue the Marketing Authorization.

In actual clinical practice, more patients are treated and events which are missed in the clinical study may come to notice. Hence the drug effects are to be observed after approval of a new drug and benefit–risk profile to be assessed.

Post-marketing assessment of new drug may be carried out in different ways as follows.

16.4.1 Stage IV/Phase IV (Post-marketing) Study

This trial is to be conducted with an approved protocol by the Regulator with inclusion and exclusion criteria, safety efficacy assessment, etc. This study may include additional drug–drug interaction, dose response and safety studies.

The study subjects are to be provided with free of cost the study drug and ethical aspects for protection of rights, safety and well-being of subjects are to be followed including that of compensation in case of clinical trial related injury or death.

16.4.2 Post-marketing Surveillance Study or Observational or Non-interventional Study for Active Surveillance

These studies with a new drug also conducted after the approval of Regulator. In such studies the treatment of the subject is based on prescribers as included in the protocol. However, as the drug is already approved for the marketing, the regulatory provisions are not applicable in such cases.

16.4.3 Post-marketing Surveillance Through Periodic Safety Update Reports

The Regulatory provisions prescribes, the periodic safety update reports (PSURs) are to be submitted in every 6 months for the first 2 years and for next 2 years are to be submitted once in a year. However, all cases involving serious unexpected adverse reactions to be reported to the regulator within 15 days on receipt of information.

The PSURs are to include all relevant new information of the new drug, patient exposure, market authorization status in different countries along with any significant safety issues and to indicate any changes in the product information to optimize the use of product.

16.5 New Drug Application Review and Approval Process by Drugs Controller General of India

The application for approval of New Drug or Investigational New Drug to be submitted to the CDSCO through online SUGAM portal along with requisite documents and application is processed as per flow chart described below and approval is given by Drugs Controller General of India.

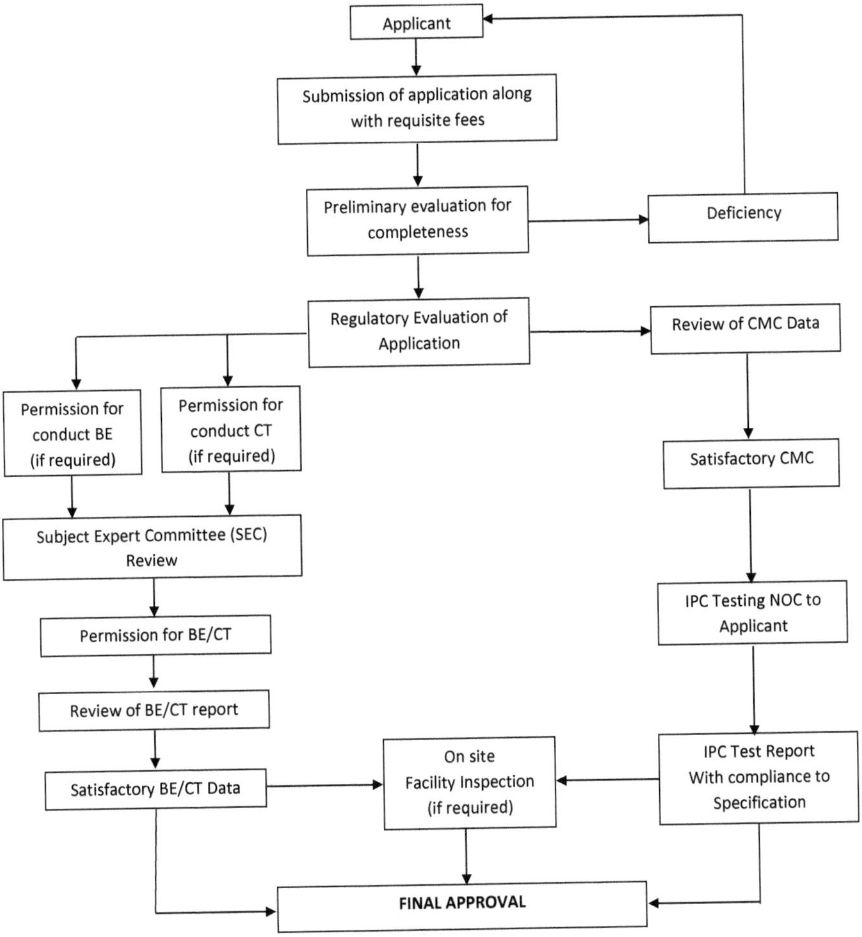

Form CT-17 – Licence to import ND or IND for CT or BA/BE

Form CT-11 – Permission to manufacture ND or IND for CT or BA/BE

Form CT-06 – Permission to conduct CT of ND or IND

Form CT-07– Permission to conduct BA/BE study of ND or IND

Acknowledgments I hereby acknowledge my sincere thanks to Ministry of Health and Family welfare, Govt. of India, Dr V.G. Somani, Drugs Controller General (India), and Dr S. Eswara Reddy, Joint Drugs Controller (India), for their support and encouragement. My sincere thanks to Mr. Ranjeet Singh Patel, Drugs Inspector, for his contribution and assistance in writing this chapter.

Annexure

Animal Toxicity Requirements for Clinical Trials and Marketing of a New Drug

Systemic toxicity studies			
Route of administration	Proposed frequency of dosing and duration of treatment in humans	Phase of clinical studies	Toxicity testing—Number of species and duration
Oral/parenteral/ transdermal	Single/several dose(s) in 1 d, up to 1 week	I, II, III	Two species; 2 weeks
	>1 week but upto 2 weeks	I, II, III	Two species; 2 weeks
	Up to 2 weeks	Marketing permission	Two species; 4 weeks
	• 2 weeks but up to 4 weeks	I, II, III	Two species; equal to duration of human exposure
		Marketing permission	Two species; 12 weeks
	>4 weeks but up to 12 weeks	I, II, III	Two species; equal to duration of human exposure
		Marketing permission	Two species; 24 weeks
	>12 weeks but up to 24 weeks	I, II, III	Two species; Equal to duration of human exposure
		Marketing permission	Two species; Rodent: 24 weeks, non-rodent 36 weeks
	>24 weeks	I, II, III	Two species; Rodent: 24 weeks, non-rodent 36 weeks
		Marketing permission	Two species; Rodent: 24 weeks, non-rodent 36 weeks
Inhalation (general anaesthetics, aerosols)	Up to 2 weeks	I, II, III	Two species; 1 month (exposure time 3 h/d, 5 d/week)
	Up to 4 weeks	I, II, III	Two species; 12 weeks, (exposure time 6 h/d, 5 d/week)
	>14 weeks	I, II, III	Two species; 24 weeks, (exposure time 6 h/d, 5 d/week)
Local toxicity studies			
Dermal	Up to 2 weeks	I, II	One species; 4 weeks
		III	Two species; 4 weeks
	>2 weeks	I, II, III	Two species; 12 weeks

(continued)

Systemic toxicity studies			
Ocular or optic or nasal	Up to 2 weeks	I, II	One species; 2 weeks
		III	One species; 2 weeks
	>2 weeks	I, II, III	Two species; 12 weeks
Vaginal or rectal	Up to 2 weeks	I, II	One species; 4 weeks
		III	Two species; 4 weeks
	>2 weeks	I, II, III	Two species; 12 weeks

Special toxicity studies
Male Fertility Study: Phase III in male volunteers or patients
Female reproduction and development toxicity studies:
Segment II studies in 2 species; Phase II, III involving female patients of childbearing age
Segment I study; Phase III involving female patients of childbearing age
Segment III study; Phase III for drugs to be given to pregnant or nursing mothers for long periods or where there are indications of possible adverse effects on foetal development
Allergenicity or hypersensitivity:
Phase I, II, III—when there is a cause of concern or for parenteral drugs (including dermal application)
Photoallergy or dermal phototoxicity:
Phase I, II, III—if the drug or a metabolite is related to an agent causing photosensitivity or the nature of action suggests such a potential
Genotoxicity:
In vitro studies—Phase I
Both in vitro and in vivo—Phase II, III
Carcinogenicity:
Phase III—when there is a cause for concern, or when the drug is to be used for more than 6 months

Abbreviations: *d* day, *h* hour, *I, II, III* phase of clinical trial
Note: Animal toxicity data generated in other countries may be accepted and may not be asked to be repeated or duplicated in India on a case-to-case basis depending upon the quality of data and the credentials of the laboratory where such data has been generated

Number of Animals Required for Repeated-Dose Toxicity Studies

2–4 weeks					12–26 weeks			
Group	Rodent (Rat)		Non-rodent (Dog or Monkey)		Rodent (Rat)		Non-rodent (Dog or Monkey)	
	♂	♀	♂	♀	♂	♀	♂	♀
Control	6–10	6–10	2–3	2–3	15–30	15–30	4–6	4–6
Dose 1	6–10	6–10	2–3	2–3	15–30	15–30	4–6	4–6
Dose 2	6–10	6–10	2–3	2–3	15–30	15–30	4–6	4–6
Dose 3	6–10	6–10	2–3	2–3	15–30	15–30	4–6	4–6

Laboratory parameters to be included in toxicity studies:

Haematological parameters			
Haemoglobin	Total red blood cell count	Haematocrit	Reticulocyte
Total white blood cell count	Differential white blood cell count	Platelet count	Terminal bone marrow examination
Erythrocyte sedimentation rate (ESR)	General blood picture: A special mention of abnormal and immature cells should be made		
Coagulation parameters (Non-rodents only): Bleeding, time, coagulation Time, prothrombin time, activated partial thromboplastin time			
Urinalysis parameters			
Colour	Appearance	Specific gravity	24 h urinary output
Reaction (pH)	Albumin	Sugar	Acetone
Bile pigments	Urobilinogen	Occult Blood	
Microscopic examination of urinary sediment			
Blood biochemical parameters			
Glucose	Cholesterol	Triglycerides	High density lipoproteins (HDL) cholesterol (Non-rodents only)
Low density lipoproteins (LDL)	Bilirubin	Serum glutamic pyruvic transaminase (SGPT) (Alanine aminotransferase (ALT)	Serum glutamic oxaloacetic transaminase (SGOT)
Cholesterol (Non-rodents only) aspartate aminotransferase (AST)			
Alkaline phosphatase (ALP)	GGT (Non-rodents only)	Blood urea nitrogen	Creatinine

Total proteins	Albumin	Globulin (Calculated values)	Sodium
Potassium	Phosphorus	Calcium	
Gross and microscopic pathology			
Brain[a]: Cerebrum, cerebellum, midbrain	(Spinal cord)	Eye	(Middle Ear)
Thyroid	(Parathyroid)	Spleen	Thymus
Adrenal[a]	(Pancreas)	(Trachea)	Lung[a]
Heart[a]	Aorta	Oesophagus	Stomach
Duodenum	Jejunum	Terminal ileum	Colon
(Rectum)	Liver[a]	Kidney[a]	Urinary bladder
Epididymis	Testis[a]	Ovary	Uterus[a]
Skin	Mammary gland	Mesenteric lymph node	Skeletal muscle

(continued)

Total proteins	Albumin	Globulin (Calculated values)	Sodium
Potassium	Phosphorus	Calcium	

Organs listed in parenthesis should be examined if indicated by the nature of the drug or observed effects
[a]Organs marked with an asterisk should be weighed

References

1. Drugs and Cosmetics Act. 1940.
2. Drugs and Cosmetics Rules. 1945.
3. New Drugs and Clinical Trials Rules. 2019.
4. Schedule Y of Drugs and Cosmetics Rules. 1945.
5. CDSCO. Good Clinical Practices guidelines.

Chapter 17
Pharmaceutical Industry, Academia, Regulatory Authorities and End User Collaboration in Successful Drug Discovery and Development

Ramarao Poduri

> *"Science is one of the very few human activities – perhaps the only one- where errors are systematically criticized and fairly often, in time, corrected"- Karl Popper* [1].

The exponential increase in the great drug (including vaccine) discoveries in the academic institutions in early part of the twentieth century resulted in newer remedies to the infectious diseases. In the later part of the twentieth century our understanding of biological and physical sciences increased and thereby how these drugs interact at the molecular level in the disease conditions [2]. From the beginning of modern drug discovery, academia played a pivotal role and the historical details of the academia and pharmaceutical industry collaborations are reported by Furman and MacGarvine [3]. More than 35 universities were collaborating with pharmaceutical industry in the east and west coast of US in 1938 (cf. Table 1 of Ref. [3]). Further it was concluded that

> *The labour markets, collaborative research and contract research were the principal mechanisms by which early university science contributed to the development of research capabilities in the emerging US pharmaceutical industry* [3].

This clearly indicates that the academia is very much in forefront and collaborated with pharmaceutical industry in establishment of R&D and drug discovery in early twentieth century. However, gradual decline in the collaboration was observed. The sulphanilamide episode [4, 5] and the catastrophic thalidomide disaster [6, 7] gave rise to implementation of stringent new drug approval procedures by the drug regulatory authorities. Regulatory sciences emerged. The time taken from the drug discovery to market increased and it is around 15 years. The number of the drugs

R. Poduri (✉)
Department of Pharmaceutical Sciences and Natural Products, Central University of Punjab, Bathinda, India
e-mail: ramaraop@yahoo.com

R. Poduri (ed.), *Drug Discovery and Development*,
https://doi.org/10.1007/978-981-15-5534-3_17

approved by USFDA per year declined. The success rate of drug approval was as low as 5% from preclinical stage to market as per the CMR report of 2014 [8]. The maximum failures were at Phase III clinical trials. Jarvis and Williams in their review reported that not able to reproduce the findings of biomedical research is the major factor in decline in the number drug approvals [9]. Several other reports indicate that biopharmaceutical factors, optimum pharmacokinetic factors [10] and unreliable preclinical data [11] are also contributing to this. The approval rate of drugs by the regulatory authorities decreased to all time low and attempts were made to cross the valley of death (the phrase used for the gap between the preclinical and clinical phase) [12] failed, and the average cost of new drug approval reached $2.87 billion in 2016 [13].

> *"The mean annual number of new drug approvals including biologics, was 34 from 1990-1999, 25 from 2000-2009, and 41 from 2010 -2018.New biologic product approvals increased from a median of 2.5 from 1990-1999, to 5 from 2000-2013, to 12 from 2014-2018.... FDA drug review times declined from more than 3 years in 1983 to less than 1 year in 2017, but total time from authorization of clinical testing to approval has remained at approximately 8 years over that period"* [14].

The recent US drug approval numbers indicate the trend is changed and the number reached 48 in year 2019 [15].

17.1 Drug Discovery and Translational Research in Academia

Researchers in academic institutions are interested in the ground-breaking discoveries by taking risks to identify novel therapeutic interventions that can save or improve the quality of life. In contrast, the counterparts in pharmaceutical industry mainly concentrate on the concept of Return on Investment (ROI) and rarely venture to take the risks to explore the complexities of the disease process. Academic institutions receive the funds from government for biomedical research, which results in capacity building and generating the opportunities to the aspiring candidates in drug discovery fields and service economy. Several examples of success are cited in the literature. The antiviral drug designed for AIDS by Robert Vince and was commercialized by GlaxoSmithKline and earned royalties to University of Minnesota [16]. Thus, Centre for Drug Design was developed at University of Minnesota. The experience and success of Drug Discovery in academic institutes leads to equip the next generation of researchers more innovative and also fruitful in their pursuits and thereby contributing to knowledge economy. The academic researchers do care for publication of their work, must teach and train the researchers, and have the academic freedom. On the other hand, pharmaceutical R&D scientists concentrate on the applied research, patenting their work, carry out the research based on the market needs and retain the know-how [17]. The quite contrasting research cultures prevailing in academia and pharma industry resulted in very minimal interaction between the two. Further the academic researchers are lacking not only the managerial but also the funds required to successfully carry out the translational research. The main reason is failure to reproduce the preclinical data or proof of concept which are mainly developed in the academic ecosystem in industrial set up.

The funding agencies (governmental and non-governmental organization) give the funds to carry out basic research for the improvement of the quality of life. The industrial organizations translate the proof of concepts into products by developing methods and maintain the quality of the products as per the guidelines issued by the regulatory agencies. Drug discovery involve several steps in sequence starting from target selection and validation, choosing the best candidate based on efficacy, safety, and toxicological data prior to clinical trials. Early stages of drug discovery (preclinical studies) can be conducted in academic institutions depending upon availability of resources [18].

Pharmaceutical industry nowadays is coming forward to interact with the academia again keep in view of new therapeutic options which can be explored at the academic research institutions since academic institutions venture into creative research. Recently, the funding agencies are encouraging the academic institutions by funding the projects to carry out the translational research [16].

17.1.1 Methods to Reduce the Hurdles in Academic Drug Discovery

Number of Academic Drug Discover Organizations increased in the recent past. This provides new vistas to academia which until now mainly concentrate on "blue sky research". However, to increase the success rate and to play pivotal role in drug discovery it is axiomatic researchers understand the hurdles in the preclinical stage of drug discovery. The following are the reasons for the increase in the academic Drug Discovery Organizations:

1. Decrease in the Preclinical Drug Discovery in big pharmaceutical companies due to prohibitive cost and high attrition.
2. Paradigm shift in the funding agencies policies as more emphasis on the evidence for the product potential.
3. Academic Drug Discovery Organizations ability to capacity building in drug discovery and can provide more trained personnel.
4. Drastic cuts by the funding agencies for the "blue sky research".

Though the industry is willing to interact with the academia, the academic organization must curtail its inertia and reduce bureaucratic, administrative and organization structure to do successful translational research. To achieve the objective confidentiality of the data needs to be maintained. The essential requirement to obtain the success in translational research is to maintain harmony among the team members coming from different disciplines. The multidisciplinary nature of drug discovery requires intensive interaction between the various teams participating in the program. The biggest hurdle is in drafting the agreements and sharing the intellectual properties. The intellectual properties can be an asset if the drug discovery is successful to academic institutions. The Bayh-Dole Act of 1980 safeguards the academic organizations' patent rights and the products developed can be commercialized [19]. Prior to the submission of IND or NDA application both academia and industry people should consult the regulatory authorities.

The reproducibility of the biomedical research carried out in the academic institutions is one of the major hurdles which results in variable end results in the efficacy and safety of the new drug developed and this leads to growing distrust in the industry regarding the research carried by the academia [9]. Several reports are available in the literature regarding not able to reproduce the preclinical data submitted by the academia in clinical trials. [20, 21]. Jarvis and Williams (2014) discussed in detail the various issues related to the magnitude and pervasiveness of the problem of reproducibility, animal models, statistics, misconduct and their resultant effect on the target validation and clinical translation [9]. To raise the quality and increase the reproducibility of preclinical research carried out by the academic institutions several guidelines and concepts are introduced in the biomedical science training [22–25]. For further guidelines and publication checklists please refer to Table 1 of Jarvis and Williams [9]. To increase the level of the confidence of pharmaceutical industry in the biomedical research carried out by the academia, the scientists working in research institutes should concentrate on the replicability and reproducibility of their results.

17.2 Regulatory Authorities and Patients

Throughout the world, health care system is undergoing rapid changes due to disruptive technologies. Academia and industrial scientists involved in regarding the drug discovery and development should be in constant interaction with the regulatory authorities and the end users, i.e. patients, and obtain the feedback to assess benefit–risk analysis. Discovery scientists should value the involvement of patients throughout the product lifecycle and improve the drug discovery and development [26]. The valuable involvement and contribution of the patients in all the stages of the drug discovery and development and its benefits are being recognized throughout the world. European Patients Academy on Therapeutic Involvement (EUPATI) developed four separate guidelines. These guideline documents cover patient involvement in (a) pharmaceutical industry led drug discovery and development, (b) ethics committees, (c) regulatory authorities and (d) health technology assessments [27]. Patients with their experience and perspective can contribute a lot in the drug discovery. Further, patient involvement leads to fairness, equity and capacity building in the drug discovery and development process.

Unbiased assessment and transparency are important considerations for drug regulatory agencies since public health is of utmost important. EMA (European Medical Agency) and TGA (Therapeutic Good Administration, Australia) publish information regarding the evaluation of medicines by public. Regulatory Agencies consider the European Public Assessment Reports and Australian Public Assessment Reports in the final approval of the drugs [27].

Increase in the awareness and transparency in US regulatory assessments of benefits–risk evaluation and involvement of patients in FDA approval was reported [28]. In a Patient preferences in Benefit and Risk Assessment during Treatment Life Cycle (PREFER) study various stakeholders give importance to patients' preferences to improve the outcome of the treatment [29]. The benefit–risk evaluation is transforming from "*subjective and inconsistent to a more structural, transparent approach with a number of quantitative methods to incorporate preference weight from various stakeholders*" [30]. The greater involvement of end users and the regulatory processes are evolving and undergoing several changes to provide the maximum benefit to patients.

17.3 Conclusion

The academic drug discovery scientists who carry out blue sky research in the biomedical areas using innovative and disruptive technologies to identify selective, safe and efficacious ligands by studying the downstream pathways of different targets can carry out translational studies by better understanding of management of it. The credibility of the basic research carried out in the academic organizations will depend on the reliability and reproducibility of the data. The awareness and implementation of the guidelines and checklists issued by the various professional bodies and the authorities need to be adhered by the researchers. A recent report indicates that the success rates of academic drug discovery and developments were more than 60% in different Phases of clinical trials and nearing 90% in NDA and BLA phase [31]. The disease foundations are also playing pivotal role by interacting with academia and industry in preclinical stages and phase I of clinical trials resulted in advancement of drug discovery and development [32]. A greater coordination between academia, pharmaceutical industry, regulatory authorities, funding agencies (public agencies and disease foundations), patient feedback and participation is required to develop successful and affordable drugs.

References

1. Popper K. Conjectures and refutations. In: The growth of scientific knowledge. London: Routledge; 1963. p. 216.
2. Fogt JS, Kitzmiller JP. Building the relationship between industry and academia: benefits and the general landscape of accomplishing research. J Transl Sci. 2018;4:1–3.
3. Furman JL and MacGarvie M. Academic collaboration and organizational innovations: the development of research capabilities in the US pharmaceutical industry 1927-1946. In Industrial and corporate Change. Pp 1-33, Oxford Press, 2009. doi:https://doi.org/10.1093/icc/dtp035.

4. Anonymous. Editorial. J Am Med. , n.s. 1937;12:185.
5. Wordley E. Diethyl glycol poisoning. Report of two cases. J Clin Pathol. 1947;1:44–6.
6. Mcbride WG. Thalidomide and congenital abnormalities. Lancet. 1961;II:1358.
7. Lenz W. Thalidomide and congenital abnormalities. Lancet. 1962;I:45.
8. 2014 CMR International Pharmaceutical R&D Factbook.
9. Javris ML, Williams M. Irreproducibility in preclinical biomedical research: perceptions, uncertainties, and knowledge gaps. Trends Pharmacol Sci. 2014;37:290–302.
10. Kola I, Landis J. Can the pharmaceutical industry reduce attrition rates? Nat Rev Drug Discov. 2004;3:711–6.
11. Bergren R, Mollar M, Moss R, Poda P, Smietana K. Outlook for the next 5 years in drug innovation. Nat Rev Drug Discov. 2012;11:435–6.
12. Butler D. Translational research: crossing the valley of death. Nature. 2008;453:840–2.
13. DiMasi JA, Grabowski HG, Hansen RW. Innovation in the pharmaceutical industry: new estimates of R& D cost. J Health Econ. 2016;47:20–33.
14. Darrow JJ, Avorn J, Kesselheim AS. FDA approval and regulation of pharmaceuticals, 1983-2018. JAMA. 2020;323:164–76.
15. https://www.fda.gov/drugs/new-drugsGrime-fda-cde rs-new-molecular-entities-and-new-therapeutic-biological-products/novel-drug-approvals-2019
16. Parrish MC, Tan YJ, Grimes KV, Mochlet-Rosen D. Surviving the valley of death: opportunities and challenges in translating academic drug discoveries. Annu Rev Pharmacol Toxicol. 2019;59:405–21.
17. Dahling JL, Inglese J, Walters MA. Mitigating risk in academic preclinical discovery. Nat Rev Drug Discov. 2015;14:279–94.
18. Roy A. Early probe and drug discovery in academia: a mini review. High Throughput. 2018;7: E4. https://doi.org/10.3390/ht7010004.
19. US Copyright Off. Title 37—patents, trademarks, and copyrights. Washington, DC: Code Fed Regul; 2013.
20. Prinz F, Schlange T, Asadullah K. Believe it or not: how much can we rely on published data on potential drug targets? Nat Rev Drug Discov. 2011;10:712–3.
21. Begley CG, Ellis LM. Drug development: raise standards for preclinical cancer research. Nature. 2012;483(7391):531–3.
22. Curtis MJ, Bond RA, Spina D, Ahluwalia A, Alexander SPA, Giembycz MA, Gilchrist MA, Hoyer D, Insel PA, Izzo AA, Lawrence AJ, MacEwan DJ, Moon LDF, Wonnacott S, Weston AH, McGrath JC. Experimental design and analysis and their reporting: new guidance guidance for publication in BJP. Br J Pharmacol. 2015;172:3461–71.
23. Mullane K, Williams M. Unknown unknowns in biomedical research: does an inability to deal with ambiguity contribute to issues of irreproducibility? Biochem Pharmacol. 2015;97:133–6.
24. McGrath JC, Drummond GB, McLachlan EM, Kilkenny C, Wainwright CL. Guidelines for reporting experiments involving animals: the ARRIVE guidelines. Br J Pharmacol. 2010;160:1573–6.
25. Kilkenny C, Parsons N, Kadyszewski E, Festing MF, Cuthill IC, Fry D, Hutton J, Altman DG. Survey of the quality of experimental design, statistical analysis and reporting of research using animals. PLoS One. 2009;4:e7824.
26. Lowe MM, Blaster DA, Cone L, Arcona S, KO J, Sasane R, Wicks P. Increasing patient involvement in drug development. Value Health. 2016;19:869–78.
27. Warnner K, See W, Haerry D, Klingmann I, Hunter A, May M. EUPATI guidance for patient involvement in medicines research and development (R&D); guidance for pharmaceutical industry-led medicines R&D. Front Med. 2018;5:270. https://doi.org/10.3389/fmed.2018.00270.
28. Johnson FR, Zhou M. Patient preferences in regulatory benefit- risk assessments: a US perspective. Value Health. 2016;19:741–5.

29. De Bekker-Grob EW, Berlin C, Levitan B, Raza K, Christoforidi K, Cleemput I, Pelouchova J, Enzmann H, Cook N, Hansson MG. Giving patients' preferences a voice in medical treatment life cycle: PREFER public-private project. Patient. 2017;10:263–6.
30. Juhaeri J. Benefit—risk evaluation: the past, present, and future. Ther Adv Drug Saf. 2019;10:1–10.
31. Takebe T, Imai R, Ono S. The status of drug discovery and development as originated in United States academia: the influence of industrial and academic collaboration on drug discovery and development. Clin Transl Sci. 2018;11:597–606.
32. Ramsey BW, Nepom GT, Lonial S. Academic, foundation, and industry collaboration in finding new therapies. N Engl J Med. 2017;376:1761–9.

Chapter 18
Statistics in Pharmacology and Toxicology

Mohammad Atiar Rahman and Karl Lin

18.1 Introduction

This chapter gives an introduction to basic statistics. Sects. 18.2 and 18.3 provide definitions of some basic statistical terminologies, Sect. 18.4 describes ways to display data, Sect. 18.5 discusses some descriptive statistics, Sects. 18.6 and 18.7 contain topics related to point and interval estimations, Sects. 18.8–18.10 discuss topics of parametric and nonparametric statistical inferences, Sect. 18.11 presents descriptions of designs of experiments and their associated statistical analyses, Sect. 18.12 contains methods of analyzing repeated measures data, Sect. 18.13 contains topic of adjustment for multiple testing, Sect. 18.14 presents regression analysis, Sect. 18.15 discusses the topic of transformation of variables, and Sect. 18.16 gives some applications of statistical methods in toxicity studies.

18.2 Random Variables

A random variable or simply a variable is any characteristics, number, or quantity that can be measured or counted. Examples of variables are age, gender, business income, capital expenditure, country of birth, class grades, eye color, vehicle type, etc. It is called a variable because the value may vary between data units or over time.

The opinions expressed herein are those of the authors and do not necessarily reflect those of the US Food and Drug Administration.

M. A. Rahman (✉) · K. Lin
Division of Biostatistics-6, Office of Biostatistics, Center for Drug Evaluation and Research, Food and Drug Administration, Silver Spring, MD, USA
e-mail: mohammad.rahman@fda.hhs.gov; Karl.Lin@fda.hhs.gov

R. Poduri (ed.), *Drug Discovery and Development*,
https://doi.org/10.1007/978-981-15-5534-3_18

The variables are generally symbolized by X, Y, Z, etc. Outcomes of a scientific experiment or observational study are assumed to come from, an inherent space, defined as a hypothetical and potentially infinite group of objects, e.g., the set of all possible hands in a game of poker. This space is known as a population. The exact specification of this population is generally unknown. However, the population is assumed to follow a probability law known as the density or distribution function. A mathematical function that may represent the density is known as density function or distribution function. Examples of such density function include Binomial, Poisson, Normal, t-distribution, etc. The outcome set of the experiment is known as the sample. A good sample is assumed to imitate the population. Therefore, a good sample can be analyzed to study the properties of a population. To accurately study the population, it is necessary that the experiment is well designed depending on the experimental conditions and aim, the number of observations (sample size) is enough based on appropriate power requirement and the data are analyzed using appropriate statistical methods. Data analyses may include descriptive statistics, graphical presentation, and statistical inference.

18.3 Parameters and Statistics

As mentioned above, the population is assumed to follow a probability law known as the distribution. A distribution is specified by some constants (generally unknown). These constants are known as parameters. For example, Binomial distribution for variable r, denoted as $B(n,p) = \binom{n}{r} p^r (1-p)^{n-r}$. This distribution provides probability rule for the realization of r success out of n trials, e.g., getting $r = 4$ successes in $n = 10$ throws of a coin. The constant p is the parameter of this distribution, which is the actual probability of one success in one through. The real value of p is generally unknown. A second example is a normal distribution for variable y, denoted as $N(\mu, \sigma^2) = \frac{1}{\sigma\sqrt{2\pi}} e^{-\frac{(y-\mu)^2}{2\sigma^2}}$. This distribution provides probability rule for the realization of a continuous variable to fall within a range of value, e.g., the probability that a randomly chosen person from a certain age group will have a systolic blood pressure Y between 100 and 120 mm Hg. The constants μ and σ^2 are the parameters of this distribution, which represent the mean and the variance of Y. The real values of μ and σ^2 are also generally unknown. Since a sample is assumed to imitate the population, some estimation of these parameters is possible using the observed sample values. These estimates of parameters are known as statistics. The estimates may be only a single number describing a possible value of the associated parameter (point estimate), or an interval describing the probability that the parameter will be in that interval (interval estimation).

Every distribution has two main characteristics, namely, centrality and dispersion. Centrality is the center value around which all realization of a variable happens, and dispersion is an extent to which values of a variable can go.

18.4 Displaying Data

Observed data are often presented in tabular or graphical form. A variable from which data come may be qualitative or quantitative. Qualitative variables include gender, color, ethnicity, etc., while quantitative variables include litter size, number of hospital visits, height, weight, blood pressure, etc. The qualitative variable may be ordinal or nominal. Ordinal variables can be ordered, e.g., patient condition "critical," "poor," "fair." Nominal variables cannot be ordered, e.g., ethnicity "black," "white," "Asian," "Hispanic." The quantitative variables may be discrete (litter size, number of hospital visits) or continuous (height, weight, heart rate, blood pressure). Depending on its type, the presentation method may also vary.

If the data set is very small, simple listing of the data is OK. For example, if five observations are taken on time (h) to performing brain operation using a new neurological procedure, then a simple listing of 5 numbers as 9.5, 10.3, 8.8, 9.5, and 9.9 may suffice to display the data. On the other hand, if 600 of these values are observed, simple listing may be difficult to comprehend. In this situation, a frequency table may be used. Table 18.1 shows the frequency of every single observation. This table also may be very big. Table 18.2 shows the frequency of intervals in which each observation falls. Since Table 18.2 can be manipulated to fit any requirements of the investigator, this kind of frequency table is preferred. Other columns like cumulative frequency, proportion, cumulative proportion, percent, and cumulative percent can be added to these tables. These tables can be prepared for both qualitative and quantitative variables. However, for nominal variables, the cumulative values may be ambiguous (Table 18.3).

Table 18.1 Individual data

Value	Frequency
9.5	10
10.3	13
.	.
.	.
.	.
16.7	1
Total	600

Table 18.2 Frequency table

Interval	Frequency
7.0–8.9	34
9.0–10.9	102
11.0–12.9	224
13.0–14.9	221
15.0–16.9	29
Total	600

Table 18.3 Frequency, cumulative frequency, proportion, cumulative proportion, percent, and cumulative percent

Interval	Freq	Cum	Freq proportion	Cum proportion	Percent	Cum percent	h
7.0–8.9	24	24	0.040	0.040	4.00	4.000	0.02000
9.0–10.9	102	126	0.170	0.210	17.00	21.000	0.08500
11.0–12.9	224	350	0.373	0.583	37.33	58.333	0.18667
13.0–14.9	221	571	0.368	0.952	36.83	95.167	0.18417
15.0–16.9	29	600	0.048	1.000	4.83	100.000	0.02417

Histogram: Frequencies, proportions, and percentages can be expressed in figures

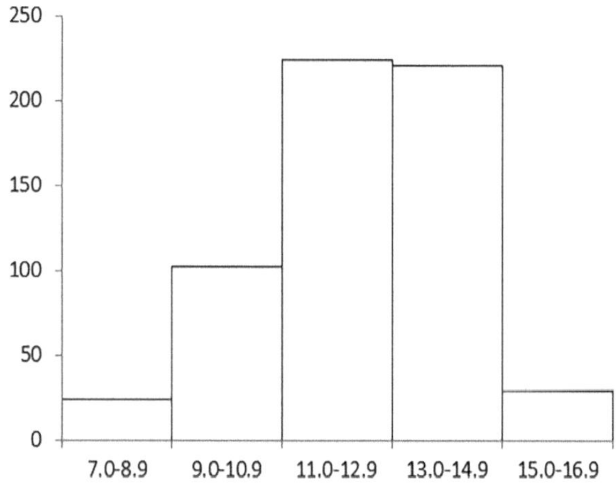

Fig. 18.1 Frequency histogram

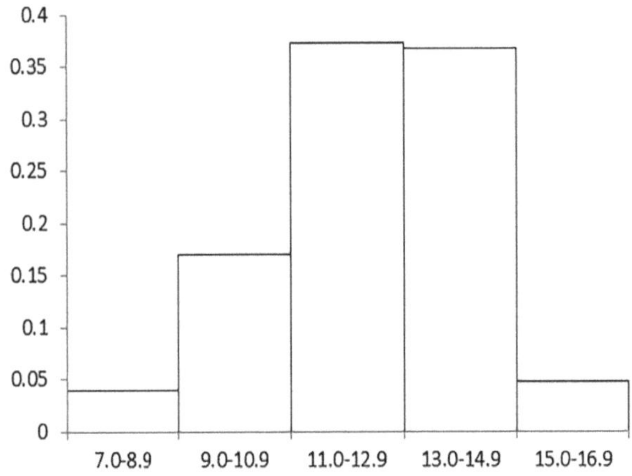

Fig. 18.2 Proportion histogram

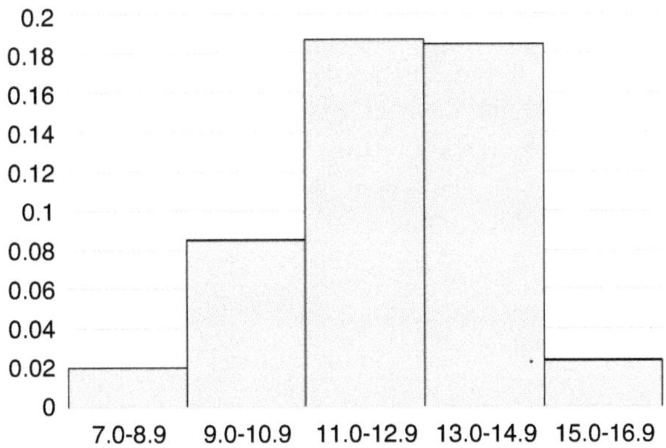

Fig. 18.3 Density histogram

Histogram Frequencies, proportions, and percentages can be expressed in figures known as frequency histogram, proportion histogram, and percentage histogram. The following are examples of histograms.

One special histogram is drawn in which the area of each rectangle represents the proportion. For this histogram, the height of each rectangle is calculated as h=proportion /width of the rectangle. Therefore, the area of the rectangle $= h \times$ width of the rectangle = proportion. This histogram is an estimate of population density. A similar histogram using the cumulative proportion is known as cumulative distribution (Table 18.3) (Figs. 18.1–18.3).

Notice that the total area under the density histogram is 1. Often this density histogram is smoothed, and a mathematical density function (e.g., normal density) is fitted to it for mathematical rigor. Following is an example of such fitting of normal density:

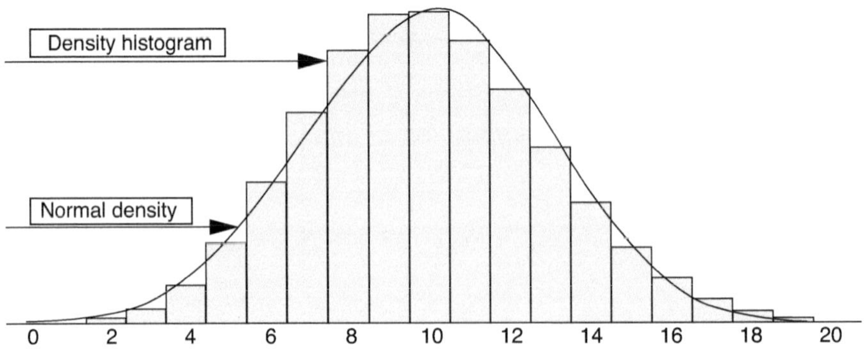

Scatter plot If two variables are observed from each experimental unit, then a convenient graphical presentation of the data is to draw a two-dimensional plot. This plot is known as a scatter plot and each observation is denoted by a dot. Consider the following data set, obtained for the ability of liver slices from guinea pigs of different ages to conjugate phenolphthalein with glucuronic acid. In notations, n=Number of animals, X= Age, and Y= Milli-micromoles conjugated.

Consider the data set with 11 pairs of observations. (X, Y): (1, 5.6), (1, 8.8), (3, 12.0), (5, 18.0), (6, 31.0), (10, 38.0), (10, 44.0), (11, 22.0), (14, 37.0), (15, 46.0), and (21, 54.0). The scatter plot of this data is as follows.

Scatter plot Y on X

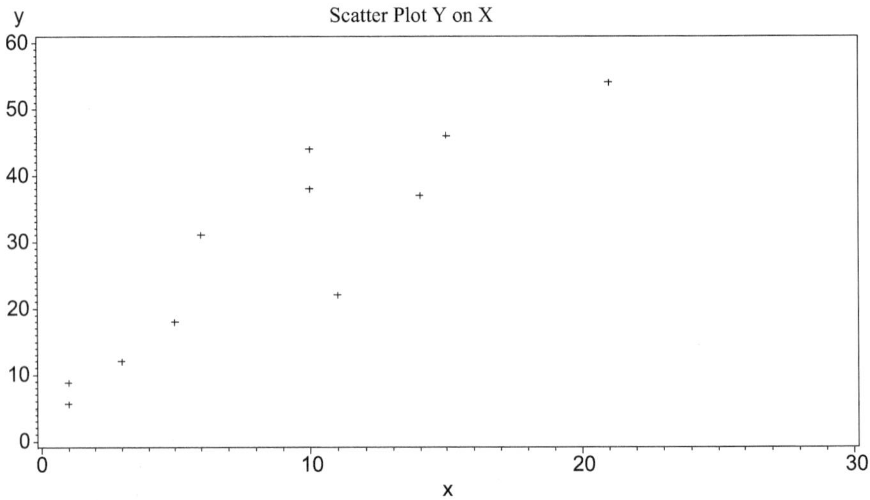

A scatter plot provides a visual aid to guess on a possible relationship between two observed variables. For example, the above scatter plot suggests a possible linear relationship between y and x with a positive slope.

There is a host of other graphical representations in the literature including stem and leaf, dot diagram, bar diagram, pie diagram, and box and whisker plot. Due to the limitation of space, these graphical methods are not discussed further. In SAS, "proc chart" command can draw all these and other important plots and diagrams.

18.5 Descriptive Statistics

Descriptive statistics are summary values of the observations, describing the characteristics of the associated population like centrality and dispersion. Common descriptive statistics are mean, median, mode, moment (more specifically m^{th} moment), percentiles, variance, standard deviation, standard error, mean deviation, etc. In this list mean, median, and mode estimate the centrality of distribution, while variance, standard deviation, standard error, mean deviation, describe the dispersion of the distribution. In this discussion the symbol $\sum_{i=1}^{n} x_i$ will represent sum of n observations. For example, if a sample of variable X has 5 values as $x_1 = 5$, $x_2 = 4$, $x_3 = 5$, $x_4 = 6$, and $x_5 = 10$, then $\sum_{i=1}^{5} x_i = 5 + 4 + 5 + 6 + 10 = 30$. Using this notation, we define Mean $= \frac{\sum_{i=1}^{n} x_i}{n}$. The population mean is generally denoted by μ and sample mean is denoted by putting a bar sign over the variable, e.g., $\bar{x}$ in this case. The population means of a variable X is also known as the expected value of X and is denoted as $E(X)$. Median = the middle value of the observed sample values in magnitude. For its calculation, we first arrange the observed sample values in ascending order and pick the middle value in case of an odd number of observations and mean of middle two observations in case of an even number of observations. Median is generally denoted by M for population and m for a sample. A mode is an observation that is repeated the most number of times. For its calculation, find how many times each observation is repeated. Pick the one that has the highest repetition. An observed sample may have no mode or more than one mode. The pth moment about A is defined as $\frac{\sum_{i=1}^{n} (x_i - A)^p}{n}$. The pth moment is generally denoted by $\mu_p$for population and by m_p for sample. Moments about mean is of special interest. Variance is defined as $\frac{\sum_{i=1}^{n} (x_i - \bar{x})^2}{n}$, where $\bar{x}$ is the mean of observed x values. Notice that variance is the 2nd moment about mean. Variance is generally denoted by V or σ^2 for population and by v or s^2 for sample. Standard deviation $= \sqrt{\text{Variance}}$. Standard deviation is generally denoted by SD or σ for population and by sd or s for sample. Mean deviation $= \frac{\sum_{i=1}^{n} |x_i - \bar{x}|}{n}$, generally denoted by MD for population and md for sample. The mth percentile, sometimes denoted by p^m, is defined as a value of the observed sample so that $m\%$ of the observations is less than this value and $(100 - m)\%$ observations are greater than this value. The value p^{25} is also known as the first quartile (Q_1) and the value p^{75} is also known as the third quartile (Q_3). Therefore, there are 25% observations smaller than Q_1 and 75% of observations are greater than Q_1. Similarly, there are 75% observations smaller than Q_3 and 25% of observations are greater than Q_3. The difference between Q_3 and Q_1 is known as the interquartile range. $Q_2 = p^{50}$ is known as the second quantile. There are 50% observations smaller than Q_2 and 50% of observations are greater than Q_2. Note that Q_2 is the same as the median. The difference between the maximum and the minimum value is known as the range. In SAS, "PROC UNIVARIATE AND PROC MEANS" provide some of the summary statistics.

Example Suppose we have blood plasma level of drug X in 10 rats with following values x: {3, 2, 4, 3, 5, 1, 5, 4, 2, 4}in ng/ml. The calculated summary statistics (using SAS command "PROC UNIVARIATE") are as follows: $\overline{x} = 3.3$ ng/ml, median = 3.5, mode = 4.0, minimum = 1.0, maximum = 5, $Q_1 = p^{25} = 2$, $Q_3 = p^{75} = 4$, sd=1.34, mean deviation =1.1, range =4.0, and interquartile range =2.

For experiments where more than one variable such as heart rate and blood pressure is measured from the same experimental unit, two important summary statistics are covariance (Cov) and correlation coefficient (r). Both statistics measure the strength of relations between variables. The correlation coefficient is the standardized value of covariance and is unit free with a value ranging in −1 and 1. A value of $r = 1$ implies a perfect correlation between two variables X and Y say, indicating perfect linear relation between X and Y, i.e., Y increases linearly as X increases and vice versa. A value of $r = -1$ also implies a perfect correlation between X and Y with reverse linear relation between X and Y, i.e., Y decreases linearly as X increases and vice versa. A value of $r = 0$ indicates no linear relation between X and Y. Other values indicate a lack of perfect linear relationship. For a given value of x_i, y_i, $i = 1, 2, \ldots, n$,of X and Y the covariance and correlation coefficient are calculated as follows:

$$\text{Covariance, Cov}(X, Y) = \frac{\sum_i^n (x_i - \overline{x}) \times (y_i - \overline{y})}{n} = \frac{\sum x_i y_i}{n} - \frac{\sum x_i}{n} \frac{\sum y_i}{n},$$

$$\text{Correlation coefficient, } r(X, Y) = \frac{\text{Cov}(X, Y)}{\sqrt{V(X) \times V(Y)}}$$

$$= \frac{\sum x_i y_i - \frac{\sum x_i}{n} \frac{\sum y_i}{n}}{\sqrt{\left(\sum x_i^2 - \frac{\left(\sum x_i\right)^2}{n}\right)\left(\sum y_i^2 - \frac{\left(\sum y_i\right)^2}{n}\right)}}.$$

Example The following data were obtained for the ability (Y) of liver slices from guinea pigs of different ages (X) to conjugate phenolphthalein with glucuronic acid.

X	Y	X^2	Y^2	XY
1	5.6	1	31.4	5.6
1	8.8	1	77.4	8.8
3	12.0	9	144.0	36.0
5	18.0	25	324.0	90.0

(continued)

X	Y	X^2	Y^2	XY
6	31.0	36	961.0	186.0
10	38.0	100	1444.0	380.0
10	44.0	100	1936.0	440.0
11	22.0	121	484.0	242.0
14	37.0	196	1369.0	518.0
15	46.0	225	2116.0	690.0
21	54.0	441	2916.0	1134.0
Total 97	316.4	1255	11803	3730.4

$$\mathrm{Cov}(X,Y) = \frac{3730.4}{11} - \frac{97}{11} \times \frac{316.4}{11} = 85.45, \text{and } r(X,Y)$$

$$= \frac{3730.4 - \frac{97}{11} \times \frac{316.4}{11}}{\sqrt{(1255 - \frac{(97)^2}{11})(11803 - \frac{(316.4)^2}{11})}} = 0.90$$

SAS command "PROC CORR" can calculate the correlation coefficient.

For a variable r with Binomial distribution $B(n,p) = \binom{n}{r} p^r (1-p)^{n-r}$ the population mean is np and variance is npq. For variable y with normal distribution $N(\mu, \sigma^2) = \frac{1}{\sigma\sqrt{2\pi}} e^{-\frac{(y-\mu)^2}{2\sigma^2}}$ the population mean is μ and variance is σ^2. There are a host of other distributions whose population mean and variance can be expressed in terms of their parameters. Due to space limitations, the proof and discussion of this topic are not included here.

18.6 Estimations of Parameters

As mentioned, the values of parameters are not generally known. Statistics can provide estimations of parameters. For example, sample proportion can estimate the population proportion, sample mean can estimate population mean, sample variance can estimate population variance, etc. More accurately they are known as point estimates. For example, for a normal distribution $N(\mu, \sigma^2)$ the estimate of mean μ is $\bar{x}$ and that of variance $\sigma^2$is s^2. Sample estimates should have some desired properties, e.g., unbiasedness, consistency, efficiency, and sufficiency. Due to the limitation of space of this article, these properties will not be discussed further. In general, an estimate with more of these properties is assumed to be better.

Since a statistic is a function of the observed sample, it may vary from sample to sample. Therefore, a statistic is a random variable. It is possible that a point estimate may be close or far from the real value of the associated parameter. This difference cannot be determined by such a single value. An alternative to point estimate is an interval estimate. This estimate calculates two values known as lower and the upper limits in such a way that the actual value of the parameter will be within these limits with a high probability. The interval defined by these two limits is known as the confidence interval (C.I.). For example, if X has a normal distribution as $N(\mu, \sigma^2)$, then $l = \overline{x} - 1.96 \times SD(\overline{x})$ and $u = \overline{x} + 1.96 \times SD(\overline{x})$ are lower and the upper limits for a 95% confidence interval for mean μ. In other words, C.I. = $(u,\ l)$ is the 95% confidence interval for μ. The interpretation of this confidence interval is, if the calculation of confidence interval is repeated 100 times by taking 100 sets of independent samples using the above formula, then 95 times the actual value of μ is expected to be inside the calculated intervals. The number 1.96 came from the 97.5% tail probability of standard normal distribution ($Z \backsim N(0, 1)$. In statistical notations $1.96 = z(\alpha/2)$, where $\alpha = 0.05$.

As mentioned above, a statistic is a random variable. Therefore, it also has a distribution with some parameters. The distribution of a statistic is known as the sampling distribution. For example, if $X \backsim N(\mu, \sigma^2)$, then $\overline{X} \backsim N(\mu, \sigma^2/_n$, i.e., $\overline{X}$ has a normal distribution with mean μ and variance $\sigma^2/_n$. The square root of this variance is the standard deviation $\overline{x}$. In the statistical literature, the standard deviation of a statistic is known as standard error (S.E.) of the statistics. Therefore, correctly $\overline{x}$ has a distribution with mean = μ and S.E. $(\overline{x}) = \sqrt{\sigma^2/_n}$. An estimate of this S.E. is $\sqrt{s^2/_n}$. Therefore, the 95% C.I. for μ is $(\overline{x} \mp 1.96 \times S.E.(\overline{X}))$. However, to be more accurate $t_{1-\frac{\alpha}{2},(n-1)}$ from t-distribution with $n - 1$ degree of freedom (d.f.) should be used instead of $z_{1 - \alpha/2}$, i.e., $(\ \overline{x} \pm t_{n-1,\alpha/2} \times S.E.(\overline{X})\ )$. The 95% C.I for σ^2 is $l = s^2\chi^2_{\alpha/2}(n-1)$ and $u = s^2\chi^2_{1-\alpha/2,(n-1)}$, where $\chi^2_{\alpha/2,(n-1)}$ is the χ^2value with $n - 1$ d.f. with $\frac{\alpha}{2}$% tail to the left, and $\chi^2_{1-\alpha/2,(n-1)}$is the χ^2 value with $n - 1$ d.f. and $(1 - \alpha/2)$% tail to the left. Theoretically, confidence intervals can be calculated for any parameter. However, for some parameters the process may be quite complicated.

Example Consider observations 3.9, 4.2, 4.2, 3.8, 3.6, 4.4, 4.1, 3.4, 4.0, 3.5, 3.7, and 4.2. Suppose that population mean μ and population variance σ^2 are unknown and we like to calculate an 80% confidence interval on the mean μ. We have $\alpha = 0.20$, $n = 12$, $\sum_i x_i = 47$, and $\sum_i x_i^2 = 185.2$. Thus, $\overline{x} = \frac{47}{12} = 3.917$ and $s = \sqrt{\frac{12(185.2) - 47^2}{12(11)}} = \sqrt{0.10152} = 0.3186$. From t-table $t_{11,\ 19} = 1.363$ and the 80% confidence interval on μ is $3.917 \mp 1.363\left(\frac{0.3186}{\sqrt{12}}\right) = 3.917 \mp 0.1254 = (3.79, 4.04)$.

Suppose $\mu_1$and $\mu_2$are population means from two populations. Also, $x_1, \ldots x_m$, and $y_1, \ldots, y_n$ be two samples for these two populations. The $(1 - \alpha)$% confidence interval on the difference of these means $\mu_1 - \mu_2$ can be found as, $(\overline{x}_1 - \overline{x}_2) \mp t_{1-\alpha/2,(m+n-2)} s_p \sqrt{\frac{1}{m} + \frac{1}{n}}$, where $s_p = \sqrt{\frac{s_1^2(n_1-1) + s_2^2(n_2-1)}{n_1+n_2-2}}$.

18.7 Central Limit Theorem

If $x_1, x_2, \ldots, x_n$ is a sample of size n from a population, not necessarily normal, with mean μ and variance σ^2. Then for large n the sample mean $\overline{X}$ has approximately a normal distribution with mean μ and variance $\sigma^2/_n$, i.e., as $n \to \infty$ $\overline{X} \approx N(\mu, \sigma^2/_n)$. This property of a large sample is known as a central limit theorem. A consequence of this theorem is that statistical tests (discussed in Sect. 18.8) originally meant for the normal population are also applicable for non-normal populations for large samples.

18.8 Statistical Hypothesis Testing

Population parameters are generally unknown. Nevertheless, they may have hypothetical values. There are statistical methods to test if such hypothetical values of parameters are true using a representative sample observation from this population. This statistical procedure is known as the test of hypothesis.

Formally, we first assume a hypothetical value of a parameter, known as the null hypothesis (H_0), and possible alternative value (H_1) if H_0 is not true. For example, consider a coin throwing example where a coin is thrown n times and count the total number of heads. This outcome is modeled by a Binomial distribution $B(n, p)$, with p as the probability of head (success) in each throw of the coin. Suppose we like to know if the used coin is a fair coin, i.e., in each throw it has a chance of 50% of showing a head (also 50% chance of showing a tail). Alternatively, if the coin is not fair, the chance of showing a head in a single throw is not 50%. Mathematically we say for a fair coin probability of a head $p = 0.5$ and alternatively for unfair coin $p \neq 0.5$. In formal notations, we write $H_0 : p = 0.5$ and $H_1 : p \neq 0.5$. In a second example for a normal distribution $N(\mu, \sigma^2)$, we may be interested in $H_0 : \mu = 100$ and $H_1 : \mu > 100$. In general notations, we write $H_0 : \theta = \theta_0$ and $H_1 : \theta > \theta_0$ *or* $H_1 : \theta < \theta_0$ (both are known as a one-sided or one tailed alternative) or $H_1 : \theta \neq \theta_0$ (known as two-sided or two tailed alternative) where θ denotes a parameter like p *or*μ.

The general idea of test of hypothesis is to calculate the difference $d = \left(\widehat{\theta} - \theta_0\right)$ or ratio $r = \frac{\widehat{\theta}}{\theta_0}$between the hypothetical value of the parameter and the corresponding estimate and make a determination on the validity of the hypothetical value of the parameter based on the magnitude of this difference or ratio. A big value of $|d|$ indicates the falsehood of the null hypothesis and a small value of d indicates the truth of the null hypothesis. Similarly, a value of r away from 1 indicates the falsehood of the null hypothesis and the value of r close to 1 indicates the truth of the null hypothesis. An important component of this process is the selection of a cutoff point on d (or r) to decide on the falsehood and truth of the null hypothesis. For this purpose the difference (or the ratio) is standardized. This standardized value is known as test statistic Z, say. Note that a test statistic is a function of sample values and has distribution. A cutoff point is then determined based on the distribution of this statistic. In the following, we will describe the process of choice of this cutoff point and examples of some test statistics.

There are two types of possible errors associated with this process, known as Type 1 and Type 2 errors. They are defined as follows:

	Accept H_0	Reject H_0
H_0 is True	Correct Decision	Type 1 Error (False Positive)
H_0 is False	Type 2 Error (False Negative)	Correct Decision

The Type 1 error is generally denoted by α and the Type 2 error is generally denoted by β. We want these errors to be as small as possible. The cutoff point for the test is generally fixed such that the Type 1 error is less than a prespecified value. This cutoff point is known as the critical value and the corresponding Type 1 error (α) is known as the test level. Generally, α is taken to be as 1% or 5%. The value $(1 - \beta)$ 100% is known as the power of the test.

There are two operational criteria to draw a conclusion in a hypothesis testing known as (a) critical value criteria and (b) p-value criteria. In the first criteria, for one-sided ($\mu > 100$) alternative, compare z with the critical value $z_{1-\alpha}$ taken from standard tables. If the calculated statistic z is greater than the critical value $z_{1-\alpha}$, reject the null hypothesis, accept otherwise. For two-sided alternative $\mu \neq 100$ compare $|z|$ with the critical value $z_{1-\frac{\alpha}{2}}$ and draw a conclusion accordingly. In the second criteria, calculate the p-value defined as $p = P(Z>|z|)$. For one-sided alternative if the calculated p-value is smaller than the test level α, reject the null hypothesis, accepts otherwise. For a two-sided alternative, compare the p-value with $\alpha/2$. Clearly, the two criteria are equivalent. Some examples are given in Sects. 18.9 and 18.10.

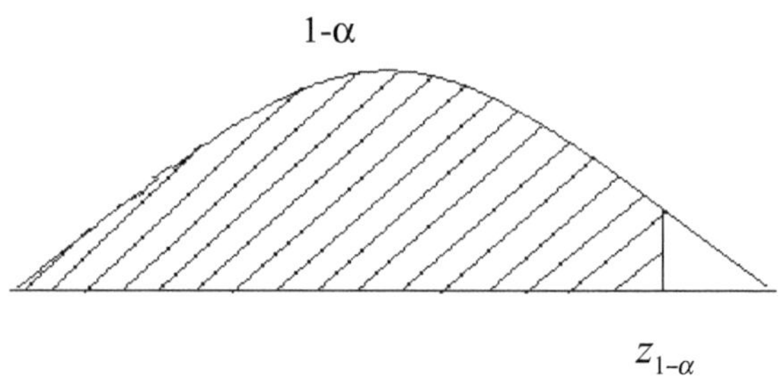

Sample size calculation Both Type 1 and Type 2 errors are sensitive to the sample size. As the sample size becomes large, both Type 1 and Type 2 errors become small. However, as the sample size becomes large, the cost, time, and complexity of conducting the experiment grow. Therefore, for any experiment, a reasonable sample size is desirable. In practice, an acceptable Type 1 error and an acceptable Type 2 error are first determined and then the sample size is calculated satisfying these two requirements. There are mathematical formulas that can calculate a sample size satisfying the required Type 1 and Type 2 errors. Due to the limitation of the space of this article, the derivation procedure of these mathematical formulas is not discussed here. An example of these mathematical formulae for testing a single mean is as follows. Consider testing the hypothesis $H_0 : \mu = \mu_0$ against the alternative $H_1 : \mu = \mu_1$. Suppose that the intended Type 1 error $=\alpha$ and Type 2 error $=\beta$, then the sample size n that satisfies these requirements is calculated using the formula,

$n = \frac{\sigma^2 \left[z_{(1-\alpha)} + z_{(1-\beta)}\right]^2}{(\mu_1 - \mu_0)^2}$, where $z_{(1-\alpha)}$ and $z_{(1-\beta)}$ are the $(1-\alpha)\%$ and $(1-\beta)\%$ tail values of z, respectively. For example, consider a test with, $H_0 : \mu = 25$ against $H_1 : \mu = 30$. Assume $\sigma = 5$, $\alpha=0.025$, and $\beta = 0.05$ or power $=95\%$. We have $z_{(1-\alpha)} = 1.96$ and $z_{(1-\beta)}=1.64$. Therefore, $n = \frac{5^2[1.96+1.64]^2}{(30-25)^2} = 12.96$. Therefore, to perform a 0.025 level test for the null hypothesis $H_0 : \mu = 25$ against $H_1 : \mu = 30$ with a power of 95% we need a sample of about 13 subjects.

Also, for the test of equality of two population means, with equal sample size from each population, we have the following formula. The sample size for each population $n = \frac{2\sigma^2 [z_{(1-\alpha)} + z_{(1-\beta)}]^2}{(\mu_1 - \mu_2)^2}$, where μ_1 and μ_2 are two population means and σ^2 is the common variance. Many statistical software are available to calculate the sample size under different requirements. Some of these software programs are nQuary, Power analysis, Sample size calculator, PASS, EAST, and SAS. The use of these software is generally simple and highly accurate.

In the following sections, we will discuss the test procedure for some common parametric and nonparametric methods for a test of hypothesis.

18.9 Parametric Test of Hypothesis

The hypotheses are regarding the parameters of a distribution. The following are some typical tests.

18.9.1 Test for Single Proportion

Consider a Binomial distribution B(n, p). We like to test the null hypothesis $H_0 : p = p_0$ with an alternative hypothesis $H_1 : p > p_0$ at test level$=\alpha$. Let x_1, x_2, . . ., x_n be a sample from this distribution of size n, where each x_i is either 0 or 1. Test statistic $z = \frac{\widehat{p} - p_0}{\sqrt{\widehat{p}(1-\widehat{p})/n}}$. Reject H_0 if $z \geq z_{1-\alpha}$, where $z_{1-\alpha}$ is the $(1-\alpha^{th})$ percentile of standard normal distribution (Z). If the alternative is two-sided, i.e., $H_1 : p \neq p_0$. Reject H_0 if $|z| \geq z_{1-\alpha/2}$ (two tail test), where $z_{1-\alpha/2}$ is the $\left(1-\frac{\alpha}{2}\right)^{th}$ percentile of z.

Example Consider an example in which coin is tossed 1000 times and it showed 667 heads. We like to test if the coin is fair or favors head, i.e., if $p_0 = 0.50$ with an alternative $p_0 > 0.5$ at test level 0.05. From the observations $\hat{p} = \frac{523}{1000} = 0.523$ and $z = \frac{0.523-0.5}{\sqrt{0.523(1-0.523)/1000}} = 1.45$. From normal probability table $z_{0.95} = 1.64$. Since $1.45 < z_{0.95}$, we accept the null hypothesis that $p = 0.5$. If the sample size is small, it is recommended to use t statistic instead of z statistic.

18.9.2 Test for Two Proportions

Consider two binomial populations $B(n_1, p_1)$ and $B(n_2, p_2)$ with observations *Sample 1:* $x_1, x_2, \ldots, x_m$ of size m, and Sample 2 $y_1, y_2, \ldots, y_n$ of size n where each x_i and y_j are either 0 or 1. We like to test the null hypothesis H$_0$: $p_1 = p_2$. With alternative hypothesis $H_1 : p_1 \neq p_2$, Test statistic: $z = \frac{\widehat{p_1} - \widehat{p_2}}{\sqrt{\widehat{p}_p\left(1-\widehat{p}_p\right)\left(\frac{1}{m}+\frac{1}{n}\right)}}$, where $\widehat{p}_p$ is the pooled estimate of the variance of p. The $\widehat{p}_p$ is calculated as $\frac{\sum x_i + \sum y_j}{m+n}$. Reject H$_0$ if $|z| \geq z_{1-\frac{\alpha}{2}}$.

Example Consider an example in which two treatments (reference drug R and test substance T) were applied on two groups of uninephrectomized rats to see their effects on urinary albumin excretion (UAE), a marker to assess the efficacy of the drug. After 4 weeks of treatment, the reference drug and test substance markedly decreased in 425 and 459 of 500 rats, respectively. We like to test the equality of the proportions of UAE decrease in the two treatment groups against not equal (two-sided) at 0.05 test level.

We have $H_0 : p_1 = p_2$, $H_1 : p_1 \neq p_2, m = 500, n = 500, \alpha = 0.05, p_1 = \frac{425}{500} = 0.85$, $\widehat{p_2} = \frac{459}{500} = 0.918$, $\widehat{p}_p = \frac{425+459}{500+500} = 0.884$, $\sqrt{\widehat{p}_p\left(1 - \widehat{p}_p\right)\left(\frac{1}{m} + \frac{1}{n}\right)}$ $=\sqrt{0.884(1 - 0.884)\left(\frac{1}{500} + \frac{1}{500}\right)} = 0.0203$, $1 - \alpha/2 = 1 - \frac{0.05}{2} = 0.975$, and $|z| = \left|\frac{0.850-0.918}{0.0203}\right| = 3.35$. From table $z_{0.975} = 1.96$. Since $|3.35| > z_{0.975}$ we reject the null hypothesis.

18.9.3 Test for Dose Response Relationship of m Proportions

Cochran-Armitage test Suppose $r + 1$ doses of a drug are applied to observe the occurrence of an event, e.g., death or occurrence of tumors. We are interested to test if there is a linear dose response in the proportions of events. Let $Y = 0, 1$ represent the occurrence or non-occurrence of the event. Also, let N_i and x_i denote the group size and dose level for the ith dose group and n_i denotes the total occurrence of the event in the ith dose group. We have the proportion of successes in the ith group as

$p_i = n_i/N_i$, the overall proportion be $P = \frac{n}{N}$, where $n = \sum n_i$ and $N = \sum N_i$. Also, let $\bar{x} = \sum N_i x_i / N$. Then the Cochran Armitage test statistic is,

$$\chi_0{}^2 = \frac{\{\sum n_i x_i - n/N \sum N_i x_i\}^2}{\frac{n(N-n)}{N^2}\left\{\sum N_i x_i{}^2 - \left(\sum N_i x_i\right)^2/N\right\}}$$

has a χ^2 distribution with 1 degree of freedom. Alternatively, the square root of this statistic

$$z = \sqrt{\chi^2} = \frac{\{\sum n_i x_i - (n/N) \sum N_i x_i\}}{\sqrt{\frac{n(N-n)}{N^2}\{\sum N_i x_i{}^2 - (\sum N_i x_i)^2/N\}}}$$

has approximately the standard normal distribution.

Example Consider the following data of thyroid gland C cell carcinoma in rats. We like to test H_0: No linear dose-response in the proportions of occurrence of C-cell carcinoma against H_1: A linear dose-response (positive or negative).

Incidence of thyroid gland C cell carcinoma in male rats (n =60)

Treatment group	Control	Low	Medium	High	Total
Dose level	0 mg	100 mg	300 mg	1000 mg	
NTBA	2	2	4	9	17
NNTBA	58	58	56	51	223
Total (N_i)	60	60	60	60	240
Proportion at risk $P_i = N_i/N$	0.25	0.25	0.25	0.25	1.00

NTBA: Number of tumor-bearing animals, *NNTBA:* Number of non-tumor-bearing animals, $N = \Sigma N_i$

We have $n = 17$, $N = 240$, $\Sigma n_i x_i = 10400$, $\Sigma N_i x_i = 84000$, $\Sigma N_i x_i^2 = 66000000$. Therefore, $z = 2.14245$, and $p = 0.0162$. Since $p < 0.025$, we reject the null hypothesis, i.e., result supports a significant dose response at two-sided 0.05 test level.

18.9.4 Test for an Association between Categorical Variables

Fisher Exact test Consider two laboratories performed a microscopic examination on N_1 and N_2 animals, respectively, to detect the existence of a tumor. Let the following be the outcome.

	Yes	No	Total
Lab 1	n_{11}	n_{10}	N_1
Lab 2	n_{21}	n_{20}	N_2
Total	n_1	n_0	N

We like to know if the laboratory and detection are independent, alternatively, there is an association between laboratory and finding.

The exact probability of this output can be calculated as (Hypergeometric)

$$P(n_1) = \frac{\binom{n_1}{n_{11}}\binom{n_0}{n_{10}}}{\binom{N}{N_1}}$$

Where (n_1/n_{11}) the combination of n_{11} animals (subjects) out of n_1 animals (subjects). The null hypothesis of no association is tested by adding probabilities of all possible tables with extreme n_1 than the observed n_1 (tail probability) keeping the marginal totals fixed.

Example Suppose the observed table is

	Yes	No	Total
Lab 1	2	3	5
Lab 2	1	4	5
Total	3	7	10

Ho: No association between laboratory and findings. Following we have four possible tables with observed margin and the associated probability.

	Yes	**No**	**Total**		**Yes**	**No**	**Total**
Lab 1	0	5	5	**Lab 1**	1	4	5
Lab 2	3	2	5	**Lab 2**	2	3	5
Total	3	7	10	**Total**	3	7	10
$p(0) = 0.0833$				$p(1) = 0.4167$			
	Yes	**No**	**Total**		**Yes**	**No**	**Total**
Lab 1	2	3	5	**Lab 1**	3	2	5
Lab 2	1	4	5	**Lab 2**	0	5	5
Total	3	7	10	**Total**	3	7	10
$p(2) = 0.4167$ (observed table)				$p(3) = 0.0833$			

Right tail $p(2) + p(3) = 0.5$. Since $0.5 > 0.05$ we accept the null hypothesis, i.e., there is no association.

Pearson Chi-Square test This is another test for association. Consider the following table with two categorical variables R and C with r and s levels, respectively.

	C_1	...	C_j	...	C_s	Total
R_1	n_{11}	...	n_{1j}		n_{1s}	N_1
.	.	.	.	.	.	.
.	.	.	.	.	.	.
.	.	.	.	.	.	.
R_i	n_{i1}	...	n_{ij}		n_{is}	N_i
.	.	.	.	.	.	.
.	.	.	.	.	.	.
.	.	.	.	.	.	.
R_r	n_{r1}		n_{rj}		n_{rs}	
Total	n_1	...	n_j	...	n_s	N

Expected value of n_{ij} is defined as $e_{ij} = (N_i \; x \; N_j)/N$. With these expected values the Pearson Chi-Square statistic is defined as

$$\chi^2 = \frac{\Sigma(o_i - e_i)^2}{e_i} \text{ with degrees of freedom d.f.}$$
$$= (\text{number of colum} - 1) \times (\text{number of rows} - 1).$$

Example Using the same data as used for Fisher Exact test we have the observed (expected) table

	Yes	No	Total
Lab 1	2 (1.5)	3 (3.5)	5
Lab 2	1 (1.5)	4 (3.5)	5
Total	3	7	10

$$\chi^2 = \frac{(2-1.5)^2}{1.5} + \frac{(3-3.5)^2}{3.5} + (1-1.5)^2 + \frac{(4-3.5)^2}{3.5}$$
$$= 0.16667 + 0.07143 + 0.16667 + 0.07143 = 0.4762$$

From table $\chi^2_{0.05\ (1)}$ =5.024. Since 0.4762 < $\chi^2_{0.05\ (1)}$ we do not reject the null hypothesis, i.e., there is no association.

18.9.5 Test for Single Mean

The inherent population is assumed to be normal $N(\mu, \sigma^2)$. The null hypothesis $H_0 : \mu = \mu_0$. Alternative hypothesis $H_1 : \mu > \mu_0$. Assume a sample $x_1, x_2, \ldots, x_n$ of size n. Test statistic is defined as $z = \frac{\overline{x} - \mu_0}{\sigma/\sqrt{n}}$ if σ^2 is known and $t = \frac{\overline{x} - \mu_0}{s/\sqrt{n}}$ if $\sigma^2$is unknown, where $s = \sqrt{\frac{\sum (x_i - \overline{x})^2}{n-1}}$. Reject H_0 if $z \geq z_{1-\alpha}$ if σ^2 is known, and $t \geq t_{1-\alpha}$,

$_{(n-1)}$ if σ^2 is unknown, where $z_{1-\alpha}$is the $(1-\alpha)^{th}$ percentile of z distribution, $t_{1-\alpha,(n-1)}$ is the $1-\alpha$ percentile of t distribution with $(n-1)$ d.f., which can be obtained from a t-table. If the alternative is two-sided, e.g., $H_1 : \mu \neq \mu_0$, reject H_0 if $|z| \geq z_{1-\alpha/2}$ if $\sigma^2$is known, and $|t| \geq t_{1-\alpha/2,(n-1)}$ if $\sigma^2$is unknown.

Example Let us consider the administration of a drug blunted the marked increase in urinary sodium excretion induced by another drug. The observed values for Na+ excretion (μmol) are: 25, 30, 28, 33, 26, 29, 36, 34, 33, 28, 22, 27. We like to test the hypothesis $H_0 : \mu = 30$ against the alternative $H_1 : \mu > 30$ at test level $\alpha = 0.05$. The population variance σ^2 is assumed to be unknown. We have $n = 12, \bar{x} = 29.25$, $\sum (x_i - \bar{x})^2 = 186.25, s = \sqrt{\frac{186.25}{12-1}} = 4.11, t = \frac{29.25-30.00}{4.11/\sqrt{12}} = -0.6314$. From the table $t_{0.95,(11)} = 2.201$. Since $t < t_{0.95,(11)}$, we accept the null hypothesis, i.e., $\mu = 30$.

For the above test normality of the population is an essential condition. Therefore, before applying this test it is necessary to check the normality assumption of the population. Several descriptive methods can be used to check for normality. A brief description of some of these methods is as follows:

1. Based on the data, construct a histogram and check the form of the graph. If the data are approximately normal, the shape of the histogram will be similar to the normal curve.
2. Calculate intervals $\bar{x} \mp s$, $\bar{x} \mp 2s$, and $\bar{x} \mp 3s$, and check the percentage of observations falling in each. If the data came from a normal population, then the percentages will be approximately 68%, 95%, and 99%, respectively.
3. Calculate the interquartile range IQR$=Q_2 - Q_1$, and standard deviation s. The ratio IQR/s should be close to 1.3 for the normal population.
4. Construct a normal probability plot. A straight or approximately straight plot indicates normality.

For constructing a normal plot, order the observations x from the smallest to the largest. Denote the ordered values of x by $x_{(m)}$. Calculate percentage $p_{(m)}$ of observations bigger than $x_{(m)}$, $m= 1, 2, \ldots$,n. Find z-score ($z_{(m)}$) such that $P(z > z_{(m)}) = p_m$. Plot $z_{(m)}$ against $x_{(m)}$.

Example An example of drawing a normal plot is as follows. Suppose we have 25 observation 20.5, 45.2, 44.1, 32.3, 50.0, 43.4, 28.2, 22.2, 59.5, 44.4, 34.8, 16.9, 56.3, 33.7, 54.3, 55.6, 36.1, 57.6, 22.9, 65.6, 54.4, 56.7, 28.5, 37.7, and 57.8. We want to plot a normal probability plot. We have the following calculations:

Ordered observations X *(m)*	Number of observations greater than *X(m)* *N(m)*	Percentage of observations greater than *X(m)* *P(m)*	Z-Score Such that *p(z>Z(m))=P(m)* *Z(m)*
16.90	24.00	100.0	.
20.50	23.00	95.83	−1.73
22.20	22.00	91.67	−1.38
22.90	21.00	87.50	−1.15
28.50	20.00	83.33	−0.97

(continued)

Ordered observations *X(m)*	Number of observations greater than *X(m)* *N(m)*	Percentage of observations greater than *X(m)* *P(m)*	Z-Score Such that *p(z>Z(m))=P(m)* *Z(m)*
28.80	19.00	79.17	−0.81
32.30	18.00	75.00	−0.67
33.70	17.00	70.83	−0.55
34.80	16.00	66.67	−0.43
36.10	15.00	62.50	−0.32
37.70	14.00	58.33	−0.21
43.40	13.00	54.17	−0.10
44.10	12.00	50.00	0.00
44.40	11.00	45.83	0.10
45.20	10.00	41.67	0.21
50.00	9.00	37.50	0.32
54.30	8.00	33.33	0.43
54.40	7.00	29.17	0.55
55.60	6.00	25.00	0.67
56.30	5.00	20.83	0.81
56.70	4.00	16.67	0.97
57.60	3.00	12.50	1.15
57.80	2.00	8.33	1.38
59.50	1.00	4.17	1.73
65.60	0.00	0.00	.

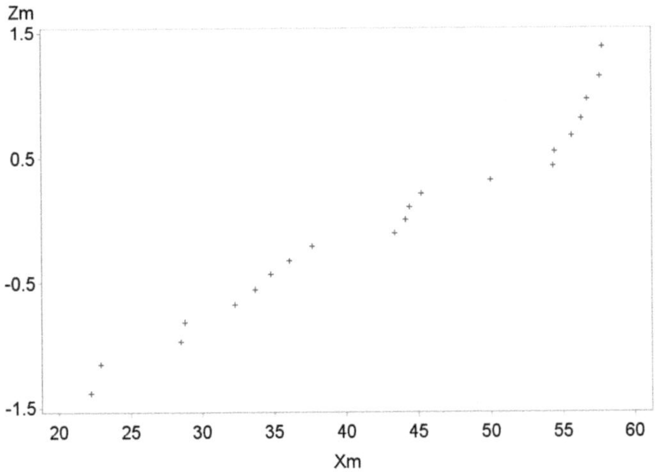

Fig. 18.4 Plot of $Z_{(m)}$ against $X_{(m)}$

The plot does not look like a straight line. Therefore, the observations do not look like coming from a normal population (Fig. 18.4).

5. There are formal statistical tests available for test for normality such as the Shapiro-Wilk test, the Kolmogorov-Smirnov test, the Anderson-Darling test, and the Cramer-von Mises test. SAS proc univariate with appropriate options can perform all these tests and draw the normal probability plot.

18.9.6 Test for Equality of Two Means from Two Population

Let $x_{11}, x_{12}, \ldots, x_{1n_1}$ be a sample of size n_1 from $N(\mu_1, \sigma^2)$ and $x_{21}, x_{22}, \ldots, x_{2n_2}$ be a sample of size n_2 from $N(\mu_2, \sigma^2)$. We like to test the null hypothesis $H_0 : \mu_1 = \mu_2$ against alternative hypothesis $H_1 : \mu_1 > \mu_2$. Test statistic is $z = \frac{\bar{x}_1 - \bar{x}_2}{\sigma\sqrt{\frac{1}{n_1} + \frac{1}{n_2}}}$ if the common variance σ^2 is known, and $t = \frac{\bar{x}_1 - \bar{x}_2}{s_p\sqrt{\frac{1}{n_1} + \frac{1}{n_2}}}$ if σ^2 is unknown, where $s_p = \sqrt{\frac{s_1^2(n_1-1) + s_2^2(n_2-1)}{n_1+n_2-2}}$. Reject H_0 if $z \geq z_{1-\alpha}$ or $t \geq t_{1-\alpha,(n_1+n_2-2)}$, where $z_{1-\alpha}$ is the $(1-\alpha)^{\text{th}}$ percentile of z distribution and $t_{1-\alpha,(n_1+n_2-2)}$ is the $(1-\alpha)^{\text{th}}$ percentile of t distribution with $n_1 + n_2 - 2$ d.f. If the alternative is two-sided, e.g., $H_1 : \mu_1 \neq \mu_2$, reject H_0 if $|z| \geq z_{1-\alpha/2}$ if the common variance σ^2 is known, and $|t| \geq t_{1-\frac{\alpha}{2},(n_1+n_2-2)}$ if σ^2 is unknown.

Example Suppose we have 40 observations each from $N(\mu_1, \sigma^2)$ and $N(\mu_2, \sigma^2)$. We like to test $H_0 : \mu_1 = \mu_2$ and $H_1 : \mu_1 \neq \mu_2$. Suppose the common population variance σ^2 is unknown. Let $n_1 = 40, n_2 = 40,\ \bar{x}_1 = 25.56, \bar{x}_2 = 33.84, s_1 = 4.63, and\ s_2 = 6.31$. We have, $s_p = \sqrt{\frac{4.63^2(40-1)+6.31^2(40-1)}{40+40-2}} = 5.53$, $t = \frac{25.56-33,84}{5.53\sqrt{1/40+1/40}}$ =-6.7317. From table $t_{0.975(78)}$= 1.99. Since $|-6.7317|>t_{0.975(78)}$ we reject $H_0 : \mu_1 = \mu_2$.

In the above discussion, the variances of the two populations were assumed to be equal. Therefore, before applying this test it is necessary to test the equality of the population variances. A test for the equality of variances is given in Sect. 18.9.9. If the test reveals that the population variances are not equal, the problem is popularly known as the Behrens-Fisher problem. In this case the test statistic is defined as $t = \frac{\bar{x}_1 - \bar{x}_2}{\sqrt{\frac{s_1^2}{n_1} + \frac{s_2^2}{n_2}}}$ and the degrees of freedom is estimated as $f = \frac{\left(\frac{s_1^2}{n_1} + \frac{s_2^2}{n_2}\right)^2}{\frac{s_1^4}{n_1^2(n_1-1)} + \frac{s_2^4}{n_2^2(n_2-1)}}$.

18.9.7 Test for Equality of Several Means

Suppose m different dose levels of a drug are tested on m groups of similar patients. The drug is expected to be effective and change the disease level y. The data may be written as,

	Level 1	Level 2	...	Level i	...	Level m
	y_{11}	y_{21}	...	y_{i1}	...	y_{m1}
	y_{12}	y_{22}	...	y_{i2}	...	y_{m2}
	.		...		...	.
	y_{1j}	y_{2j}	...	y_{ij}	...	y_{mj}
	.	.	...	.	...	.
	y_{1n_1}	y_{2n_2}	...	y_{in_i}	...	y_{mn_m}
Sample size	n_1	n_2	...	n_i	...	n_m
Sample mean	$\overline{y}_1$	$\overline{y}_2$	...	$\overline{y}_i$	...	$\overline{y}_m$

Let $\mu_1, \mu_2, \ldots, \mu_i, \ldots, \mu_m$ be the population mean disease level for dose group 1, 2, ..., m, with a common variance of σ^2.

Let $n = n_1 + n_2 + \ldots + n_m$; $\overline{y} = \frac{\sum_i \sum_j y_{ij}}{n}$; $\overline{y}_{i.} = \frac{\sum_j y_{ij}}{n_i}$, $i = 1, 2, \ldots, m$. Define the following sum of squares (SS), $SS_{Tot} = \sum_i \sum_j (y_{ij} - \overline{y})^2$; $SS_{Trt} = \sum_i n_i (\overline{y}_{i.} - \overline{y})^2$; and $SS_{Error} = \sum_i \sum_j (y_{ij} - \overline{y}_{i.})^2$. It can be shown that $SS_{Tot} = SS_{Trt} + SS_{Error}$. The terms SS_{Tot} is known as the total sum of squares which represents the total variation among all n patients, SS_{Trt} is known as treatment sum of squares which represents the variation between treatment groups, and SS_{Error} is known as error sum of squares which represents the unexplained part of the total variation. The expected value $E\left(\frac{SS_{Tot}}{n-1}\right) = \sigma^2$; $E\left(\frac{SS_{Trt}}{m-1}\right) = \sigma^2 + \left[\sum_i n_i \frac{(\mu_i - \mu)^2}{m-1}\right]$; and $E\left(\frac{SS_{Error}}{n-m}\right) = \sigma^2$. It is clear to see that if all $\mu_1, \mu_2, \ldots, \mu_i, \ldots, \mu_p$ are equal, then $E\left(\frac{SS_{Trt}}{m-1}\right) = \sigma^2$. In case all $\mu_1, \mu_2, \ldots, \mu_i, \ldots, \mu_p$ are equal, then $E\left(\frac{SS_{Error}}{n-m}\right) = E\left(\frac{SS_{Trt}}{m-1}\right)$. Therefore, a ratio of these two observed quantities $F = \frac{SS_{Trt}/(m-1)}{SS_{Error}/(n-m)}$ can be used to test the equality of all $\mu_1, \mu_2, \ldots, \mu_i, \ldots, \mu_p$. The value F (or more appropriately $F_{m-1, n-m}$) is known as the F statistics with $m - 1$ and $n - m$ degrees of freedom. The method can be summarized in the following analysis of variance (ANOVA) table.

Analysis of variance table

Source	Sum of squares	Degrees of freedom	Mean sum of squares	F
Treatment	SS_{Trt}	$m - 1$	$MS_{\text{Trt}} = \frac{SS_{\text{Trt}}}{m-1}$	$\frac{MS_{\text{Trt}}}{MS_{\text{Error}}}$
Error	SS_{Error}	$n - m$	$MS_{\text{Error}} = \frac{SS_{\text{Error}}}{n-m}$	
Total	SS_{Tot}	$n - 1$		

Therefore, to test the equality of several means reject H_0 if $F \geq F_{1-\alpha, (m-1, n-m)}$, where $F_{1-\alpha, (m-1, n-m)}$ is the $1 - \alpha$ percentile of F distribution with $(m - 1, n - m)$, degrees of freedom, which can be obtained from an F-table. If $T = \sum_i \sum_j y_{ij}$ and $T_i = \sum_j y_{ij}$, then it can be shown that $SS_{Tot} = \sum_i \sum_j y_{ij}^2 - \frac{T^2}{n}$, $SS_{Trt} = \sum_i \frac{T_i^2}{n_i} - \frac{T^2}{n}$, and $SS_{Error} = SS_{tot} - SS_{trt} = \sum_i \sum_j y_{ij}^2 - \sum_i \frac{T_i^2}{n_i}$. For practical calculations, these formulas are easy to use. For further details of this topic, the interested readers are referred to the book by Hogg and Tanis [1].

Example In a pilot study, a single effective oral dose of an antiepileptic drug was tested in 10 rats. Control animals received normal saline. Convulsions were induced by subcutaneous injection of pentylenetetrazol 1 hr after administration of test substance. All animals were observed for clonic convulsions. The duration of clonic convulsions was measured in seconds and expressed as milliseconds. We like to test the null hypothesis that the population mean duration of clonic convulsions for two groups are same, i.e., $H_0 : \mu_1 = \mu_2$ *and* $H_1 : \mu_1 \neq \mu_2$. This test can be done using the two-sided t-test. However, here we like to perform an F test using the ANOVA technique.

Control	Treated	$(\text{Control})^2$	$(\text{Treated})^2$
600	555	360,000	308,025
620	540	384,400	291,600
650	530	422,500	280,900
610	520	372,100	270,400
655	570	429,025	324,900
700	560	490,000	313,600
630	538	396,900	289,444
645	530	416,025	280,900
605	515	366,025	265,225
680	525	462,400	275,625
Total 6395	5383	4,099,375	2,900,619

$n = 10 + 10 = 20$, $\sum_j y_{1j}$=6395, $\sum_j y_{2j}$=5383, T=6395+5383=11778, $\Sigma_{1j}y^2_{1j}$=4099375, $\Sigma_{2j}y^2_{2j}$=2900619, $\sum_i\sum_j y_{ij}^2$ =4099375+2900619=6999994. Therefore, $SS_{Tot} = 6999994 - \frac{11778^2}{20} = 63929.8$, $SS_{trt} = \left(\frac{6395^2}{10}\right) + \left(\frac{5383^2}{10}\right) - \frac{11778^2}{20} =$ 51207.2, and, by subtraction, SS_{Error} =3929.8 − 51207.2 = 12722.6. The ANOVA table is

Source	Sum of squares	Degrees of freedom	Sum of squares	Mean Sum of squares	*F*
Treatment	SS_{trt}	*1*	51207.20	51207.20	72.45
Error	SS_{Error}	*18*	12722.60	706.81	
Total	SS_{Tot}	*19*	63929.80		

From table $F_{0.95,\ (1, 18)} = 4.41$. Since $72.45 > F_{0.95(1, 18)}$, we reject the null hypothesis.

18.9.8 Test for Single Variance

Let $x_1, x_2, \ldots, x_n$ be a sample of size n from $N(\mu, \sigma^2)$. We like to test the null hypothesis $H_0 : \sigma^2 = \sigma_0^2$ against alternative hypothesis $H_1 : \sigma^2 > \sigma_0^2$. Test statistic is $\chi^2 = \frac{s^2}{\sigma_0^2}$, where $s^2 = \frac{\sum (x_i - \bar{x})^2}{n-1}$. Reject H_0 if $\chi^2 \geq \chi^2_{1-\alpha,(n-1)}$, where $\chi^2_{1-\alpha,(n-1)}$ is the

$1 - \alpha$ percentile of χ^2 distribution with $n - 1$ degrees of freedom. The above test may also be two-tailed if the alternative hypothesis is $H_1 : \sigma^2 \neq \sigma_0^2$.

Example Suppose from a single sample of size $n = 20$, the estimated variance turns out to be $s^2 = 23.2$. We are interested to test the hypothesis that the population variance $\sigma^2 = 20$ against the alternative hypothesis that it is greater than 20. We have $n = 20$, H_0: $\sigma^2 = 20$, and $H_1 : \sigma^2 > 20$. From these values $\chi^2 = \frac{23.2}{20} = 1.16$. Also, from table $\chi^2_{0.95,(19)}$ =30.14. Since $1.16 < \chi^2_{0.95,(19)}$, we accept the null hypothesis $\sigma^2 = 25$.

18.9.9 Test for Equality of Two Variances

Let $x_{11}, x_{12}, \ldots, x_{1n_1}$ and $x_{21}, x_{22}, \ldots, x_{2n_2}$ be two samples from $N\left(\mu_1, \sigma_1^2\right)$ and $N\left(\mu_2, \sigma_2^2\right)$. We like to test the null hypothesis $H_0 : \sigma_1^2 = \sigma_2^2$ against alternative hypothesis $H_1 : \sigma_1^2 > \sigma_2^2$. Test statistic is $F = \frac{(n_2-1)s_2^2}{(n_1-1)s_1^2}$, where $s_1^2 = \frac{\sum_i (x_{1i}-\bar{x}_1)^2}{n_1-1}$ and $s_2^2 = \frac{\sum (x_{2i}-\bar{x}_2)^2}{n_2-1}$. Reject H_0 if $F \geq F_{1-\alpha,(n_2-1,n_1-1)}$, where $F_{1-\alpha,(n_2-1,n_1-1)}$ is the $1 - \alpha$ percentile of F distribution with $(n_1 - 1, n_2 - 1)$ degrees of freedom.

Example Consider two samples of sizes $n_1 = 20$ and $n_2 = 25$ from two populations give estimates of their respective population variance as $s_1^2 = 20.4$ and $s_2^2 = 23.1$. We like to test the null hypothesis $H_0 : \sigma_1^2 = \sigma_2^2$ against $H_1 : \sigma_1^2 > \sigma_2^2$. From these values $F = \frac{(25-1)23.1}{(20-1)20.4} = 1.43$. Also, from table $F_{0.95,\ (24,\,19)} = 2.025$. Since $1.43 < F_{0.95,\ (14,\,9)}$, we accept the null hypothesis.

18.9.10 Test for the Equality of Several Variances

The following test is known as the Bartlett test. The null hypothesis is H_0: all k population variances are equal against the alternative that at least two are different.

If n_i is the sample size from ith population, then the test statistic is defined as,

$$\chi^2 = \frac{(N-k)\ln\left(S_p^2\right) - \sum_i (n_i - 1)\ln\left(S_i^2\right)}{1 + \frac{1}{3(k-1)}\left(\sum_I \left(\frac{1}{n_i-1}\right) - \frac{1}{N-k}\right)}$$

where $N = \sum_i n_i$ and $S_p^2 = \frac{1}{N-k}\sum_i (n_i - 1)S_i^2$ is the pooled estimate for the variance.

The test statistic χ^2 has approximately a χ^2_{k-1} distribution. Bartlett's test is sensitive to departures from normality. In case the parent distribution is not normal, the Levine test and Brown-Forsythe test are two alternatives to the Bartlett test (Snedecor and Cochran 1989 [2]). These two tests are less sensitive to departures from normality.

Example The diuretic effect of furosemide was evaluated at five doses (1, 3, 10, 20, 30), mg/kg, oral for 7 days, in conscious restrained dogs (n= 6/dose). Urine output (ml) was collected 0 – 3 h on day 7. A dose-dependent diuresis with a maximal effect at 30 mg/kg was noted. Mean void volume (ml) at each dose is given in the following table.

Dose	1	3	10	20	30
n	6	6	6	6	6
Mean	200	450	900	1000	1100
Variance	5.6	8.6	6.6	6.9	8.7

H0: all variances are equal and H_1: at least one of the variances is not equal. The calculated $\chi^2 = 0.38394$. The tabulated $\chi^2_{0.05(4)} = 9.49$. Since $0.38394 < \chi^2_{0.05(4)}$, we accept the null hypothesis.

18.10 Nonparametric Method

In the parametric hypothesis testing, we tested hypothetical values of parameters of the distribution of a population. However, in practice it is not always possible to know the inherent distribution. In such cases, tests can be done on the general characteristics of the population. The related tests are known as nonparametric tests.

In the following sections, we will present some common nonparametric tests. For a detailed discussion of these tests the reader is referred to the book by Hollander and Wolfe [3].

18.10.1 Single Proportion Problem: Binomial Test

Assume that a trial has a constant probability of p of success. We like to test the hypothesis $H_0 : p = p_0$ against $H_1 : p > p_0$ (one-sided test). The test statistic is B=total number of success in n repetition of the trial. Reject H_0 if $B \geq b_{1-\alpha}$, where $b_{1-\alpha}$ is the upper $(1 - \alpha)\%$ point of the binomial distribution $B(n, p_0)$ (see book of Hollander and Wolfe). The two-sided tests can also be similarly carried out.

Example Thrombocytopenia was noted in four of 20 patients receiving vasopressin for the treatment of shock. We like to test if the true proportion of thrombocytopenia cases is more than 0.25. We have $H_0 : p = 0.25$, $H_0 : p > 0.25$. From binomial table with $p_0 = 0.25$ and $n = 20$ the critical value $b_{0.95} = 6$. Since $4 < b_{0.95}$ accept the null hypothesis, i.e., accept that the true proportion of the thromocytopenia case is 0.25.

18.10.2 One Sample Location Problem: Signed Rank Test or Wilcoxon Test

Let $y_1, y_2, \ldots y_i, \ldots, y_n$ be n observations. Assume model $y_i = \mu + e_i$, where μ is the centrality parameter and e_i are mutually independent. Assume that e_i comes from a continuous population. We like to test $H_0 : \mu = \mu_0$ against $H_1 : \mu > \mu_0$. Test statistic is defined as follows: order the n absolute values of y_i's from the lowest to highest and let R_i denote the rank of the absolute value $|y_i|$ of ith observation. Use average rank in case of ties. Also, let ψ_i=1 if $y_i > \mu_0$ and $\psi_i = 0$ otherwise. The signed rank test is defined as $T = \Sigma R_i\psi_i$. Reject H_0 if $T \geq t_{1-\alpha,\, n}$, where $t_{1-\alpha,\, n}$ is the upper $\alpha\%$ point of the distribution of T and can be obtained from the tabulated values given in standard nonparametric books. The two-sided tests can also be similarly carried out.

Example Suppose we have the following 15 observations from an experiment (arbitrary units in an increase in the force of contraction). We like to test the hypothesis $H_0 : \mu = 0$ against the alternative $H_1 : \mu > 0$ with test level $\alpha = 0.05$. We have the following:

Ordered absolute values of observations	Rank R	Original observations	ψ_i	$R_i\psi_i$
9	1.0	−9	0	0.0
10	2.0	−10	0	0.0
11	3.0	11	1	3.0
12	4.5	12	1	4.5
12	4.5	−12	0	0.0
13	6.0	−13	0	0.0
14	8.0	14	1	8.0
14	8.0	14	1	8.0
14	8.0	14	1	8.0
16	10.5	−16	0	0.0
16	10.5	16	1	10.5
17	13.0	17	1	13.0
17	13.0	17	1	13.0
17	13.0	17	1	13.0
18	15.0	18	1	15.0
				Total = 96

From the table the critical value $t_{0.95,\ 15} = 44$. Since $96 > t_{0.95(15)}$ we reject the null hypothesis.

A special case of one sample test is the case of paired observation, where observations (e.g., measuring blood pressure, blood glucose level) are made at pre- and post- treatment. Suppose x_{ib} and x_{ip} denote the before and after treatment for the ith animal. Let $y_i = x_{ip} - x_{ib}$. The above test now can be conducted for the null hypothesis $H_0 : \mu = 0$ against $H_1 : \mu > 0$.

Another test for the paired observation is called the sign test. In this test only the sign ψ_i is enumerated as +1 if $y_i > 0$ and 0 otherwise. The test statistic $B = \Sigma\,\psi_i$. Reject

H_0 if $B \geq b_{1-\alpha}$, $(n, ½)$, where $b_{1-\alpha}$, $(n, ½)$ is the critical value of binomial distribution with sample size n and parameters $p = ½$.

18.10.3 Two Sample Location Problem: Rank Sum Test (or Wilcoxon Rank Sum Test)

Let $x_1, x_2, \ldots x_i, \ldots, x_m$ be a sample from the first population and $y_1, y_2, \ldots y_i, \ldots, y_n$ from the second population. Therefore, we have a total of $N = m + n$ observations. Assume a model $x_i = \mu + e_i$, $i = 1, \ldots, m$, and $y_i = \mu + \Delta + e_m + j, j = 1, \ldots, n$, where e is an error term and Δ is the parameter of interest, known as the shift in location. The error is assumed to be mutually independent and come from a continuous population. Let $H_0 : \Delta = 0$ against $H_1 : \Delta > 0$. A test statistic is defined as follows: order the N observations from least to the greatest and let R_j denote the rank of the y_j. The rank sum test is defined as $W = \Sigma R_j$, level of significance α, rejection rule: reject H_0 if $W \geq w_{1-\alpha, (m,n)}$, where $w_{1-\alpha, (m,n)}$ is the critical value and is the upper $(1-\alpha)\%$ point of the distribution of W and can be obtained from the tabulated values given in standard nonparametric books. The two-sided tests can also be similarly carried out.

Example The following data show the tritiated water diffusion across human chorioamnion.

At term X^*	12–26 weeks gestational age Y^*
$n = 10$	$m = 5$
0.80, 0.83, 1.89, 1.04, 1.45, 1.38, 1.91, 1.64, 0.73, 1.46	1.15, 0.88, 0.99, 0.74, 1.21

*Pd (10–4 cm/s)

We like to test that H_0: the two centralities are the same, i.e., $H_0 : \Delta = 0$ against $H_1 : \Delta > 0$. After ranking the pooled sample values from x and y, we have the values of y and the related ranks as follows:

Y	Rank R
1.15	8
0.88	5
0.99	6
0.74	2
1.21	9
	Total $W = 30$

From table $w_{0.95, (10,5)} = 54$. Since $30 < w_{0.95, (10,5)}$ we accept the null hypothesis.

For two sample location problem, Mann and Whitney proposed a test based on the signs only. This test may be described as follows: Let $\varphi(a, b) = 1$ if $a < b$ and 0 otherwise. The test statistic is $U = \sum_{i=1}^{m}\sum_{j=1}^{n}\varphi(x_i, y_j)$. Reject H_0 if $U \geq u(\alpha, m, n)$, where $u(\alpha, m, n)$ is the critical value and is the upper $\alpha\%$point of the distribution of U

and can be obtained from the tabulated values given in standard nonparametric books. The two-sided tests can also be similarly carried out.

18.10.4 Two Sample Dispersion Problem: Rank Test (Ansari-Bradley Test)

Let $x_1, x_2, \ldots x_i, \ldots, x_m$ be from the first population and $y_1, y_2, \ldots y_i, \ldots, y_n$ from the second population. Therefore, we have a total of $N = m + n$ observations. Assume models $x_i = \mu + \sigma_1 e_i;\ i = 1, \ldots, m$, and $y_i = \mu + \sigma_2 e_m + j, j = 1, \ldots, n$, where e's are error terms and μ is the common mean of X and Y. The error e's are mutually independent and come from a continuous population.

Let $\gamma^2 = \frac{\sigma_2^2}{\sigma_1^2}$. We are interested in testing hypothesis $H_0 : \gamma^2 = 1$ against $H_1 : \gamma^2 > 1$. Test statistic is defined as follows. Order the N observations from the least to the greatest. Assign rank 1 to the smallest and the largest observations in the combined sample, assign rank 2 to the second smallest and second largest, and continue in this manner. Let R_i denote the rank of the x_i in this ranking. The Ansari-Bradley test is defined as $W = \Sigma\ R_i$, level of significance α, reject H_0 *if* $W \geq w_{1\ -\ \alpha,\ m,\ n}$, where $w_{1\ -\ \alpha,\ m,\ n}$ is the critical value and is the upper α% point of the distribution of W and can be obtained from the tabulated values given in standard nonparametric books. The two-sided tests can also be similarly carried out.

Example The following data represents 12 measurements of plasma concentrations (ng/ml) of test substance by two different methods. We are interested in testing $H_0 : \gamma^2 = 1$ against $H_1 : \gamma^2 > 1$ at test level α=0.05

Ramsay method[a]	Jung-Parekh method[a]
111, 107, 100, 99, 102	107, 108, 106, 98, 105, 103, 110

[a]μg/100 ml

We have $m = 5$ and $n = 7$. Also, from the ordered pooled sample we have

i	Sample	Serum value	Rank R
1	2	98	1
2	1	99	2
3	1	100	3
4	1	102	4
5	2	103	5
6	2	105	6
7	2	106	6
8	1	107	4.5
9	2	107	4.5
10	2	108	3
11	2	110	2
12	1	111	1

Summing over the ranks of the first sample $W = 14.5$. From table $w_{0.95,5,\ 7} = 23$. Since $14.5 < w_{0.95,5,\ 7}$ we accept the null hypothesis.

In case the locations are not the same, a test suggested by Moses can be used. This test will not be discussed in this article. The interested readers are referred to the book by Hollander and Wolfe for details of this test.

18.10.5 Test of Equality of Several Locations: Rank Test (Kruskal-Wallis Test)

Consider the following data,

Level 1	Level 2	...	Level j	...	Level k
x_{11}	x_{12}	...	x_{1j}	...	x_{1k}
x_{21}	x_{22}	...	x_{2j}	...	x_{2k}
.	.	...	.	...	.
x_{i1}	x_{i2}	...	x_{ij}	...	x_{ik}
.	.	...	.	...	.
	$x_{n_2 2}$	...	$x_{n_i j}$	...	$x_{n_m k}$
$x_{n_1 1}$					
Sample size n_1	n_2	...	n_j	...	n_k

Let $N = \Sigma n_j$. Assume model $x_{ij} = \mu + \tau_j + e_{ij}\ i = 1, \ldots, n_j,\ and\ j = 1, \ldots, k$, where μ is the overall mean, τ_j is the effect of treatment j with $\Sigma \tau_j = 0$. The errors are assumed as mutually independent and come from a continuous population.

The null hypothesis $H_0 : \tau_1 = \tau_2 = \ldots = \tau_{jk}$ against H_1: at least one pair of τ_i and τ_j are different. Order the N observations from least to the greatest. Let r_{ij} denote the rank of x_{ij} in the joint ranking. For ties use the average rank. Define $R_j = \sum_i r_{ij}$, $R_{.j} = \frac{R_j}{n_j}$, $R_{..} = \frac{N+1}{2}$. The Kruskal-Wallis test is defined as $H = \frac{12}{N(N+1)} \sum_j n_j \left(R_{.j} - R_{..}\right)^2 = \left(\frac{12}{N(N+1)} \sum_j \frac{R_j^2}{n_j}\right) - 3(N+1)$.

Reject H_0 if $H \geq h_{(1-\alpha,k,(n_1,\ldots,n_k))}$, where $h_{(1-\alpha,k,(n_1,\ldots,n_k))}$ is the upper $\alpha\%$point of the distribution of H and can be obtained from the tabulated values given in standard nonparametric books.

The Kruskal-Wallis test can be motivated by considering the usual analysis of variance F statistic calculated using the ranks, rather than the original observations. The F statistic can be written as $F = \frac{cSSB}{SST-SSB}$, where c is a constant depending only on sample sizes, SST is total sum of squares, and SSB is the between sum of squares. The statistic SSB reduces to $\sum_j n_j (R_{.j} - R_{..})^2$when applied to the ranks rather than the original observations, and SST becomes a fixed constant when calculated on ranks. Therefore, when F is calculated for ranks, F is an increasing function of H. It is evident from this discussion that a simple analysis of variance on ranks can also be performed to test the above hypothesis, namely, $H_0 : \tau_1 = \tau_2 = \ldots = \tau_{jk}$.

In cases where randomization is performed restricting by blocking variables (e.g., age, sex, race), a test for the equality of location is given by Friedman. The test is known as Friedman rank sums test.

Example The objective of the study is to evaluate the heart protective effects of 4 drugs (Groups 2 to 5, $n = 6$/group) in a model of established diabetic nephropathy, the uninephrectomized diabetic obese db/db mouse. After 10 weeks of treatment, animals were killed to determine the severity of disease affecting the heart. Untreated mice (Group 1) were used as disease control. Scores (collagen staining score, 0 to 5 where 0 no pathology and 1 to 5 mild to marked/severe) were based on the reduction in collagen deposition in the cardiac interstitial and perivascular areas. Data are as follows:

Group 1 (control, no treatment): 4, 5, 5, 4, 5, 5; Group 2 (Drug A): 4, 4, 4, 3, 4, 4; Group 3 (Drug B): 2, 3, 1, 2, 3, 1; Group 4 (Drug C): 0, 1, 1, 2, 1, 1; and Group 5 (Drug D): 3, 2, 2, 3, 1, 2.

The Kruskal-Wallis statistic is $= \frac{12}{30(30+1)}\left\{\frac{160^2}{6} + \frac{132^2}{6} + \frac{67^2}{6} + \frac{32.5^2}{6} + \frac{73.5^2}{6}\right\} - 3(30+1)$ =23.07. From table $H_{0.95(5,\,(6,6,6,6,6))} = 9.49$. Since, 23.07 > $H_{0.95(5,\,(6,6,6,6,6))}$ we reject the null hypothesis. An ANOVA analysis of the ranks gives the following:

Analysis of variance table

Source	Sum of squares	Degrees of freedom	Sum of squares	Mean sum of squares	*F*	*P*-value
Sample	SS_{Sample}	4	*1787.75*	*446.94*	*30.10*	<0.001
Error	SS_{Error}	*25*	*371.25*	*14.85*		
Total	SS_{Total}	*29*	*2159.00*			

Since the $p-$value is <0.001, we reject the null hypothesis.

18.11 Design of Experiment

In scientific experiments, certain independent variables are manipulated to explore their effects on one or more dependent variables. In statistical terminologies, these dependent variables are known as the response variables and independent variables are known as factors or treatments. Each of the factors generally has several levels. For example, consider an experiment in which a drug is tested at a low dose and a high dose levels to see its effect on a disease intensity in animals. In this experiment, the disease intensity is the dependent variable (response) and the applied drug is the independent variable (factor) with two levels, namely low and high doses. The drug levels are assigned to the experimental units, which are the animals in this experiment. The resulting data are statistically analyzed to draw a conclusion on the objective of the study. However, the validity of the statistical analysis and the associated conclusion depends on the validity of the observed data which include recruitment of sufficient number of experimental units (sample size), appropriate randomization of the experimental units to treatments, blocking relevant factors, taking care of the confounding factors, selection of the order in which the experiment is to be conducted, determination of the mathematical model to explain the relation between the response variables and the factors, selection of analyses methods, the fulfillment of the required assumptions, etc. To acquire a valid sample, it is important that the experiment is designed scientifically and conducted carefully. In the following section, we will briefly discuss this topic for experimental design with different conditions and objectives. For detailed discussions of these topics, the interested readers are referred to the book by Hicks [4].

18.11.1 Single Factor Completely Random Design

Assume one response variable (y) related to one factor (x) with m levels. Experimental units are assigned completely randomly to treatments without any restriction, except that no experimental units are assigned to a group that is already full. The observations can be represented as,

Sample with m levels of the factor

Level 1	Level 2	...	Level i	...	Level m
y_{11}	y_{21}	...	y_{i1}	...	y_{m1}
y_{12}	y_{22}	...	y_{i2}	...	y_{m2}
.	.	...	.	...	.
y_{1j}	y_{2j}	...	y_{ij}	...	y_{mj}
.	.	...	.	...	.
y_{1n1}	y_{2n2}	...	y_{ini}	...	y_{mnm}
Sample size n_1	n_2	...	n_i	...	n_m
Sample mean $\overline{y}_1$	$\overline{y}_2$	...	$\overline{y}_i$	...	$\overline{y}_m$

Yij can be modelled as, $y_{ij} = \mu + \tau_i + \varepsilon_{ij}$, where y_{ij} is the jth observation for i^{th} dose level, μ is known as the common effect for the whole experiment, τ_i is the extra effect due to i^{th} level of the factor, and ε_{ij} is the unexplainable random error. The error $\varepsilon_{ij}$is assumed to have a normal distribution with mean zero and a common variance σ_e^2 for all treatment groups. It is also assumed that $\sum_i \tau_i = 0$. Analysis method: One-way analysis of variance is as follows:

ANOVA table for completely randomized design

Source	Sum of squares	Degrees of freedom	Mean sum of squares	F
Treatment	SS_{Trt}	$m - 1$	$MS_{Trt} = \frac{SS_{Trt}}{m-1}$	$F = \frac{MS_{Trt}}{MS_{Error}}$
Error	SS_{Error}	$n - m$	$MS_{Error} = \frac{SS_{Error}}{n-m}$	
Total	SS_{Tot}	$n - 1$		

where $n = n_1 + n_2 + \ldots + n_m$; $\bar{y} = \frac{\sum_i \sum_j y_{ij}}{n}$; $\bar{y}_{i.} = \frac{\sum_j y_{ij}}{n_i}$ $i = 1, 2, \ldots, m$, $SS_{Tot} = \sum_i \sum_j (y_{ij} - \bar{y})^2$; $SS_{Trt} = \sum_i n_i (\bar{y}_{i.} - \bar{y})^2$; and $SS_{Error} = \sum_i \sum_j (y_{ij} - \bar{y}_{i.})^2$.

The SAS proc ANOVA or proc GLM can perform this analysis.

Example In a clinical trial comparing weight loss in obese healthy subjects of 5 different age groups (n= 8 per group), the effect of test substance was assessed after 12 weeks of treatment. The difference (in kg) in body weight between initial weight and the weight after 12 weeks of treatment is given in the following table. We are interested in testing the equality of mean weight loss with test level $\alpha = 0.05$.

Weight loss data

Group									
A	B	C	D	E	A^2	B^2	C^2	D^2	E^2
6	5	7	8	6	36	25	49	64	36
5	7	6	6	8	25	49	36	36	64
7	8	6	8	7	49	64	36	64	49
6	9	5	9	8	36	81	25	81	64
5	6	8	9	6	25	36	64	81	36
7	7	7	7	6	49	49	49	49	36
8	8	9	6	5	64	64	81	36	25
6	9	7	8	6	36	81	49	64	36
Total 50	59	55	61	52	320	449	389	475	346

$n_A = 8, n_B = 8, n_C = 8, n_D = 8, n_E = 8, n = 40, T = 277, \sum_i \sum_j y_{ij}^2 = 1979,$ - $T_1 = 50, T_2 = 59, T_3 = 55,\ T_4 = 61,\ T5 = 52$. Therefore, $SS_{\text{Tot}} = 1979 - \frac{277^2}{40} = 60.78$, $SS_{\text{Regimen}} = (\frac{50^2}{8}) + (\frac{59^2}{8}) + (\frac{55^2}{8}) + (\frac{61^2}{8}) + (\frac{52^2}{8}) - \frac{277^2}{40} = 10.65$, and, by subtraction, $SS_{\text{Error}} = 60.78 - 10.65 = 50.13$. The ANOVA table is given below.

Analysis of variance table

Source	Sum of squares	Degrees of freedom	Sum of squares	Mean sum of squares	F
Treatment	SS_{Regimen}	*4*	*10.65*	*2.66*	*1.86*
Error	SS_{Error}	*35*	*50.13*	*1.43*	
Total	SS_{Tot}	*39*	60.78		

There are many conveniently usable software programs available to perform this analysis. SAS Proc ANOVA with treatment as class; model y = treatment produced the following ANOVA table.

Source	DF	Sum of squares	Mean square	*F* value	Pr > *F*
Regimen	4	10.65	2.66	1.86	0.1396
Error	35	50.13	1.43		
Total	39	60.76			

Since p-value of the test is 0.1396 > 0.05, we accept the null hypothesis, i.e., we conclude that there is no difference in mean weight loss among treatment groups.

In the following section, we will discuss a few more commonly used designs. However, due to the limitations of the space of this article, most of the mathematical details will be skipped.

18.11.2 Single Factor Randomized Block Design

Suppose a drug is applied to a group of animals at two levels. It is possible that the drug may work differently on young animals (<1/2 years old), middle-aged animals (1/2≤ years <1 year) and old animals (≥ 1 years old). Therefore, for meaningful comparisons, it is desirable that animals of all ages are assigned to each treatment group. A completely random assignment may end up assigning all young animals to one treatment group and all old animals to another treatment group making the comparison less meaningful. The randomized block design puts a restriction on the randomization so that animals from all age groups are included in each dose group. Animals are randomized to dose groups in each age group separately. The restricting variable (age) is known as a blocking variable.

Operationally animals are randomly assigned to treatments in each block (age group) separately. For the experiment with p blocks, animals are assessed to treatment groups separately so that each group gets animals balanced by age.

Sample with m levels of treatment and p blocks

	Trt. Level 1	.	Trt. Level i	.	Trt. m
Block 1	$y_{111}, \cdots y_{11\text{r}11}$	.	$y_{i11}, \cdots y_{i1\text{r}i1}$	.	$y_{m11}, \cdots y_{m1\text{rm}1}$
.	.	.	.	.	.
Block j	$y_{1j1}, \cdots y_{1jr1\text{j}}$	.	$y_{ij1}, \cdots y_{ijrij}$	.	$y_{mj1}, \cdots y_{11rmj}$
.	.	.		.	.
Block p	$y_{1p1}, \cdots y_{1pr1p}$	.	$y_{ip1}, \cdots y_{iprip}$	.	$y_{mp1}, \cdots y_{mprmp}$

The analysis model is $y_{ijk} = \mu + \beta_j + \tau_i + \varepsilon_{ijk}$, where y_{ijk} is the kth observation for i^{th} dose level in j^{th} block, μ is known as the common effect for the whole experiment, $\tau_i$is the effect due to treatment level i, β_j is the effect due to block j, and ε_{ijk} is a random error. The error $\varepsilon_{ijk}$is assumed to follow a normal distribution with mean zero and a common variance σ_e^2 for all treatment groups. It is also assumed that $\sum_j \tau_j = 0$ and $\sum_i \tau_i = 0$. The analysis of variance table is as follows:

ANOVA table for one factor with one blocking variable

Source	Sum of squares	Degrees of freedom	Mean sum of squares	F
Treatment	SS_{Trt}	$m - 1$	$MS_{Trt} = \frac{SS_{Trt}}{m-1}$	$F_1 = \frac{MS_{Trt}}{MS_{Error}}$
Block	SS_{Block}	$p - 1$	$MS_{Block} = \frac{SS_{Blick}}{m-1}$	$F_2 = \frac{MS_{Block}}{MS_{Error}}$
Error	SS_{Error}	$n - m - p$	$MS_{Error} = \frac{SS_{Error}}{n-m-p}$	
Total	SS_{Tot}	$n - 1$		

In SAS “PROC ANOVA” or “PROC GLM” command can perform this analysis.

Example In an experiment, a group of 20 mice were given 4 different diets. At the end of 3 months their body weights were collected. Since the body weight gain was suspected to be affected by the litter, the randomization was blocked by litter, i.e., diets were randomly assigned within a litter. The following table shows the data. We want to test the null hypothesis of equality of body weight means of four groups at significance level $\alpha = 0.05$.

Litter	Diet			
	A	B	C	D
1	22	20	18	22
2	18	20	20	23
3	19	24	19	25
4	19	18	16	24
5	18	16	18	24

A complete analysis can be done using the formulas given above. The ANOVA table given by SAS proc ANOVA, with model bw = litter diet, is presented below.

Analysis of Variance

Source	DF	Sum of squares	Mean square	F value	Pr > F
Litter	4	19.300	4.825	1.42	0.2854
Diet	3	84.550	28.183	8.31	0.0029
Error	12	40.700	3.392		
Total	19	144.550			

The results show statistically significant difference in bodyweight gain among diets (p-value $0.0029 < 0.05$). Results do not show statistically significant difference among litters (p-value $0.2854 > 0.05$).

18.11.3 *Single Factor Randomized Block Design for more than One Blocking Variable (Latin Square Design)*

Experiments may be designed for a single factor blocked by more than one blocking factor. In this section, we will discuss a special design with two blocking factors, where the number of treatment groups and levels of each blocking factor are equal. This type of experiment can be conveniently carried out using a special design, called Latin Square design. Such design with 5 levels for each blocking factor and 5 levels of treatment may be expressed as follows. The letters A, B, C, D, and E represent the levels of the treatment. There may be more than one observation for each combination of the factors. The main feature of this allocation is that each row and each column contain each treatment only once.

	Blocking factor 1					
		Block 1	Block 2	Block 3	Block 4	Block 5
Blocking factor 2	Block 1	A	B	C	D	E
	Block 2	B	C	D	E	A
	Block 3	C	D	E	A	B
	Block 4	D	E	A	B	C
	Block 5	E	A	B	C	D

Each animal is randomly assigned to one of these cells until the cell is full. The analysis of variance table is as follows.

ANOVA table for one factor with two three blocking variable				
Source	Sum of squares	Degrees of freedom	Mean sum of squares	F
Treatment	SS_{Trt}	$m-1$	$MS_{Trt}=\frac{SS_{Trt}}{m-1}$	$F_1=\frac{MS_{Trt}}{MS_{Error}}$
Blocking Var 1	SS_{Tlock}	$m-1$	$MS_{Block}=\frac{SS_{Blick}}{m-1}$	$F_2=\frac{MS_{Block}}{MS_{Error}}$
Blocking Var 2	SS_{Tlock}	$m-1$	$MS_{Block}=\frac{SS_{Blick}}{m-1}$	$F_2=\frac{MS_{Block}}{MS_{Error}}$
Error	SS_{Error}	$n-m-p$	$MS_{Error}=\frac{SS_{Error}}{n-m-p}$	
Total	SS_{Tot}	$n-1$		

The SAS proc ANOVA or proc GLM can perform this analysis.

Example In an experiment, three dose levels of a drug were tested for the tumorigenic effect on three strains (Sprague-Dawley, Wistar, Fisher) of rats. However, the difference in body weight gains was a concern. Therefore, it was of interest to analyze the body weight gain data as a function of diet for a valid conclusion of tumorigenicity. The data may be considered as coming from a 3x3 Latin square design with Strain and Diet as Blocking factors and Dose as a treatment variable, each with three levels. Three observations per combination of diet strain and treatment. The following table shows the data.

Body weight by strain, diet, and treatment level			
	Strain 1	Strain 2	Strain 3
Diet 1	Ttr 1 369, 385, 382	Ttr 2 335, 351, 333	Ttr 3 338, 327, 367
Diet 2	Ttr 2 367, 385, 358	Ttr 3 338, 329, 324	Ttr 1 330, 337, 349
Diet 3	Ttr 3 339, 359, 370	Ttr 1 340, 358, 355	Ttr 2 356, 365, 348

We have H_{01}: no effect of strain on body weight gain, H_{02}: no effect of diet on body weight gain, and H_{03}: no effect of treatment dose on body weight gain. We use a significant level of $\alpha = 0.05$ for all tests. Using the SAS, "PROC ANOVA" command with model *y*= Strain Diet Treatment, we have the following ANOVA table.

Analysis of variance table

Source	DF	SS	Mean square	F value	Pr > F
Strain	2	3878.74	1939.37	112.88	<.0001
Diet	2	379.19	189.59	11.03	0.0006
Treatment	2	3878.74	1939.37	112.88	<.0001
Error	20	343.63	17.18		
Corrected Total	26	8480.30			

Using the p-values criteria we reject all null hypotheses $H_{10}, H_{20}, and\ H_{30}$, i.e., the strain, diet, and treatment dose have statistically significant effects on body weight gain.

18.11.4 Factorial Design

In our discussion so far, we had only one factor (independent variable) on one response (dependent variable), e.g., drug and its effect. However, in some experiments, we want to study the effect of two (or more) factors on a response variable. For example, consider the effects of the combination of two drugs each at some prespecified levels. One possibility is running separate single factor experiments by varying the level of one factor and keeping the level of the other factors fixed. Repeat the process of running a new experiment until all combinations of the remaining factors are complete. This process is very tedious, lengthy, and expensive. Alternatively, it is possible running experiments simultaneously to study the effect of more than one factor. This design is known as factorial design. In this design, all possible combinations of all factors are first enumerated. Subjects are randomly assigned to these factors following a randomization scheme, like completely random or randomized by blocking, as appropriate. One important difference between one factor and multi factor experiments is that in the latter experiment there is an interaction to be studied between the factors of interest. Interaction is defined as the extra effect of a combination of the factors over their individual effect separately. In other words, at different levels of two factors reacts differently for same level of treatment.

For example, consider an experiment where we like to study the effects of a combination of two drugs on a disease. Suppose the first drug has two levels, low and high denoted by 0 and 1, and the second drug has three levels, low, medium, and high denoted by 0, 1, and 2, respectively. Therefore, there are 2 × 3=6 total combinations (0,0), (0,1), (0,2), (1,0), (1,1), and (1,2).

The analysis model is assumed to be $y_{ijk} = \mu + \beta_j + \tau_i + \gamma_{ij} + \varepsilon_{ijk}$, where y_{ijk} is the kth observation for i^{th} level of Drug 1, j^{th} level of drug 2. The term μ is known as the common effect for the whole experiment, τ_i is the effect due to i^{th} level of Drug 1, β_j is the effect due to j^{th} level of Drug 2, γ_{ij} is the interaction effect due to i^{th} level of Drug 1 and j^{th}level of Drug 2, and ε_{ijk} is the random error. The error $\varepsilon_{ijk}$is assumed to have a normal distribution with mean zero and a common variance σ_e^2 for all treatment groups. It is also assumed that $\sum_i \tau_i = 0$, $\sum_j B_j = 0$, and $\sum_i \sum_j \gamma_{ij} = 0$. The analysis of variance table is as follows.

ANOVA table for factorial design with two factors

Source	Sum of squares	Degrees of freedom	Mean sum of squares	F
Treatment 1	SS_{Trt1}	$m-1$	$MS_{Trt1}=\frac{SS_{Trt1}}{m-1}$	$F_1=\frac{MS_{Trt1}}{MS_{Error}}$
Treatment 2	SS_{Trt2}	$m-1$	$MS_{Trt2}=\frac{SS_{Trt2}}{m-1}$	$F_2=\frac{MS_{Trt2}}{MS_{Error}}$
Interaction				
$T_1 x\ T_2$	$SS_{Trt1xTrt2}$	$(m-1)(m-1)$	$MS_{Trt1xTrt2}=\frac{SS_{Trt1xTrt2}}{(m-1)(m-1)}$	$F_3=\frac{MS_{Block}}{MS_{Error}}$
Error	SS_{Error}	$n-m-p$	$MS_{Error}=\frac{SS_{Error}}{n-m-p}$	
Total	SS_{Tot}	$n-1$		

where $n = \sum_i \sum_j \sum_k n_{ijk}$; $n_i = \sum_k \sum_j n_{ijk}$; $n_j = \sum_k \sum_i n_{ijk}$; $n_{ij} = \sum_k n_{ijk}$; $\bar{y} = \frac{\sum_i \sum_j \sum_k y_{ijk}}{n}$; $\bar{y}_i = \frac{\sum_j \sum_k y_{ijk}}{n_i}$; $\bar{y}_j = \frac{\sum_i \sum_k y_{ijk}}{n_j}$, $SS_{Tot} = \sum_i \sum_j \sum_k \left(y_{ijk} - \bar{y}\right)^2$; $SS_{Treat1} = \sum_i n_i (\bar{y}_i - \bar{y})^2$; $SS_{Treat2} = \sum_j n_j \left(\bar{y}_j - \bar{y}\right)^2$; $SS_{Treat1xTreat2} = \sum_i \sum_j n_{ij} \left(\bar{y}_{ij} - \bar{y}\right)^2$; and $SS_{Error} = \sum_i \sum_j \sum_k \left(y_{ijk} - \bar{y}_i - \bar{y}_j + \bar{y}\right)^2$.

In SAS, "proc ANOVA" or "proc GLM" can perform this analysis.

Example Two drugs and a placebo were tested for their ability to deplete the adrenal glands of ascorbic acid in control rats and in rats anesthetized with ether. Six rats were randomly assigned to each of the three treatments. Therefore, for each treatment, we observe two effects, one in control rats and one in anesthetized rats. In this experiment, we like to investigate the simultaneous effects of two factors, namely, treatment and anesthetic status. There could be an interaction between treatment and anesthetic status. Therefore, an interaction term would be included in the model. It should be noted that for an estimate of the interaction term it is necessary that more than one observation is taken in each cell. The data is as follows:

Effect of treatments in the absence and presence of anesthesia on the depletion of ascorbic acid in adrenal glands

	Drug		
Status	Placebo	A	B
Absence	4, 5, 4	−2, −1, −3	−20, −26, −23
Anesthesia	−3, −4, −5	−10, −11, −14	−32, −37, −38

Using the SAS proc ANOVA, with model y= drug status drug*status, we have the following ANOVA table.

Analysis of variance table					
Source	DF	Sum of squares	Mean square	*F* value	Pr > *F*
Drug	2	2851.00	1425.50	328.96	<.0001
Status	1	470.22	470.22	108.51	<.0001
Drug*status	2	14.78	7.39	1.71	0.2230
Error	12	52.00	4.33		
Corrected total	17	3388.00			

The result shows statistically significant effects of both treatment and anesthetic status. However, the interaction term is not statistically significant, indicating that the adrenal ascorbic acid content is significantly lowered by Treatment irrespective of anesthetic status.

18.11.5 Nested or Hierarchical design

In a certain experiment, a blocking factor may be embedded or nested within another blocking factor. For example, technicians may be nested within a laboratory. Further, different test techniques may be nested within each technician. This is called a nested design. Symbolically, the jth level of Factor 2 nested within the ith level of Factor 1 factor is written as $F1F2_{j(i)}$. For a nested experiment with two blocking factors F1 and F2, with *m* levels of F1 and *n* levels of F2, and F2 nested in F1, the model for k^{th} observation is written as $y_{ijk} = \mu + F1_i + F2_{i(j)} + \varepsilon_{k(ij)}$, where $i = 1, 2, \ldots, m$, $i = 1, 2, \ldots, n$, and $k = 1, 2, \ldots, p$. The ANOVA table is as follows:

Analysis of variance table				
Source	Sum of squares	Degrees of freedom	Mean sum of squares	*F*
F1	SS_{F1}	$m - 1$	$MS_{F1} = \frac{SS_{F1}}{m-1}$	$F_1 = \frac{MS_{F1}}{MS_{Error}}$
$F2_{(F1)}$	SS_{F2F1}	$m(n - 1)$	$MS_{F2F1} = \frac{SS_{F1_{F2}}}{m(n-1)}$	$F_2 = \frac{MS_{F2_{F2}}}{MS_{Error}}$
Error	SS_{Error}	$mnk - mn + 1$	$MS_{Error} = \frac{SS_{Error}}{mnk-mn+1}$	
Total	SS_{Tot}	$mnk - 1$		

Example To assess the reliability of a new method and to exclude the influence of laboratory and personnel involved, serum was sent to 4 laboratories for analyses. Two different technicians in each laboratory were asked to analyze serum alkaline phosphatase (AP). The reported results are entered in the table and analyzed for variation between laboratories and between technicians within a laboratory.

Laboratory	A		B		C		D	
Technician	T1	T2	T1	T2	T1	T2	T1	T2
AP, U/L	90	89	100	99	90	92	110	111
	91	94	102	98	92	94	105	107
	95	93	99	101	94	91	102	102

Using the SAS proc ANOVA with model $y=$ lab tech (lab), we have the following ANOVA table.

Analysis of variance table					
Source	DF	Squares	Mean square	*F* value	Pr > *F*
Lab	3	835.46	278.49	36.13	<.0001
Tech(lab)	4	3.17	0.79	0.10	0.9799
Error	16	123.33	7.708		
Total	23	961.94			

The result shows statistically significant effects of the laboratory but not of technicians within any laboratory, i.e., the accuracy of measurement differs significantly among at least one pair of laboratories, but no significant pairwise difference between technicians within a laboratory.

18.11.6 Split Plot Design

This design is useful if an experiment is conducted in a certain order. In design, the experimental space is first divided into blocks (also known as plots), the blocks are then divided into subblocks (or subplots), these subblocks may then be subdivided into smaller subblocks. The experiment is conducted by blocks. Design can be represented as,

Plot 1			Plot2		
Sub Plot 1	Sub Plot 2	Sub Plot 3	Sub Plot 1	Sub Plot 2	Sub Plot 3

The experiment is conducted in the order of subplots.

Data can be represented as,

		Treatment groups		
		(Subblock 2)		
(Block) Replication	(Subblock 1) Dose Levels	Group 1 By Time	Group 2 By Time	Group 3 By Time
I	Level 1	Y_{111}	Y_{121}	Y_{131}
		Y_{112}	Y_{122}	Y_{132}
		Y_{113}	Y_{123}	Y_{133}
	Level 2	Y_{211}	Y_{221}	Y_{231}
		Y_{212}	Y_{222}	Y_{232}
		Y_{213}	Y_{223}	Y_{233}
II	Level 1	Y_{111}	Y_{121}	Y_{131}
		Y_{112}	Y_{122}	Y_{132}
		Y_{113}	Y_{123}	Y_{133}
	Level 2	Y_{211}	Y_{221}	Y_{231}
		Y_{212}	Y_{222}	Y_{232}
		Y_{213}	Y_{223}	Y_{233}

The hypotheses of interest are H_0: No dose effect, H_0: No time effect, and H_0: No dose by time interaction etc. The ANOVA table is as follows:

Analysis of variance table

Source	Sum of squares	Degrees of freedom	Mean sum of squares	F
Replication R	SS_{Rep}	$m-1$	$MS_{Rep}=\frac{SS_{Rep}}{m-1}$	$F_1=\frac{MS_{Rep}}{MS_{Error}}$
Dose D	SS_{Doe}	$n-1$	$MS_{Dos}=\frac{SS_{Dos}}{n-1}$	$F_2=\frac{MS_{Dos}}{MS_{Error}}$
Interaction $R \times D$	SS_{RxD}	$(m-1)(n-1)$	$MS_{RxD}=\frac{SS_{RxD}}{(m-1)(n-1)}$	$F_5=\frac{MS_{RxD}}{MS_{Error}}$
Time T	SS_{Time}	$p-1$	$MS_T=\frac{SS_T}{p-1}$	$F_4=\frac{MS_T}{MS_{Error}}$
Interaction $R \times T$	SS_{RxT}	$(m-1)(p-1)$	$MS_{RxT}=\frac{SS_{RxT}}{(m-1)(p-1)}$	$F_5=\frac{MS_{RxT}}{MS_{Error}}$
Interaction $D \times T$	SS_{DxT}	$(n-1)(p-1)$	$MS_{DxT}=\frac{SS_{DxT}}{(n-1)(p-1)}$	$F_6=\frac{MS_{DxT}}{MS_{Error}}$
Error	SS_{Error}	$N-m-n-p+2$	$MS_{Error} \doteq \frac{SS_{Error}}{N-n-m-p}$	
Total	SS_{Tot}	$N-1$		

18.11.7 *Cross Over Design*

In a crossover experiment subjects in all treatment groups receive each treatment including placebo in different periods. Thus, each subject becomes control for itself. The order in which each subject receives the treatments depends on the design chosen for the experiment. The simplest design is a 2×2 design. In this design, each subject receives two different treatments, A (normally control/placebo) and B (drug treatment). Half the subjects receive A first and then crossover to B. The remaining subjects receive B first and then crossover to A. A potential disadvantage of this design lingering effect of a previously applied treatment known as the carry-over effect. Carry-over effect contaminates the real effect of the treatment in the next period(s). To reduce the carry-over effect completely or to an acceptable level, often a washout period is introduced in between treatment change. Thus, the design is not suitable for drugs with a long half-life. One advantage of this design is that it needs a fewer number of subjects compared to parallel design study. Common notations are s treatment groups, t treatment groups, and p periods. For example,

	Periods						
	Group	Period 1	...	Period 2	...	Period $p-1$	Period p
	Group 1	T_1		T_2	...	T_{t-1}	T_t
Group 2	T_1	T_4 .	...	T_t		T_{t-7}	
.							
.							
.							
Group s	T_t	T_1	...	T_2		T_{t-s}	

Note that some of these periods may be washout periods.
Model: $Y_{ijk} = \mu + s_{ik} + \pi_j + \tau_{T[i,j]} + \lambda_{T[i,j-1]} + \varepsilon_{ijk}$
where
Y_{ijk} = the response observed on the k^{th}subject in period j of group i.
μ = the general mean
s_{ik} = the effect of subject k in group i, $i = 1, 2, \ldots, s$, $k = 1, 2, \ldots, n_i$
π_j = the effect of period j, $j = 1, 2, \ldots, p$
$\tau_{T[i,j]}$ = the direct effect of the treatment administered in period j of group i
$\lambda_{T[i,j-1]}$ = the effect of the carry-over of the treatment administered in period $j-1$ of group i, where $\lambda_{[i,0]} = 0$
ε_{ijk} = a random error for subject k in period j in group i
The general null hypothesis is,
$H_0 : \tau_{1[i,j]}=\tau_{2[i,j]}=\ldots=\tau_{d[i,j]}$, i.e., no treatment effect.
Skeleton analysis of variance table is as follows:

Analysis of variance table				
Source	d.f.	S.S.	M.S.S.	F
Between subjects				
Within subjects Due to period Due to treatment Carry-Over effect Residual				
Total				

For 2 × 2 crossover design equality of carry-over effect and direct treatment effect for two groups of subjects can be tested using the t-test. To test the carry-over effect H$_0$: $\lambda_1=\lambda_2$. The test statistic $t_\lambda = \frac{\widehat{\lambda_d}}{\sqrt{\widehat{\sigma_t^2} m}}$ with $n_1 + n_2 - 2$ d.f., where $m = \frac{1}{n_1} + \frac{1}{n_2}$ and $\sigma_t^2 = 2(2\sigma_s^2 + \sigma^2)$, $E\left[\widehat{\lambda_d}\right] = \lambda_d$, $V\left[\widehat{\lambda_d}\right] = \sigma_t^2 m$. The estimate of σ_t^2 is $\widehat{\sigma_t^2} = \sum_{i=1}^{2}\sum_{k=1}^{n_i}(t_{ik} - \bar{t}_i)^2/(n_1 + n_2 - 2)$.

To test the direct treatment effect $H_0 : \tau_1 = \tau_2$. The test statistic $t_\tau = \frac{\widehat{\tau_d}}{\sqrt{\widehat{\sigma_d^2} m/4}}$ with $n_1 + n_2 - 2$ d.f., where $m = \frac{1}{n_1} + \frac{1}{n_2}$, $\sigma_d^2 = 2\sigma^2$, $E[\widehat{\tau_d}] = \tau_d$, $V[\widehat{\tau_d}] = \frac{\sigma_d^2}{4} m$,. The estimate of σ_d^2 is $\widehat{\sigma_d^2} = \sum_{i=1}^{2}\sum_{k=1}^{n_i}\left(d_{ik} - \bar{d}_i\right)^2/(n_1 + n_2 - 2)$. For a detailed discussion of this topic, the readers are referred to the book by Jones and Kenward (1989) [5]. For analysis SAS proc mixed may be used.

Example Effect of test substance on QT interval prolongation in dogs. The experiments were performed on 5 dogs. A single dose of test substance (A) and normal saline (control, B) was administered according to a Latin square design. ECG parameters (QT, QRS, PR) were recorded. The duration of the QT interval (uncorrected, and corrected with Bazett, Fridericia, and Van de Water formulas) was determined 1 h before and 2 h after dosing. In the following example, active drug A and saline B were applied to two groups of animals in two different sequences. The first group received A first and B second with a washout period (3 days) in the middle, while the second group received A first and B second with a washout period in the middle. The outcome variable is measured QT interval duration (corrected) expressed as % increase above pretest value.

Group 1				Group 2			
Animal #	Period 1 Y_1	Period 2 Y_2	Sequence	Animal #	Period 1 Y_2	Period 2 Y_1	Sequence
1	17.0	2.5	AB	2	2.7	15.3	BA
2	15.2	1.3	AB	3	1.5	14.2	BA
3	18.0	2.1	AB	4	1.6	18.1	BA
4	16.1	1.5	AB	5	2.2	14.5	BA
5	14.3	2.4	AB	6	1.9	19.7	BA

Using SAS proc mixed data = aa; class Subject Period Treatment sequence; model y=Treatment Period Treatment*Period/noint; random subject; we have the following analysis of variance table:

Analysis of variance table					
Source	d.f.	S.S.	M.S.S.	*F*	*P*-value
Subjects	8	16.490000	2.061250	0.9417	0.5328
Due to treatment	1	1018.164500	1018.165	465.18	<0.0001
Due to period	1	0.060500	0.060500	0.0276	0.8721
Period*Treatment	1	0.087550	0.087550	0.04	0.8446
Residual	8	17.510000	2.188750	.	.
Total	19				

The results show a highly significant treatment effect with no significant difference in periods or subjects.

18.12 Analysis of Data with Repeated Measures

In some experimental studies, observations are taken on each subject repeatedly at different time points. Repeated measures on n subjects with m observations on each subject may be presented as an mxn matrix. If p variables are observed for each subject at each time point, the data can be presented as mxnxp matrix.

Repeated measure data can be modeled as $y_{ij} = \mu_j + \alpha_{ij} + \varepsilon_{ij}$, where y_{ij} denotes the jth observation of the ith subject, μ_j= the mean level of y_{ij}, a subject at j^{th} time point, α_{ij}= departure of $y_{ij}$from $\mu_j$for i^{th} subject and ε_{ij}= error term, representing the random departure of y_{ij} from $\mu_j + \alpha_{ij}$

The above model is known as a mixed effect model with μ_j assumed to be fixed and α_{ij} assumed to be random from subject to subject. The general assumptions are, for given i and j E(α_{ij}) = 0, V(α_{ij})= $\sigma^2_{\alpha j}$, E(ε_{ij}) = 0, V(α_{ij})= σ^2_j, Cov(α_{ij},$\alpha_{ij'}$) = $\sigma_{\alpha j j'}$ and Cov(α_{ij},$\alpha_{i'j'}$) =0 for $i \neq i$, Cov(ε_{ij},$\varepsilon_{ij'}$) =0 for $i \neq i'$or $j \neq j'$, Cov(α_{ij}, $\varepsilon_{i'j'}$) = 0 for all i, j, i',and j', and the $\alpha_{ij}$and $\varepsilon_{i'j'}$are normally distributed.

A complete discussion of this topic is more extensive and is outside the scope of this article. Crowder and Hand [6] provide a good discussion of this topic. A brief description of this ANOVA procedure will be presented in the following examples.

Example In an experiment in which changes in blood glucose levels were measured at 1, 2, 3, 4, 6, 8 h following the administration of a single oral dose of test substance. There were 6 animals per dose group. The variable of interest is the % change in blood glucose.

	Time after dosing					
Dose (mg/kg)	1 h	2 h	3 h	4 h	6 h	8 h
Control	1, 2, −2 0, 2, 1	1, 1, 2 −1, 0, 0	−1, 1, −1 2, 1, 2	−2, 1, 2 3, 2, 1	0, 3, 1 3, 1, −2	2, 1, 0 2, 3, −1
1	−2, −1, 0 1, 0, 1	−2, −1, 1 0.−2,−1	−2, −1, 0 1, −2, 1	−2, 0, −1 1, 2, 3	2, 0, 0 −1, 0, 1	−1, 1, 2 2, 0, 1
3	−5, −6, −1 0, −2, 1	−4, −2, 0 −1, −2, 0	−4, −2, −2 1, 1, −2	−4, −3, 2 2, −1, −2	−3, −3, 2 0, 2, 0	−1, 1, 1 2, 0, −2
10	−6, −5, −7 −5, −4, −5	−7, −9, −8 −4, −6, −8	−6, −5, −9 −4, −7, −8	−6, −9, −6 −5, −6, −8	−4, −5, −2 −2, 0, −2	−1, −2, 1 −2, 0, 0
30	−11, −9,−12 −12, −8,−11	−14,−15, −14 −18,−20, −15	−14,−16,−15 −19,−21,−16	−14, −16, −18, −20,−21,−20	−16, −15, −18 −17,−10, −14	−7, −4, −3 −6, −5, −5
100	−25,−20,−12 −16,−18,−22	−25,−28,−28 −22,−25,−22	−20,−20,−18 −28,−28,−30	−18,−20,−22 −29,−30,−30	−17,−15,−17 −20,−17,−18	−10,−14,−12 −10,−11,−10

The objective is to evaluate any difference in the profile of change in blood glucose among treatment groups. The design is considered as a two-way layout (dose × time) experiment. The hypothesis is H_0: no difference in an overall change in blood glucose among dose groups. There may be an interaction between factors dose and time.

Using the SAS commands, proc mixed; class id dose; model value = dose time dose*time; random intercept, where id represents the observation number within a dose group, we have the following ANOVA table.

Analysis of variance table

Effect	DF	DF	F Value	Pr > F
Dose	5	204	68.21	<.0001
Time	1	204	30.90	<.0001
Time*dose	5	204	3.37	0.0061

The result shows statistically significant effects of both doses and time. The results also show that there is significant interaction between factors dose and time, indicating that the mean blood glucose does not change similarly overtime for all dose levels.

18.13 Adjustment for Multiple Testing

Making a conclusion on if a drug has the pharmacological or toxicological effect on tested subjects based on results of simultaneously testing multiple null hypotheses on data of different variables is a common problem in statistical analyses in pharmacology and toxicology studies. One requirement for the test of hypothesis is that the Type-1 error remains less than the significance level α. For example, in an experiment, a high and a low dose are tried to two randomly selected animals. The drug is assumed to be effective if either the high dose reduces the disease symptom by at least 50% or the low dose reduces the disease symptom by at least 30%. In this experiment, two tests are performed simultaneously and there are two ways to make a favorable conclusion. In this case, the overall Type-1 error (generally known as the overall false positive rate or family wise error rate FWER) is probability of at least one false positive. Mathematically, this is 1-probability of no false positive error or $1 - (1 - \alpha)^2 > \alpha$. Another example of multiple testing is the repeated measure experiment. In this experiment, tests are performed at several time points to compare treatment with control. A significant difference is concluded if a statistical significance is found at any time point. In general, if m simultaneous tests are performed for concluding on one endpoint, then the overall false positive rate is $1 - (1 - \alpha)^p > \alpha$, where m is the number of independent hypothesis tested. The FWER increases as the number of simultaneous tests of independent hypotheses increases. For example, if $\alpha = 0.05$ and $m = 10$, i.e., 10 independent null hypotheses are simultaneously tested, each at 0.05 level. The overall false positive rate is $1 - (1 - 0.05)^{10} = 0.40$ or 40% as shown in the graph below. This is not desirable. Therefore, some kind of adjustment is needed for simultaneous tests. This process is known as an adjustment for multiple testing.

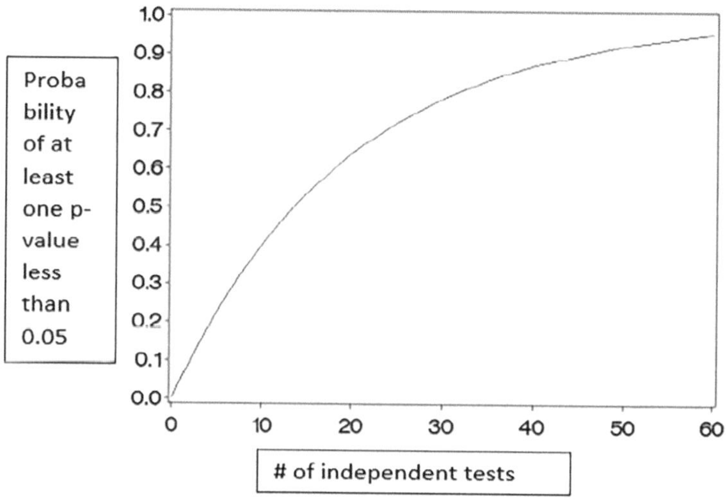

There are various methods available for this adjustment. One popular method for this purpose is the Bonferroni method. In this method, for m simultaneous tests, each hypothesis is tested at $\alpha_a = \frac{\alpha}{m}$. Sidak suggested another method like Bonferroni. He suggested the use of $\alpha_a = 1 - \sqrt[m]{1-\alpha}$ to test each hypothesis. The other popular methods are Dunnett method, Scheffe method, Tukey method, Simes method, Hommel method, Hochberg method, Rom method, Holm method, Gate keeping method, Closed testing method, Benjamini and Hochberg method, etc. There are other methods based on Bayesian and bootstrap-resampling methodologies. A complete discussion of this topic is outside the scope of this chapter. The readers are referred to Westfall et al. [7] for a detailed discussion of many of these methods.

Bonferroni method and Sidak method Suppose we want to test three simultaneous null hypotheses and with an overall false positive rate of 5% (i.e., FWER $\alpha = 0.05$). For this case, the Bonferroni adjustment method suggests using a test level of $\alpha_a = \frac{0.05}{3} = 0.01667$ for each test and the Sidak method suggests using a test level of $\alpha_a = 1 - \sqrt[3]{1-0.05} = 0.01695$.

Dunnett's test Dunnett suggested a special method for the adjustment for multiple testing in situations in which several treatment groups are simultaneously compared with the control to see the efficacy in any of the selected doses. A brief description of his method is as follows.

Suppose we have an experiment with two or more treatment groups and one control group and we want to compare each treatment group with the control. This involves multiple testing. Dunnett test is conducted by computing a t-statistic between each treatment group with the control group using the formula, $t_d = \frac{\bar{y}_i - \bar{y}_c}{\sqrt{2MSE/n_h}}$, where $\bar{y}_i$ is the mean of the ith treatment group, $\bar{y}_c$ is the mean of the control group, MSE is the mean square error as computed from the analysis of variance, and n_h is the harmonic mean of the total sample size of the treatment group and the control group. The degree of freedom for the (each) test is $N - r$, where N is the total number of subjects in all groups and r is the number of groups (including the control). The critical value of t_d depends on the number of treatment groups in the experiment and can be obtained from a table of Dunnett's test. The readers are referred to Westfall et al. [8] for a detailed discussion of many of these methods.

18.14 Regression Analysis

As mentioned earlier, sometimes we measure more than one variable from an individual experimental subject. Some of these variables may be dependent on the values of the other variables known as independent variables. For example, body weight may depend on age, height, and waist size. It is of interest in data analysis to investigate possible mathematical relationships between dependent and independent variables. In simple case where there are only one dependent variable and one or few independent variables, this investigation starts with a scatter plot that provides a visual aid to guess on possible relationship between the dependent and independent variables. For example, the scatter plot in Fig. 18.4 suggests a possible linear relationship between Y and X with a positive slope. In the next step, we look for a mathematical model that may explain the relationship between the dependent and the independent variables. It is worth mentioning that in observational data the mathematical model that looks most suitable to the data may not describe it exactly. It may only be the most reasonable model to approximately fit the data. In the above example, a linear relationship may be expressed as $y_i = \alpha + \beta x_i$. However, since the scatter plot does not suggest a perfect linear relation, the actual realization of y may be modeled as $y_i = \alpha + \beta x_i + \varepsilon_i$, where ε_i is a random fluctuation around the straight line, known as the error. The error term ε_i is assumed to follow a normal distribution with mean zero and variance σ^2, i.e., $\varepsilon_i \sim N(0, \sigma^2)$. Alpha ($\alpha$) and beta ($\beta$) are the two parameters of the model. The actual values of these parameters are unknown. Estimates of the parameters can be done from sample values using statistical procedures like least square method, maximum likelihood method, Bayesian method, etc. Due to the limitation of space, in this article we will discuss only the least square method. The model we get after the estimation of parameters is known as the fitted model and is denoted as $\widehat{y}_i = \widehat{\alpha} + \widehat{\beta} x_i$. The y_i value coming from the fitted model is known as the fitted or estimated value of y_i, denoted by $\widehat{y}_i$.

18.14.1 The Least Square Method

The difference between actual and the estimated values $e_i = (y_i - \widehat{y}_i)$ is known as residual. The least square method suggests, choose $\widehat{\alpha}$ and $\widehat{\beta}$ such that sum of the square of all residuals $\varphi = \sum e_i^2$ has a minimum value. Alpha hat ($\widehat{\alpha}$) and beta hat ($\widehat{\beta}$) are known as the lease square estimate of α and β. Avoiding the mathematical detail for limitation of space, we mention that the least square estimates $\widehat{\alpha}$ and $\widehat{\beta}$ of α and β, respectively, are

$$\widehat{\alpha} = \frac{\sum y_i}{n} - \widehat{\beta}\frac{\sum x_i}{n} \text{ and } \widehat{\beta} = \frac{\sum x_i y_i - \frac{(\sum x_i)(\sum y_i)}{n}}{\sum x_i^2 - \frac{(\sum x_i)^2}{n}}$$

Models with more than one independent and dependent variables can also be fitted based on the least square method. In SAS, "Proc reg" command can fit such a model.

Example For the above data ability (Y) of liver slices from guinea pigs of different ages (X) to conjugate phenolphthalein with glucuronic acid, we have the following calculations.

X	Y	X^2	XY	$\widehat{Y}$
1	5.6	1	5.6	10.39
1	8.8	1	8.8	10.39
3	12.0	9	36.0	15.09
5	18.0	25	90.0	19.79
6	31.0	36	186.0	22.14
10	38.0	100	380.0	31.54
10	44.0	100	440.0	31.54
11	22.0	121	242.0	33.89
14	37.0	196	518.0	40.94
15	46.0	225	690.0	43.29
21	54.0	441	1134.0	57.39
Total 97	316.4	1255	3730.4	316.4

Therefore, $\widehat{\beta} = \frac{3730.4 - \frac{97x316.4}{11}}{1255 - \frac{97^2}{11}} = 2.35$ and $\widehat{\alpha} = \frac{316.4}{11} - 2.35x\frac{97}{11} = 8.04$. Therefore, the fitted model is $\widehat{y}_i = 8.04 + 2.35x_i$. If $x_1 = 1$, then $\widehat{y} = 8.04 + 2.35x_1 = 10.39$.

Figure 18.5 displays this data along with the fitted regression line.

Clearly, the estimates $\widehat{\alpha}$ and $\widehat{\beta}$ are statistics. Further mathematical details show that the mean of $\widehat{\alpha}$ is $E(\widehat{\alpha}) = \alpha$ and the variance of $\widehat{\alpha}$ is $V(\widehat{\alpha}) = s^2\left(\frac{1}{n} + \frac{\overline{x}^2}{\sum(x_i - \overline{x})^2}\right)$, also the mean of $\widehat{\beta}$ is $E\left(\widehat{\beta}\right) = \beta$ and the variance of $\widehat{\beta}$ is $V\left(\widehat{\beta}\right) = \frac{s^2}{\sum(x_i - \overline{x})^2}$, where $s^2 = \frac{\sum\left(y_i - \widehat{y}_i\right)^2}{n-2}$.

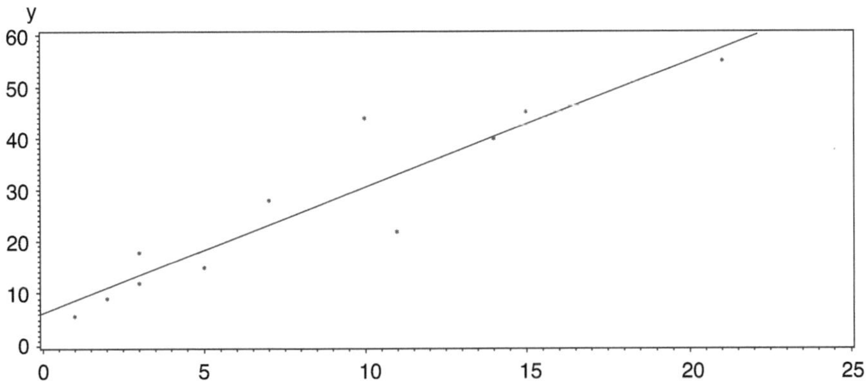

Fig. 18.5 Least square fit of straight line

18.14.2 Estimation and Prediction

Using the fitted model $\widehat{y}_i = \widehat{\alpha} + \widehat{\beta}x_i$, the value of $\widehat{y}_i$ can be calculated for a given value of x_i. For example, if $x_i = 3$, the estimated value is $\widehat{y}_i = 8.04 + 2.35 \times 3 = 15.09$. This $\widehat{y}_i$ value is called the estimate of y_i if the value of x_i is within the data range and is called prediction if the value of x_i is beyond the data range, in which case x_i is also known as the predictor of y_i. For our example, the estimates of y_i using the fitted model are given in the last column of the above table.

Clearly, the estimate $\widehat{y}_i$ is also a statistic. Some mathematical details show that for a given value of $x_i = x_p$ say, the mean of $\widehat{y}_p$ is $E(\widehat{y}_p) = E(y_p)$=Mean of y_p, and the variance of $\widehat{y}_p$ is $V(\widehat{y}_p) = \sigma^2\left(\frac{1}{n} + \frac{(x_p-\bar{x})^2}{\sum(x_i-\bar{x})^2}\right)$. An estimate of this variance is $V(\widehat{y}_p) = s^2\left(\frac{1}{n} + \frac{(x_p-\bar{x})^2}{\sum(x_i-\bar{x})^2}\right)$.

18.14.3 Inference about the Model Parameters

Null hypothesis is $H_{10} : \alpha = \alpha_0$ against the alternative like $H_{11} : \alpha > \alpha_0$ (or $\alpha < \alpha_0$ *or* $\alpha \neq \alpha_0$), and $H_{20} : \beta = \beta_0$ against the alternative like $H_{21} : \beta > \beta_0$ (or $\beta < \beta_0$ *or* $\beta \neq \beta_0$). The test statistic for testing $H_{10} : \alpha = \alpha_0$ is $t = \frac{\widehat{\alpha}-\alpha_0}{\sqrt{V(\widehat{\alpha})}}$ with $(n-2)$ d.f., and that for testing $H_{20} : \beta = \beta_0$ is $t = \frac{\widehat{\beta}-\beta_0}{\sqrt{V(\widehat{\beta})}}$ with $(n-2)$ d.f. For data in the above example, using SAS proc reg; model y=x; we have the following results for $H_{10} : \alpha = 0$ against the alternative $H_{11} : \alpha \neq 0$, and $H_{20} : \beta = 0$ against alternative $H_{21} : \beta \neq 0$.

Parameter estimates					
Variable	DF	Parameter estimate	Standard error	t-Value	Pr > | t|
Intercept α	1	8.01483	3.94021	2.03	0.0725
Slope β	1	2.35296	0.36889	6.38	0.0001
Error	9	489.43560	54.38173		
Total	10	2701.98545			

Results show a significant slope (i.e., β is not zero), but not a significant intercept (i.e., α is not different from zero).

The $(1-\alpha)\%$ confidence intervals for α, β, and y_p can be calculated using the formulas, $\widehat{\alpha} \mp t_{1-\alpha/2,(n-2)}\sqrt{V(\widehat{\alpha})}$, $\widehat{\beta} \mp t_{1-\alpha/2,(n-2)}\sqrt{V\left(\widehat{\beta}\right)}$, and $\widehat{y}_p \mp t_{1-\alpha/2,(n-2)}\sqrt{V\left(\widehat{y}_p\right)}$, respectively. Interested readers are referred to Draper and Smith [7] for advanced discussions on this topic.

18.15 Transformation of Variables

For convenient and meaningful analysis some time variables are transformed. In this process the original variable is replaced by a function of that variable: for example, replace a variable x by its square root or its logarithm. New regressions or relations are considered among transformed variables. Good transformation may reduce skewness, equalize variance among treatment levels (homoscedasticity), and linearize relationships.

Commonly used transformations are reciprocal ($1/x$), logarithm ($\log(x)$ *or* $\ln(x)$), cube root $\left(x^{\frac{1}{3}}\right)$, square root ($\sqrt{x}$), square, and exponential (e^x *or* e^{-x}). For example, a multiplicative relationship like $y = a * \exp(bx)$ becomes linear $\ln(y) = \ln(a) + bx$ by a log transformation. It is easier to deal with and understand this linear regression compared to the multiplicative relation. The main criterion in choosing a transformation is: what works with the data and what makes physical (biological, economic, whatever) sense, for example, for extremely skewed data suggest using log transformation. If possible, we prefer measurement scales that are easy to think about.

On the downside, the selection of a type of transformation is difficult. The choice of appropriate transformations can cause debate even among experienced data analysts. Some use them routinely, others much less.

Data that are proportions (between 0 and 1) often benefit from special transformations. The most common is the logit (or logistic) transformation, which is $logit\ p = log\ (p/(1-p))$ for proportions. This way of writing brings out a symmetrical way in which very high and very low values are treated. (If p is small, $1-p$ is large, and vice versa).

18.16 Some Application of Statistics in Toxicity Studies

For initial dose range-finding or exploratory safety studies the sample size is generally small, typically 5 or less. For such studies, a statistical inference may not make much sense. Descriptive statistics may be enough to proceed to a further research program. For toxicity studies using slightly large group sizes of 20–25 animals, statistical inferences can be performed. In the analysis of continuous variable data, the global null hypothesis of equal means or medians of the variable of treatment groups can be tested by analysis of variance (ANOVA) or analysis of covariance (ANCOVA), or their corresponding nonparametric procedures such as Kruskal-Wallis test. Dose responses in those continuous variables collected in a toxicity study can be tested either by using the regression procedure or by testing the special contrasts of the means in ANOVA developed for testing trends. For binary and other categorical data, pairwise differences in proportion between pairs of treatment groups can be tested using Pearson's Chi-Square test and Fisher's exact test1test. The Cochran-Armitage test and the permutation trend test can be used to test the dose-response in proportion in binary variables. For data with more than two categories, the Wilcoxon rank sum test and Jonckheere's test can be used for the evaluation of pairwise differences and dose-responses in those categorical variables. The parametric methods of logit analysis are widely used to estimate LD50 and confidence interval for LD50.

For genotoxicity studies, the dose response and pairwise differences of data proportions of cells with chromosomal aberration can be analyzed by the Cochran-Armitage trend test and Fisher's exact test. Logistic regression using $\mu_i = exp\ (\beta_0 + \beta_1 d_i)/(1 + exp\ [\beta_0 + \beta_1 d_i])$ can also be used to test the dose-response effect in the proportion of cells with a chromosomal aberration. For Reproductive and Developmental Toxicity (DART) Studies the binary data can also be analyzed by the Cochran-Armitage trend test, an exact permutation trend test for dose-response in group proportion, and by the Fisher's exact test for pairwise comparisons in proportion between the control group and each of the treated groups

For long-term carcinogenicity studies the Cox test, the generalized Wilcoxon or Kruskal-Wallis test, and the Tarone trend tests can be used to test for heterogeneity in survival distributions and significant dose-response relationships in survival. For tumor data sets, dose response and pairwise comparisons of the proportions of tumor-bearing animals can be tested using the Cochran-Armitage test. Survival adjusted dose-response test and pairwise comparisons are suggested, e.g., combination of prevalence and death rate method by Peto et al. [9], and Poly-K method by Bailer and Portier [10]. Both Peto and Poly-K use the Cochran Armitage test with a modified denominator to calculate proportions.

Acknowledgment The authors express their thanks to Dr. G. Jagadeesh, for offering the opportunity of contributing this chapter to this very important book. The authors are grateful to the Division of Biometrics-6 director Dr. Yi Tsong and other colleagues for their continuous support.

References

1. Hogg RV, Tanis EA. *Probability and statistical inference*. New York: Macmillan Publishing Company; 1983.
2. Snedecor GW, Cochran WG. Statistical methods. 8th ed: Iowa State University Press; 1989.
3. Hollander M, Wolfe DA. Nonparametric statistical methods. New York: John Wiley & Sons; 1973.
4. Hicks CR. Fundamental concepts in the design of experiments 5th edn, Oxford University Press
5. Jones B, Kenward MG. Design and analysis of cross over trials. New York: Chapman and Hall; 1989.
6. Crowder MJ, Hand DJ. Analysis of repeated measures. New York: Chapman and Hall; 1990.
7. Draper NR, Smith. Applied regression analysis. 2nd ed. New York: John Wiley and Sons; 1966.
8. Westfall PH, Tobias RD, Rom D, Wolfinger RD, Hochberg Y. Multiple comparisons and multiple tests using the SAS system. Cary, NC: SAS Institute Inc.; 1999.
9. Peto R, Pike MC, Day NE, Gray RG, Lee PN, Parish S, Peto J, Richards S, and Wahrendorf J. Guidelines for simple, sensitive significance tests for carcinogenic effects in long-term animal experiments," *Long-term and Short-term Screening Assays for Carcinogens: An Critical Appraisal, World Health Organization* 1980.
10. Bailer AJ, Portier CJ. Effects of treatment-induced mortality and tumor-induced mortality on tests for carcinogenicity in small samples. Biometrics. 1988;44:417–31.